AN INTRODUCTION TO
BOUNDARY LAYER METEOROLOGY

ATMOSPHERIC SCIENCES LIBRARY

ROLAND B. STULL

*Department of Meteorology,
University of Wisconsin, Madison, U.S.A.*

An Introduction to
Boundary Layer Meteorology

KLUWER ACADEMIC PUBLISHERS
DORDRECHT / BOSTON / LONDON

Library of Congress Cataloging in Publication Data

Stull, Roland B., 1950-
    An introduction to boundary layer meteorology / Roland B. Stull.
        p.   cm. -- (Atmospheric sciences library)
    Includes bibliographies and index.
    ISBN 9027727686. ISBN 9027727694 (pbk.)
    1. Boundary layer (Meteorology)   I. Title. II. Series.
QC880.4.B65S784 1988
551.5--dc19   88-15564
                                                        CIP

ISBN 90-277-2768-6
ISBN 90-277-2769-4 (pbk.)

Published by Kluwer Academic Publishers,
P.O. Box 17, 3300 AA Dordrecht, The Netherlands.

Kluwer Academic Publishers incorporates
the publishing programmes of
D. Reidel, Martinus Nijhoff, Dr W. Junk and MTP Press.

Sold and distributed in the U.S.A. and Canada
by Kluwer Academic Publishers,
101 Philip Drive, Norwell, MA 02061, U.S.A.

In all other countries, sold and distributed
by Kluwer Academic Publishers Group,
P.O. Box 322, 3300 AH Dordrecht, The Netherlands.

Printed in The Netherlands

Contents

14 Geographic Effects 587

Appendices 619

Subject Index 649

Preface

Part of the excitement of boundary-layer meteorology is the challenge in studying and understanding turbulent flow — one of the unsolved problems of classical physics. Additional excitement stems from the rich diversity of topics and research methods that we collect under the umbrella of boundary-layer meteorology. That we live our lives within the boundary layer makes it a subject that touches us, and allows us to touch it. I've tried to capture some of the excitement, challenges and diversity within this book.

I wrote this book with a variety of goals in mind. First and foremost, this book is designed as a *textbook*. Fundamental concepts and mathematics are presented prior to their use, physical interpretations of equation terms are given, sample data is shown, examples are solved, and exercises are included. Second, the book is organized as a *reference*, with tables of parameterizations, procedures, field experiments, useful constants, and graphs of various phenomena in a variety of conditions. Third, the last several chapters are presented as a *literature review* of the current ideas and methods in boundary layer meteorology.

It is assumed that this book will be used at the beginning graduate level for students with an undergraduate background in meteorology. However, a diversity in the background of the readers is anticipated. Those with a strong mathematical background can skip portions of Chapter 2 on statistics, and those with experience with time series can skip Chapter 8. These two mathematics chapters were separated to offer the reader a chance to apply the first dose of statistics to boundary layer applications before delving back into more math. Some students might have had a course on geophysical turbulence or statistical fluid mechanics, and can skim through the first 5 chapters to get to the boundary-layer applications. By excluding a few chapters, instructors can easily fit the remaining material into a one-semester course. With supplemental readings, the book can serve as a two-semester sequence in atmospheric turbulence and boundary-layer meteorology.

Notational diversity proved to be the greatest difficulty. Each subdiscipline appeared to have its own set of notation, which often conflicted with the notation of other subdisciplines. To use the published notation would have lead to confusion. I was therefore forced to select a consistent set of notation to use throughout the book. In most cases I've tried to retain the notation most frequently used in the literature, or to add subscripts or make logical extensions to existing notation. In other cases, I had to depart from previously published notation. Readers are referred to Appendix B for a comprehensive list of notation.

I certainly cannot claim to be an expert in all the myriad subdisciplines of boundary-layer meteorology, yet I knew the book should be comprehensive to be useful. My interest and enthusiasm in writing this book motivated many trips to the library, and stimulated my analysis of many research papers to learn the underlying themes and common tools used in the diverse areas of boundary layer meteorology. Unfortunately, the limited space within this book necessitated some difficult decisions regarding the amount and level of material to include. Hopefully I've presented sufficient background to lay the building blocks upon which more advanced concepts can be built by other instructors and researchers. Certain topics such as atmospheric dispersion and agricultural micrometeorology are not covered here, because there are other excellent books on these subjects.

Many colleagues and friends helped with this book and contributed significantly to its final form and quality. Michelle Vandall deserves special recognition and thanks for drawing most of the figures, designing the page and chapter headers, and for her overall dedication to this project. Colleagues Steve Stage, Larry Mahrt, George Young, Jacq Schols, Chandran Kaimal, Steve Silberberg, Beth Ebert, and Bruce Albrecht reviewed various chapters and provided valuable suggestions. Information regarding field experiments was provided by Anton Beljaars, Ad Driedonks, Jean-Claude André, Jean-Paul Goutorbe, Anne Jochum, Steve Nelson, Bob Murphy, Ruwim Berkowicz, Peter Hildebrand, Don Lenschow, and others. Eric Nelson proofread the manuscript, and helped with the list of notation. Sam Zhang compiled the index. Some of the equations were typeset by Camille Riner and Michelle Vandall (her name keeps reappearing). Four years of students in my micrometeorology courses at Wisconsin graciously tolerated various unfinished drafts of the book, and caught many mistakes. The patient editors of Kluwer Academic Publishers (formerly D. Reidel) provided constant encouragement, and are to be congratulated for their foresight and advice. The American Meteorological Society is acknowledged for permission to reproduce figures 10.7, 12.10 and 12.19. To all of these people and the many more to whom I apologize for not naming here, I thank you.

To my wife, Linda, I give my gratitude for her devotion during this exciting and exhausting episode of my life.

Roland B. Stull
Boundary Layer Research Team
Madison, Wisconsin

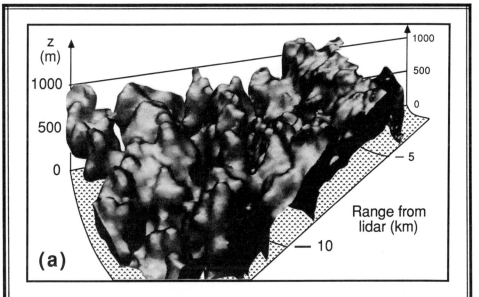

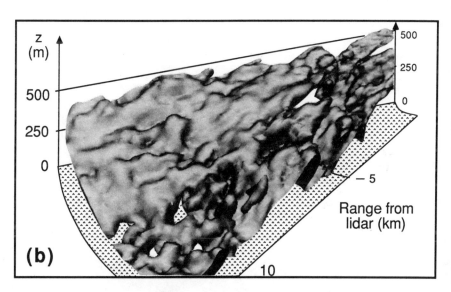

Frontispiece Lidar images of the aerosol-laden boundary layer, obtained during the FIFE field experiment in Kansas. (a) Convective mixed layer observed at 1030 local time on 1 July 1987, when winds were generally less than 2 m/s. (b) Slightly-stable boundary layer with shear-generated turbulence, observed at 530 local time on 7 July 1987. Winds ranged from 5 m/s near the surface to 15 m/s near the top of the boundary layer. Photographs from the Univ. of Wisconsin lidar are courtesy of E. Eloranta, Boundary Layer Research Team.

1 Mean Boundary Layer Characteristics

From our first breath, we spend most of our lives near the earth's surface. We feel the warmth of the daytime sun and the chill of the nighttime air. It is here where our crops are grown, our dwellings are constructed, and much of our commerce takes place. We grow familiar with our local breezes and microclimates, and we sense the contrasts when we travel to other places.

Such near-earth characteristics, however, are not typical of what we observe in the rest of the atmosphere. One reason for this difference is the dominating influence of the earth on the lowest layers of air.

The earth's surface is a boundary on the domain of the atmosphere. Transport processes at this boundary modify the lowest 100 to 3000 m of the atmosphere, creating what is called the *boundary layer* (Fig 1.1). The remainder of the air in the troposphere is loosely called the *free atmosphere*. The nature of the atmosphere as perceived by most individuals is thus based on the rather peculiar characteristics found in a relatively shallow portion of the air.

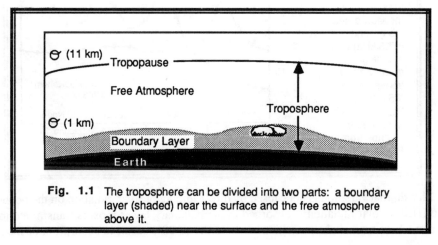

Fig. 1.1 The troposphere can be divided into two parts: a boundary layer (shaded) near the surface and the free atmosphere above it.

1

This chapter is meant to provide a descriptive overview of the boundary layer. Emphasis is placed on mid-latitude boundary layers over land, because that is where much of the world's population resides and where many boundary layer measurements have been made. In this region the diurnal cycle and the passage of cyclones are the dominant forcing mechanisms. Tropical and maritime boundary layers will also be briefly reviewed. Theories and equations presented in later chapters will make more sense when put into the context of the descriptive nature of the boundary layer as surveyed here.

1.1 A Boundary-Layer Definition

The troposphere extends from the ground up to an average altitude of 11 km, but often only the lowest couple kilometers are directly modified by the underlying surface. We can *define the boundary layer as that part of the troposphere that is directly influenced by the presence of the earth's surface, and responds to surface forcings with a timescale of about an hour or less.* These forcings include frictional drag, evaporation and transpiration, heat transfer, pollutant emission, and terrain induced flow modification. The boundary layer thickness is quite variable in time and space, ranging from hundreds of meters to a few kilometers.

An example of temperature variations in the lower troposphere is shown in Fig 1.2. These time-histories were constructed from rawinsonde soundings made every several hours near Lawton, Oklahoma. They show a *diurnal variation* of temperature near the ground that is not evident at greater altitudes. Such diurnal variation is one of the key characteristics of the boundary layer over land. The free atmosphere shows little diurnal variation.

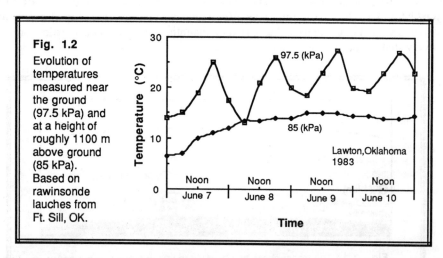

Fig. 1.2
Evolution of temperatures measured near the ground (97.5 kPa) and at a height of roughly 1100 m above ground (85 kPa). Based on rawinsonde lauches from Ft. Sill, OK.

This diurnal variation is not caused by direct forcing of solar radiation on the boundary layer. Little solar radiation is absorbed in the boundary layer; most is transmitted to the ground where typical absorptivities on the order of 90% result in absorption of much of the solar energy. It is the ground that warms and cools in response to the radiation, which in turn forces changes in the boundary layer via transport processes. *Turbulence* is one of the important transport processes, and is sometimes also used to define the boundary layer.

Indirectly, the whole troposphere can change in response to surface characteristics, but this response is relatively slow outside of the boundary layer. Hence, our definition of the boundary layer includes a statement about one-hour time scales. This does not imply that the boundary layer reaches an equilibrium in that time, just that alterations have at least begun.

Two types of clouds are often included in boundary-layer studies. One is the *fair-weather cumulus cloud*. It is so closely tied to thermals in the boundary layer that it is difficult to study the dynamics of this cloud type without focusing on the triggering boundary-layer mechanisms. The other type is the **stratocumulus cloud**. It fills the upper portion of a well-mixed, humid boundary layer where cooler temperatures allow condensation of water vapor. *Fog*, a stratocumulus cloud that touches the ground, is also a boundary-layer phenomenon.

Thunderstorms, while not a surface forcing, can modify the boundary layer in a matter of minutes by drawing up boundary-layer air into the cloud, or by laying down a carpet of cold downdraft air. Although thunderstorms are rarely considered to be boundary layer phenomena, their interaction with the boundary layer will be reviewed in this book.

1.2 Wind and Flow

Air flow, or wind, can be divided into three broad categories: *mean wind*, *turbulence*, and *waves* (Fig 1.3). Each can exist separately, or in the presence of any of the others. Each can exist in the boundary layer, where transport of quantities such as moisture, heat, momentum, and pollutants is dominated in the horizontal by the mean wind, and in the vertical by turbulence.

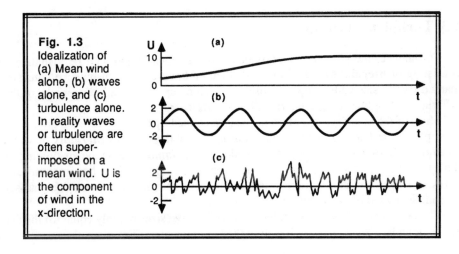

Fig. 1.3 Idealization of (a) Mean wind alone, (b) waves alone, and (c) turbulence alone. In reality waves or turbulence are often superimposed on a mean wind. U is the component of wind in the x-direction.

Mean wind is responsible for very rapid horizontal transport, or *advection*. Horizontal winds on the order of 2 to 10 m/s are common in the boundary layer. Friction causes the mean wind speed to be slowest near the ground. Vertical mean winds are much smaller, usually on the order of millimeters to centimeters per second.

Waves, which are frequently observed in the nighttime boundary layer, transport little heat, humidity, and other scalars such as pollutants. They are, however, effective at transporting momentum and energy. These waves can be generated locally by mean-wind shears and by mean flow over obstacles. Waves can also propagate from some distant source, such as a thunderstorm or an explosion.

The relatively high frequency of occurrence of turbulence near the ground is one of the characteristics that makes the boundary layer different from the rest of the atmosphere. Outside of the boundary layer, turbulence is primarily found near the jet stream where strong wind shears can create clear air turbulence (CAT).

Sometimes atmospheric waves may enhance the wind shears in localized regions, causing turbulence to form. Thus, wave phenomena can be associated with the turbulent transport of heat and pollutants, although waves without turbulence would not be as effective.

A common approach for studying either turbulence or waves is to split variables such as temperature and wind into a *mean part* and a *perturbation part*. The mean part represents the effects of the mean temperature and mean wind, while the perturbation part can represent either the wave effect or the turbulence effect that is superimposed on the mean wind.

As will be seen in later chapters of this book, such a splitting technique can be applied to the equations of motion, creating a number of new terms. Some of these terms, consisting of products of perturbation variables, describe *nonlinear* interactions between variables and are associated with turbulence. These terms are usually neglected when wave motions are of primary interest. Other terms, containing only one perturbation variable, describe *linear* motions that are associated with waves. These terms are neglected when turbulence is emphasized.

1.3 Turbulent Transport

Turbulence, the gustiness superimposed on the mean wind, can be visualized as consisting of irregular swirls of motion called *eddies*. Usually turbulence consists of many different size eddies superimposed on each other. The relative strengths of these different scale eddies define the *turbulence spectrum*.

Much of the boundary layer turbulence is generated by forcings from the ground. For example, *solar heating* of the ground during sunny days causes *thermals* of warmer air to rise. These thermals are just large eddies. *Frictional drag* on the air flowing over the ground causes *wind shears* to develop, which frequently become turbulent. *Obstacles* like trees and buildings deflect the flow, causing *turbulent wakes* adjacent to, and downwind of the obstacle.

The largest boundary layer eddies scale to (ie, have sizes roughly equal to) the depth of the boundary layer; that is, 100 to 3000 m in diameter. These are the most intense eddies because they are produced directly by the forcings discussed above. "Cats paws" on lake surfaces and looping smoke plumes provide evidence of the larger eddies.

Smaller size eddies are apparent in the swirls of leaves and in the wavy motions of the grass. These eddies feed on the larger ones. The smallest eddies, on the order of a few millimeters in size, are very weak because of the dissipating effects of molecular viscosity.

Turbulence is several orders of magnitude more effective at transporting quantities than is molecular diffusivity. It is turbulence that allows the boundary layer to respond to changing surface forcings. The frequent lack of turbulence above the boundary layer means that the rest of the free atmosphere cannot respond to surface changes. Stated more directly, the free atmosphere behaves as if there were no boundary to contend with, except in sense of mean wind flowing over the boundary-layer-top height contours.

1.4 Taylor's Hypothesis

We often need information on the size of eddies and on the scales of motions in the boundary layer. Unfortunately, it is difficult to create a snapshot picture of the boundary layer. Instead of observing a large region of space at an instant in time, we find it easier to make measurements at one point in space over a long time period. For example, meteorological instruments mounted on a tower can give us a time record of the boundary layer as it blows past our sensors.

In 1938, G. I. Taylor suggested that for some special cases, turbulence might be considered to be *frozen* as it advects past a sensor. Thus, the wind speed could be used to translate turbulence measurements as a function of time to their corresponding measurements in space. We must keep in mind that turbulence is not really frozen. Taylor's simplification is thus useful for only those cases where the turbulent eddies evolve with a timescale longer than the time it takes the eddy to be advected past a sensor (Powell and Elderkin, 1974).

Let U and V represent the eastward-moving and northward-moving Cartesian wind components, and let M represent the total wind magnitude (speed) given by $M^2 = U^2 + V^2$. If an eddy of diameter λ is advected at mean wind speed M, then the time period $\mathbb{P}$ for it to pass by a stationary sensor is given by $\mathbb{P} = \lambda / M$.

Suppose that some variable like temperature varies from one side of the eddy to the other. Then the temperature measured at our sensors would vary with time as the eddy advects past. For example, Fig 1.4a shows an initial condition where an eddy of 100 m in diameter is beginning to advect past our sensor tower. The leading side of the eddy, at a temperature of 10°C, is warmer than the trailing side, at only 5°C. This is our spatial description of the eddy at an instant in time. Namely, the temperature gradient across the eddy is $\partial T/\partial x_d = 0.05$ K/m, where x_d is in a direction parallel to the mean wind.

At that initial instant, our sensor on the tower measures a temperature of 10°C. If the wind speed were 10 m/s, then the sensor would measure a temperature of 5°C ten seconds later, assuming that the eddy has not changed as it advected by (Fig 1.4b). The local change of temperature with time measured at our sensor is $\partial T/\partial t = -0.5$ K/s. We see that:

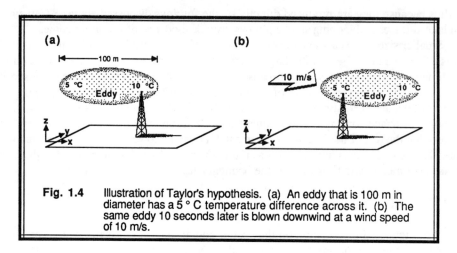

Fig. 1.4   Illustration of Taylor's hypothesis. (a) An eddy that is 100 m in diameter has a 5 ° C temperature difference across it. (b) The same eddy 10 seconds later is blown downwind at a wind speed of 10 m/s.

$$\frac{\partial T}{\partial t} = -M \frac{\partial T}{\partial x_d} \tag{1.4a}$$

which is an expression of Taylor's hypothesis for temperature in one dimension.

For any variable ξ, Taylor's hypothesis states that turbulence is frozen when $d\xi/dt = 0$. But the total derivative is defined by: $d\xi/dt = \partial\xi/\partial t + U \, \partial\xi/\partial x + V \, \partial\xi/\partial y + W \, \xi/\partial z$. Thus, the general form of Taylor's hypothesis is

$$\frac{\partial\xi}{\partial t} = -U \frac{\partial\xi}{\partial x} - V \frac{\partial\xi}{\partial y} - W \frac{\partial\xi}{\partial z} \tag{1.4b}$$

This hypothesis can also be stated in terms of a wavenumber, κ, and frequency, f:

$$\kappa = f / M \tag{1.4c}$$

where $\kappa = 2\pi/\lambda$, and $f = 2\pi/\mathbb{P}$, for wavelength λ and wave period $\mathbb{P}$ (Wyngaard and Clifford, 1977). The dimensions of κ are radians per unit length, while f has dimensions of radians per unit time.

To satisfy the requirements that the eddy have negligible change as it advects past a sensor, Willis and Deardorff (1976) suggest that

$$\sigma_M < 0.5 \, M \tag{1.4d}$$

where σ_M, the standard deviation (see chapter 2 for a review of statistics) of wind speed, is a measure of the intensity of turbulence. Thus, Taylor's hypothesis should be satisfactory when the turbulence intensity is small relative to the mean wind speed.

1.5 Virtual Potential Temperature

Buoyancy is one of the driving forces for turbulence in the BL. Thermals of warm air rise because they are less dense than the surrounding air, and hence *positively buoyant*. *Virtual temperature* is a popular variable for these studies because it is the temperature that dry air must have to equal the density of moist air at the same pressure. Thus, variations of virtual temperature can be studied in place of variations in density.

Water vapor is less dense than dry air; thus, moist unsaturated air is more buoyant than dry air of the same temperature. The virtual temperature of unsaturated moist air is therefore always greater than the absolute air temperature, T. Liquid water, however, is more dense than dry air, making cloudy air heavier or less buoyant than the corresponding cloud-free air. The suspension of cloud droplets in an air parcel is called *liquid water loading*, and it always reduces the virtual temperature.

Virtual potential temperatures are analogous to potential temperatures in that they remove the temperature variation caused by changes in pressure altitude of an air parcel. Turbulence includes vertical movement of air, making a variable such as virtual potential temperature not just attractive but almost necessary.

1.5.1 Definitions

For saturated (cloudy) air, the virtual potential temperature, θ_v, is defined by:

$$\theta_v = \theta \cdot (1 + 0.61 \cdot r_{sat} - r_L) \qquad (1.5.1a)$$

where r_{sat} is the water-vapor saturation mixing ratio of the air parcel, and r_L is the liquid-water mixing ratio. In (1.5.1a) the potential temperatures are in units of K, and the mixing ratios are in units of g/g. For unsaturated air with mixing ratio r, the virtual potential temperature is:

$$\theta_v = \theta \cdot (1 + 0.61 \cdot r) \qquad (1.5.1b)$$

A derivation of the virtual temperature is given in Appendix D.

As usual, the potential temperature, θ, is defined as

$$\theta = T \left(\frac{P_o}{P} \right)^{0.286} \qquad (1.5.1c)$$

where P is air pressure and P_o is a reference pressure. Usually, P_o is set to 100 kPa

(1000 mb), but sometimes for boundary layer work the surface pressure is used instead.
To first order, we can approximate the potential temperature by

$$\theta \cong T + (g/C_p) \cdot z \qquad (1.5.1d)$$

where z is the height above the 100 kPa (1000 mb) level, although sometimes height above ground level (agl) is used instead. The quantity $g/C_p = 0.0098$ K/m is just the negative of the dry adiabatic lapse rate (9.8 °C per kilometer), where g is the gravitational acceleration and C_p is the specific heat at constant pressure for air. Sometimes the quantity $C_p \cdot \theta$ is called the *dry static energy*.

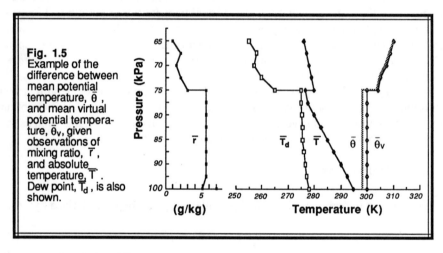

Fig. 1.5
Example of the difference between mean potential temperature, $\bar{\theta}$, and mean virtual potential temperature, $\bar{\theta}_v$, given observations of mixing ratio, $\bar{r}$, and absolute temperature, $\bar{T}$. Dew point, $\bar{T}_d$, is also shown.

An example of the difference between potential temperature and virtual potential temperature is shown in Fig 1.5 for a case of moist unsaturated air. The difference is small, but not negligible. Only when the air is very dry can we neglect the difference.

1.5.2   Example

Problem. Given a temperature of 25°C and a mixing ratio of 20 g/kg measured at a pressure of 90 kPa (900 mb), find the virtual potential temperature.

Solution. First, we must find the potential temperature:

$$\theta = T (P_o/P)^{0.286} = 298.16 \cdot (100/90)^{0.286} = 307.28 \text{ K}$$

The air is unsaturated, allowing us to find the virtual potential temperature from:

$$\theta_v = \theta \cdot (1 + 0.61 \cdot r) = 307.28 \cdot [1 + 0.61 \cdot (0.020)] = 311.03 \text{ K}$$

Discussion. Even though the virtual potential temperature is only about 4 K warmer than the potential temperature, this difference is on the same order as the difference between the warm air rising in thermals and the surrounding environment. Thus, neglect of the humidity in buoyancy calculations could lead to erroneous conclusions regarding convection and turbulence.

1.6 Boundary Layer Depth and Structure

Over oceans, the boundary layer depth varies relatively slowly in space and time. The sea surface temperature changes little over a diurnal cycle because of the tremendous mixing within the top of the ocean. Also, water has a large heat capacity, meaning that it can absorb large amounts of heat from the sun with relatively little temperature change. Thus, a slowly varying sea surface temperature means a slowly varying forcing into the bottom of the boundary layer.

Most changes in boundary layer depth over oceans are caused by synoptic and mesoscale processes of vertical motion and advection of different air masses over the sea surface. An air mass with a temperature different than that of the ocean will undergo a modification as its temperature equilibrates with that of the sea surface. Once equilibrium is reached, the resulting boundary layer depth might vary by only 10% over a horizontal distance of 1000 km. Exceptions to this gentle variation can occur near the borders between two ocean currents of different temperatures (Stage and Weller, 1976).

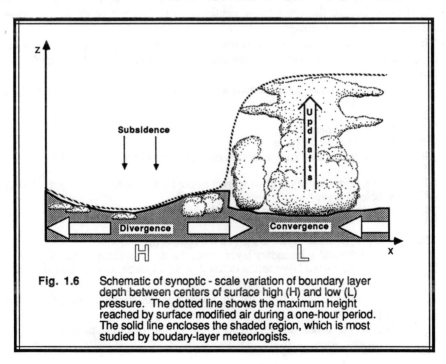

Fig. 1.6   Schematic of synoptic - scale variation of boundary layer depth between centers of surface high (H) and low (L) pressure. The dotted line shows the maximum height reached by surface modified air during a one-hour period. The solid line encloses the shaded region, which is most studied by boudary-layer meteorlogists.

Over both land and oceans, the general nature of the boundary layer is to be thinner in high-pressure regions than in low-pressure regions (Fig 1.6). The subsidence and low-level horizontal divergence associated with synoptic high pressure moves boundary layer air out of the high towards lower pressure regions. The shallower depths are often associated with cloud-free regions. If clouds are present, they are often fair-weather cumulus or stratocumulus clouds.

In low pressure regions the upward motions carry boundary-layer air away from the ground to large altitudes throughout the troposphere. It is difficult to define a boundary-layer top for these situations. Cloud base is often used as an arbitrary cut-off for boundary layer studies in these cases. Thus, the region studied by boundary layer meteorologists may actually be thinner in low-pressure regions than in high pressure ones (see Fig 1.6).

Over land surfaces in high pressure regions the boundary layer has a well defined structure that evolves with the diurnal cycle (Fig 1.7). The three major components of this structure are the *mixed layer*, the *residual layer*, and the *stable boundary layer*. When clouds are present in the mixed layer, it is further subdivided into a *cloud layer* and a *subcloud layer*.

The *surface layer* is the region at the bottom of the boundary layer where turbulent fluxes and stress vary by less than 10% of their magnitude. Thus, the bottom 10% of the boundary layer is called the surface layer, regardless of whether it is part of a mixed layer or stable boundary layer. Finally, a thin layer called a *microlayer* or *interfacial layer* has been identified in the lowest few centimeters of air, where molecular transport dominates over turbulent transport.

The following shorthand notation is often used for the various parts of the boundary layer. For the sake of completeness, some additional terms are listed here that will not be discussed until later:

BL   Boundary layer (also known as the planetary boundary layer, PBL, or the atmospheric boundary layer, ABL)
CL   Cloud layer
FA   Free atmosphere
IBL   Internal boundary layer
ML   Mixed layer (also known as the convective boundary layer, CBL)
RL   Residual layer
SBL   Stable boundary layer (also known as the nocturnal boundary layer, NBL)
SCL   Subcloud layer
SL   Surface layer (the bottom 10% of the boundary layer)

The tops of four of these layers are given the following symbols:
h   Top of the stable boundary layer (often defined as the top of the NBL)
z_i Top of the mixed layer (often defined as the average base of the overlying stable layer)
z_r Top of the residual layer (often defined as the average base of the overlying stable layer)
z_b Top of the subcloud layer (this is the height of cloud base, usually near the *lifting condensation level*, LCL)

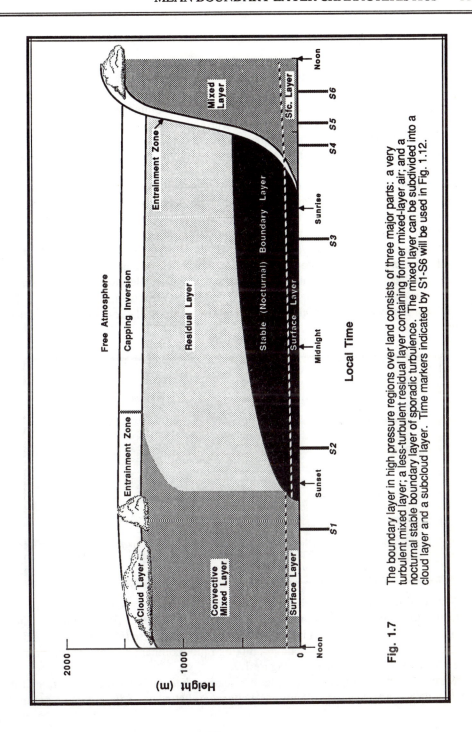

Fig. 1.7   The boundary layer in high pressure regions over land consists of three major parts: a very turbulent mixed layer; a less-turbulent residual layer containing former mixed-layer air; and a nocturnal stable boundary layer of sporadic turbulence. The mixed layer can be subdivided into a cloud layer and a subcloud layer. Time markers indicated by S1-S6 will be used in Fig. 1.12.

1.6.1 Mixed Layer

The turbulence in the mixed layer is usually *convectively driven*, although a nearly well-mixed layer can form in regions of strong winds. Convective sources include heat transfer from a warm ground surface, and radiative cooling from the top of the cloud layer. The first situation creates thermals of warm air rising from the ground, while the second creates thermals of cool air sinking from cloud top. Both can occur simultaneously, particularly when a cool stratocumulus topped mixed layer is being advected over warmer ground.

Even when convection is the dominant mechanism, there is usually *wind shear* across the top of the ML that contributes to the turbulence generation. This free-shear situation is more akin to CAT, and is thought to be associated with the formation and breakdown of waves in the air known as *Kelvin-Helmholtz* waves.

On initially cloud-free days, however, ML growth is tied to solar heating of the ground. Starting about a half hour after sunrise, a turbulent ML begins to grow in depth. This ML is characterized by intense mixing in a statically unstable situation where thermals of warm air rise from the ground (Fig 1.8). The ML reaches its maximum depth in late afternoon. It grows by *entraining*, or mixing down into it, the less turbulent air from above.

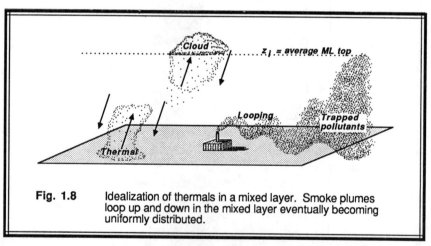

Fig. 1.8   Idealization of thermals in a mixed layer. Smoke plumes loop up and down in the mixed layer eventually becoming uniformly distributed.

The resulting turbulence tends to mix heat, moisture, and momentum uniformly in the vertical. Pollutants emitted from smoke stacks exhibit a characteristic *looping* as those portions of the effluent emitted into warm thermals begin to rise (Fig 1.8). The resulting profiles of virtual potential temperature, mixing ratio, pollutant concentration, and wind speed frequently are as sketched in Figure 1.9.

Virtual potential temperature profiles are nearly *adiabatic* in the middle portion of the ML. In the surface layer one often finds a *superadiabatic* layer adjacent to the ground.

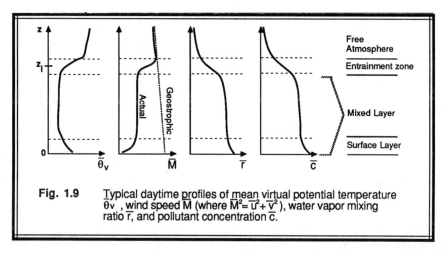

Fig. 1.9   Typical daytime profiles of mean virtual potential temperature $\overline{\theta}_v$, wind speed $\overline{M}$ (where $\overline{M}^2 = \overline{u}^2 + \overline{v}^2$), water vapor mixing ratio $\overline{r}$, and pollutant concentration $\overline{c}$.

A stable layer at the top of the ML acts as a lid to the rising thermals, thus restraining the domain of turbulence. It is called the **entrainment zone** because entrainment into the ML occurs there.

At times this capping stable layer is strong enough to be classified as a temperature inversion; that is, the absolute temperature increases with height. In fact, it is frequently called an **inversion layer** regardless of the magnitude of the stability. The most common symbol for ML depth is z_i, which represents the average height of the inversion base.

Wind speeds are **subgeostrophic** throughout the ML, with wind directions **crossing the isobars** at a small angle towards low pressure. The middle portion of the ML frequently has nearly constant wind speed and direction. Wind speeds decrease towards zero near the ground, resulting in a wind speed profile that is nearly **logarithmic** with height in the surface layer. Wind directions cross the isobars at increasingly large angles as the ground is approached, with 45 degree angles not uncommon near the surface.

Mixing ratios tend to decrease with height, even within the center portion of the ML. This reflects the evaporation of soil and plant moisture from below, and the entrainment of drier air from above. The moisture decrease across the top of the ML is very pronounced, and is often used together with potential temperature profiles to identify the ML top from rawinsonde soundings.

Most pollutant sources are near the earth's surface. Thus, pollutant concentrations can build up in the ML while FA concentrations remain relatively low. Pollutants are transported by eddies such as thermals; therefore, the inability of thermals to penetrate very far into the stable layer means that the stable layer acts as a lid to the pollutants too. **Trapping** of pollutants below such an "inversion layer" is common in high-pressure regions, and sometimes leads to pollution alerts in large communities.

As the tops of the highest thermals reach greater and greater depths during the course of the day, the highest thermals might reach their *lifting condensation level*, LCL, if sufficient moisture is present. The resulting fair-weather clouds are often targets for soaring birds and glider pilots, who seek the updraft of the thermals.

High or middle overcast can reduce the insolation at ground level. This, in turn, reduces the intensity of thermals. On these days the ML may exhibit slower growth, and may even become nonturbulent or neutrally-stratified if the clouds are thick enough.

1.6.2 Residual Layer

About a half hour before sunset the thermals cease to form (in the absence of cold air advection), allowing turbulence to decay in the formerly well-mixed layer. The resulting layer of air is sometimes called the residual layer because its initial mean state variables and concentration variables are the same as those of the recently-decayed mixed layer.

For example, in the absence of advection, passive tracers dispersed into the daytime mixed layer will remain aloft in the RL during the night. The RL is neutrally stratified, resulting in turbulence that is nearly of equal intensity in all directions. As a result, smoke plumes emitted into the RL tend to disperse at equal rates in the vertical and lateral directions, creating a cone-shaped plume. Figure 1.10 shows a sketch of *coning*.

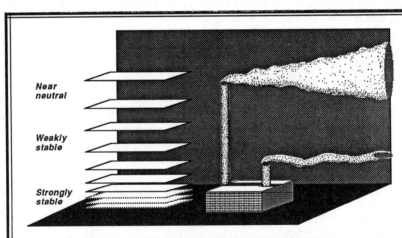

Fig. 1.10 The static stability decreases with height in the nocturnal boundary layer, gradually blending into the neutrally-stratified residual layer aloft, as indicated by the isentropic surfaces sketched on the left. Smoke emissions into the stable air fan out in the horizontal with little vertical dispersion other than wavelike oscillations. Smoke emissions in the neutral residual-layer air spread with an almost equal rate in the vertical and horizontal, allowing the smoke plume to assume a cone-like shape.

Nonpassive pollutants may react with other constituents during the night to create compounds that were not originally emitted from the ground. Sometimes gaseous chemicals may react to form aerosols or particulates which can precipitate out. The RL often exists for a while in the mornings before being entrained into the new ML. During this time solar radiation may trigger photochemical reactions among the constituents in the RL.

Moisture often behaves as a passive tracer. Each day, more moisture may be evaporated into the ML and will be retained in the RL. During succeeding days, the re-entrainment of the moist air into the ML might allow cloud formation to occur where it otherwise might not.

Variables such as virtual potential temperature usually decrease slowly during the night because of radiation divergence. This cooling rate is on the order of 1 °C/d. The cooling rate is more-or-less uniform throughout the depth of the RL, thus allowing the RL virtual potential temperature profile to remain nearly adiabatic. When the top of the next day's ML reaches the base of the RL, the ML growth becomes very rapid.

The RL does not have direct contact with the ground. During the night, the nocturnal stable layer gradually increases in thickness by modifying the bottom of the RL. Thus, the remainder of the RL is not affected by turbulent transport of surface-related properties and hence does not really fall within our definition of a boundary layer. Nevertheless, we will include the RL in our studies as an exception to the rule.

1.6.3 Stable Boundary Layer

As the night progresses, the bottom portion of the residual layer is transformed by its contact with the ground into a stable boundary layer. This is characterized by statically stable air with weaker, sporadic turbulence. Although the wind at ground level frequently becomes lighter or calm at night, the winds aloft may accelerate to *supergeostrophic speeds* in a phenomenon that is called the *low-level jet* or *nocturnal jet*.

The statically stable air tends to suppress turbulence, while the developing nocturnal jet enhances wind shears that tend to generate turbulence. As a result, turbulence sometimes occurs in relatively short bursts that can cause mixing throughout the SBL. During the nonturbulent periods, the flow becomes essentially decoupled from the surface.

As opposed to the daytime ML which has a clearly defined top, the SBL has a poorly-defined top that smoothly blends into the RL above (Fig 1.10 and 1.11). The top of the ML is defined as the base of the stable layer, while the SBL top is defined as the top of the stable layer or the height where turbulence intensity is a small fraction of its surface value.

Pollutants emitted into the stable layer disperse relatively little in the vertical. They disperse more rapidly, or "fan out", in the horizontal. This behavior is called *fanning*, and is sketched as the bottom smoke plume in Fig 1.10. Sometimes at night when winds are lighter, the effluent *meanders* left and right as it drifts downwind.

Winds exhibit a very complex behavior at night. Just above ground level the wind speed often becomes light or even calm. At altitudes on the order of 200 m above ground, the wind may reach 10-30 m/s in the nocturnal jet. Another few hundred meters above

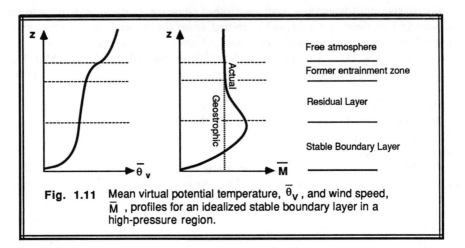

Fig. 1.11 Mean virtual potential temperature, $\overline{\theta}_v$, and wind speed, $\overline{M}$, profiles for an idealized stable boundary layer in a high-pressure region.

that, the wind speed is smaller and closer to its geostrophic value. The strong shears below the jet are accompanied by a rapid change in wind direction, where the lower level winds are directed across the isobars towards low pressure.

Touching the ground, however, is a thin (order of a few meters) layer of *katabatic* or *drainage* winds. These winds are caused by the colder air, adjacent to the ground, flowing downhill under the influence of gravity. Wind speeds of 1 m/s at a height of 1 m are possible. This cold air collects in the valleys and depressions and stagnates there. Unfortunately, many weather stations are located in or near valleys, where the observed surface winds bear little relationship to the synoptic-scale forcings at night.

Wave motions are a frequent occurrence in the SBL. The strongly stable NBL not only supports gravity waves, but it can trap many of the higher-frequency waves near the ground. Vertical wave displacements of 100 m have been observed, although the associated wind and temperature oscillations are relatively small and difficult to observe without sensitive instruments.

SBLs can also form during the day, as long as the underlying surface is colder than the air. These situations often occur during warm-air advection over a colder surface, such as after a warm frontal passage or near shorelines.

1.6.4 Virtual Potential Temperature Evolution

Given the virtual potential temperature profiles from the previous subsections, it is useful to integrate these profiles into our concept of how the boundary layer evolves. If rawinsonde soundings were made at the times indicated by flags S1 through S6 in Fig 1.7, then Fig 1.12 shows the resulting virtual potential temperature profile evolution.

We see from these soundings that knowledge of the virtual potential temperature profile is usually sufficient to identify the parts of the boundary layer. The structure of the BL is clearly evident.

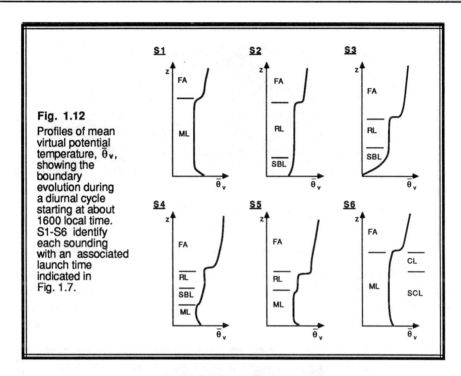

Fig. 1.12
Profiles of mean virtual potential temperature, $\bar{\theta}_v$, showing the boundary evolution during a diurnal cycle starting at about 1600 local time. S1-S6 identify each sounding with an associated launch time indicated in Fig. 1.7.

Stated another way, knowledge of the virtual potential temperature lapse rate is usually sufficient for determining the static stability. An exception to this rule is evident by comparing the lapse rate in the middle of the RL with that in the middle of the ML. Both are adiabatic; yet, the ML corresponds to statically unstable air while the RL contains statically neutral air.

One way around this apparent paradox for the classification of adiabatic layers is to note the lapse rate of the air immediately below the adiabatic layer. If the lower air is superadiabatic, then both that superadiabatic layer and the overlying adiabatic layer are statically unstable. Otherwise, the adiabatic layer is statically neutral. A more precise definition of static stability is presented later.

It is obvious that as the virtual potential temperature profile evolves with time, so must the behavior of smoke plumes. For example, smoke emitted into the top of the NBL or into the RL rarely is dispersed down to the ground during the night because of the limited turbulence. These smoke plumes can be advected hundreds of kilometers downwind from their sources during the night.

Smoke plumes in the RL may disperse to the point where the bottom of the plume hits the top of the NBL. The strong static stability and frequent reduction in turbulence reduces the downward mixing into the NBL. The top of the smoke plume sometimes can continue to rise into the neutral air. This is called *lofting* (see Fig 1.13).

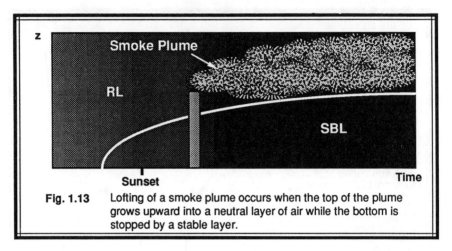

Fig. 1.13   Lofting of a smoke plume occurs when the top of the plume grows upward into a neutral layer of air while the bottom is stopped by a stable layer.

After sunrise a new ML begins to grow, eventually reaching the height of the elevated smoke plume from the previous night. At this time, the elevated pollutants are mixed down to the ground by ML entrainment and turbulence in a process that is called *fumigation*. A sketch of this process is shown in Fig 1.14. An analogous process is

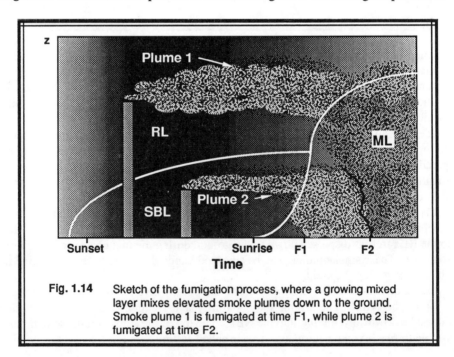

Fig. 1.14   Sketch of the fumigation process, where a growing mixed layer mixes elevated smoke plumes down to the ground. Smoke plume 1 is fumigated at time F1, while plume 2 is fumigated at time F2.

often observed near shorelines, where elevated smoke plumes in stable or neutral air upstream of the shoreline are continuously fumigated downstream of the shoreline after advecting over a warmer bottom boundary that supports ML growth.

1.7 Micrometeorology

Compared to the other scales of meteorological motions, turbulence is on the small end. Figure 1.15 shows a classification scheme for meteorological phenomena as a function of their time and space scales. Phenomena such as turbulence with space scales smaller than about 3 km and with time scales shorter than about 1 h are classified as *microscale*. *Micrometeorology* is the study of such small-scale phenomena.

It is evident that the study of the boundary layer involves the study of microscale processes. For this reason, boundary layer meteorology and micrometeorology are virtually synonymous. Since many of the early micrometeorological measurements were made with sensors on short stands and towers, micrometeorology was often associated with surface-layer phenomena.

Regardless of what you call it, the small-scale phenomena being studied here are so transient in nature that the deterministic description and forecasting of each individual eddy is virtually impossible. As a result, micrometeorologists have developed three primary avenues for exploring their subject:

- stochastic methods
- similarity theory
- phenomenological classifications.

Stochastic methods deal with the average statistical effects of the eddies. *Similarity theory* involves the apparent common-behavior exhibited by many empirically-observed phenomena, when properly scaled. In the *phenomenological* methods, the largest size structures such as thermals are classified and sometimes approached in a partially deterministic manner.

Micrometeorology has always relied heavily on *field experiments* to learn more about the boundary layer. Unfortunately, the large variety of scales involved and the tremendous variability in the vertical require a large array of sensors including airborne platforms and remote sensors. The relatively large costs have limited the scope of many field experiments. Only a few general-purpose, large-scale boundary layer experiments have been conducted.

Alternative studies have used *numerical* and *laboratory simulations*. Much of the turbulence work has been performed in laboratory tanks, usually using liquids such as water as the working medium. Although there have been many successful laboratory studies of small-scale turbulence, there have been only a few simulations of larger phenomena such as thermals. Wind tunnel studies have been used to observe the flow of neutral boundary layers over complex terrain and buildings, although the difficulty of stratifying the air has meant that typical daytime and nighttime boundary layers could not be adequately simulated.

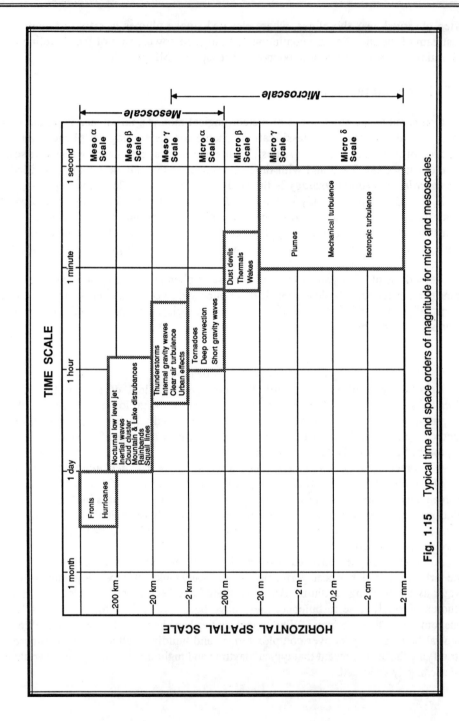

Fig. 1.15   Typical time and space orders of magnitude for micro and mesoscales.

Numerical simulation using digital computers has been very popular since the 1960's, with many discoveries that were verified years later with field studies. Most of these computer simulations employ the stochastic method of modeling fluid flow. Unfortunately, a difficulty known as the *closure problem* has meant that each of these models has at best been able to only approximate the governing equations, with uncertainties introduced via a necessary parameterization of the unknowns.

That same closure problem has limited the avenues for *theoretical studies* involving analytical solutions. For the most part, only highly simplified approximations to the boundary layer have been amenable to direct solution.

If anything, these difficulties have stimulated, rather than stifled, the work of micrometeorologists. There is an underlying assumption in meteorology that subsynoptic-scale phenomena such as turbulence might be responsible, in part, for the difficulty in making quality weather forecasts beyond a few days. Thus, part of the effort in boundary layer meteorology involves the search for adequate turbulence parameterization schemes for larger-scale numerical forecast models.

Additional motivation has come from concern over our environment. Every species of animal and plant modifies its environment; the human species, however, is in a position to recognize the consequences of its pollution and take appropriate action. Since most of the anthropogenic effluents are emitted from near-surface sources, the resulting dispersion of the pollutants is tied to boundary layer processes. As a result, *air-pollution meteorology* is an applied form of micrometeorology.

Other applications include *agricultural meteorology*, where airborne transport of chemicals necessary to plant life is governed by turbulence. Nocturnal processes such as frost formation warrant improved study and forecast methods for crop protection. Fog and low stratocumulus, which inhibit aviation operations, are essentially boundary layer phenomena. Wind-generated power, a popular energy source for centuries, has had a recent increase in interest as wind turbines have been designed to extract energy more efficiently from the boundary layer wind. Other structures such as bridges and buildings must be designed to withstand wind gusts appropriate to their sites.

1.8 Significance of the Boundary Layer

The role of the boundary layer on our lives is put into perspective when we compare the characteristics of the boundary layer and free atmosphere (Table 1-1). A taste of the importance of the BL is given in the following summary:
 • People spend most of their lives in the BL.
 • Daily weather forecasts of dew, frost, and maximum and minimum temperatures are really BL forecasts.
 • Pollution is trapped in the BL.
 • Fog occurs within the BL.
 • Some aviation, shipping, and other commerce activities conducted within it.
 • Air masses are really boundary layers in different parts of the globe that have equilibrated with their underlying surface. Baroclinicity is generated this way.

Table 1-1. Comparison of boundary layer and free atmosphere characteristics.

Property	Boundary Layer	Free Atmosphere
Turbulence	• Almost continuously turbulent over its whole depth.	• Sporadic CAT in thin layers of large horizontal extent.
Friction	• Strong drag against the earth's surface. Large energy dissipation.	• Small viscous dissipation.
Dispersion	• Rapid turbulent mixing in the vertical and horizontal.	• Small molecular diffusion. Often rapid horizontal transport by mean wind.
Winds	• Near logarithmic wind speed profile in the surface layer. Subgeostrophic, cross-isobaric flow common.	• Winds nearly geostrophic.
Vertical Transport	• Turbulence dominates.	• Mean wind and cumulus-scale dominate
Thickness	• Varies between 100 m to 3 km in time and space. Diurnal oscillations over land.	• Less variable. 8-18 km. Slow time variations.

- The primary energy source for the whole atmosphere is solar radiation, which for the most part is absorbed at the ground and transmitted to the rest of the atmosphere by BL processes. About 90% of the net radiation absorbed by oceans causes evaporation, amounting to the evaporation of about 1m of water per year over all the earth's ocean area. The latent heat stored in water vapor accounts for 80% of the fuel that drives atmospheric motions.
- Crops are grown in the BL. Pollen distributed by boundary layer circulations.
- Cloud nuclei are stirred into the air from the surface by BL processes.
- Virtually all water vapor that reaches the FA is first transported through the BL by turbulent and advective processes.
- Thunderstorm and hurricane evolution are tied to the inflow of moist BL air.
- Turbulent transport of momentum down through the BL to the surface is the most important momentum sink for the atmosphere.
- About 50% of the atmosphere's kinetic energy is dissipated in the BL.
- Turbulence and gustiness affects architecture in the design of structures.

• Wind turbines extract energy from the BL winds.
• Wind stress on the sea surface is the primary energy source for ocean currents.
• Turbulent transport and advection in the BL move water and oxygen to and from immobile life forms like plants.

Obviously the list could go on, but the main point is BL processes affect our lives directly, and indirectly via its influence on the rest of the weather. In this book we examine some of the processes that occur in the BL, develop some schemes for coping with turbulence, and show how they can be applied to benefit mankind.

For additional information on boundary layers, turbulence and micrometeorology, the General References section below lists books and other secondary sources. A reference section at the end of each chapter lists the specific articles sited in the chapter.

A summary table of frequently-used scaling variables and dimensionless groups is given in Appendix A. Appendix B lists notation, including abbreviations, symbols and acronyms. Appendix C gives values of frequently-used parameters and constants. Additional appendices are specifically referenced in the text.

1.9 General References

Turbulence & Waves:
Frost, W. and T.H. Moulden, 1977: *Handbook of Turbulence, Vol.1, Fundamentals and Applications.* Plenum Press, NY. 498 pp.
Gossard, E.E. and W.H. Hooke, 1975: *Waves in the Atmosphere, Atmospheric Infrasound and Gravity Waves - their Generation and Propagation.* Elsevier Scientific Publ. Co., NY. 456pp.
Hinze, J.O., 1975: *Turbulence* (2nd ed). McGraw-Hill Series in Mechanical Engineering. McGraw-Hill Book Co., NY 790pp.
Lumley, J.L. and H.A. Panofsky, 1964: *The Structure of Atmospheric Turbulence.* Monographs and Texts in Physics and Astronomy. Vol XII. Interscience Publ., John Wiley & Sons, NY 239pp.
Monin, A.S. and A.M. Yaglom, 1973: *Statistical Fluid Mechanics*, Vols 1 & 2. Edited by John Lumley. The MIT Press, Cambridge, MA. 769pp.
Panofsky, H.A., and J.A. Dutton, 1984: *Atmospheric Turbulence, Models and Methods for Engineering Applications.* Wiley-Interscience, John Wiley & Sons, NY. 397pp.
Scorer, R.S., 1978: *Environmental Aerodynamics.* Ellis Horwood, Halsted Press, John Wiley & Sons. London. 488pp.
Stanisic, M.M., 1985: *The Mathematical Theory of Turbulence.* Universitext, Springer-Verlag, NY. 429pp.
Tennekes, H. and J.L. Lumley, 1982: *A First Course in Turbulence* (2nd Ed). The MIT Press, Cambridge, MA 300pp.
Townsend, A.A., 1976: *The Structure of Turbulent Shear Flow* (2nd Ed). Cambridge University Press, Cambridge, England. 429pp.
Tritton, D.J., 1977: *Physical Fluid Dynamics*, Van Nostrand Reinhold. NY. 362pp.
Turner, J.S., 1973: *Buoyancy Effects in Fluids*, Cambridge Univ. Press. 367pp.

Van Dyke, M., 1982: *An Album of Fluid Motion*. The Parabolic Press, Stanford.176 pp.

Boundary Layers:
Bhumralkar, C.M., 1975: *A Survey of Parameterization Techniques for the Planetary Boundary Layer in Atmospheric Circulation Models*. Report R-1653-ARPA July 1975. ARPA order no. 189-1. 6P10 Information Processing Techniques Office, Rand Corp, Santa Monica, CA 90406. 84pp.
Boundary Layer Meteorology, a journal published by Reidel, is devoted to boundary layer topics.
Coantic, M.F., 1978: *An Introduction to Turbulence in Geophysics, and Air-Sea Interactions*. NATP-AGARD. Available from NTIS, Springfield, VA 22161. 242pp.
Kraus, E.B., 1972: *Atmosphere-Ocean Interaction*. Clarendon Press, Oxford, England. 271pp.
Nieuwstadt, F.T.M. and H. van Dop, 1982: *Atmospheric Turbulence and Air Pollution Modeling*. D. Reidel Publ. Co., P.O. Box 17, 3300 AA Dordrecht, The Netherlands. 358pp.
Plate, E.J., 1971: *Aerodynamic Characteristics of Atmospheric Boundary Layers*. AEC Critical Review Series, US Atomic Energy Commission, Office of Information Services. Available as TID-25465 from NTIS, Springfield, VA 22151. 190pp.
Roll, H.U., 1965: *Physics of the Marine Atmosphere*, Academic Press, NY. 426pp.
Schlichting, H., 1968: *Boundary Layer Theory*, 6th Ed. McGraw-Hill Series in Mechanical Engineering, McGraw-Hill Book Co., NY 747pp.
Sorbjan, Z., 1988: *Structure of the Atmospheric Boundary Layer*. Prentice-Hall, NY. 300pp.
Stull, R.B., 1986: *Boundary Layer Basics, A Survey of Boundary Layer Meteorology*. (available from the author), 51pp. Also published as "Atmospheric Boundary Layer", *The Encyclopedia of Physical Science and Technology* (edited by R.A. Meyers). Academic Press, Inc., NY.
Wyngaard, J.C., 1980: *Workshop on the Planetary Boundary Layer*. Am. Meteor. Soc., 45 Beacon St., Boston, MA 02108. 322pp.

Boundary Layer Experimental Techniques:
Lenschow, D.H. (Ed.), 1986: *Probing the Atmospheric Boundary Layer*. Am. Meteor. Soc., 45 Beacon St., Boston, MA 02108. 269pp.
Vinnichenko, N.K., N.Z. Pinus, S.M. Shmeter, and G.N. Shur, 1980: *Turbulence in the Free Atmosphere* (2nd Ed.). Consultants Bureau, Plenum Publ., NY. 310 pp.

Micrometeorology:
Brutsaert, W., 1982: *Evaporation into the Atmosphere*, Theory, History, and Applications. D. Reidel Publ. Co., P.O. Box 17, 3300 AA Dordrecht, The Netherlands. 258pp.
Geiger, R., 1965: *The Climate Near the Ground*. Harvard Univ. Press, Cambridge,

MA. 611pp.

Haugen, D.A., (Ed.), 1973: *Workshop on Micrometeorology*. Amer. Meteor. Soc., 45 Beacon St., Boston, MA 02108. 392pp.

Oke, T.R., 1978. *Boundary Layer Climates*. Halsted Press, NY 372pp.

Sutton, O.G., 1953: *Micrometeorology*. McGraw-Hill (1977 reprint by Kreiger Publ. Co., Inc., 645 NY Ave., Huntington, NY 11743). 333pp.

1.10 References for this Chapter

Powell, D.C. and C.E. Elderkin, 1974: An investigation of the application of Taylor's hypothesis to atmospheric boundary layer turbulence. *J. Atmos. Sci.*, **31**, 990-1002.

Stage, S.A. and R.A. Weller, 1986: The frontal air-sea interaction experiment (FASINEX); part II: experimental plan. *Bull. Amer. Meteor. Soc.*, **67**, 16-20.

Taylor, G.I., 1938: The spectrum of turbulence. *Proc. R. Soc.*, **A164**, 476-490.

Willis, G.E. and J.W. Deardorff, 1976: On the use of Taylor's translation hypothesis for diffusion in the mixed layer. *Quart. J. Roy. Meteor. Soc.*, **102**, 817-822.

Wyngaard, J.C. and S.F. Clifford, 1977: Taylor's hypothesis and high-frequency turbulence spectra. *J. Atmos. Sci.*, **34**, 922-929.

1.11 Exercises

1) a) Resketch figure 1.7 without any shading.
b) Next, shade on your figure those regions where the lapse rate is superadiabatic. (A color like red is appropriate.)
c) In a different color (like dark blue), shade those regions that have a strongly stable lapse rate.
d) With a different color (like light blue), shade those regions that have a weakly stable lapse rate.
e) Leave the nearly adiabatic lapse rate regions unshaded. (Hint, look at the virtual potential temperature graphs shown after figure 1.7.)

2) From your answer above, we see that a large portion of the mixed layer and the residual layer have nearly adiabatic lapse rates. Nevertheless, there is a difference between the two. Comment on this difference. How would smoke disperse in these regions?

3) If a very tall smoke stack could inject a smoke plume into the free atmosphere, then what characteristic dispersion pattern would you expect? Options include looping, fanning, coning, lofting, fumigation, and trapping within a layer.

4) Given the boundary layer evolution shown in Fig 1.7 and 1.12, sketch the modified evolution that would occur if the sun was turned off at time S5. Also comment on the processes that could effectively turn off the effects of the sun on the boundary layer.

5) If you observed the following measurement of wind speed on a strip chart, then what diameter (in meters) are the major eddies? (Hint: First find the typical time duration of an eddy, and then employ Taylor's Hypothesis.)

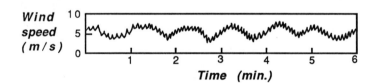

6) Suppose that the boundary layer can be approximated by a 1 km constant thickness layer over the whole continental United States. Above that, assume that a constant thickness (10 km) free atmosphere exists. Assume that each of these layers is well mixed for simplicity. (You may need to consult synoptics textbooks or study a number of weather maps to answer the following questions.)
a) On the average in spring, what fraction of the area of the continental United States is covered by cyclones (low pressure regions)?
b) What is the average vertical velocity magnitude out of the top of the boundary layer in those cyclonic regions?
c) What is the average subsidence velocity acting on the top of the boundary layer in anticyclonic (high pressure) regions? Remember that the boundary layer is of constant

thickness, so the total inflow of air must equal the outflow. (Normally, subsidence would not be able to cause air to penetrate down into the top of the boundary layer, but for simplicity we will assume that the turbulent entrainment rate of air into the top of the boundary layer is exactly equal to the subsidence velocity. Thus, all of the subsiding air would enter the boundary layer.)

d) What are the average residence times, P , of air within the boundary layer and the free atmosphere? Residence time is defined as P=V/F, where F = volumetric flow rate (volume/time) and V = volume. Neglect horizontal advection out of the United States.

e) Compare these residence times with the typical Rossby wave time period (that is, the average time period between cyclone passage over a fixed point). Comment on the significance of these times.

7) Would there be a boundary layer on a planet that had an atmosphere, but that did not experience a diurnal variation of net radiation at the ground?

8) Define the boundary layer in your own terms.

9) Add five items to the list in section 1.8 regarding the significance of the boundary layer. Hint, look for things close to home, school, or work.

10) a) Given air at a pressure height of 90 kPa (900 mb) with a temperature of 30 °C and a mixing ratio of 20 g/kg, find the virtual potential temperature.

b) Given saturated air at 85 kPa (850 mb) with a temperature of 20 °C and a total water mixing ratio (i.e., sum of vapor and liquid mixing ratios) of 20 g/kg find the virtual potential temperature.   (Hint, you might want to employ a thermodynamic diagram.)

11) Given the following virtual potential temperature sounding, identify each layer (example, ML, RL, SBL, FA). Estimate what time of day that sounding was made, and then sketch the virtual potential temperature profile that you might expect four hours later.

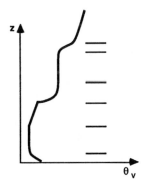

2 Some Mathematical & Conceptual Tools: Part 1. Statistics

Turbulence is an intrinsic part of the atmospheric boundary layer that must be quantified in order to study it. The randomness of turbulence makes deterministic description difficult. Instead, we are forced to retreat to the use of statistics, where we are limited to average or expected measures of turbulence. In this chapter we review some basic statistical methods and show how measurements of turbulence can be put into a statistical framework. Usually, this involves separating the turbulent from the nonturbulent parts of the flow, followed by averaging to provide the statistical descriptor.

The role of spectra in this separation of parts is also described. Although at first glance we see a labyrinth of motions, turbulence may be idealized as consisting of a variety of different-size swirls or eddies. These eddies behave in a well-ordered manner when displayed in the form of a spectrum.

Statistical descriptors such as the variance or covariance are of limited usefulness unless we can physically interpret them. Variances are shown to be measures of turbulence intensity or turbulence kinetic energy, and covariances are shown to be measures of flux or stress. Flux and stress concepts are explored further, and Einstein's summation notation is introduced as a shorthand way to write these variables.

The concepts developed in this chapter are used extensively in the remainder of the book to help describe the turbulent boundary layer.

2.1 The Signature of Turbulence and Its Spectrum

Since we have lived most of our lives within the turbulent boundary layer, we have developed feelings or intuitions about the nature of turbulence that can be refined to help us classify and describe this phenomenon. Consider figure 2.1 for example, which shows the near-surface wind speed measured during a 2.5 hour period. A number of features stand out.

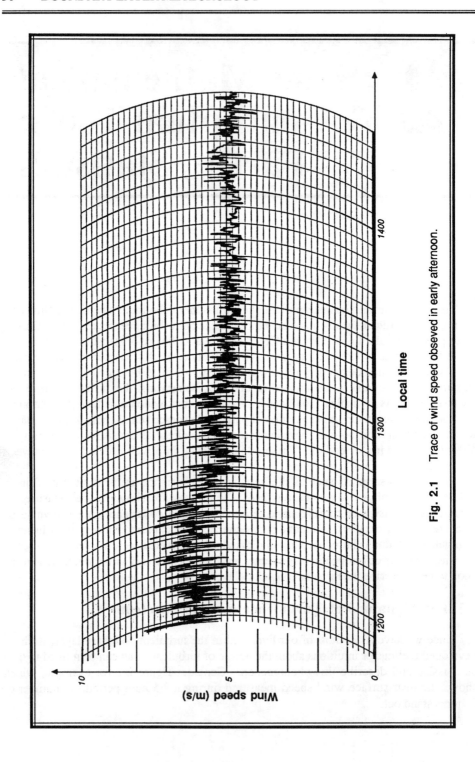

Fig. 2.1   Trace of wind speed obseved in early afternoon.

- The wind speed varies in an irregular pattern — a characteristic signature of turbulence. This quasi-randomness is what makes turbulence different from other motions, like waves.
- We can visually pick out a mean, or typical, value of the wind speed. For example, between noon and 1230 local time the average wind speed is about 6 m/s, while a bit later (between 1400 and 1430) the winds have decreased to about 5 m/s on the average. The ability to find a statistically-stable mean value suggests that turbulence is not completely random.
- The wind speeds do not vary from 0 to 100 m/s in this graph, but rather vary over a limited range of speeds. In other words, there is a measurable and definable intensity to the turbulence that shows up on this graph as the vertical spread of wind speed. The turbulence intensity appears to decrease between noon and 1400 local time.

  Near noon the instantaneous wind speed is often 1 m/s faster or slower than the mean, while at 1400 the wind speed varies by only about 0.5 m/s about its mean. Such a bounded characteristic of the wind speed means that we can use statistics such as the variance or standard deviation to characterize the turbulence intensity.
- There appears to be a wide variety of time-scales of wind variation superimposed on top of each other. If we look closely we see that the time period between each little peak in wind speed is about a minute. The larger peaks seem to happen about every 5 min. There are other variations that indicate a 10 min time period. The smallest detectable variations on this chart are about 10 s long.

  If each of these time variations is associated with a different size turbulent eddy (Taylor's hypothesis. See exercise 5 in Chapter 1), then we can conclude that we are seeing the signature of eddies ranging in size from about 50 m to about 3000 m. In other words, we are observing evidence of the spectrum of turbulence.

The turbulence spectrum is analogous to the spectrum of colors that appears when you shine a light through a prism. White light consists of many colors (i.e., many wavelengths or frequencies) superimposed on one another. The prism is a physical device that separates the colors. We could measure the intensity of each color to learn the magnitude of its contribution to the original light beam. We can perform a similar analysis on a turbulent signal using mathematical rather than physical devices to learn about the contribution of each different size eddy to the total turbulence kinetic energy.

Figure 2.2 shows an example of the spectrum of wind speed measured near the ground. The ordinate is a measure of the portion of turbulence energy that is associated with a particular size eddy. The abscissa gives the eddy size in terms of the time period and frequency of the wind-speed variation. Small eddies have shorter time periods than large eddies (again, using Taylor's hypothesis).

Peaks in the spectrum show which size eddies contribute the most to the turbulence kinetic energy. The leftmost peak with a period of near 100 h corresponds to wind speed variations associated with the passage of fronts and weather systems. In other words, there is evidence of the Rossby-wave cycle in our wind speed record. The next peak, at 24 h, shows the diurnal increase of wind speed during the day and decrease at night. The rightmost peak is the one we will study in this book. It indicates the microscale eddies having durations of 10 s to 10 min, just what we would have guessed from examining Fig 2.1 by eye.

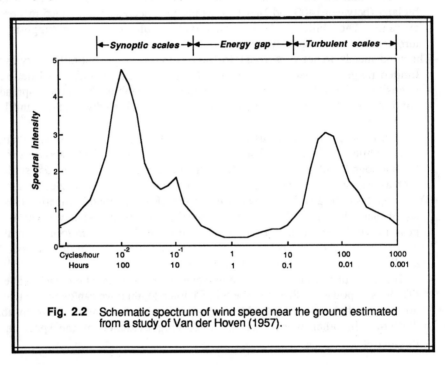

Fig. 2.2   Schematic spectrum of wind speed near the ground estimated from a study of Van der Hoven (1957).

Within the rightmost peak we see that the largest eddies are usually the most intense. The smaller, high frequency, eddies are very weak, as previously discussed. Large-eddy motions can create eddy-size wind-shear regions, which can generate smaller eddies. Such a net transfer of turbulence energy from the larger to the smaller eddies is known as the energy cascade. At the smallest size eddies, this cascade of energy is dissipated into heat by molecular viscosity. The flavor of this energy cascade was captured by Lewis Richardson in his 1922 poem:

> Big whorls have little whorls,
> Which feed on their velocity;
> And little whorls have lesser whorls,
> And so on to viscosity
> (in the molecular sense).

In a later chapter we will describe the mathematical tools necessary to calculate spectra.

2.2 The Spectral Gap

There appears to be a distinct lack of wind-speed variation in Fig 2.1 having time periods of about 30 min to 1 h. The slow variation of the mean wind speed from 6 to 5 m/s over the 2 h period was already discussed. Shorter time periods were what we associated with microscale turbulence. The lack of variation at the intermediate time or space scales has been called the *spectral gap*.

A separation of scales is evident in Fig 2.2, where the spectral gap appears as the large valley separating the microscale from the synoptic scale peaks. Motions to the left of the gap are said to be associated with the *mean flow*. Motions to the right constitute *turbulence*. The center of the gap is near the one-hour time period.

It is no accident that the response time used in Chapter 1 to define the BL is one hour. Implicit in the definition of the BL is the concept that turbulence is the primary agent for effecting changes in the BL. Hence, the spectral gap provides a means to separate the turbulent from the nonturbulent influences on the BL (in the microscale sense).

For some flows there might not be a spectral gap. For example, larger cumulus clouds act like large eddies with time scales on the order of an hour. Consequently, a spectrum of wind speed made in the cloud layer might not exhibit a vivid separation of scales. Most analyses of turbulence rely on the separation of scales to simplify the problem; hence, cloud-filled flow regimes might be difficult to properly describe.

Many of the operational numerical weather prediction models use grid spacings or wavelength cutoffs that fall within the spectral gap. This means that larger-scale motions can be explicitly resolved and deterministically forecast. The smaller-scale motions, namely turbulence, are not modeled directly. Rather, the effects of those subgrid scales on the larger scales are approximated. These smaller-size motions are said to be parameterized by subgrid-scale stochastic (statistical) approximations or models.

2.3 Mean and Turbulent Parts

There is a very easy way to isolate the large-scale variations from the turbulent ones. By averaging our wind speed measurements over a period of 30 minutes to one hour, we can eliminate or "average out" the positive and negative deviations of the turbulent velocities about the mean. Once we have the mean velocity, $\overline{U}$, for any time period, we can subtract it from the actual instantaneous velocity, U, to give us just the turbulent part, u':

$$u' = U - \overline{U} \tag{2.3a}$$

The existence of a spectral gap allows us to partition the flow field in this manner.

We can think of u' as the gust that is superimposed on the mean wind. It represents the part of the flow that varies with periods shorter than about one hour. The mean, $\overline{U}$, represents the part that varies with a period longer than about one hour.

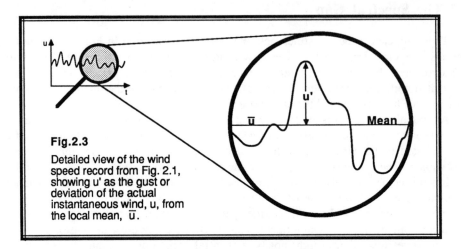

Fig.2.3

Detailed view of the wind speed record from Fig. 2.1, showing u' as the gust or deviation of the actual instantaneous wind, u, from the local mean, $\bar{u}$.

Fig 2.3 shows an expanded view of just a small portion of the wind trace from Fig 2.1. The straight line represents the mean wind over that portion of the record, while the wiggly line represents the actual instantaneous wind speed. The gust part, u', is sketched as the distance between those two lines. At some times the gust is positive, meaning the actual wind is faster than average. At other times the gust is negative, indicative of a slower than average wind.

Microscale turbulence is a three-dimensional phenomenon. Therefore, we expect that gusts in the x-direction might be accompanied by gusts in the y- and z-directions. Turbulence, by definition, is a type of motion. Yet motions frequently cause variations in the temperature, moisture, and pollutant fields if there is some mean gradient of that variable across the turbulent domain. Hence, we can partition each of the variables into mean and turbulent parts:

$$U = \overline{U} + u'$$
$$V = \overline{V} + v'$$
$$W = \overline{W} + w' \qquad (2.3b)$$
$$\theta_v = \overline{\theta}_v + \theta_v'$$
$$q = \overline{q} + q'$$
$$c = \overline{c} + c'$$

Each of these terms varies in time and space (see Appendix B for a list of symbols).

2.4 Some Basic Statistical Methods

Because one of the primary avenues for studying turbulent flow is the stochastic approach, it is desirable to have a good working knowledge of statistics. This section will survey some of the basic methods of statistics, including the mean, variance, standard deviation, covariance, and correlation. Those readers having an adequate background on statistics might wish to skim this section.

2.4.1 The Mean

Time $\overset{t}{(\ )}$, space $\overset{s}{(\ )}$, and ensemble $\overset{e}{(\ )}$ averages are three ways to define a mean. The *time average* applies at one specific point in space, and consists of a sum or integral over time period P. For any variable, $A(t, s)$, that is a function of time, t, and space, s:

$$\overset{t}{\overline{A}}(s) = \frac{1}{N} \sum_{i=0}^{N-1} A(i, s) \quad \text{or} \quad \overset{t}{\overline{A}}(s) = \frac{1}{P} \int_{t=0}^{P} A(t, s) \, dt \qquad (2.4.1a)$$

where $t = i \, \Delta t$ for the discrete case.

The *spatial average*, which applies at some instant in time, is given by a sum or integral over spatial domain S:

$$\overset{s}{\overline{A}}(t) = \frac{1}{N} \sum_{j=0}^{N-1} A(t,j) \quad \text{or} \quad \overset{s}{\overline{A}}(t) = \frac{1}{S} \int_{t=0}^{S} A(t,s) \, ds \qquad (2.4.1b)$$

where $s = j\Delta s$ in the discrete case.

An *ensemble average* consists of the sum over n identical experiments:

$$\overset{e}{\overline{A}}(t,s) = \frac{1}{N} \sum_{i=0}^{N-1} A_i(t,s) \qquad (2.4.1c)$$

In the equations above, $\Delta t = P/N$ and $\Delta s = S/N$, where N is the number of data points.

For laboratory experiments, the ensemble average is the most desirable, because it allows us to reduce random experimental errors by repeating the basic experiment. Unlike laboratory experiments, however, we have little control over the atmosphere, so we are rarely able to observe reproducible weather events. We are therefore unable to use the ensemble average.

Spatial averages are possible by deploying an array of meteorological sensors covering a line, area, or volume. If the turbulence is *homogeneous* (statistically the same at every point in space, see Fig 2.4) then each of the sensors in the array will be measuring the same phenomenon, making a spatial average meaningful. The real atmosphere, however, is horizontally homogeneous in only limited locations, meaning that most spatial means are averaged over a variety of different phenomena. By proper choice of sensor-array domain size as well as intra-array spacing, one can sometimes isolate scales of phenomena for study, while averaging out the other scales.

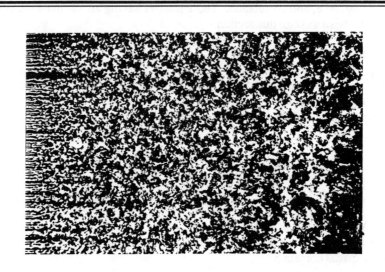

Fig. 2.4   Laboratory generation of homogeneous turbulence behind a grid. Using a finer grid than Fig 2.4, the merging unstable wakes quickly form a homogeneous field. As it decays downstream, it provides a useful approximation of homogeneous turbulence.(From Van Dyke, 1982)

Volume averaging is virtually impossible using direct sensors such as thermometers because of the difficulty of deploying these sensors at all locations and altitudes throughout the BL. Remote sensors such as radars, lidars, and sodars, however, can scan volumes of the atmosphere, making volume averages of selected variables possible. Details of these sensors are discussed in chapter 10.

Area averaging in the surface layer is frequently performed within small domains by deploying an array of small instrumented masts or instrument shelters on the ground. *Line averages* are similarly performed by erecting sensors along a road, for example.

Sensors mounted on a moving platform, such as a truck or an aircraft, can provide quasi-line averages. These are not true line (spatial) averages because the turbulence state of the flow may change during the time it takes the platform to move along the desired path. As a result, most measurement paths are designed as a compromise between long length (to increase the statistical significance by observing a larger number of data points) and short time (because of the diurnal changes that occur in the mean and turbulent state over most land surfaces).

Time averages are frequently used, and are computed from sensors mounted on a single, fixed-location platform such as a mast or tower. The relative ease of making observations at a fixed point has meant that time averaging has been the most popular in the lower BL. Some vertically-looking remote sensors also use this method to observe the middle and top of the BL. For turbulence that is both homogeneous and *stationary* (statistically not changing over time), the time, space and ensemble averages should all be equal. This is called the *ergodic* condition, which is often assumed to make the turbulence problem more tractable:

$$\overline{(\ )}^{e} = \overline{(\ )}^{t} = \overline{(\ )}^{s} \equiv \overline{(\ )} \tag{2.4.1d}$$

This book will use the overbar $\overline{(\ )}$ as an abbreviation for a generic average, not specifying whether it is a time, space, or ensemble average. Because of the popularity of the time average, however, many of our examples will use time as the independent variable.

2.4.2 Rules of Averaging

Let A and B be two variables that are dependent on time, and let c represent a constant. To find the average of the sum of A and B, we can employ the equations of the previous section with some basic rules of summation or integration to show that:

$$\overline{(A+B)} = \overline{A} + \overline{B} \tag{2.4.2a}$$

In terms of discrete sums, the average is:

$$
\begin{aligned}
\overline{(A+B)} &= \frac{1}{N} \sum_{i=0}^{N-1} (A_i + B_i) \\
&= \frac{1}{N} \left(\sum_i A_i + \sum_i B_i \right) \\
&= \frac{1}{N} \sum_i A_i + \frac{1}{N} \sum_i B_i \\
&= \overline{A} + \overline{B}
\end{aligned}
$$

In terms of continuous integrals:

$$\overline{(A + B)} = \frac{1}{P} \int_{t=0}^{P} (A + B)\, dt$$

$$= \frac{1}{P} \left(\int_t A\, dt + \int_t B\, dt \right)$$

$$= \frac{1}{P} \int_t A\, dt + \frac{1}{P} \int_t B\, dt$$

$$= \overline{A} + \overline{B}$$

Both the sum and integral approaches give the same answer, as expected.
We can use similar methods to show that:

$$\overline{(c\,A)} = c\,\overline{(A)} \tag{2.4.2b}$$

$$\overline{c} = c \tag{2.4.2c}$$

An important consequence of averaging is that an average value acts like a constant when averaged a second time over the same time period, P:

Define

$$\frac{1}{P} \int_{t=0}^{P} A(t,s)\, dt \equiv \overline{A}(P,s)$$

Therefore

$$\frac{1}{P} \int_{t=0}^{P} \overline{A}(P,s)\, dt \equiv \overline{A}(P,s)\, \frac{1}{P} \int_{t=0}^{P} dt$$

$$= \overline{A}(P,s)$$

Leaving

$$\overline{(\overline{A})} = \overline{A} \tag{2.4.2d}$$

Similarly, it can be shown that:

$$(\overline{A\ B}) = \overline{A}\ \overline{B} \tag{2.4.2e}$$

Often we need to find the average of a derivative of a dependent variable. For example, let A be dependent on both t and s, where s is an independent variable such as x, y, or z. In this case, we must use **Leibniz' theorem**:

$$\frac{d}{dt}\left[\int_{S_1(t)}^{S_2(t)} A(t,s)\,ds\right] = \int_{S_1(t)}^{S_2(t)}\left[\frac{\partial A(t,s)}{\partial t}\right]ds + A(t, S_2)\frac{d S_2}{dt} - A(t, S_1)\frac{d S_1}{dt} \tag{2.4.2f}$$

where S_1 and S_2 are the limits of integration.

For the special case where S_1 and S_2 are constant with time, we can simply interchange the order of integration and differentiation:

$$\frac{d}{dt}\left[\int_s A\,ds\right] = \int_s\left[\frac{\partial A}{\partial t}\right]ds$$

Multiplying both sides by 1/S gives:

$$\frac{d\left(^s\overline{A}\right)}{dt} = {}^s\overline{\left(\frac{\partial A}{\partial t}\right)} \tag{2.4.2g}$$

This special case is not always valid for variable depth boundary layers.

Suppose we wish to find the time-rate-of-change of a BL-averaged mixing ratio, $\overline{r}$, where the BL average is defined by integrating over the depth of the BL; i.e., from z=0 to $z=z_i$. Since z_i varies with time, we can use the full Leibniz' theorem to give:

$$\frac{d}{dt}\left[z_i \, ^s\overline{r}\right] = z_i \, {}^s\overline{\left[\frac{\partial r}{\partial t}\right]} + r(t, z_i^+)\frac{d z_i}{dt} \tag{2.4.2h}$$

Finally, let's re-examine the spectral gap. If our averaging time is 30 minutes to 1 hour, turbulent fluctuations will be eliminated, leaving the longer-period time variations. As we saw in Fig 2.1, the 30-minute mean wind speed changes over the period of a few hours. Thus, we can take the 30-minute average of the time-derivative of variable A to find how $\overline{A}$ varies over longer periods:

$$\overline{\left(\frac{dA}{dt}\right)}^{t} = \frac{d\overline{A}^{t}}{dt} \tag{2.4.2i}$$

In other words, the average of the local slopes (slope = rate of change with time) equals the slope of the averages (see Fig 2.5).

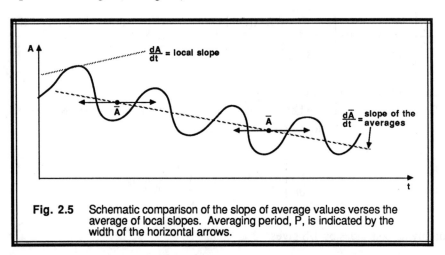

Fig. 2.5   Schematic comparison of the slope of average values verses the average of local slopes. Averaging period, P, is indicated by the width of the horizontal arrows.

This is a difficult concept that deserves some thought on the part of the reader. It is an important consequence of the spectral gap because it allows us to make a deterministic forecast of a mean variable such as $\overline{A}$ using simplified, stochastic, representations of the turbulence. Otherwise, operational forecasts of seemingly simple variables such as temperature or wind would be much more difficult.

To summarize the rules of averaging:

$$\overline{c} = c$$

$$\overline{(c\,A)} = c\,\overline{A}$$

$$\overline{(\overline{A})} = \overline{A}$$

$$\overline{(\overline{A}\,B)} = \overline{A}\,\overline{B} \tag{2.4.2k}$$

$$\overline{(A + B)} = \overline{A} + \overline{B}$$

$$\overline{\left(\frac{dA}{dt}\right)} = \frac{d\,\overline{A}}{dt}$$

2.4.3 Reynolds Averaging

The averaging rules of the last section can now be applied to variables that are split into mean and turbulent parts. Let $A = \overline{A} + a'$ and $B = \overline{B} + b'$. Starting with the instantaneous value, A, for example, we can find its mean using the fifth and third rules of the previous section:

$$\overline{(A)} = \overline{(\overline{A} + a')} = \overline{(\overline{A})} + \overline{a'} = \overline{A} + \overline{a'}$$

The only way that the left and right sides can be equal is if

$$\overline{a'} = 0 \qquad\qquad (2.4.3a)$$

This result is not surprising if one remembers the definition of a mean value. By definition, the sum of the positive deviations from the mean must equal the sum of the negative deviations. Thus the deviations balance when summed, as implied in the above average.

Another example: start with the product $\overline{B} a'$ and find its average. Employing the above result together with the fourth averaging rule, we find that

$$\overline{(\overline{B} \, a')} = \overline{B} \, \overline{a'} = \overline{B} \cdot 0 = 0 \qquad\qquad (2.4.3b)$$

Similarly, $\overline{A} \, b' = 0$. One should not become too lax about the average of primed variables, as is demonstrated next.

The average of the product of A and B is

$$\overline{(A \cdot B)} = \overline{(\overline{A} + a')(\overline{B} + b')}$$
$$= \overline{(\overline{A}\,\overline{B} + a'\overline{B} + \overline{A}\,b' + a'\,b')}$$
$$= \overline{(\overline{A}\,\overline{B})} + \overline{(a'\,\overline{B})} + \overline{(\overline{A}\,b')} + \overline{(a'\,b')}$$
$$= \overline{A}\,\overline{B} + 0 + 0 + \overline{a'\,b'}$$
$$= \overline{A}\,\overline{B} + \overline{a'\,b'} \qquad\qquad (2.4.3c)$$

The nonlinear product $\overline{a'\,b'}$ is NOT necessarily zero. The same conclusion holds for other nonlinear variables such as:

$$\overline{a'^2} \ , \ \overline{a'b'^2} \ , \ \overline{a'^2 b'^2} \ .$$

In fact, these nonlinear terms must be retained to properly model turbulence. This is a dramatic difference from many linear theories of waves, where the nonlinear terms are often neglected as a first-order approximation.

2.4.4 Variance, Standard Deviation and Turbulence Intensity

One statistical measure of the dispersion of data about the mean is the variance, σ^2, defined by:

$$\sigma_A^2 = \frac{1}{N} \sum_{i=0}^{N-1} (A_i - \overline{A})^2 \qquad (2.4.4a)$$

This is known as the **biased variance**. It is a good measure of the dispersion of a sample of BL observations, but not the best measure of the dispersion of the whole population of possible observations. A better estimate of the variance (an **unbiased variance**) of the population, given a sample of data, is

$$\sigma_A^2 = \frac{1}{(N-1)} \sum_{i=0}^{N-1} (A_i - \overline{A})^2 \qquad (2.4.4b)$$

When N is large, as it often is for turbulence measurements, $1/N \cong 1/(N-1)$. As a result, the biased definition is usually used in BL meteorology for convenience.

Recall that the turbulent part (or the perturbation or gust part) of a turbulent variable is given by a' = A - $\overline{A}$. Substituting this into the biased definition of variance gives

$$\sigma_A^2 = \frac{1}{N} \sum_{i=0}^{N-1} a'^2_i = \overline{a'^2} \qquad (2.4.4c)$$

Thus, whenever we encounter the average of the square of a turbulent part of a variable, such as $\overline{u'^2}$, $\overline{v'^2}$, $\overline{w'^2}$, $\overline{\theta'^2}$, $\overline{r'^2}$, or $\overline{q'^2}$,we can interpret these as variances.

The standard deviation is defined as the square root of the variance:

$$\sigma_A = \left(\overline{a'^2} \right)^{1/2} \qquad (2.4.4d)$$

The standard deviation always has the same dimensions as the original variable. Fig 2.6 shows the relationship between a turbulent trace of wind speed and the corresponding standard deviation. It can be interpreted as a measure of the magnitude of the spread or dispersion of the original data from its mean. For this reason, it is used as a measure of the intensity of turbulence. In figure 2.1, for example, we might guess the standard deviation to be about 0.5 - 0.6 m/s at noon, dropping to about 0.3 m/s by 1400 local time.

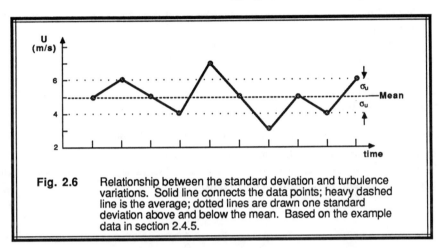

Fig. 2.6   Relationship between the standard deviation and turbulence variations. Solid line connects the data points; heavy dashed line is the average; dotted lines are drawn one standard deviation above and below the mean. Based on the example data in section 2.4.5.

Near the ground, the turbulence intensity might be expected to increase as the mean wind speed, M, increases. For this reason a dimensionless measure of the *turbulence intensity*, I, is often defined as

$$I = \sigma_M / \overline{M} \qquad (2.4.4e)$$

For mechanically generated turbulence, one might expect σ_M to be a simple function of M. As we learned in Chapter 1, $I < 0.5$ is required for Taylor's hypothesis to be valid.

2.4.5 Covariance and Correlation

In statistics, the covariance between two variables is defined as

$$\text{covar}(A,B) \equiv \frac{1}{N} \sum_{i=0}^{N-1} (A_i - \overline{A}) \cdot (B_i - \overline{B}) \qquad (2.4.5a)$$

Using our Reynolds averaging methods, we can show that:

$$\text{covar}(A,B) \equiv \frac{1}{N} \sum_{i=0}^{N-1} a_i' b_i'$$

$$= \overline{a' b'} \tag{2.4.5b}$$

Thus, the nonlinear turbulence products that were introduced in section 2.4.3 have the same meaning as covariances.

The covariance indicates the degree of common relationship between the two variables, A and B. For example, let A represent air temperature, T, and let B be the vertical velocity, w. On a hot summer day over land, we might expect the warmer than average air to rise (positive T' and positive w'), and the cooler than average air to sink (negative T' and negative w'). Thus, the product w'T' will be positive on the average, indicating that w and T vary together. The covariance $\overline{w'T'}$ is indeed found to be positive throughout the bottom 80% of the convective mixed layer.

Sometimes, one is interested in a normalized covariance. Such a relationship is defined as the *linear correlation coefficient*, r_{AB}:

$$r_{AB} \equiv \frac{\overline{a' b'}}{\sigma_A \sigma_B} \tag{2.4.5c}$$

This variable ranges between -1 and +1 by definition. Two variables that are perfectly correlated (i.e., vary together) yield r = 1. Two variables that are perfectly negatively correlated (i.e., vary oppositely) yield r = -1. Variables with no net variation together yield r = 0. Fig 2.7 shows typical correlation coefficients in the ML.

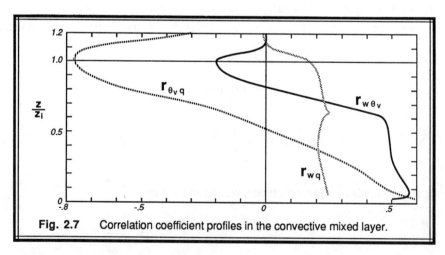

Fig. 2.7   Correlation coefficient profiles in the convective mixed layer.

2.4.6 Example

Problem. Suppose that we erect a short mast instrumented with anemometers to measure the U and W wind components. We record the instantaneous wind speeds every 6 s for a minute, resulting in the following 10 pairs of wind observations:

U (m/s):	5	6	5	4	7	5	3	5	4	6
W (m/s):	0	-1	1	0	-2	1	2	-1	1	-1

Find the mean, biased variance, and standard deviation for each wind component. Also, find the covariance and correlation coefficient between U and W.

Solutions.

$$\overline{U} = 5 \text{ m·s}^{-1} \qquad \sigma_U^2 = 1.20 \text{ m}^2\text{·s}^{-2} \qquad \sigma_U = 1.10 \text{ m·s}^{-1}$$

$$\overline{W} = 0 \text{ m·s}^{-1} \qquad \sigma_W^2 = 1.40 \text{ m}^2\text{·s}^{-2} \qquad \sigma_W = 1.18 \text{ m·s}^{-1}$$

$$\overline{u'w'} = -1.10 \text{ m}^2\text{·s}^{-2} \qquad r_{UW} = -0.85 \quad \text{(dimensionless)}$$

Discussion. Thus, the turbulent variations of W are more intense than those of U, even though the mean wind speed for W is zero in this example. U and W tend to vary in opposite directions on the average, as indicated by the negative values for the covariance and the correlation coefficient. The magnitude of the correlation coefficient is fairly high (close to one), meaning that there are just a few observations where U and W vary in the same direction, but many more observations where they vary oppositely.

2.5 Turbulence Kinetic Energy

The usual definition of kinetic energy (KE) is KE $= 0.5 \text{ m M}^2$, where m is mass. When dealing with a fluid such as air it is more convenient to talk about kinetic energy per unit mass, which is just 0.5 M^2.

It is enticing to partition the kinetic energy of the flow into a portion associated with the mean wind (MKE), and a portion associated with the turbulence (TKE). By taking advantage of the mean and turbulent parts of velocity introduced in section 2.3, we can immediately write the desired equations:

$$\text{MKE/m} = \frac{1}{2}\left(\overline{U}^2 + \overline{V}^2 + \overline{W}^2\right) \tag{2.5a}$$

$$e = \frac{1}{2}\left(\overline{u'^2} + \overline{v'^2} + \overline{w'^2}\right) \tag{2.5b}$$

where e represents an instantaneous turbulence kinetic energy per unit mass. There is an additional portion of the total KE consisting of mean-turbulence products, but this disappears upon averaging.

Rapid variations in the value of e with time can be expected as we measure faster and slower gusts. By averaging over these instantaneous values, we can define a mean turbulence kinetic energy (TKE) that is more representative of the overall flow:

$$\frac{\text{TKE}}{\text{m}} = \frac{1}{2}\left(\overline{u'^2} + \overline{v'^2} + \overline{w'^2}\right) = \bar{e} \tag{2.5c}$$

We can immediately see the relationship between TKE/m and the definition of variance defined in the last section. It is apparent that statistics will play an important role in our quantification of turbulence.

The turbulence kinetic energy is one of the most important quantities used to study the turbulent BL. We have already discussed in Chapter 1 that turbulence can be generated by buoyant thermals and by mechanical eddies. It is suppressed by statically stable lapse rates and dissipated into heat by the effects of molecular viscosity. By writing a budget equation for TKE, we can balance the production terms against the loss terms to determine whether the BL will become more turbulent, or whether turbulence will decay in the BL. This will be done in Chapter 5.

A typical daytime variation of TKE in convective conditions is shown in Fig 2.8. Examples of the vertical profile of TKE for various boundary layers are shown in Fig 2.9. During the daytime, buoyancy allows air parcels to accelerate in the middle of the ML, allowing $\overline{w'^2}$ to be large there and contributing to the total TKE (Fig 2.9a).

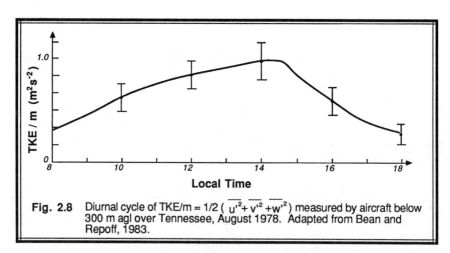

Fig. 2.8   Diurnal cycle of TKE/m = 1/2 ($\overline{u'^2} + \overline{v'^2} + \overline{w'^2}$) measured by aircraft below 300 m agl over Tennessee, August 1978. Adapted from Bean and Repoff, 1983.

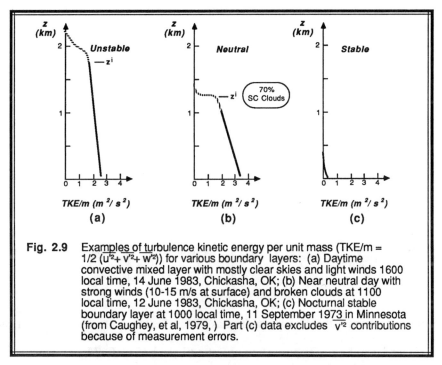

Fig. 2.9   Examples of turbulence kinetic energy per unit mass (TKE/m = $1/2 \, (\overline{u'^2} + \overline{v'^2} + \overline{w'^2})$) for various boundary layers: (a) Daytime convective mixed layer with mostly clear skies and light winds 1600 local time, 14 June 1983, Chickasha, OK; (b) Near neutral day with strong winds (10-15 m/s at surface) and broken clouds at 1100 local time, 12 June 1983, Chickasha, OK; (c) Nocturnal stable boundary layer at 1000 local time, 11 September 1973 in Minnesota (from Caughey, et al, 1979,) Part (c) data excludes $\overline{v'^2}$ contributions because of measurement errors.

On overcast days when there is little heating of the ground, wind shears and flow over obstacles create turbulence near the ground that gradually decreases intensity with height

(Fig 2.9b). This turbulence is produced in primarily the $\overline{u'^2}$ and $\overline{v'^2}$ components. Days of both strong winds and strong heating will have both sources of turbulence.

For night, Fig 2.9c shows how the static stability suppresses the TKE, causing it to decrease rapidly with height. Turbulence is produced primarily near the ground by wind shears, although the enhanced shears near the nocturnal jet can also generate turbulence. Not apparent in this figure is the observation that nocturnal turbulence is sometimes sporadic: happening in turbulent bursts followed by quiescent periods.

2.6 Kinematic Flux

2.6.1 Definitions

Flux is the transfer of a quantity per unit area per unit time. In BL meteorology, we are often concerned with mass, heat, moisture, momentum and pollutant fluxes. The dimensions of these fluxes are summarized below, using SI units as the example:

Flux	Symbol	Units
mass	$\tilde{M}$	$\left[\dfrac{kg_{air}}{m^2 \cdot s}\right]$
heat	$\tilde{Q}_H$	$\left[\dfrac{J}{m^2 \cdot s}\right]$
moisture	$\tilde{R}$	$\left[\dfrac{kg_{water}}{m^2 \cdot s}\right]$
momentum	$\tilde{F}$	$\left[\dfrac{kg \cdot (m \cdot s^{-1})}{m^2 \cdot s}\right]$
pollutant	$\tilde{\chi}$	$\left[\dfrac{kg_{pollutant}}{m^2 \cdot s}\right]$ or $\left[\dfrac{kg_{pollutant}}{m^3} \cdot \dfrac{m}{s}\right]$

Sometimes the moisture flux is rewritten as a latent heat flux, $\tilde{Q}_E$, where $\tilde{Q}_E = L_v \tilde{R}$ and L_v is the latent heat of vaporization of water ($L_v \cong 2.45 \times 10^6$ J/kg at a summertime BL temperature of 20 °C).

As a reminder, momentum is mass times velocity (kg·m/s); thus, a momentum flux is (kg·m/s)/(m²·s). These units are identical to N/m², which are the units for stress. The nature of stress is reviewed in section 2.9.

Unfortunately, we rarely measure quantities such as heat or momentum directly. Instead we measure things like temperature or wind speed. Therefore, for convenience the above fluxes can be redefined in **kinematic form** by dividing by the density of moist air, ρ_{air}. In the case of sensible heat flux, we also divide by the specific heat of air. In fact, the term $\rho \, C_p = 1.216 \times 10^3$ (W/m²) / (K·m/s) allows us to easily convert between kinematic heat fluxes and normal heat fluxes.

Kinematic Flux	Symbol	Equation	Units	
mass		$M = \dfrac{\tilde{M}}{\rho_{air}}$	$\left[\dfrac{m}{s}\right]$	(2.6.1a)
heat		$Q_H = \dfrac{\tilde{Q}_H}{\rho_{air} C_{p_{air}}}$	$\left[K\dfrac{m}{s}\right]$	(2.6.1b)

$$\text{moisture} \quad R = \frac{\tilde{R}}{\rho_{air}} \quad \left[\frac{kg_{water}}{kg_{air}} \cdot \frac{m}{s}\right] \quad (2.6.1c)$$

$$\text{momentum} \quad F = \frac{\tilde{F}}{\rho_{air}} \quad \left[\frac{m}{s} \cdot \frac{m}{s}\right] \quad (2.6.1d)$$

$$\text{pollutant} \quad \chi = \frac{\tilde{\chi}}{\rho_{air}} \quad \left[\frac{kg_{pollutant}}{kg_{air}} \cdot \frac{m}{s}\right] \quad (2.6.1e)$$

The above definitions are viable because the boundary layer is usually so thin that the density change across it can be neglected in comparison to changes of the other meteorological variables. For example, the standard atmospheric air density is 1.225 kg/m³ at sea level and 1.112 kg/m³ at 1000m, a difference of only 10%.

These kinematic fluxes are now expressed in units that we can measure directly: wind speed for mass and momentum fluxes; temperature and wind speed for heat flux; and specific humidity (q) and wind speed for moisture flux. The pollutant flux is frequently expressed in either form: concentration and wind speed, or mass ratio (like parts per million, ppm) and wind speed.

Each of these fluxes can be split into three components. For example, there might be a vertical component of heat flux, and two horizontal components of heat flux, as sketched in Fig 2.10. Similar fluxes could be expected for mass, moisture, and pollutants. Hence, we can picture these fluxes as vectors.

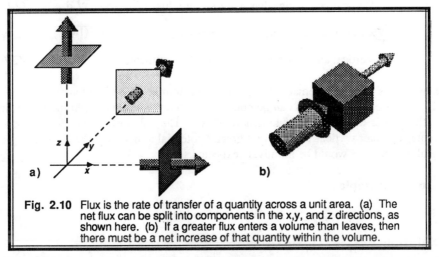

Fig. 2.10 Flux is the rate of transfer of a quantity across a unit area. (a) The net flux can be split into components in the x,y, and z directions, as shown here. (b) If a greater flux enters a volume than leaves, then there must be a net increase of that quantity within the volume.

For momentum, we have the added dimension that the flux in any one direction might be the flux of U, V or W momentum (see Fig 2.11). This means that there are nine components of this flux to consider: each of the three momentum components can pass through a plane normal to any of the three cartesian directions. The momentum flux is thus said to be a **tensor**. Just for the record, this kind of tensor is known as a second order tensor. A vector is a first order tensor, and a scalar is a zero order tensor.

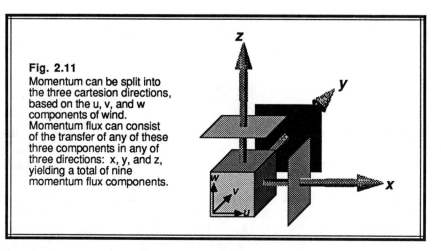

Fig. 2.11
Momentum can be split into the three cartesion directions, based on the u, v, and w components of wind. Momentum flux can consist of the transfer of any of these three components in any of three directions: x, y, and z, yielding a total of nine momentum flux components.

As you might guess, we can split the fluxes into mean and turbulent parts. For the flux associated with the mean wind (i.e., advection), it is easy to show, for example, that

Vertical kinematic advective heat flux	$= \overline{W} \cdot \overline{\theta}$	(2.6.1f)
Vertical kinematic advective moisture flux	$= \overline{W} \cdot \overline{q}$	(2.6.1g)
x-direction kinematic advective heat flux	$= \overline{U} \cdot \overline{\theta}$	(2.6.1h)
Vertical kinematic advective flux of U-momentum $= \overline{W} \cdot \overline{U}$		(2.6.1i)

The last flux is also the kinematic flux of W-momentum in the x-direction.

Fluxes in other directions can be constructed in an analogous fashion. These fluxes have the proper dimensions for kinematic fluxes. They also make physical sense. For example, a greater vertical velocity or a greater potential temperature both create a greater vertical heat flux, as would be intuitively expected.

2.6.2   Example

Problem.   Given $\tilde{Q}_H = 365$ W·m^{-2}. Find Q_H.

Solution. $Q_H = \tilde{Q}_H / (\rho \, C_p)$

$= (365 \text{ Wm}^{-2}) / [(1.21 \text{ kg·m}^{-3}) \cdot (1005 \text{ J·kg}^{-1} \cdot \text{K}^{-1})] = 0.30 \text{ K·m/s}$

Discussion. This is a typical daytime kinematic heat flux during strong convection.

2.7 Eddy Flux

We saw in the last section that fluid motion can transport quantities, resulting in fluxes. Turbulence also involves motion. Thus we expect that turbulence transports quantities too.

2.7.1 Concepts

A term like $\overline{w'\theta'}$ looks similar to the kinematic flux terms of the last section, except that the perturbation values are used instead of the mean values of W and θ. If turbulence is completely random, then a positive $w'\theta'$ one instant might cancel a negative $w'\theta'$ at some later instant, resulting in a near zero value for the average turbulent heat flux. As is shown below, however, there are situations where the average turbulent flux might be significantly different from zero.

As a conceptual tool, suppose we examine a small idealized eddy near the ground on a hot summer day (see Fig 2.12a). The average potential temperature profile is usually superadiabatic in such surface layers. If the eddy is a swirling motion, then some of the air from position 1 will be mixed downward (i.e., w' is negative) , while some air from position 2 will mix up (i.e., w' is positive) to take its place. The average motion caused by turbulence is $\overline{w'} = 0$, as expected (from section 2.4.3).

The downward moving air parcel (negative w') ends up being cooler than its surroundings (negative θ', assuming that θ' was conserved during its travel), resulting in an instantaneous product $w'\theta'$ that is positive. The upward moving air (positive w') is warmer than its surroundings (positive θ'), also resulting in a positive instantaneous product $w'\theta'$. Both the upward and downward moving air contribute positively to the flux, $w'\theta'$; thus, the average kinematic eddy heat flux $\overline{w'\theta'}$ is positive for this small-eddy mixing process.

This important result shows that turbulence can cause a net transport of a quantity such as heat ($\overline{w'\theta'} \neq 0$), even though there is no net transport of mass ($\overline{w'} = 0$). Turbulent eddies transport heat upward in this case, tending to make the lapse rate more adiabatic.

Next, let's examine what happens on a night where a statically stable lapse rate is present (Fig 2.12b). Again, picture a small eddy moving some air up and some back down. An upward moving parcel ends up cooler than its surrounding (negative $w'\theta'$), while a downward moving parcel is warmer (negative $w'\theta'$). The net effect of the small eddy is to cause a negative $\overline{w'\theta'}$, meaning a downward transport of heat.

Fruit growers utilize this process on cold nights as one method to prevent their fruit from freezing. They run motor-driven fans throughout the orchard to generate turbulent eddies. These eddies mix the warmer air down towards the fruit, and mix the cooler near-surface air upward out of the orchard, thereby potentially saving the crop.

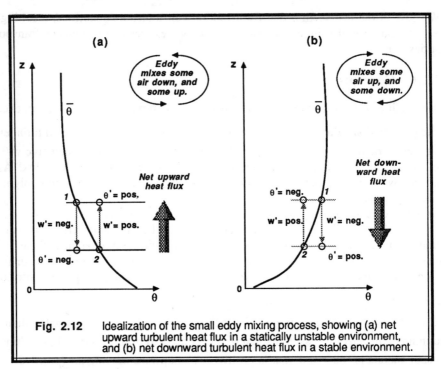

Fig. 2.12   Idealization of the small eddy mixing process, showing (a) net upward turbulent heat flux in a statically unstable environment, and (b) net downward turbulent heat flux in a stable environment.

Again we see the statistical nature of our description of turbulence. A kinematic flux such as $\overline{w'\theta'}$ is nothing more than a statistical covariance. We will usually leave out the word "kinematic" in future references to such fluxes.

As before, we can extend our arguments to write various kinds of eddy flux:

Vertical kinematic eddy heat flux	$= \overline{w'\theta'}$	(2.7.1a)
Vertical kinematic eddy moisture flux	$= \overline{w'q'}$	(2.7.1b)
x-direction kinematic eddy heat flux	$= \overline{u'\theta'}$	(2.7.1c)
Vertical kinematic eddy flux of U-momentum	$= \overline{u'w'}$	(2.7.1d)

The last flux is also the x-direction kinematic eddy flux of W-momentum.

Comparing the advective fluxes to the eddy fluxes, it is important to recognize that $\overline{W} \cong 0$ throughout most of the boundary layer. As a result, the vertical advective fluxes are usually negligible compared to the vertical turbulent fluxes. No such statement can be made about the horizontal fluxes, where strong mean horizontal winds and strong turbulence can cause fluxes of comparable magnitudes.

Finally, it is important to note that turbulence in the real atmosphere usually consists of many large positive and negative values of the instantaneous fluxes, such as heat flux $w'\theta'$. Only after averaging does a smaller, but significant, net flux $\overline{w'\theta'}$ become apparent.

An example of the instantaneous heat flux values measured by an aircraft flying near the ground during the 1983 Boundary Layer Experiment (BLX83, see Stull & Eloranta, 1984) is shown in Fig. 2.13. The net flux for this case was $\overline{w'\theta'} = 0.062$ K·m·s^{-1}. We see from the figure that most of the time there are small positive and negative values of $w'\theta'$ that average to near zero. Occasional large positive spikes associated with convective plumes cause the net average value of $\overline{w'\theta'}$ to be positive for this case. Figs 2.14 show the corresponding histograms of frequency of occurrence of w', θ', and $w'\theta'$. The aircraft was flying at about 75 m/s, so it is easy to convert the time axis of Fig 2.13 to a distance axis. Statistics for this afternoon case are shown in table 2-1.

Table 2-1. Statistics for the 100 s segment time series shown in Fig 2.14. This segment is extracted from a 4 min flight leg near the surface, during the BLX83 field experiment. The 4 min mean values were used as the reference from which the perturbation values were calculated, which explains why $\overline{w'}$ and $\overline{\theta'}$ are not exactly zero for this 100 s segment.

Statistic	w' (m/s)	θ' (K)	$w'\theta'$ (K m/s)
Average	0.017	-0.017	0.062
σ	0.67	0.18	0.14
Maximum	2.51	0.87	1.93
Minimum	-2.07	-0.44	-0.38

2.7.2 Turbulent Flux Profiles

Fig 2.15 shows idealizations of the turbulent heat, momentum, and moisture fluxes for both the daytime and nighttime BLs. During the daytime, the fluxes are large, and usually change linearly with height over the ML. At night, the fluxes are much weaker.

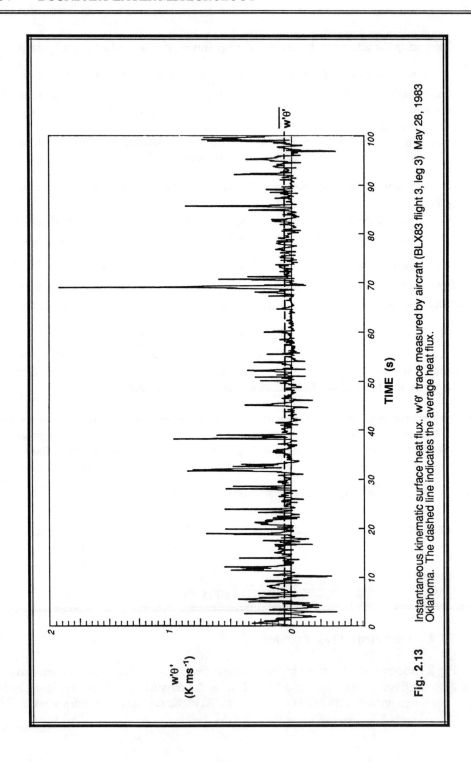

Fig. 2.13   Instantaneous kinematic surface heat flux. w'θ' trace measured by aircraft (BLX83 flight 3, leg 3) May 28, 1983 Oklahoma. The dashed line indicates the average heat flux.

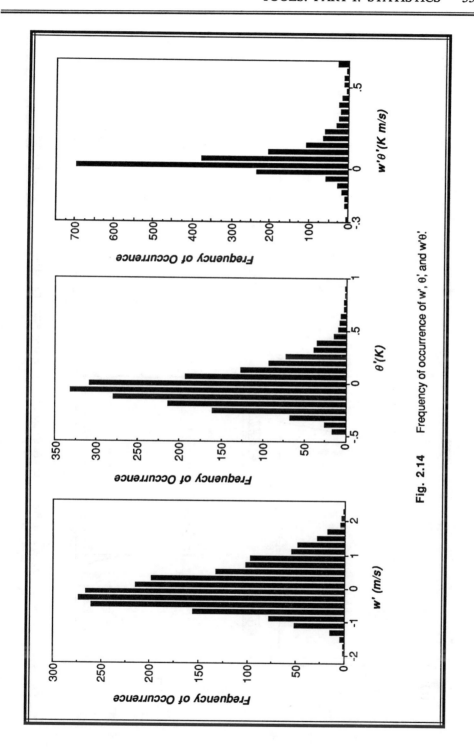

Fig. 2.14   Frequency of occurrence of w', θ', and w'θ'.

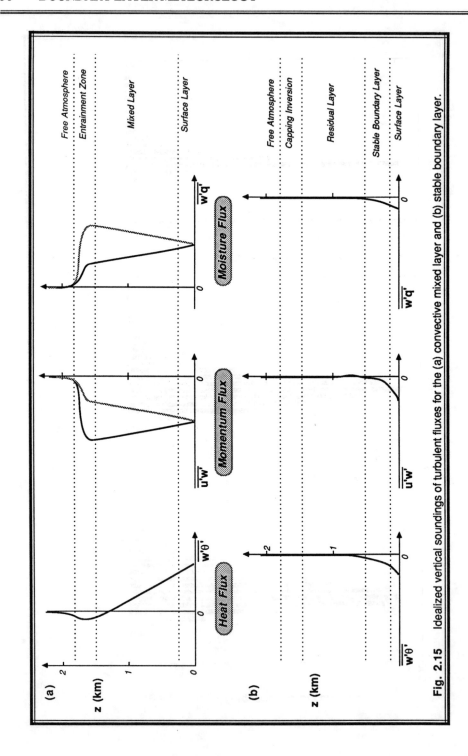

Fig. 2.15   Idealized vertical soundings of turbulent fluxes for the (a) convective mixed layer and (b) stable boundary layer.

2.8 Summation Notation

In the last section we encountered heat fluxes with three components and momentum fluxes with nine components. It is very laborious to write a separate forecast equation for each of these nine fluxes, but seemingly necessary if we want to better understand the boundary layer.

To ease the burden, we can employ a shorthand notation known as Einstein's summation notation. With just one term, we can represent all nine of the momentum fluxes. In this section, we first define some terms, then state rules with some examples, and finally show how summation notation and vector notation are related.

2.8.1 Definitions and Rules

Let m, n, and q be integer variable indices that can each take on the values of 1, 2, or 3. Let A_m represent a generic velocity vector, X_m represent a generic component of distance, and δ_m represent a generic unit vector (a vector of length unity and direction in one of the three Cartesian directions). By using indices as subscripts to these generic variables, we can define:

$$m = 1, 2, \text{ or } 3 \qquad A_1 = u \qquad X_1 = x$$
$$n = 1, 2, \text{ or } 3 \qquad A_2 = v \qquad X_2 = y$$
$$q = 1, 2, \text{ or } 3 \qquad A_3 = w \qquad X_3 = z$$

A variable with: no free indices   = scalar
                   1 free index   = vector
                   2 free indices   = tensor

Unit vectors: $\delta_1 = \mathbf{i}$ $\delta_2 = \mathbf{j}$ $\delta_3 = \mathbf{k}$

Physically, we expect that some forces act in all directions while others might act in just one or two. To be able to isolate such directional dependence, we must define two new terms with unusual characteristics:

Kronecker Delta (a scalar quantity even though it has two indices):

$$\delta_{mn} = \begin{cases} +1 & \text{for } m = n \\ 0 & \text{for } m \neq n \end{cases} \qquad (2.8.1a)$$

Alternating Unit Tensor (a scalar even though it has three indices):

$$\varepsilon_{mnq} = \begin{cases} +1 & \text{for } mnq = 123, 231, \text{ or } 312 \\ -1 & \text{for } mnq = 321, 213, \text{ or } 132 \\ 0 & \text{for any two or more indices alike} \end{cases} \qquad (2.8.1b)$$

The unit vector, δ_m, and the Kronecker delta, δ_{mn}, can easily be confused. They represent distinctly different quantities that are not interchangeable. To help distinguish between these two quantities, remember that the Kronecker delta is a scalar and always has two subscripts, while the unit vector is a vector and always has just one subscript.

Two fundamental rules apply within summation notation: one concerns repeated indices within any one term, and the other concerns nonrepeated (free) indices.

Rule (a): Whenever two identical indices appear in the same one term, it is implied that there is a sum of that term over each value (1, 2, and 3) of the repeated index.

Rule (b): Whenever one index appears unsummed (free) in a term, then that same index must appear unsummed in all terms in that equation. Hence, that equation effectively represents 3 equations, one for each value of the unsummed index. This insures that all terms are tensorally consistent with the other terms in the equation.

2.8.2 Examples

Problem 1 and Solution, demonstrating Rule (a).

$$A_n \frac{\partial B_m}{\partial X_n} = A_1 \frac{\partial B_m}{\partial X_1} + A_2 \frac{\partial B_m}{\partial X_2} + A_3 \frac{\partial B_m}{\partial X_3}$$

$$= u \frac{\partial B_m}{\partial x} + v \frac{\partial B_m}{\partial y} + w \frac{\partial B_m}{\partial z}$$

Problem 2 and Solution, demonstrating Rule (a).

$$\delta_{2n} A_n = \delta_{21} A_1 + \delta_{22} A_2 + \delta_{23} A_3$$

$$= 0 + A_2 + 0$$

$$= v$$

The latter example leads to an important general conclusion:

$$\delta_{mn} A_n = A_m \qquad (2.8.2a)$$

namely, the Kronecker delta changes the index of A from n to m.

Problem 3, demonstrating Rule (b):   Given the following equation, expand it:

$$A_m = B_m + \delta_{mn} C_n$$

Solution to 3. In each term is the same unrepeated index, m.   Thus, this equation represents three equations:

$$
\begin{cases}
A_1 = B_1 + \delta_{1n}\, C_n \\
A_2 = B_2 + \delta_{2n}\, C_n \\
A_3 = B_3 + \delta_{3n}\, C_n
\end{cases}
$$

The last term in each equation has the Kronecker delta, which means we can use the general conclusion above to yield:

$$
\begin{cases}
A_1 = B_1 + C_1 \\
A_2 = B_2 + C_2 \\
A_3 = B_3 + C_3
\end{cases}
$$

Problem 4, demonstrating Both Rules: One form of the equation of motion is written here in summation notation.   For now, just accept this equation as a given example; we will discuss the physics of it in more detail in the next chapter. This equation employs repeated and nonrepeated indices, the Kronecker delta, the alternating unit tensor, and the stress tensor τ (to be discussed in the next section).   Let both A and B represent velocities.   This is quite a complex example, which you should study carefully:

$$
\frac{\partial A_m}{\partial t} + B_n \frac{\partial A_m}{\partial X_n} = -\delta_{m3}\, g + f_c\, \varepsilon_{mn3}\, B_n - \frac{1}{\rho}\frac{\partial p}{\partial X_m} + \frac{1}{\rho}\left[\frac{\partial \tau_{mn}}{\partial X_n}\right] \qquad (2.8.2b)
$$

Using the previous rules and definitions, we can step-by-step expand the shorthand equation above to discover the equivalent set of equations written in more conventional form.

Solution to 4.   First, sum over repeated indices:

$$
\frac{\partial A_m}{\partial t} + B_1 \frac{\partial A_m}{\partial X_1} + B_2 \frac{\partial A_m}{\partial X_2} + B_3 \frac{\partial A_m}{\partial X_3} = -\delta_{m3}\, g + f_c\, \varepsilon_{m13}\, B_1 + f_c\, \varepsilon_{m23}\, B_2
$$

$$
+ f_c\, \varepsilon_{m33}\, B_3 - \frac{1}{\rho}\frac{\partial p}{\partial X_m} + \frac{1}{\rho}\left[\frac{\partial \tau_{m1}}{\partial X_1} + \frac{\partial \tau_{m2}}{\partial X_2} + \frac{\partial \tau_{m3}}{\partial X_3}\right]
$$

The term with ε_{m33} becomes zero because of the repeated index in the alternating unit tensor.

Next, write a separate equation for each value of the free index, m:

For m = 1:

$$\frac{\partial A_1}{\partial t} + B_1 \frac{\partial A_1}{\partial X_1} + B_2 \frac{\partial A_1}{\partial X_2} + B_3 \frac{\partial A_1}{\partial X_3} = -\delta_{13}\, g + f_c\, \varepsilon_{113}\, B_1 + f_c\, \varepsilon_{123}\, B_2$$

$$-\frac{1}{\rho}\frac{\partial p}{\partial X_1} + \frac{1}{\rho}\left[\frac{\partial \tau_{11}}{\partial X_1} + \frac{\partial \tau_{12}}{\partial X_2} + \frac{\partial \tau_{13}}{\partial X_3}\right]$$

In this equation the terms with δ_{13} and ε_{113} are both zero. We will leave out similar terms in the equations for the remaining components. The factor $\varepsilon_{123} = 1$.

For m = 2:

$$\frac{\partial A_2}{\partial t} + B_1 \frac{\partial A_2}{\partial X_1} + B_2 \frac{\partial A_2}{\partial X_2} + B_3 \frac{\partial A_2}{\partial X_3} = + f_c\, \varepsilon_{213}\, B_1$$

$$-\frac{1}{\rho}\frac{\partial p}{\partial X_2} + \frac{1}{\rho}\left[\frac{\partial \tau_{21}}{\partial X_1} + \frac{\partial \tau_{22}}{\partial X_2} + \frac{\partial \tau_{23}}{\partial X_3}\right]$$

The factor ε_{213} in the equation above equals -1.

For m = 3:

$$\frac{\partial A_3}{\partial t} + B_1 \frac{\partial A_3}{\partial X_1} + B_2 \frac{\partial A_3}{\partial X_2} + B_3 \frac{\partial A_3}{\partial X_3} = -\delta_{33}\, g$$

$$-\frac{1}{\rho}\frac{\partial p}{\partial X_3} + \frac{1}{\rho}\left[\frac{\partial \tau_{31}}{\partial X_1} + \frac{\partial \tau_{32}}{\partial X_2} + \frac{\partial \tau_{33}}{\partial X_3}\right]$$

The factor $\delta_{33} = 1$ in the equation above.

After substituting u for A_1, y for X_2, τ_{zx} for τ_{31}, etc., we finally get:

$$\frac{\partial u}{\partial t} + u\frac{\partial u}{\partial x} + v\frac{\partial u}{\partial y} + w\frac{\partial u}{\partial z} = +f_c v - \frac{1}{\rho}\frac{\partial p}{\partial x} + \frac{1}{\rho}\left[\frac{\partial \tau_{xx}}{\partial x} + \frac{\partial \tau_{xy}}{\partial y} + \frac{\partial \tau_{xz}}{\partial z}\right]$$

$$\frac{\partial v}{\partial t} + u\frac{\partial v}{\partial x} + v\frac{\partial v}{\partial y} + w\frac{\partial v}{\partial z} = -f_c u - \frac{1}{\rho}\frac{\partial p}{\partial y} + \frac{1}{\rho}\left[\frac{\partial \tau_{yx}}{\partial x} + \frac{\partial \tau_{yy}}{\partial y} + \frac{\partial \tau_{yz}}{\partial z}\right]$$

$$\frac{\partial w}{\partial t} + u\frac{\partial w}{\partial x} + v\frac{\partial w}{\partial y} + w\frac{\partial w}{\partial z} = -g - \frac{1}{\rho}\frac{\partial p}{\partial z} + \frac{1}{\rho}\left[\frac{\partial \tau_{zx}}{\partial x} + \frac{\partial \tau_{zy}}{\partial y} + \frac{\partial \tau_{zz}}{\partial z}\right]$$

$$(2.8.2c)$$

Discussion of Problem 4. Comparing the above set of equations to the original shorthand version, we begin to appreciate the power of Einstein's summation notation. This tool will be used throughout the remaining chapters. It is analogous to vector notation, which is examined in the next section.

Usually, the shorthand (summation) form of (2.8.2b) is written as follows:

$$\frac{\partial u_i}{\partial t} + u_j\frac{\partial u_i}{\partial x_j} = -\delta_{i3}\, g + f_c\, \varepsilon_{ij3}\, u_j - \frac{1}{\rho}\frac{\partial p}{\partial x_i} + \frac{1}{\rho}\left[\frac{\partial \tau_{ij}}{\partial x_j}\right] \qquad (2.8.2d)$$

where vectors like u_i have three components (u, v, w), and where (i, j, k) are indices, not unit vectors. This is the form that will be used for the remainder of the text.

Finally, we should recognize that a whole term is a scalar, vector, or tensor if the term has zero, one, or two unsummed (free) variable indices, respectively. For example, the $f_c\, \varepsilon_{ij3}\, u_j$ term in the equation above is a vector, because there is only one unsummed variable index, i.

2.8.3 Comparison with Vector Notation

Vectors can represent three Cartesian components, and tensors can represent nine. There is a one-to-one correspondence between vector definitions and Einstein's summation notation, as might be expected. The following is an explanation of how vectors can be rewritten in summation notation, and how vector operations such as the dot and cross product can be represented. In these examples, vector operations apply only to the vector parts of each term; scalar parts can be separated and interpreted as simple products.

The definitions below relate basic vector forms (especially, unit vectors) and operators to summation notation:

Vector:$\qquad \qquad \mathbf{A} \equiv A_m \, \delta_m$ $\qquad \qquad$ (2.8.3a)

Dot Product:$\qquad \delta_m \cdot \delta_n \equiv \delta_{mn}$ $\qquad \qquad$ (2.8.3b)

Cross Product: $\delta_m \times \delta_n \equiv \varepsilon_{mnq} \, \delta_q$ $\qquad \qquad$ (2.8.3c)

Del Operator:$\quad \nabla(\ ) \quad \equiv \delta_m \dfrac{\partial(\ )}{\partial X_m}$ $\qquad$ (2.8.3d)

Examples are presented here of other vector operations using the definitions above. First, consider the dot product between two vectors:

$$\mathbf{A} \cdot \mathbf{B} \;=\; (\delta_m \, A_m) \cdot (\delta_n \, B_n)$$

$$=\; (\delta_m \cdot \delta_n) \, A_m \, B_n$$

$$=\; \delta_{mn} \, A_m \, B_n$$

$$=\; A_m \, B_m$$

$$(2.8.3e)$$

On the first line, we substituted each vector by its summation notation as three Cartesian components times their respective component magnitudes. On the second line, the vector (boldface) terms were grouped together, leaving the product of the magnitudes remaining at the end. Then, the definition of a vector dot product was used to substitute the Kronecker delta. Finally, the Kronecker delta was used to change one subscript to equal the other. The end result, $A_m \, B_m = A_1 \, B_1 + A_2 \, B_2 + A_3 \, B_3$, is indeed a scalar that is equal in value to the scalar result of the vector dot product.

A similar development can be made for the cross product of two vectors:

$$\mathbf{A} \times \mathbf{B} \;=\; (\delta_m \, A_m) \times (\delta_n \, B_n)$$

$$=\; (\delta_m \times \delta_n) \, A_m \, B_n$$

$$=\; \varepsilon_{mnq} \, A_m \, B_n \, \delta_q$$

$$(2.8.3f)$$

The result is a vector, as is required for a cross product between two vectors. The reader can perform the implied sums to verify that the expected terms are obtained.

As a final example, we will look at the divergence of a vector:

$$\nabla \cdot \mathbf{A} = \left(\delta_m \frac{\partial}{\partial X_m} \right) \cdot (\delta_n A_n)$$

$$= (\delta_m \cdot \delta_n) \frac{\partial A_n}{\partial X_m}$$

$$= (\delta_{mn}) \frac{\partial A_n}{\partial X_m}$$

$$= \frac{\partial A_m}{\partial X_m} \qquad\qquad (2.8.3g)$$

We will have little use for vector notation in the remainder of this book, because summation notation is frequently easier to use. This section was presented only because most meteorologists are familiar with vector notation from their studies of atmospheric dynamics.

2.9 Stress

We have seen that the covariance statistic describes a turbulent flux. But a momentum flux is analogous to a stress. In this section, we review the nature of stress and relate it to various turbulence statistics.

Stress is the force tending to produce deformation in a body. It is measured as a force per unit area. Three types of stress appear frequently in studies of the atmosphere: pressure, Reynolds stress, and viscous shear stress.

2.9.1 Pressure

Pressure is a type of stress that can act on a fluid at *rest*. For an infinitesimally small fluid element, such as idealized as the cube sketched in Fig 2.16a, pressure acts equally in all directions. *Isotropic* is the name given to characteristics that are the same in all directions (see Figs 2.4 and 2.5).

If we consider just one face of this cube, as in Fig. 2.16b, we see that the isotropic nature of pressure tends to counteract itself in all directions except in a direction normal to (perpendicular to) the surface of the cube. Forces acting normal to all faces of the cube tend to compress or expand the cube, thereby deforming it (Fig 2.16c).

At sea level, the standard atmospheric pressure is 1.013×10^5 N/m^2. A pascal (Pa) is defined as 1 N/m^2, thus standard atmospheric pressure at sea level is 101.3 kPa. Historically, millibars (mb) have also been used as a pressure unit, where 1 mb = 100 pascals. In kinematic units, standard sea-level pressure is 82714 m^2/s^2. Although this value is much larger than the other stresses to be discussed next, it is almost totally counteracted by the influence of gravity, as described by the hydrostatic approximation.

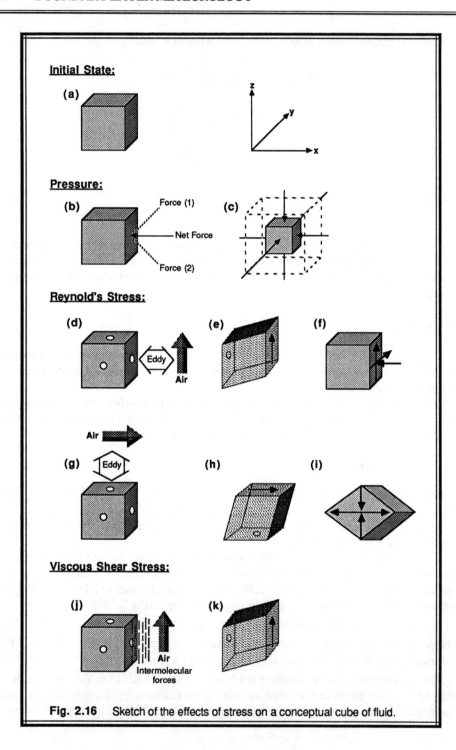

Fig. 2.16   Sketch of the effects of stress on a conceptual cube of fluid.

Pressure is a concept that is familiar to most meteorologists; therefore, we will not dwell on it here. Because pressure is not dependent on direction, we need just one number to describe it at any point in space and time. Thus, pressure is a scalar.

2.9.2 Reynolds stress

Reynolds stress exists only when the fluid is in *turbulent motion*. A turbulent eddy can mix air of different wind speeds into our cube of interest (Fig 2.16d). When this different-speed air is incorporated into one face of the cube and not the opposite face, the cube deforms because of the velocity differences between those two faces (Fig 2.16e).

The rate that air of different speeds is transported across any face of the cube is just the momentum flux, by definition (see section 2.6). The effect of this flux on the cube is identical to what we would observe if we applied a force on the face of the cube; namely, the cube would deform. Thus, turbulent momentum flux acts like a stress, and is called the *Reynolds stress*. In this example, air moving upward (possessing w'), was mixed towards the cube (at rate u'), resulting in a Reynolds stress component described by

$-\rho \overline{u'w'}$. The magnitude of this component of Reynolds stress or momentum flux in

kinematic units is thus $|\overline{u'w'}|$. Sometimes the symbol $\tau_{Reynolds}$ is used for the Reynolds stress.

Even if we consider just one face of the cube (Fig 2.16f), air moving in any one of the three Cartesian directions could be mixed into it, resulting in a variety of deformations.

For that one face, we must consider $\overline{u'u'}$, $\overline{u'v'}$, and $\overline{u'w'}$. Since the same number of combinations holds for faces normal to the other two cartesian directions, we have a total of nine components of the Reynolds stress to account for. Based on what we learned in section 2.6, this is expected because there are nine components of the momentum flux.

Let's consider one other example. Suppose a u' wind gust is being turbulently mixed towards the top face of a cube at rate w', as is sketched in Fig 2.16g. The resulting

deformation shown in Fig 2.16h is associated with a $\overline{w'u'}$ momentum flux. The nature of the deformation shown in Fig 2.16h is identical to that of Fig 2.16e; the only difference is that (e) is rotated relative to the orientation of (h). Based on the nature of the deformation

alone, we must then conclude that $\overline{u'w'} = \overline{w'u'}$. Fig 2.16i shows the type of deformation that is exhibited by both Figs 2.16 e and h, without regard to the rotational differences.

Similar arguments can be made for the other faces. We can conclude that the Reynolds stress tensor (represented by the elements in the matrix below) is symmetric:

$$\begin{bmatrix} \overline{u'u'} & \overline{u'v'} & \overline{u'w'} \\ \overline{v'u'} & \overline{v'v'} & \overline{v'w'} \\ \overline{w'u'} & \overline{w'v'} & \overline{w'w'} \end{bmatrix} = \begin{bmatrix} \overline{u'u'} & \overline{u'v'} & \overline{u'w'} \\ \overline{u'v'} & \overline{v'v'} & \overline{v'w'} \\ \overline{u'w'} & \overline{v'w'} & \overline{w'w'} \end{bmatrix} \qquad (2.9.2)$$

Hence, we need only be concerned with six independent stress components. The convenience of Einstein's summation notation is apparent here. We could easily represent any one of the components of the matrix on the left by $\overline{u_i' u_j'}$. The kinematic Reynolds stress in typical atmospheric surface layers is on the order of 0.05 m^2/s^2.

Momentum flux and Reynolds stress are properties of the *flow*, not of the fluid. This stress is fully described by the matrices above, which contain products of velocities (a flow characteristic) that could apply to any fluid. Such is not the case with viscous shear stresses. Although the effect of $\overline{u_i' u_j'}$ is like a stress, the Reynolds stress is not a true stress (force per unit area) as is the viscous shear stress.

2.9.3 Viscous Shear Stress

Viscous shear stress exists only if there are **shearing motions** in the fluid. The motion can be laminar or turbulent. When one portion of a fluid moves, the intermolecular forces tend to drag adjacent fluid molecules in the same direction (Fig 2.16j). The strength of these intermolecular forces depend on the nature of the fluid: molasses has stronger forces than water, which in turn has stronger forces than air. A measure of these forces is the viscosity. The result of this stress is a deformation of the fluid (Fig. 2.16k).

These viscous forces can act in any of the three cartesian directions on any of the three faces of our conceptual cube (Fig. 2.16f). Thus, the viscous stress is also a tensor with nine components. Like the Reynolds stress, the shear stress tensor is symmetric, leaving six independent components.

A fluid for which the viscous shear stress is linearly dependent on the shear is said to be a *Newtonian fluid*. The stress, τ_{ij}, for a Newtonian fluid is usually given by:

$$\tau_{ij} = \mu \left(\frac{\partial u_i}{\partial x_j} + \frac{\partial u_j}{\partial x_i} \right) + \left(\mu_B - \frac{2}{3} \mu \right) \frac{\partial u_k}{\partial x_k} \delta_{ij} \qquad (2.9.3)$$

where μ_B is the bulk viscosity coefficient (near zero for most gases) and μ is the dynamic viscosity coefficient. We can interpret τ_{ij} as the force per unit area in the x_i-direction acting on the face that is normal to the x_j-direction.

The viscous shear stress can be put into kinematic form by dividing by the mean density of the fluid. A corresponding kinematic viscosity is defined by $\nu = \mu / \rho$. The standard atmospheric sea-level value for kinematic viscosity of air is 1.4607×10^{-5} m^2/s. For a mean wind shear of 0.5 s^{-1} (typical for atmospheric surface layers), the resulting viscous shear stress is 7.304×10^{-6} m^2/s^2.

This value is so much smaller than the Reynolds stresses in the BL that the viscous stress is usually neglected in mean wind forecasts. Turbulent eddies, however, can have much larger values of shear in localized eddy-size regions. We thus can not neglect viscosity when forecasting turbulence.

2.10 Friction Velocity

During situations where turbulence is generated or modulated by wind shear near the ground, the magnitude of the surface Reynolds' stress proves to be an important scaling variable. The total vertical flux of horizontal momentum measured near the surface is

$$\tau_{xz} = -\rho \, \overline{u'w'}_s \quad \text{and} \quad \tau_{yz} = -\rho \, \overline{v'w'}_s$$

$$\left| \tau_{Reynolds} \right| = \left[\tau_{xz}^{\,2} + \tau_{yz}^{\,2} \right]^{1/2} \tag{2.10a}$$

Based on this relationship, a velocity scale called the *friction velocity*, u_*, is defined as

$$u_*^2 \equiv \left[\overline{u'w'}_s^{\,2} + \overline{v'w'}_s^{\,2} \right]^{1/2}$$

$$= \left| \tau_{Reynolds} \right| / \overline{\rho} \tag{2.10b}$$

For the special case where the coordinate system is aligned so that the x-axis points in the direction of the surface stress, we can rewrite (2.10b) as $u_*^2 = |\overline{u'w'}_s| = |\tau_{Reynolds}| / \rho$. Examples of u_* evolution are shown in Figs 2.17 and 4.1.

While we are talking about surface scales, we can also introduce surface layer temperature (θ_*^{SL}) and humidity (q_*^{SL}) scales that are defined by:

$$\theta_*^{SL} = \frac{-\overline{w'\theta'}_s}{u_*} \tag{2.10c}$$

$$q_*^{SL} = \frac{-\overline{w'q'}_s}{u_*} \tag{2.10d}$$

These scales will be used in later chapters dealing with surface-layer similarity theory.

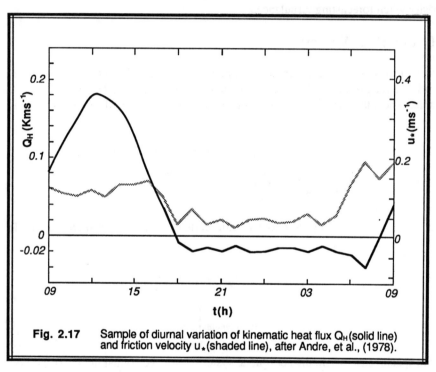

Fig. 2.17   Sample of diurnal variation of kinematic heat flux Q_H (solid line) and friction velocity u_* (shaded line), after Andre, et al., (1978).

2.11 References

André, J.-C., G. De Moor, P. Lacarrère, G. Therry, and R. du Vachat, 1978: Modeling the 24-hour evolution of the mean and turbulent structure of the planetary boundary layer. *J. Atmos. Sci.*, **35**, 1861-1883.

Bean, B.R. & T.P Repoff, 1983: A study of turbulent energy over complex terrain (STATE, 1978). *Bound.-Layer Meteor.*, **25**, 17-23.

Caughey, S.J., J.C. Wyngaard and J.C. Kaimal, 1979: Turbulence in the evolving stable boundary layer. *J. Atmos. Sci.*, **36**, 1041-1052.

Richardson, L.F., 1922: *Weather Prediction by Numerical Process*. Cambridge Univ. Press.

Stull, R.B. and E.W. Eloranta, 1984: Boundary layer experiment - 1983. *Bull. Am. Meteor. Soc.*, **65**, 450-456.

Van der Hoven, I., 1957: Power spectrum of horizontal wind speed in the frequency range from 0.0007 to 900 cycles per hour. *J. Meteor.*, **14**, 160.

Van Dyke, M., 1982: *An Album of Fluid Motion*. The Parabolic Press, Stanford.176 pp.

2.12 Exercises

1) It has been suggested that in regions of strong static stability, the lower (long wavelength, small wavenumber) end of the inertial subrange occurs at a wavenumber, κ_B, given by $\kappa_B \cong N_{BV}^{3/2}\, \varepsilon^{-1/2}$, where N_{BV} is the Brunt-Vaisala frequency, and ε is the turbulence dissipation rate. Between this wavenumber and lower wavenumbers is a region called the buoyancy subrange, where the gravitational effects (i.e., buoyancy) are important. Within the buoyancy subrange sketched below, would you expect turbulence to be isotropic?

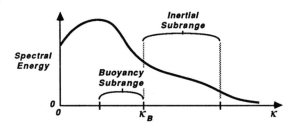

2) Given the following instantaneous measurements of potential temperature (θ) and vertical velocity (w) in this table, fill in all the remaining blanks in the table. Also, verify with the answers from above that $\overline{w\theta} = \overline{w}\,\overline{\theta} + \overline{w'\theta'}$.

	Measurements:		Calculations:					
index	w	θ	w'	θ'	$(w')^2$	$(\theta')^2$	wθ	w'θ'
0	0.5	295						
1	-0.5	293						
2	1.0	295						
3	0.8	298						
4	0.9	292						
5	-0.2	294						
6	-0.5	292						
7	0.0	289						
8	-0.9	293						
9	-0.1	299						
Average:								

3) Given the data in problem (2), find the biased standard deviation for w and θ, and find the linear correlation coefficient between w and θ.

4) Using your results from problems (2) and (3), is the data characteristic of a stable, neutral, or unstable boundary layer?

5) Let: c = constant, $s \neq$ function of time, and $A = \overline{A} + a'$, $B = \overline{B} + b'$, and $E = \overline{E} + e'$.

Expand the following terms into mean and turbulent parts, and apply Reynold's averaging rules to simplify your expression as much as possible:

a) $\overline{(c\,A\,B)}$ = ?

d) $\overline{\left(\dfrac{\partial A}{\partial s}\right)\cdot\left(\dfrac{\partial B}{\partial s}\right)}$ = ?

b) $\overline{(A\,B\,E)}$ = ?

c) $\overline{\left(A\cdot\dfrac{\partial B}{\partial s}\right)}$ = ?

e) $\overline{\left(c\,\nabla^2 A\right)}$ = ?

6) The following terms are given in summation notation. Expand them (that is, write out each term of the indicated sums).

a) $\dfrac{\partial(\overline{u_i'\,u_j'})}{\partial x_j}$

d) $\overline{u_i'\,u_j'}\ \dfrac{\partial \overline{U}_k}{\partial x_j}$

b) $u_i'\,\dfrac{\partial \theta'}{\partial x_i}$

e) $\dfrac{\partial(\overline{u_i'\,u_j'\,u_k'})}{\partial x_j}$

c) $\overline{U}_j\,\dfrac{\partial(\overline{u_i'\,u_k'})}{\partial x_j}$

f) $\left(\dfrac{\partial u_i}{\partial x_j}\right)\left(\dfrac{\partial u_k}{\partial x_j}\right)$

7) Express the following terms in summation notation. They are given to you in vector notation.

a) Gradient of a scalar: ∇s

c) Total derivative: dV/dt

b) Curl of a vector: $\nabla \times V$

d) Laplacian: ∇^2

8) Consider a 100 m thick layer of air at sea level, with an initial potential temperature of 290 K. If the kinematic heat flux into the bottom of this layer is 0.2 K m/s and the flux out of the top is 0.1 K m/s, then what is the potential temperature of that layer 2 hours later? Assume that the potential temperature is constant with height in the layer. (This is a thought question, meant to stimulate the student's ability to interpret a physical situation in an acceptable mathematical framework.)

9) Given the typical variation of wind speed with height within the surface layer (see Chapt 1), and using a development similar to that in section 2.7:

a) determine whether the net kinematic momentum flux, $\overline{u'w'}$ is positive, negative, or zero within the surface layer.

b) Does you answer mean the momentum is being transported up or down, on the average?

c) This momentum that is transported up or down, where does it go or where does it come from, and how would that alter the mean state of the atmosphere?

10) Suppose we define an average wind speed by

$$\overline{U}(P) \;=\; \frac{1}{N} \sum_{i=0}^{N-1} U(t_i)$$

where $t_0 = 0$, $t_N = P$, and where the averaging time, P, is the interval between t=0 and t=P. When P is small, we will find one value for $\overline{U}$, and when P is larger, we might find a different value. In fact, $\overline{U}$ (P) is probably a smoothly varying function of P. Given the instantaneous measurements of U(t) shown in the table below:

a) Plot U vs t.

b) Plot $\overline{U}$ vs P for P = 0 to 60 min. (A calculator or computer might make the job easier.)

c) Comment on the relationship between the spectral gap and your answer from part (b).

Data:

i	0	1	2	3	4	5	6	7	8	9	10	11	12	13	14	15	16	17	18	19	20
t (min)	0	3	6	9	12	15	18	21	24	27	30	33	36	39	42	45	48	51	54	57	60
U (m/s)	3	4	3	6	6	2	5	3	7	8	5	4	3	5	2	2	4	5	3	6	5

11) Suppose that the following wind speed traces were observed on different days using an anemometer mounted on a 10m mast:

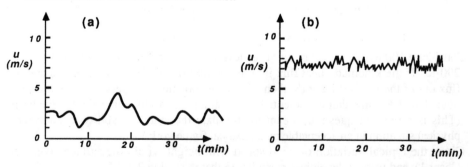

a) Sketch (both on the same graph) your best guess of the frequency spectrum of a) and of b). Do not digitize the data from graphs a) and b), but visually estimate the

a) Sketch (both on the same graph) your best guess of the frequency spectrum of a) and of b). Do not digitize the data from graphs a) and b), but visually estimate the contribution of various size eddies to the total turbulence kinetic energy when answering this question.

b) If the mean wind speed at anemometer height was 5 m/s, then reanswer part a) in terms of a wavenumber spectrum rather than a frequency spectrum.

12) For selected operational weather forecast models, determine whether the grid spacing (for finite-difference models) or the smallest wavelength (for spectral models) falls within the spectral gap. Find out how the subgrid-scale motions (turbulence) are parameterized in those models. (Hint: The instructor should select one or two current models for this project. It involves outside reading on the part of the students.)

13) When might the ergodic assumption fail for boundary layer studies?

14) Turbulence is usually anisotropic, inhomogeneous, and nonstationary in the boundary layer, although we often limit our studies to situations where one or more of these special cases are approximately valid. Comment on the difference in meaning between the words isotropic, homogeneous, and stationary. Can turbulence be homogeneous but anisotropic? Can turbulence be isotropic but inhomogeneous? How are homogeniety and stationarity related?

15) What would be the dimensions of kinematic pressure flux? Of kinematic vorticity flux?

16) The solar constant is about 1380. W/m^2. Rewrite this in kinematic units.

17) What is the difference between $\partial U_j / \partial x_j$ and $\partial U_k / \partial x_k$?

18) If $\partial U_k / \partial x_k = 0$ (valid for incompressible flow), then write the resulting equation for $\partial \tau_{ij} / \partial x_j$ in the simplest form possible. [Hint, use what you learned from question (17).]

19) Given the typical value for kinematic viscosity from section 2.9, find what value of shear is necessary to make the viscous shear stress as large as the previously quoted typical value of Reynolds stress. Could such a shear really be found within eddies? If so, what size eddies? (Hint, assume the viscous shear stress is equal to the viscosity times the shear).

20) How is the pressure stress different from a stress like $\overline{u' u'}$ or τ_{22} ?

21) Suppose that on the planet Krypton turbulent motions are affected by a strange form of viscosity that dissipates turbulent energy in only the vertical direction. How would the average turbulence kinetic energy differ from that found on earth?

22) How, if at all, are terms A and B below related to each other. Answer with equations or words.

$$A = f \, \varepsilon_{ij3} \, \overline{U_j} \quad , \quad\quad B = f \, \varepsilon_{ij3} \, \overline{U_i}$$

23) Given the term $U \, \partial(V^2) / \partial x$, which represents the U-advection of the V-component of total kinetic energy. Expand the U and V variables into mean and turbulent parts, Reynolds average, and simplify as much as possible.

24) Expand the Coriolis term [-2 $\varepsilon_{ijk} \Omega_j U_k$] using summation notation for the case i=1. Assume that $\Omega_j = (0, \omega \cos \lambda, \omega \sin \lambda)$ are the three components of the angular velocity vector, where λ is latitude and ω is the speed of rotation of the earth ($\omega = 360$ degrees in 24 hours). Assume that $W = 0$ for simplicity.

25) For each separate term in problem 6, count the number of indices and determine if each term is a scalar, vector, tensor, etc.

26) Given the following variances in m^2/s^2:

Where:	Location A		Location B	
When (UTC):	1000	1100	1000	1100
$\overline{u'^2}$	0.50	0.50	0.70	0.50
$\overline{v'^2}$	0.25	0.50	0.25	0.25
$\overline{w'^2}$	0.70	0.50	0.70	0.25

Where, when and for which variables is the turbulence:
a) Stationary?
b) Homogeneous?
c) Isotropic?

27) What boundary layer flow phenomena or characteristics have scale sizes on the order of:
a) 1 mm ?
b) 10 m ?
c) 1 km ?

28) Simplify the following term (assume horizontal homogeneity).

$$\delta_{k1} \, \varepsilon_{ijk} \, \frac{\partial \overline{u_i' \theta'}}{\partial x_j}$$

3 Application of the Governing Equations to Turbulent Flow

To quantitatively describe and forecast the state of the boundary layer, we turn to the equations of fluid mechanics that describe the dynamics and thermodynamics of the gases in our atmosphere. Motions in the boundary layer are slow enough compared to the speed of light that the Galilean/Newtonian paradigm of classical physics applies. These equations, collectively known as the *equations of motion*, contain time and space derivatives that require initial and boundary conditions for their solution.

Although the equations of motion together with other conservation equations can be applied directly to turbulent flows, rarely do we have sufficient initial and boundary condition information to resolve all turbulent scales down to the smallest eddy. We often don't even care to forecast all eddy motions. For simplicity, we instead pick some cut-off eddy size below which we include only the statistical effects of turbulence. In some mesoscale and synoptic models the cutoff is on the order of 10 to 100 km, while for some boundary layer models known as *large eddy simulation models* the cutoff is on the order of 100 m.

The complete set of equations as applied to the boundary layer are so complex that no analytical solution is known. As in other branches of meteorology, we are forced to find approximate solutions. We do this by either finding exact analytical solutions to simplified subsets of the equations, or by finding approximate numerical solutions to a more complete set of equations. Both approximations are frequently combined to allow boundary layer meteorologists to study particular phenomena.

In this chapter we start with the basic governing equations and statistically average over the smaller eddy sizes. Along the way we demonstrate simplifications based on boundary layer scaling arguments. Numerical methods for solving the resulting set of equations are not covered.

75

3.1 Methodology

Because the upcoming derivations are sometimes long and involved, it is easy "to lose sight of the forest for the trees". The following summary gives the steps that will be taken in the succeeding sections to develop prognostic equations for mean quantities such as temperature and wind:

Step 1. Identify the basic governing equations that apply to the boundary layer.

Step 2. Expand the total derivatives into the local and advective contributions.

Step 3. Expand dependent variables within those equations into mean and turbulent (perturbation) parts.

Step 4. Apply Reynolds averaging to get the equations for mean variables within a turbulent flow.

Step 5. Add the continuity equation to put the result into flux form.

Additional steps take us further towards understanding the nature of turbulence itself:

Step 6. Subtract the equations of step 5 from the corresponding ones of step 3 to get equations for the turbulent departures from the mean.

Step 7. Multiply the results of step 6 by other turbulent quantities and Reynolds average to yield prognostic equations for turbulence statistics such as kinematic flux or turbulence kinetic energy.

Section 3.2 covers steps 1 and 2. Section 3.3 takes a side road to look at some simplifications and scaling arguments. In section 3.4 we get back on track and utilize steps 3-5 to derive the desired prognostic equations. After a few more simplifications in section 3.5, a summary of the governing equations for mean variables in turbulent flow is presented.

Steps 6 and 7 are addressed in Chapters 4 and 5.

3.2 Basic Governing Equations

Five equations form the foundation of boundary layer meteorology: the equation of state, and the conservation equations for mass, momentum, moisture, and heat. Additional equations for scalar quantities such as pollutant concentration may be added. It is assumed that the reader has already been exposed to these equations; hence, the derivations are not given here.

3.2.1 Equation of State (Ideal Gas Law)

The ideal gas law adequately describes the state of gases in the boundary layer:

$$P = \rho_{air} \, \Re \, T_v \qquad (3.2.1)$$

where P is pressure, ρ_{air} is the density of moist air, T_v is the virtual absolute temperature, and $\Re$ is the gas constant for **dry** air ($\Re = 287$ J·K^{-1} kg^{-1}). Sometimes, the density of moist air is abbreviated as ρ for simplicity.

3.2.2 Conservation of Mass (Continuity Equation)

Two equivalent forms of the continuity equation are

$$\frac{\partial \rho}{\partial t} + \frac{\partial (\rho U_j)}{\partial x_j} = 0 \tag{3.2.2a}$$

and

$$\frac{d\rho}{dt} + \rho \frac{\partial U_j}{\partial x_j} = 0 \tag{3.2.2b}$$

where the definition of the total derivative is used to convert between these forms.

If V and L are typical velocity and length scales for the boundary layer, then it can be shown (Businger, 1982) that $(d\rho/dt)/\rho \ll \partial U_j/\partial x_j$ if the following conditions are met: (1) $V \ll 100$ m/s;   (2) $L \ll 12$ km;   (3) $L \ll C_s^2/g$;   and (4) $L \ll C_s/f$, where C_s is the speed of sound and f is frequency of any pressure waves that might occur. Since these conditions are generally met for all turbulent motions smaller than mesoscale, (3.2.2b) reduces to

$$\frac{\partial U_j}{\partial x_j} = 0 \tag{3.2.2c}$$

This is the *incompressibility* approximation.

3.2.3 Conservation of Momentum (Newton's Second Law)

As presented at the end of section 2.8.2, one form for the momentum equation is

$$\frac{\partial U_i}{\partial t} + U_j \frac{\partial U_i}{\partial x_j} = - \delta_{i3} g - 2 \varepsilon_{ijk} \Omega_j U_k - \frac{1}{\rho} \frac{\partial p}{\partial x_i} + \frac{1}{\rho} \frac{\partial \tau_{ij}}{\partial x_j} \tag{3.2.3a}$$

$$\quad \text{I} \qquad \text{II} \qquad\quad \text{III} \qquad \text{IV} \qquad\quad \text{V} \qquad\quad \text{VI}$$

Term I   represents storage of momentum (inertia).
Term II   describes advection.
Term III allows gravity to act vertically.
Term IV describes the influence of the earth's rotation (Coriolis effects).
Term V   describes pressure-gradient forces.
Term VI represents the influence of viscous stress.

In term IV, the components of the angular velocity vector of the earth's rotation Ω_j are $[0, \omega \cos(\phi), \omega \sin(\phi)]$ where ϕ is latitude and $\omega = 2\pi$ radians/24h $= (7.27 \times 10^{-5} \text{ s}^{-1})$ is the angular velocity of the earth. Often term IV is written as $+ f_c \, \varepsilon_{ij3} \, U_j$, where the *Coriolis parameter* is defined as $f_c = 2 \, \omega \sin \phi = (1.45 \times 10^{-4} \text{ s}^{-1}) \sin \phi$. For a latitude of about $44°$ (e.g., southern Wisconsin), $f_c = 10^{-4} \text{ s}^{-1}$.

To a close approximation, air in the atmosphere behaves like a Newtonian fluid. Thus, the expression for viscous shear stress from section 2.9.3 allows us to write term VI as:

$$\text{Term VI} = \left(\frac{1}{\rho}\right) \frac{\partial}{\partial x_j} \left\{ \mu \left[\frac{\partial U_i}{\partial x_j} + \frac{\partial U_j}{\partial x_i} \right] - \left(\frac{2}{3}\right) \mu \left[\frac{\partial U_k}{\partial x_k} \right] \delta_{ij} \right\}$$

where the bulk viscosity coefficient μ_B was assumed to be near zero. Upon applying the derivative to each term, assuming that the viscosity μ is not a function of position, and rearranging, this expression can be written as:

$$\text{Term VI} = \left(\frac{\mu}{\rho}\right) \left\{ \frac{\partial^2 U_i}{\partial x_j^2} + \frac{\partial}{\partial x_i} \left[\frac{\partial U_j}{\partial x_j} \right] - \left(\frac{2}{3}\right) \frac{\partial}{\partial x_i} \left[\frac{\partial U_k}{\partial x_k} \right] \right\}$$

By assuming incompressibility, this reduces to

$$\text{Term VI} = \nu \frac{\partial^2 U_i}{\partial x_j^2}$$

where the kinematic viscosity, ν, has been substituted for μ/ρ.

Substituting this back into (3.2.3a) gives the form for the momentum equation that is most often used as a starting point for turbulence derivations:

$$\frac{\partial U_i}{\partial t} + U_j \frac{\partial U_i}{\partial x_j} = - \delta_{i3} \, g \, - \, f_c \, \varepsilon_{ij3} \, U_j \, - \, \frac{1}{\rho} \frac{\partial p}{\partial x_i} \, + \, \nu \frac{\partial^2 U_i}{\partial x_j^2} \tag{3.2.3b}$$

$$\text{I} \qquad \text{II} \qquad \text{III} \qquad \text{IV} \qquad \text{V} \qquad \text{VI}$$

where each term represents the same process as before.

3.2.4 Conservation of Moisture

Let q_T be the total specific humidity of air; namely, the mass of water (all phases) per unit mass of moist air. The conservation of water substance can be written, assuming incompressibility, as

$$\frac{\partial q_T}{\partial t} + U_j \frac{\partial q_T}{\partial x_j} = v_q \frac{\partial^2 q}{\partial x_j^2} + \frac{S_{q_T}}{\rho_{air}}$$

(3.2.4a)

$$\quad\text{I}\qquad\qquad\text{II}\qquad\qquad\text{VI}\qquad\text{VII}$$

where v_q is the molecular diffusivity for water vapor in the air. S_{q_T} is a net moisture source term (sources - sinks) for the remaining processes not already included in the equation. Its units are: mass of total water per unit volume per unit time.

By splitting the total humidity into vapor (q) and non-vapor (q_L) parts using $q_T = q + q_L$ and $S_{q_T} = S_q + S_{q_L}$, (3.2.4a) can be rewritten as a pair of coupled equations

$$\frac{\partial q}{\partial t} + U_j \frac{\partial q}{\partial x_j} = v_q \frac{\partial^2 q}{\partial x_j^2} + \frac{S_q}{\rho_{air}} + \frac{E}{\rho_{air}}$$

(3.2.4b)

and

$$\frac{\partial q_L}{\partial t} + U_j \frac{\partial q_L}{\partial x_j} = \qquad + \frac{S_{q_L}}{\rho_{air}} - \frac{E}{\rho_{air}}$$

(3.2.4c)

$$\quad\text{I}\qquad\qquad\text{II}\qquad\qquad\text{VI}\qquad\text{VII}\qquad\text{VIII}$$

where E represents the mass of water vapor per unit volume per unit time being created by a phase change from liquid or solid. The convergence of falling liquid or solid water (e.g., precipitation) that is not advecting with the wind is included as part of term VII. It has been assumed in (3.2.4c) that molecular diffusion has a negligible effect on liquid and solid precipitation or cloud particles.

Terms I, II, and VI are analogous to the corresponding terms in the momentum equation. Term VII is a net body source term, and term VIII represents the conversion of solid or liquid into vapor.

3.2.5 Conservation of Heat   (First Law of Thermodynamics)

The First Law of Thermodynamics describes the conservation of enthalpy, which includes contributions from both sensible and latent heat transport. In other words, the water vapor in air not only transports sensible heat associated with its temperature, but it has the potential to release or absorb additional latent heat during any phase changes that might occur. To simplify the equations describing enthalpy conservation, micrometeorologists often utilize the phase change information, E, contained in the moisture conservation equations. Thus, an equation for θ can be written

$$\frac{\partial \theta}{\partial t} + U_j \frac{\partial \theta}{\partial x_j} = \nu_\theta \frac{\partial^2 \theta}{\partial x_j^2} - \frac{1}{\rho\, C_p}\left(\frac{\partial Q_j^*}{\partial x_j}\right) - \frac{L_p\, E}{\rho\, C_p} \qquad (3.2.5)$$

$$\text{I} \qquad\quad \text{II} \qquad\quad \text{VI} \qquad\quad \text{VII} \qquad\quad \text{VIII}$$

where ν_θ is the thermal diffusivity, and L_p is the latent heat associated with the phase change of E. The values for latent heat at 0°C are $L_v = 2.50 \times 10^6$ J/kg (gas:liquid), $L_f = 3.34 \times 10^5$ J/kg (liquid:solid), and $L_s = 2.83 \times 10^6$ J/kg of water (gas:solid).

Q_j^* is the component of net radiation in the jth direction. The specific heat for *moist* air at constant pressure, C_p, is approximately related to the specific heat for dry air, $C_{pd} = 1004.67$ J kg-1 K-1, by $C_p = C_{pd} (1 + 0.84\, q)$. Given typical magnitudes of q in the boundary layer, it is important not to neglect the moisture contribution to C_p.

Terms I, II, and VI are the storage, advection, and molecular diffusion terms, as before. Term VII is the "body source" term associated with radiation divergence. Term VIII is also a "body source" term associated with latent heat released during phase changes. These body source terms affect the whole volume, not just the boundaries.

3.2.6 Conservation of a Scalar Quantity

Let C be the concentration (mass per volume) of a scalar such as a tracer in the atmosphere. The conservation of tracer mass requires that

$$\frac{\partial C}{\partial t} + U_j \frac{\partial C}{\partial x_j} = \nu_C \frac{\partial^2 C}{\partial x_j^2} + S_C \qquad (3.2.6)$$

$$\text{I} \qquad\quad \text{II} \qquad\quad \text{VI} \qquad\quad \text{VII}$$

where ν_C is the molecular diffusivity of constituent C. S_C is the body source term for the remaining processes not already in the equation, such as chemical reactions. The physical interpretation of each term is analogous to that of (3.2.4c).

3.3 Simplifications, Approximations, and Scaling Arguments

Under certain conditions the magnitudes of some of the terms in the governing equations become smaller than the other terms and can be neglected. For these situations the equations become simpler — a fact that has allowed advances to be made in atmospheric dynamics that would otherwise have been more difficult or impossible.

One simplification is called the **shallow motion approximation** (Mahrt, 1986). This approximation is valid if all of the following conditions are true:

1) the vertical depth scale of density variations in the boundary layer is much shallower than the scale depth of the lower atmosphere. (This latter scale depth $= \rho \, (\partial \rho / \partial z)^{-1} \cong 8$ km.);

2) advection and divergence of mass at a fixed point approximately balance, leaving only slow or zero variations of density with time.

3) the perturbation magnitudes of density, temperature, and pressure are much less than their respective mean values; and

A more stringent simplification, called the **shallow convection approximation**, requires all of the conditions above plus:

4) the mean lapse rate $(\partial T / \partial z)$ can be negative, zero, or even slightly positive. For the statically stable positive case, $(\partial T / \partial z) \ll g/\Re$, where $g/\Re = 0.0345$ K/m; and

5) the magnitude of the vertical perturbation pressure gradient term must be of the same order or less than the magnitude of the buoyancy term in the equation of motion.

This latter condition says that vertical motion is limited by buoyancy, which is origin of the term "shallow convection".

We have already employed conditions (1) and (2) to yield the incompressible form of the continuity equation. The other conditions will be applied below to yield further simplifications.

3.3.1 Equation of State

Start with the equation of state (3.2.1) and split the variables into mean and turbulent parts: $\rho = \bar{\rho} + \rho'$, $T_v = \bar{T}_v + T_v'$, $p = \bar{P} + p'$. The result can be rearranged to be

$$\frac{\bar{P}}{\Re} + \frac{p'}{\Re} = (\bar{\rho} + \rho') \cdot (\bar{T}_v + T_v')$$

or

$$\frac{\bar{P}}{\Re} + \frac{p'}{\Re} = \bar{\rho} \cdot \bar{T}_v + \rho' \, \bar{T}_v + \bar{\rho} \, T_v' + \rho' T_v' \qquad (3.3.1a)$$

Upon Reynolds averaging, we are left with

$$\frac{\bar{P}}{\Re} = \bar{\rho} \, \bar{T}_v + \overline{\rho' T_v'}$$

The last term is usually much smaller in magnitude than the others, allowing us to neglect it. As a result, the equation of state holds in the mean:

$$\frac{\overline{P}}{\mathcal{R}} = \overline{\rho}\ \overline{T}_v \qquad (3.3.1b)$$

This is a reasonable approximation because the equation of state was originally formulated from measurements made with crude, slow-response sensors that were essentially measuring mean quantities. As we shall see in section 3.4, however, we can't make similar assumptions for the other governing equations.

Subtracting (3.3.1b) from (3.3.1a) leaves

$$\frac{p'}{\mathcal{R}} = \rho'\ \overline{T}_v + \overline{\rho}\ T_v' + \rho' T_v'$$

Finally, dividing by (3.3.1b) gives

$$\frac{p'}{\overline{P}} = \frac{\rho'}{\overline{\rho}} + \frac{T_v'}{\overline{T}_v} + \frac{\rho' T_v'}{\overline{\rho}\ \overline{T}_v}$$

Using condition (3) above and the data below, one can show that the last term is smaller than the others, leaving the *linearized perturbation ideal gas law*:

$$\frac{p'}{\overline{P}} = \frac{\rho'}{\overline{\rho}} + \frac{T_v'}{\overline{T}_v} \qquad (3.3.1c)$$

Static pressure fluctuations are associated with variations in the mass of air from column to column in the atmosphere. For the larger eddies and thermals in the boundary layer, these fluctuations may be as large as 0.01 kPa (0.1 mb), while for smaller eddies the effect is smaller. Dynamic pressure fluctuations associated with wind speeds of up to about 10 m/s also cause fluctuations of about 0.01 kPa. Thus, for most boundary layer situations, $p'/\overline{P} = 0.01\ \text{kPa}\ /\ 100\ \text{kPa} = 10^{-4}$, which is smaller than $T_v'/\overline{T}_v = 1\ \text{K}\ /\ 300\ \text{K} = 3.33 \times 10^{-3}$. For these cases we can make the shallow convection approximation [conditions (4) & (5)] to neglect the pressure term, yielding:

$$\frac{\rho'}{\overline{\rho}} = -\frac{T_v'}{\overline{T}_v} \qquad (3.3.1d)$$

Using Poisson's relationship with the same scaling as above yields:

$$\frac{\rho'}{\bar{\rho}} = -\frac{\theta_v'}{\bar{\theta}_v} \qquad (3.3.1e)$$

Physically, (3.3.1e) states that air that is warmer than average is less dense than average. Although not a surprising conclusion, these equations allow us to substitute temperature fluctuations, easily measurable quantities, in place of density fluctuations, which are not so easily measured.

3.3.2 Flux Form of Advection Terms

All of the conservation equations of section 3.2 include an advection term of the form

$$\text{Advection Term} = U_j \, \partial\xi/\partial x_j$$

where ξ denotes any variable, such as a wind component or humidity. If we multiply the continuity equation (3.2.2c) by ξ, we get $\xi \, \partial U_j/\partial x_j = 0$. Since this term is equal to zero, adding it to the advection term will cause no change (other than the mathematical form). Performing this addition gives

$$\text{Advection Term} = U_j \, \partial\xi/\partial x_j + \xi \, \partial U_j/\partial x_j$$

By using the product rule of calculus, we can combine these two terms to give

$$\text{Advection Term} = \partial(\xi U_j)/\partial x_j \qquad (3.3.2)$$

This is called the *flux form* of the advection term, because as was demonstrated in section 2.6 the product of $(\xi \, U_j)$ is nothing more than a kinematic flux.

3.3.3 Conservation of Momentum

Vertical Component. By setting i=3 in (3.2.3b), we can focus on just the vertical component of momentum to study the role of gravity, density, and pressure on turbulent motions. Utilizing U_3=W and the definition of the total derivative, $dU_i/dt = \partial U_i/\partial t + U_j \, \partial U_i/\partial x_j$, gives

$$\frac{dW}{dt} = -g - \frac{1}{\rho}\left(\frac{\partial P}{\partial z}\right) + \nu \frac{\partial^2 W}{\partial x_j^2}$$

In the following development, we will treat viscosity as a constant. Multiply the above equation by ρ and let $\rho = \bar{\rho} + \rho'$, $W = \bar{W} + w'$ and $p = \bar{P} + p'$:

$$(\bar{\rho} + \rho') \frac{d(\bar{W}+w')}{dt} = -(\bar{\rho} + \rho') g \;-\; \frac{\partial(\bar{P}+p')}{\partial z} + \mu \frac{\partial^2(\bar{W}+w')}{\partial x_j^2}$$

Dividing by $\bar{\rho}$ and rearranging gives:

$$\left(1 + \frac{\rho'}{\bar{\rho}}\right) \frac{d(\bar{W}+w')}{dt} = -\frac{\rho'}{\bar{\rho}} g - \frac{1}{\bar{\rho}} \frac{\partial p'}{\partial z} + \nu \frac{\partial^2(\bar{W}+w')}{\partial x_j^2} - \frac{1}{\bar{\rho}}\left[\frac{\partial \bar{P}}{\partial z} + \bar{\rho} g\right]$$

If we assume that the mean state is in *hydrostatic equilibrium* ($\partial \bar{P}/\partial z = -\bar{\rho} g$), then the term in square brackets is zero. Furthermore, if we remember from section 3.3.1 that $\rho'/\bar{\rho}$ is on the order of 3.33 x 10^{-3} , then we see that the factor on the left hand side of the equation is approximated by $(1 + \rho'/\bar{\rho}) \cong 1$. We can't neglect, however, the first term on the right hand side of the equal sign, because the product $[\rho'/\bar{\rho}\, g\,]$ is as large as the other terms in the equation. The process of neglecting density variations in the inertia (storage) term, but retaining it in the buoyancy (gravity) term is called the *Boussinesq approximation*. These two approximations leave

$$\frac{d(\bar{W}+w')}{dt} = -\frac{\rho'}{\bar{\rho}} g - \frac{1}{\bar{\rho}} \frac{\partial p'}{\partial z} + \nu \frac{\partial^2(\bar{W}+w')}{\partial x_j^2}$$

A prerequisite for the Boussinesq approximation is that the shallow convection conditions be satisfied.

By comparing the original equations to the scaled equations above, one finds differences in the terms involving ρ and g. Thus, a simple way to apply the Boussinesq approximation without performing the complete derivation is stated here:

Practical Application of the Boussinesq Approximation:
Given any of the original governing equations, replace every occurrence

of ρ with $\bar{\rho}$, and replace every occurrence of g with $\left[g - (\theta_v'/\bar{\theta}_v)g \right]$.

Although subsidence, $\overline{W}$, is important in mass conservation and in the advection of material (moisture, pollutants, etc) from aloft, we see that it is less important in the momentum equation because it is always paired in a linear manner with w'. In fair-weather boundary layers, subsidence can vary from zero to 0.1 m/s, which is considered a relatively large value. This is small compared to the vertical velocity fluctuations, which frequently vary over the range 0 to 5 m/s. Thus, for **only** the momentum equation for fair-weather conditions can we usually *neglect subsidence:*

$$\overline{W} \cong 0 \tag{3.3.3a}$$

This leaves the vertical component of the momentum equation as

$$\frac{dw'}{dt} = -\left(\frac{\rho'}{\overline{\rho}}\right)g - \frac{1}{\overline{\rho}}\frac{\partial p'}{\partial z} + v\frac{\partial^2 w'}{\partial x_j^2}$$

Using (3.3.1) to replace the density variations with temperature variations gives

$$\frac{dw'}{dt} = \left(\frac{\theta_v'}{\overline{\theta_v}}\right)g - \frac{1}{\overline{\rho}}\frac{\partial p'}{\partial z} + v\frac{\partial^2 w'}{\partial x_j^2} \tag{3.3.3b}$$

The physical interpretation of the first two terms in (3.3.3b) is that warmer than average air is accelerated upward (i.e., hot air rises). The last two terms describe the influences of pressure gradients and viscous shear stress on the motion. This equation therefore plays an important role in the evolution of convective thermals.

Horizontal Component. Although the BL winds are rarely geostrophic, we can use the definition of the *geostrophic wind* as a substitute variable for the horizontal pressure gradient terms:

$$f_c U_g = -\frac{1}{\rho}\frac{\partial P}{\partial y} \qquad \text{and} \qquad f_c V_g = +\frac{1}{\rho}\frac{\partial P}{\partial x} \tag{3.3.3c}$$

Thus, the horizontal components of (3.2.3b) become

$$\frac{dU}{dt} = -f_c (V_g - V) + \nu \frac{\partial^2 U}{\partial x_j^2} \qquad (3.3.3d)$$

$$\frac{dV}{dt} = +f_c (U_g - U) + \nu \frac{\partial^2 V}{\partial x_j^2} \qquad (3.3.3e)$$

$$\text{I} \qquad\qquad \text{II} \qquad\qquad \text{III}$$

Term I is the *inertia* or storage term.   Term II is sometimes called the *geostrophic departure* term, because it is zero when the actual winds are geostrophic. As we stated before, however, the winds are rarely geostrophic in the BL.   Term III describes viscous shear stress.

Combined Momentum Equation. Combining the results from the previous two subsections yields

$$\frac{\partial U_i}{\partial t} + U_j \frac{\partial U_i}{\partial x_j} = -\varepsilon_{ij3} f_c (U_{gj} - U_j) + \delta_{i3} \left[\frac{\theta_v{}'}{\overline{\theta}_v} g - \frac{1}{\overline{\rho}} \frac{\partial p'}{\partial z} \right] + \nu \frac{\partial^2 U_i}{\partial x_j^2}$$

$$(3.3.3f)$$

where we have applied the shallow convection, incompressibility, hydrostatic and Boussinesq approximations, and where $U_{gj} = (U_g, V_g, 0)$.

3.3.4 Horizontal Homogeneity

Expanding the total derivative of any mean variable, $\overline{\xi}$, yields

$$\frac{d\overline{\xi}}{dt} = \frac{\partial \overline{\xi}}{\partial t} + U \frac{\partial \overline{\xi}}{\partial x} + V \frac{\partial \overline{\xi}}{\partial y} + W \frac{\partial \overline{\xi}}{\partial z} \qquad (3.3.4)$$

$$\text{I} \qquad\quad \text{II} \qquad \text{III} \qquad \text{IV}$$

From examples like Figs 1.12 and 2.9 we saw that averaged variables such as potential temperature or turbulence kinetic energy exhibit large vertical variations over the 1 to 2 km of boundary layer depth. Those same variables, however, usually exhibit a much smaller horizontal variation over the same 1 to 2 km scale. Counteracting this disparity of gradients is a disparity of velocities. Namely, U and V are often on the order of m/s while

W is on the order of mm/s or cm/s. The resulting terms I through IV in the above equation are thus nearly equal in magnitude for many cases.

The bottom line is that we usually can **not** neglect horizontal advection (terms II & III), and we can **not** neglect subsidence (term IV) as it affects the movement of conserved variables.

Sometimes micrometeorologists wish to focus their attention on turbulence effects at the expense of neglecting mean advection. By assuming *horizontal homogeneity*, we can set $\partial \overline{\xi} / \partial x = 0$ and $\partial \overline{\xi} / \partial y = 0$,   and *neglecting subsidence* gives $\overline{W} = 0$. Although these assumptions are frequently made by theorists to simplify their derivations, they are rarely valid in the real atmosphere. When they are made, they cause the advection terms of only mean variables (like $\overline{\xi}$) to disappear; the turbulent flux terms do **not** disappear, and in fact are very important.

3.3.5 Reorientating and Rotating the Coordinate System

Although we usually use a Cartesian coordinate system aligned such that the (x, y, z) axes point (east, north, up), sometimes it is convenient to rotate the Cartesian coordinate system about the vertical (z) axes to cause x and y to point in other directions. Some examples include aligning the x-axis with:

> the mean wind direction,
> the geostrophic wind direction
> the direction of surface stress, or
> perpendicular to shorelines or mountains.

The only reason for doing this is to simplify some of the terms in the governing equations. For example, by choosing the x-axis aligned with the mean wind, we find U=M and V=0. In such a system, the x-axis is called the *along-wind direction* and the y-direction is called the *crosswind direction*.

3.4 Equations for Mean Variables in a Turbulent Flow

3.4.1 Equation of State

As was already stated in section 3.3.1, the equation of state is assumed to hold in the mean, and is rewritten here for the sake of organization:

$$\frac{\overline{P}}{\mathfrak{R}} = \overline{\rho}\ \overline{T}_v \tag{3.4.1}$$

3.4.2 Continuity Equation

Start with the continuity equation (3.2.2c) and expand the velocities into mean and turbulent parts to give:

$$\frac{\partial\ (\overline{U}_j\ +\ u_j')}{\partial x_j}\ =\ 0$$

or

$$\frac{\partial \overline{U}_j}{\partial x_j}\ +\ \frac{\partial u_j'}{\partial x_j}\ =\ 0 \qquad\qquad (3.4.2a)$$

Next average over time

$$\overline{\frac{\partial \overline{U}_j}{\partial x_j}\ +\ \frac{\partial u_j'}{\partial x_j}}\ =\ 0$$

Upon applying Reynold's averaging rules, the last term becomes zero, leaving

$$\frac{\partial \overline{U}_j}{\partial x_j}\ =\ 0 \qquad\qquad (3.4.2b)$$

Thus, the continuity equation holds in the mean. Subtracting this from (3.4.2a) gives the continuity equation for turbulent fluctuations:

$$\frac{\partial u_j'}{\partial x_j}\ =\ 0 \qquad\qquad (3.4.2c)$$

This equation will allow us to put turbulent advection terms into flux form, in the same manner as was demonstrated for (3.3.2).

3.4.3 Conservation of Momentum

Starting with the conservation of momentum expressed by (3.2.3b), make the Boussinesq approximation:

$$\frac{\partial U_i}{\partial t}\ +\ U_j\frac{\partial U_i}{\partial x_j}\ =\ -\delta_{i3}\left[g\ -\ \left(\frac{\theta_v'}{\overline{\theta_v}}\right)g\right]\ +\ f_c\varepsilon_{ij3}U_j\ -\ \frac{1}{\rho}\frac{\partial P}{\partial x_i}\ +\ \frac{v\partial^2 U_i}{\partial x_j^2}$$

Next, expand the dependent variables into mean and turbulent parts (except for the $\overline{\theta_v'}/\overline{\theta_v}$ term, for which the expansion has previously been made):

$$\frac{\partial(\overline{U}_i + u_i')}{\partial t} + \frac{(\overline{U}_j + u_j')\partial(\overline{U}_i + u_i')}{\partial x_j} = -\delta_{i3}\left[g - \left(\frac{\theta_v'}{\overline{\theta}_v}\right)g\right] + f_c\varepsilon_{ij3}(\overline{U}_j + u_j')$$

$$- \frac{1}{\overline{\rho}}\frac{\partial(\overline{P}+p')}{\partial x_i} + \nu\partial^2\frac{(\overline{U}_i + u_i')}{\partial x_j^2}$$

Upon performing the indicated multiplications, and separating terms, we find

$$\frac{\partial\overline{U}_i}{\partial t} + \frac{\partial u_i'}{\partial t} + \frac{\overline{U}_j\partial\overline{U}_i}{\partial x_j} + \frac{\overline{U}_j\partial u_i'}{\partial x_j} + \frac{u_j'\partial\overline{U}_i}{\partial x_j} + \frac{u_j'\partial u_i'}{\partial x_j} =$$

$$-\delta_{i3}g + \delta_{i3}\left(\frac{\theta_v'}{\overline{\theta}_v}\right)g + f_c\varepsilon_{ij3}\overline{U}_j + f_c\varepsilon_{ij3}u_j' - \frac{1}{\overline{\rho}}\frac{\partial\overline{P}}{\partial x_i} - \frac{1}{\overline{\rho}}\frac{\partial p'}{\partial x_i} + \nu\frac{\partial^2\overline{U}_i}{\partial x_j^2} + \nu\frac{\partial^2 u_i'}{\partial x_j^2}$$

$$(3.4.3a)$$

Next, average the whole equation:

$$\overline{\frac{\partial\overline{U}_i}{\partial t}} + \overline{\frac{\partial u_i'}{\partial t}} + \overline{\frac{\overline{U}_j\partial\overline{U}_i}{\partial x_j}} + \overline{\frac{\overline{U}_j\partial u_i'}{\partial x_j}} + \overline{\frac{u_j'\partial\overline{U}_i}{\partial x_j}} + \overline{\frac{u_j'\partial u_i'}{\partial x_j}} =$$

$$- \overline{\delta_{i3}g} + \overline{\delta_{i3}\left(\frac{\theta_v'}{\overline{\theta}_v}\right)g} + \overline{f_c\varepsilon_{ij3}\overline{U}_j} + \overline{f_c\varepsilon_{ij3}u_j'} - \overline{\frac{1}{\overline{\rho}}\frac{\partial\overline{P}}{\partial x_i}} - \overline{\frac{1}{\overline{\rho}}\frac{\partial p'}{\partial x_i}} + \overline{\nu\frac{\partial^2\overline{U}_i}{\partial x_j^2}} + \overline{\nu\frac{\partial^2 u_i'}{\partial x_j^2}}$$

By applying Reynolds averaging rules the second, fourth, fifth, eighth, tenth, twelfth and fourteenth terms become zero. We are left with:

$$\frac{\partial\overline{U}_i}{\partial t} + \frac{\overline{U}_j\partial\overline{U}_i}{\partial x_j} + \frac{\overline{u_j'\partial u_i'}}{\partial x_j} = -\delta_{i3}g + f_c\varepsilon_{ij3}\overline{U}_j - \frac{1}{\overline{\rho}}\frac{\partial\overline{P}}{\partial x_i} - \frac{\nu\partial^2\overline{U}_i}{\partial x_j^2} \quad (3.4.3b)$$

Finally, multiply the continuity equation for turbulent motions (3.4.2c) by u_i', average it, and add it to (3.4.3b) to put the turbulent advection term into flux form:

$$\frac{\partial \overline{U}_i}{\partial t} + \frac{\overline{U}_j \partial \overline{U}_i}{\partial x_j} + \frac{\partial (\overline{u_i' u_j'})}{\partial x_j} = -\delta_{i3} g + f_c \varepsilon_{ij3} \overline{U}_j - \frac{1}{\overline{\rho}} \frac{\partial \overline{P}}{\partial x_i} + \frac{v \partial^2 \overline{U}_i}{\partial x_j^2}$$

By moving this flux term to the right hand side of the equation, we see something very remarkable; namely, the following forecast equation for mean wind is very similar to the basic conservation equation we started with (3.2.3b), except for the addition of the turbulence term at the end.

$$\frac{\partial \overline{U}_i}{\partial t} + \overline{U}_j \frac{\partial \overline{U}_i}{\partial x_j} = -\delta_{i3} g + f_c \varepsilon_{ij3} \overline{U}_j - \frac{1}{\overline{\rho}} \frac{\partial \overline{P}}{\partial x_i} + \frac{v \partial^2 \overline{U}_i}{\partial x_j^2} - \frac{\partial (\overline{u_i' u_j'})}{\partial x_j}$$

$$\text{(3.4.3c)}$$

$$\quad\text{I} \qquad\quad \text{II} \qquad\quad \text{III} \qquad \text{IV} \qquad\quad \text{V} \qquad\quad \text{VI} \qquad\quad \text{X}$$

Term I   represents storage of mean momentum (inertia).
Term II   describes advection of mean momentum by the mean wind.
Term III allows gravity to act in the vertical direction only.
Term IV   describes the influence of the earth's rotation (Coriolis effects).
Term V   describes the mean pressure-gradient forces.
Term VI   represents the influence of viscous stress on the mean motions.
Term X   represents the influence of Reynolds' stress on the mean motions (see
          section 2.9.2). It can also be described as the divergence of turbulent
          momentum flux.

Term X can also be written as $(1 / \overline{\rho}) \, \partial \tau_{ij \, \text{Reynolds}} / \partial x_j$ where $\tau_{ij \, \text{Reynolds}} = - \overline{\rho} \, \overline{u_i' u_j'}$.
The implication of this last term is that **turbulence must be considered** in making forecasts in the turbulent boundary layer, **even if we are trying to forecast only mean quantities.** Term X can often be as large in magnitude, or larger, than many other terms in the equation. Sometimes term X is labled as "F" by large-scale dynamists to denote friction.

3.4.4 Conservation of Moisture

For total specific humidity, start with (3.2.4a) and split the dependent variables into mean and turbulent parts:

$$\frac{\partial \overline{q}_T}{\partial t} + \frac{\partial q_T'}{\partial t} + \frac{\overline{U}_j \, \partial \overline{q}_T}{\partial x_j} + \frac{\overline{U}_j \, \partial q_T'}{\partial x_j} + \frac{u_j' \, \partial \overline{q}_T}{\partial x_j} + \frac{u_j' \, \partial q_T'}{\partial x_j} =$$

$$\frac{\nu_q \partial^2 \overline{q}}{\partial x_j^2} + \frac{\nu_q \partial^2 q'}{\partial x_j^2} + \frac{S_{qT}}{\overline{\rho}_{air}} \tag{3.4.4a}$$

where the net remaining source term, S_{qT}, is assumed to be a mean forcing. Next, average the equation, apply Reynolds' averaging rules, and use the turbulent continuity equation to put the turbulent advection term into flux form:

$$\frac{\partial \overline{q_T}}{\partial t} + \frac{\overline{U}_j \partial \overline{q_T}}{\partial x_j} = \frac{\nu_q \partial^2 \overline{q}}{\partial x_j^2} + \frac{\overline{S_{qT}}}{\overline{\rho}_{air}} - \frac{\partial (\overline{u_j' q_T'})}{\partial x_j} \tag{3.4.4b}$$

$$\quad\ \text{I} \qquad\ \ \text{II} \qquad\qquad \text{VI} \qquad \text{VII} \qquad \text{X}$$

Term I   represents the storage of mean total moisture.
Term II   describes the advection of mean total moisture by the mean wind.
Term VI represents the mean molecular diffusion of water vapor.
Term VII is the mean net body source term for additional moisture processes.
Term X   represents the divergence of turbulent total moisture flux.

As before, this equation is similar to the basic conservation equation (3.2.4a), except for the addition of the turbulence term at the end. Similar equations can be written for the vapor and non-vapor parts of total specific humidity.

3.4.5 Conservation of Heat

Start with the basic heat conservation equation (3.2.5) and expand the dependent variables into mean and turbulent parts

$$\frac{\partial \overline{\theta}}{\partial t} + \frac{\partial \theta'}{\partial t} + \frac{\overline{U}_j \partial \overline{\theta}}{\partial x_j} + \frac{\overline{U}_j \partial \theta'}{\partial x_j} + \frac{u_j' \partial \overline{\theta}}{\partial x_j} + \frac{u_j' \partial \theta'}{\partial x_j} =$$

$$\frac{\nu_\theta \partial^2 \overline{\theta}}{\partial x_j^2} + \frac{\nu_\theta \partial^2 \theta'}{\partial x_j^2} - \frac{1}{\overline{\rho} C_p} \frac{\partial \overline{Q_j^*}}{\partial x_j} - \frac{1}{\overline{\rho} C_p} \frac{\partial Q_j^{*'}}{\partial x_j} - \frac{L_v E}{\overline{\rho} C_p} \tag{3.4.5a}$$

Next, Reynolds average and put the turbulent advection term into flux form to give:

$$\frac{\partial \overline{\theta}}{\partial t} + \frac{\overline{U}_j}{\partial x_j}\frac{\partial \overline{\theta}}{} = \frac{v_\theta \partial^2 \overline{\theta}}{\partial x_j^2} - \frac{1}{\overline{\rho}\,C_p}\frac{\partial \overline{Q}_j^*}{\partial x_j} - \frac{L_v E}{\overline{\rho}\,C_p} - \frac{\partial(\overline{u_j'\theta'})}{\partial x_j} \qquad (3.4.5b)$$

$$\;\;\;\;\;\;\text{I}\;\;\;\;\;\;\;\;\;\;\text{II}\;\;\;\;\;\;\;\;\;\;\text{VI}\;\;\;\;\;\;\;\;\;\;\text{VII}\;\;\;\;\;\;\;\;\;\;\text{VIII}\;\;\;\;\;\;\;\;\;\;\text{X}$$

Term I   represents the mean storage of heat.
Term II   describes the advection of heat by the mean wind.
Term VI represents the mean molecular conduction of heat.
Term VII is the mean net body source associated with radiation divergence.
Term VIII is the body source term associated with latent heat release.
Term X   represents the divergence of turbulent heat flux.

3.4.6 Conservation of a Scalar Quantity

Start with the basic conservation equation (3.2.6) of tracer C and expand into mean and turbulent parts:

$$\frac{\partial \overline{C}}{\partial t} + \frac{\partial c'}{\partial t} + \frac{\overline{U}_j \partial \overline{C}}{\partial x_j} + \frac{\overline{U}_j \partial c'}{\partial x_j} + \frac{u_j' \partial \overline{C}}{\partial x_j} + \frac{u_j' \partial c'}{\partial x_j} =$$

$$\frac{v_c \partial^2 \overline{C}}{\partial x_j^2} + \frac{v_c \partial^2 c'}{\partial x_j^2} + S_c \qquad (3.4.6a)$$

where the net remaining source term, S_c, is assumed to be a mean forcing. Next, Reynolds average and use the turbulent continuity equation to put the turbulent advection term into flux form:

$$\frac{\partial \overline{C}}{\partial t} + \frac{\overline{U}_j \partial \overline{C}}{\partial x_j} = \frac{v_c \partial^2 \overline{C}}{\partial x_j^2} + S_c - \frac{\partial(\overline{u_j'c'})}{\partial x_j} \qquad (3.4.6b)$$

$$\;\;\;\;\;\;\text{I}\;\;\;\;\;\;\;\;\;\;\text{II}\;\;\;\;\;\;\;\;\;\;\text{VI}\;\;\;\;\;\;\;\;\text{VII}\;\;\;\;\;\;\;\;\text{X}$$

Term I   represents the mean storage of tracer C.
Term II   describes the advection of the tracer by the mean wind.
Term VI represents the mean molecular diffusion of the tracer.
Term VII is the mean net body source term for additional tracer processes.
Term X   represents the divergence of turbulent tracer flux.

3.5 Summary of Equations, with Simplifications

To simplify usage of the equations, we have collected them in this section and organized them in a way that similarities and differences can be more easily noted. Before we list these equations, however, we can make one additional simplification based on the scale of viscous effects vs. turbulent effects on the mean fields.

3.5.1 The Reynolds Number

The *Reynolds number*, Re, is defined as

$$Re \equiv VL/\nu = \rho VL/\mu \qquad (3.5.1)$$

where V and L are velocity and length scales in the boundary layer. Given $\nu_{air} \cong$ 1.5 x 10^{-5} m^{2}s^{-1} and the typical scaling values $V = 5$ m/s and $L = 100$ m in the surface layer, we find that Re = 3 x 10^7.   In the atmospheric mixed layer, the Reynolds number is even larger. The Reynolds number can be interpreted as the ratio of inertial to viscous forcings.

3.5.2 Neglect of Viscosity for Mean Motions

In each of the conservation equations except mass conservation, there are molecular diffusion/viscosity terms.   Observations in the atmosphere indicate that the molecular diffusion terms are several order of magnitudes smaller than the other terms and can be neglected.

For example, after making the hydrostatic assumption, the momentum conservation equation for mean motions in turbulent flow (3.4.3c) can be rewritten as

$$\left[\frac{\partial \overline{U}_i}{\partial t}\right] + \left[\overline{U}_j \frac{\partial \overline{U}_i}{\partial x_j}\right] = \left[f_c \, \epsilon_{ij3} \, \overline{U}_j\right] - \left[\frac{1}{\overline{\rho}}\frac{\partial \overline{P}}{\partial x}\right] - \left[\frac{1}{\overline{\rho}}\frac{\partial \overline{P}}{\partial y}\right] - \left[\frac{\partial \overline{u_i' u_j'}}{\partial x_j}\right] + \frac{1}{Re}\left[VL\frac{\partial^2 \overline{U}_i}{\partial x_j^2}\right]$$

$$(3.5.2)$$

Each of the terms in square brackets is roughly the same order of magnitude. The last term, however, is multiplied by (1/Re) -- a very small number (on the order of 10^{-7}). Hence, the last term can be neglected compared to the rest, except in the lowest few centimeters above the surface.

3.5.3 Summary of Equations for Mean Variables in Turbulent Flow

Neglecting molecular diffusion and viscosity, and making the hydrostatic and Boussinesq approximations to the governing equations leaves:

$$\frac{\overline{P}}{\mathfrak{R}} = \overline{\rho}\,\overline{T}_v \tag{3.5.3a}$$

$$\frac{\partial \overline{U}_j}{\partial x_j} = 0 \tag{3.5.3b}$$

$$\frac{\partial \overline{U}}{\partial t} + \overline{U}_j \frac{\partial \overline{U}}{\partial x_j} = -f_c(\overline{V}_g - \overline{V}) - \frac{\partial(\overline{u_j'u'})}{\partial x_j} \tag{3.5.3c}$$

$$\frac{\partial \overline{V}}{\partial t} + \overline{U}_j \frac{\partial \overline{V}}{\partial x_j} = +f_c(\overline{U}_g - \overline{U}) - \frac{\partial(\overline{u_j'v'})}{\partial x_j} \tag{3.5.3d}$$

$$\frac{\partial \overline{q}_T}{\partial t} + \overline{U}_j \frac{\partial \overline{q}_T}{\partial x_j} = +S_{qT}/\overline{\rho}_{air} - \frac{\partial(\overline{u_j'q_T'})}{\partial x_j} \tag{3.5.3e}$$

$$\frac{\partial \overline{\theta}}{\partial t} + \overline{U}_j \frac{\partial \overline{\theta}}{\partial x_j} = -\frac{1}{\overline{\rho}\,C_p}\left[L_v E + \frac{\partial \overline{Q_j^*}}{\partial x_j}\right] - \frac{\partial(\overline{u_j'\theta'})}{\partial x_j} \tag{3.5.3f}$$

$$\frac{\partial \overline{C}}{\partial t} + \overline{U}_j \frac{\partial \overline{C}}{\partial x_j} = +S_c - \frac{\partial(\overline{u_j'c'})}{\partial x_j} \tag{3.5.3g}$$

$$\text{I} \qquad \text{II} \qquad\qquad \text{VII} \qquad\qquad \text{X}$$

The similarity between the last five equations reflects that the same forcings are present in each conservation equation:

> Term I represents storage.
> Term II represents advection.
> Term VII represent sundry body forcings.
> Term X describes the turbulent flux divergence.

The covariances appearing in term X reinforce the earlier assertion that statistics play an important role in the study of turbulent flow.

In the two momentum equations above, the *mean geostrophic wind components* were defined using the mean horizontal pressure gradients:

$$\overline{U}_g = -\frac{1}{f_c \overline{\rho}} \frac{\partial \overline{P}}{\partial y} \quad \text{and} \quad \overline{V}_g = +\frac{1}{f_c \overline{\rho}} \frac{\partial \overline{P}}{\partial x} \qquad (3.5.3h)$$

Sometimes the left hand side of equations c thru g are simplified using

$$\frac{d(\ )}{dt} = \frac{\partial(\ )}{\partial t} + \frac{\overline{U}_j \partial(\ )}{\partial x_j} \qquad (3.5.3i)$$

where the total derivative d( )/dt is inferred to include only **mean** advective effects, and not the turbulent effects.

3.5.4   Examples

Many applications will have to wait until more realistic PBL initial and boundary conditions have been covered. For now, just a few artificial sample exercises showing the use of equations (3.5.3) will be presented.

Problem 1. Suppose that the turbulent heat flux decreases linearly with height according to $\overline{w'\theta'} = a - b\,z$, where a = 0.3 (K ms^{-1}) and b = 3x10^{-4} (K s^{-1}) . If the initial potential temperature profile is an arbitrary shape (i.e., pick a shape), then what will be the shape of final profile one hour later? Neglect subsidence, radiation, latent heating, and assume horizontal homogeneity.

Solution. Neglecting subsidence, radiation, and latent heating leaves (3.5.3f) as

$$\frac{\partial \overline{\theta}}{\partial t} + \frac{\overline{U}\partial \overline{\theta}}{\partial x} + \frac{\overline{V}\partial \overline{\theta}}{\partial y} = -\frac{\partial(\overline{u'\theta'})}{\partial x} - \frac{\partial(\overline{v'\theta'})}{\partial y} - \frac{\partial(\overline{w'\theta'})}{\partial z}$$

By assuming horizontal homogeneity, the x and y derivatives drop out, giving

$$\frac{\partial \overline{\theta}}{\partial t} = -\frac{\partial(\overline{w'\theta'})}{\partial z}$$

Plugging in the expression for $\overline{w'\theta'}$ gives $\partial\overline{\theta}/\partial t = +b$.   This answer is not a function of z; hence, air at each height in the sounding warms at the same rate. Integrating over time from $t = t_0$ to t gives

$$\overline{\theta}\,\big|_T = \overline{\theta}\,\big|_{to} + b(t - t_0)$$

The warming in one hour is b (t - t$_0$) = [3x10^{-4} (K/s)]·[3600 (s)] = 1.08 K.

Discussion. This scenario frequently occurs in daytime mixed layers. Thus, given an adiabatic ML initially, the potential temperature profile a bit later will also be adiabatic because air at all heights is warming at the same rate. In fact, *anytime the heat flux changes linearly with height, the shape of the potential temperature profile will be preserved while it warms, regardless of its initial shape.*

Problem 2. If a horizontal wind of 10 m/s is advecting drier air into a region, where the horizontal moisture gradient is $(5 \, g_{water}/kg_{air})/100$ km, then what vertical gradient of turbulent moisture flux in the BL is required to maintain a steady-state specific humidity? Assume all the water is in vapor form, and that there is no body source of moisture. Be sure to state any additional assumptions you make.

Solution. A *steady-state* situation is defined as one where there are no local changes of a variable with time (i.e., where $\partial(\ )/\partial t = 0$). Choose the x-axis to be aligned with the mean wind direction for simplicity. Equation (3.5.3e) becomes

$$\frac{\overline{U}\partial \overline{q}}{\partial x} + \frac{\overline{W}\partial \overline{q}}{\partial z} = -\frac{\partial(\overline{u'q'})}{\partial x} - \frac{\partial(\overline{v'q'})}{\partial y} - \frac{\partial(\overline{w'q'})}{\partial z}$$

No information was given in the problem about subsidence, or about horizontal flux gradients; therefore, for simplicity let's assume they are zero. This leaves

$$\frac{\overline{U}\partial \overline{q}}{\partial x} = -\frac{\partial(\overline{w'q'})}{\partial z}$$

$$[10 \ (m/s)] \cdot [5x10^{-5} \ (g_{water} \ (kg_{air})^{-1} m^{-1})] = -\frac{\partial(\overline{w'q'})}{\partial z}$$

Thus

$$\frac{\partial(\overline{w'q'})}{\partial z} = -5 \times 10^{-4} \ g_{water} \ (kg_{air})^{-1} s^{-1}$$

Discussion. A gradient of this magnitude corresponds to a 0.5 (g/kg)(m/s) decrease of $\overline{w'q'}$ over a vertical distance of 1 km. Also, we see from both sample problems that a decrease of turbulence flux with height (i.e., flux convergence) results in an increase of the mean variable (e.g., temperature or moisture) with time. For the latter example, the potential increase was balanced by advective drying.

Problem 3: Assume a turbulent BL at a latitude of 44°N, where the mean wind is 2 m/s slower than geostrophic (i.e., the wind is subgeostrophic). Neglect subsidence, and assume horizontal homogeneity and steady state.
  a) Find the Reynolds stress divergence necessary to support this velocity deficit.
  b) If that stress divergence was related to molecular viscosity instead of turbulence, what curvature in the mean wind profile would be necessary?

Solution. a) Pick a coordinate system aligned with the stress, for simplicity. Therefore, use equation (3.5.3c). Assuming horizontal homogeneity, steady state, and neglecting subsidence leaves

$$0 = -f_c(\overline{V}_g - \overline{V}) - \frac{\partial \overline{u'w'}}{\partial z}$$

or

$$-\frac{\partial \overline{u'w'}}{\partial z} = f_c(\overline{V}_g - \overline{V}) = [10^{-4}(s^{-1})]\,[2\,(m/s)] = 2 \times 10^{-4} m\,s^{-2}$$

b) Looking back at equation (3.4.3c), we see that the viscous stress term is expressed by $\nu\partial^2 U/\partial z^2$. Thus $\nu\partial^2 U/\partial z^2 = 2\times10^{-4}$ m s^{-2}. Using the value of ν from appendix C, we can solve for the wind profile curvature $\partial^2 U/\partial z^2$:

$$\frac{\partial^2 \overline{U}}{\partial z^2} = \frac{[2 \times 10^{-4}(m\,s^{-2})]}{[1.5 \times 10^{-5}(m^2 s^{-1})]} = 13.33\,(m\,s)^{-1}$$

Discussion. This is a tremendously large value for curvature. If we assume that such a profile was observed within the the middle of the BL (z=0.5z$_i$), where the wind speed is, say 5 m/s, then we can integrate the above equation to find the mean wind at any other height z' away from the middle of the BL: $U(z = 0.5z_i+z') = 5 + 6.67\,z'^2$.

For example, at a height of $0.5z_i + 10$m, the wind speed would be 672 m/s, assuming no shear at z$_i$. Since realistic wind speeds and shears are several orders of magnitude smaller over most of the PBL, it is apparent that viscous stress plays a much smaller role than turbulent Reynolds' stress in the mean wind equation. As we shall see later, however, *viscosity is very important for turbulent motions, and can not be neglected.*

3.6 Case Studies

3.6.1 Daytime Cases

The following cases are meant to acquaint the reader with typical observations of some of the terms in the equations of this chapter. They are analyses of real data, most of which are based on the BLX83 field experiment near Chickasha, Oklahoma (Stull and Eloranta, 1984). This data set was taken in fair-weather anticyclonic conditions during the daytime when deep convective mixed layers formed.

Figs 3.1a through 3.3a show heat and moisture fluxes as measured by an instrumented Queen Air aircraft, flying at about 72 m/s along level flight paths of about 30 km long. Measurements of w, T, and q were taken 20 times per second [i.e., two measurements per 7.2 m (this is the *Nyquist wavelength*, discussed in chapter 8)]. From this data, average values over each flight leg were found, and linear trends were calculated. These were subtracted from the observed values to give w', T', and q'.

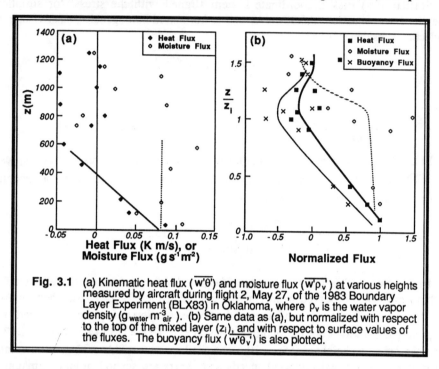

Fig. 3.1 (a) Kinematic heat flux ($\overline{w'\theta'}$) and moisture flux ($\overline{w'\rho_v'}$) at various heights measured by aircraft during flight 2, May 27, of the 1983 Boundary Layer Experiment (BLX83) in Oklahoma, where ρ_v is the water vapor density ($g_{water}\ m_{air}^{-3}$). (b) Same data as (a), but normalized with respect to the top of the mixed layer (z_i), and with respect to surface values of the fluxes. The buoyancy flux ($\overline{w'\theta_v'}$) is also plotted.

An *FFT* (Fast Fourier Transform) filter was used to eliminate all wavelengths longer than 6.25 km. from these space series. This was necessary to reduce the effect of unresolved long (mesoscale) waves that would otherwise contaminate the data. The resulting filtered values were used to calculate kinematic fluxes $\overline{w'T'}$ and $\overline{w'q'}$ using the *eddy correlation method* (i.e., the method of exercise (2) in section 2.12; also see chapter 10). Thus, the averages are line (spatial) averages, not time averages. Although each level flight leg took less than 5 min to fly, the many legs making up any one flight took from 2 to 4 hours to complete. The following table list the flight times, where CDT denotes Central Daylight Time (CDT = UTC - 5 h):

Table 3-1. Flight information for selected flights during the BLX83 field experiment.

Flight	Date	Start (CDT)	Duration (hr)	Boundary Layer
2	27 May 1983	1034	2.5	ML
3	28 May 1983	1425	3.6	ML
13	14 June 1983	1406	3.3	ML

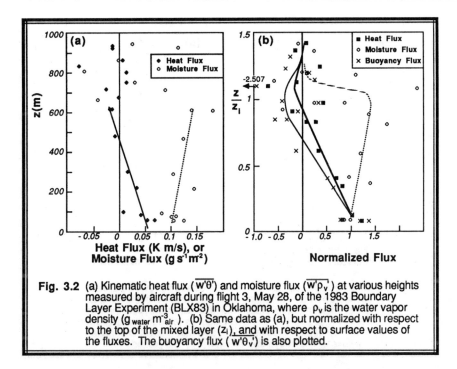

Fig. 3.2 (a) Kinematic heat flux ($\overline{w'\theta'}$) and moisture flux ($\overline{w'\rho_v}$) at various heights measured by aircraft during flight 3, May 28, of the 1983 Boundary Layer Experiment (BLX83) in Oklahoma, where ρ_v is the water vapor density ($g_{water}\ m^{-3}_{air}$). (b) Same data as (a), but normalized with respect to the top of the mixed layer (z_i), and with respect to surface values of the fluxes. The buoyancy flux ($\overline{w'\theta_v'}$) is also plotted.

Each data point in Figs 3.1a to 3.3a represent a flight leg average flux. In general, we see that the heat flux decreases with height, starting at a large positive value near the surface, and becoming negative near the top of the mixed layer. The positive heat flux near the surface is associated with solar heating of the earth's surface, which transfers its heat to the atmosphere. The negative heat flux near the ML top is associated with the entrainment of warmer FA air down into the ML (warm air mixed down causes a negative heat flux). This slope of the heat flux profile causes the temperature to become warmer with time (see eq 3.5.3f).

There is much more scatter in the moisture flux values in these figures. In general, they are positive near the surface, implying evaporation of moisture from the ground into the air. The values just below the top of the ML are also positive, which in this case is related to dry air being entrained down into the ML (note that moist air moving up and dry air moving down both yield a positive moisture flux — see section 2.7). Thus, the moisturizing from the surface and drying from aloft nearly counteract each other in the cases studied, as indicated by the nearly vertical profile of $\overline{w'q'}$ with height (i.e., zero slope implies zero humidity increase, according to eq 3.5.3e). Notice that on Flight 3, the moisture flux increases slightly with height, implying a net drying of the ML.

One problem with these figures is that sufficient time elapses between the low altitude flights and the high altitude flights that non-stationarity of the ML comes into play. In particular, the diurnal cycle causes changes in solar heating with time. Also, the top of the

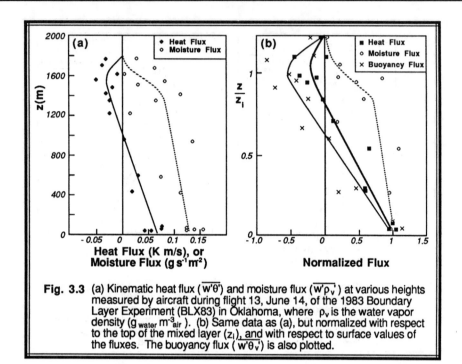

Fig. 3.3 (a) Kinematic heat flux ($\overline{w'\theta'}$) and moisture flux ($\overline{w'\rho_v'}$) at various heights measured by aircraft during flight 13, June 14, of the 1983 Boundary Layer Experiment (BLX83) in Oklahoma, where ρ_v is the water vapor density ($g_{water}\ m^{-3}_{air}$). (b) Same data as (a), but normalized with respect to the top of the mixed layer (z_i), and with respect to surface values of the fluxes. The buoyancy flux ($\overline{w'\theta_v'}$) is also plotted.

ML can increase substantially during that time interval. To remove these effects, micrometeorologist often normalize their plots. Height is normalized by dividing by the depth of the ML, z_i. Flux is normalized by the concurrent surface value of flux observed (or estimated).

The resulting normalized flux profiles are shown in Figs 3.1b - 3.3b. In addition, the

buoyancy flux $\overline{w'\theta_v}'$ is shown. The buoyancy flux has much less scatter than the

other fluxes. It is largest at the surface, and decreases linearly with height in the ML.

Figs 3.4a-c show the estimated evolution of the potential temperature profile for the three cases (Crum, et al., 1987). Each data point corresponds to a flight-leg average. The two profiles shown for flight 3 were started near the surface, and each ended an hour later above the top of the ML. Thus, the warming that took place during each hour contaminates the profiles, causing them to appear tilted. The top of each profile is tilted towards the warmer tempeartures. Just the opposite tilt is observed for the flight 13, because the flights above the top of the ML were flown first. In spite of these tilts, it is obvious that the ML becomes warmer and deeper with time.

Better examples of ML evolution for flight 3 are shown in Figs 3.5 and 3.6. The data set in these figures was taken while the aircraft climbed or descended, thereby making soundings. Given typical descent rates of the aircraft, measurements were made with about 0.5 m resolution in the vertical (extremely high resolution soundings).

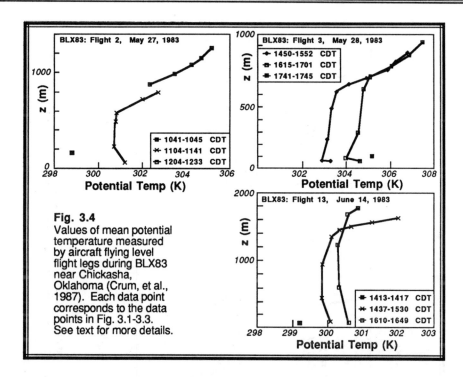

Fig. 3.4
Values of mean potential temperature measured by aircraft flying level flight legs during BLX83 near Chickasha, Oklahoma (Crum, et al., 1987). Each data point corresponds to the data points in Fig. 3.1-3.3. See text for more details.

It took about 10 minutes to complete a sounding, so we can essentially consider the soundings to be instantaneous. These sounding legs were started at the following times:

Leg 1   -   1438 CDT
Leg 12 -   1604
Leg 22 -   1731

The top of the ML, z_i, is very evident by the strong temperature inversion and drop in humidity. ML growth stands out, as does the warming of the ML. There is little change in the humidity with time, however.

From Figs 3.5a-c, we can observe the rate of warming, at any height. This can be compared with the slope of the heat flux profile from Fig 3.2. It is left as an exercise to use (3.5.3f) to see what percentage of the warming within the ML is associated with turbulent flux divergence (convergence) and what percentage is associated with other forcings (radiative, latent heating, advective, etc.).

A similar study can be made for moisture, using the specific humidity evolution shown in Figs 3.6a-c, and comparing that to the expected moisturizing using the moisture flux profiles of Fig 3.2 along with (3.5.3e). For both moisture and temperature, it is evident that the turbulence term in equations (3.5.3) plays a very important role during daytime conditions over land, when vigorous convective mixing is occurring.

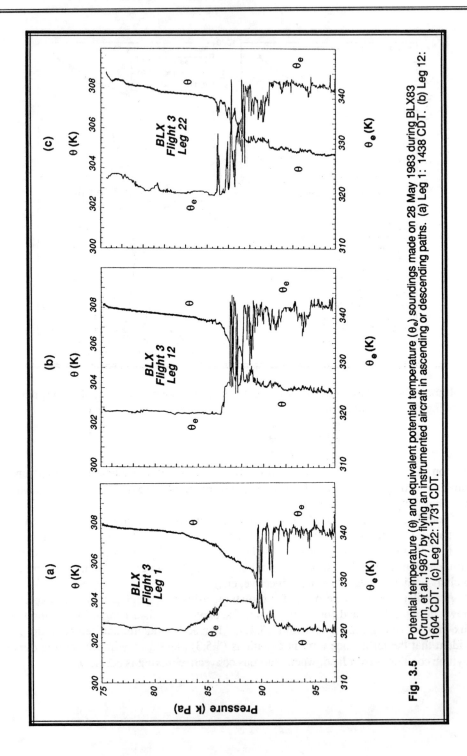

Fig. 3.5 Potential temperature (θ) and equivalent potential temperature (θ$_e$) soundings made on 28 May 1983 during BLX83 (Crum, et al.,1987) by flying an instrumented aircraft in ascending or descending paths. (a) Leg 1: 1438 CDT. (b) Leg 12: 1604 CDT. (c) Leg 22: 1731 CDT.

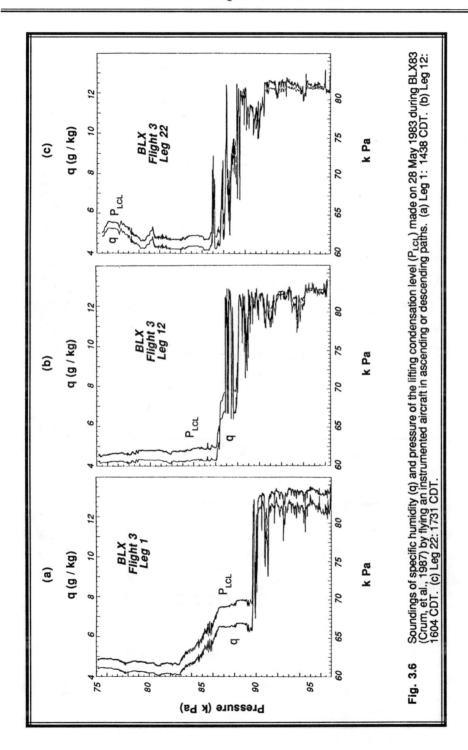

Fig. 3.6   Soundings of specific humidity (q) and pressure of the lifting condensation level (P_{LCL}) made on 28 May 1983 during BLX83 (Crum, et al., 1987) by flying an instrumented aircraft in ascending or descending paths. (a) Leg 1: 1438 CDT. (b) Leg 12: 1604 CDT. (c) Leg 22: 1731 CDT.

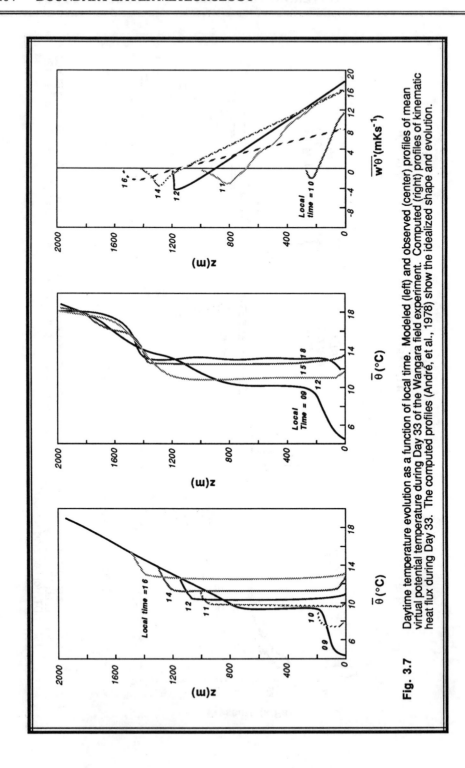

Fig. 3.7   Daytime temperature evolution as a function of local time. Modeled (left) and observed (center) profiles of mean virtual potential temperature during Day 33 of the Wangara field experiment. Computed (right) profiles of kinematic heat flux during Day 33. The computed profiles (André, et al., 1978) show the idealized shape and evolution.

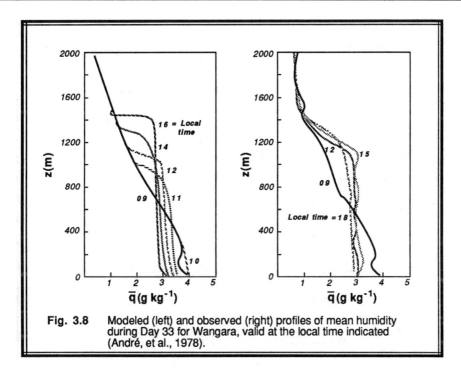

Fig. 3.8   Modeled (left) and observed (right) profiles of mean humidity during Day 33 for Wangara, valid at the local time indicated (André, et al., 1978).

The 1967 Wangara field experiment in Australia (Clarke, et al., 1971) also yielded much useful boundary layer data. André, et al., (1978) have used Day 33 from that experiment as the basis for a numerical simulation of boundary layer evolution. Fig 3.7

shows the modeled $\overline{\theta}$ evolution and the corresponding verification soundings. Modeled heat fluxes are shown in Fig 3.7c. The nearly uniform potential temperature with height is apparent, as are the linear heat flux profiles that are a characteristic signature of convective ML turbulence.

Modeled and observed humidity profiles are shown in Fig 3.8 for the Wangara experiment. The mean specific humidity decreases slightly with height. This slight slope occurs when dry air is entraining into the top of the ML, while moisture is evaporating into the bottom.

Evolution of the observed wind speeds are shown in Fig 3.9. During the afternoon hours when the mixed layer is over 1000 m thick, the winds within the interior of the ML have approximately constant wind speed with height. In the surface layer the winds must decrease towards zero at the ground. Across the entrainment zone at the top of the ML the winds change to their geostrophic values. For these cases, baroclinicity caused the geostrophic wind speed in the mixed layer to be faster than those higher above the ground. As a result, the wind speed above the ML is less than the winds within the ML, even though the ML winds are subgeostrophic and the FA winds are close to geostrophic.

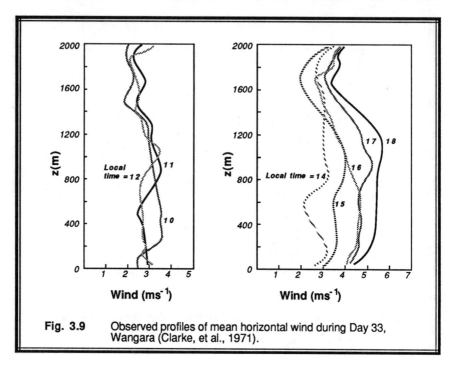

Fig. 3.9   Observed profiles of mean horizontal wind during Day 33,
Wangara (Clarke, et al., 1971).

3.6.2 Nighttime

At nighttime the turbulence is often less vigorous. As a result, other effects such as
advection, radiation, and subsidence become as important or more important than
turbulence in causing changes in temperature and humidity.   For example, Fig 3.10
shows BLX83 field experiment data taken during the night of 18 June 1983 near Canton,
Oklahoma (Carlson and Stull, 1986). Part (a) shows the temperature evolution between
2100 CDT (plotted as circles) and 2230 CDT (plotted as squares), as observed by special
high resolution rawinsonde balloon soundings.   Cooling is evident near the surface
during this 1.5 hour period, while there is warming aloft.

Also during the night, measurements were made of radiation budgets and subsidence.
Computer models were then used to estimate the contributions of the terms in (3.5.3f)
towards the total cooling/heating.   These contributions are shown in part (b), where the
grey lines represent the turbulence part, the dotted line represents radiation divergence,
and the solid lines represent subsidence contributions.   For the grey and the solid lines,
two curves are shown to indicate how they evolved with time between the initial and the
final soundings. It is apparent that subsidence and radiation dominate in the upper part of
these sounding, but turbulence becomes more important near the ground.

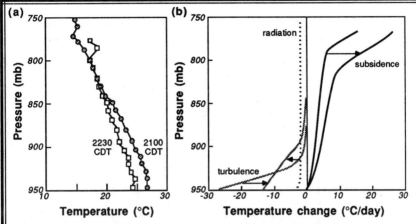

Fig. 3.10   (a) Temperature profiles measured by high resolution rawinsonde soundings launched near Canton, Oklahoma on 18 June 1983 during the BLX 83 field experiment. The circles show data from the 2100 CDT lauch. Squares show those of the 2230 CDT launch. (b) Contributions of turbulence (shaded lines), radiation (dotted line), and subsidence (solid lines) to the cooling rate are modeled for that 1.5 hour period. The two shaded and two solid lines show the range of the respective contributions during the period (Carlson and Stull, 1986).

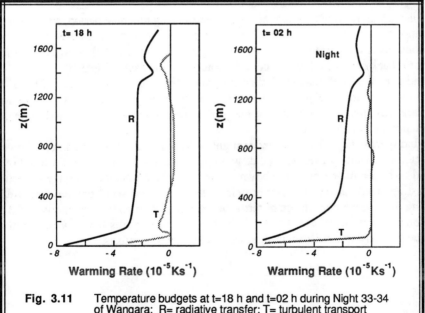

Fig. 3.11   Temperature budgets at t=18 h and t=02 h during Night 33-34 of Wangara: R= radiative transfer; T= turbulent transport (André, et al., 1978).

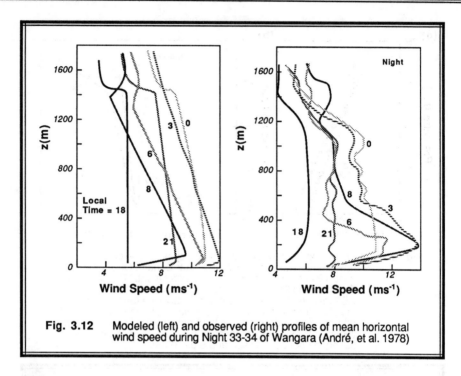

Fig. 3.12   Modeled (left) and observed (right) profiles of mean horizontal wind speed during Night 33-34 of Wangara (André, et al. 1978)

Although there were no latent heating effects during this time period, advection might be important. There were insufficient measurements to calculate the advective contributions to temperature change on this night. Instead, it is left as an exercise for the reader to "back out" the advective contribution, given the data in Fig 3.10.

André, et al. (1978) made a similar analysis of the relative importance of terms in the heat budget equation. Fig 3.11 shows that radiation played a much larger role for the Wangara SBL just after sunset, but that turbulence just above the ground increased in importance later in the night.

Wind speeds (Fig 3.12) observed at night in the Wangara field experiment showed the characteristic nocturnal jet with peak wind speed of about 14 m/s at 200 m above the ground. The simulated profiles shown in the same figure demonstrate the difficulty of forecasting winds at night. Nevertheless, the simulated wind profiles are useful in studying the relative importance of terms in the momentum budget equations (3.5.3 c and d). For this particular case, Fig 3.13 shows that the Coriolis terms were much more important in causing accelerations than were the turbulence effects.

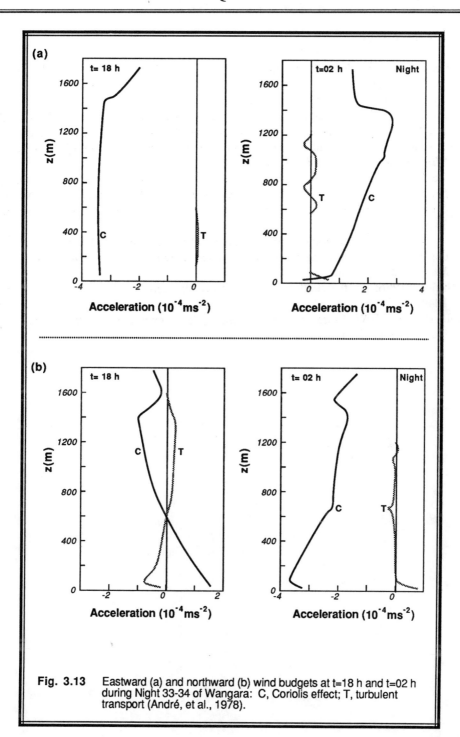

Fig. 3.13   Eastward (a) and northward (b) wind budgets at t=18 h and t=02 h during Night 33-34 of Wangara: C, Coriolis effect; T, turbulent transport (André, et al., 1978).

3.7 References

André, J.-C., G. De Moor, P. Lacarrère, G. Therry, and R. du Vachat, 1978: Modeling the 24-hour evolution of the mean and turbulent structure of the planetary boundary layer. *J. Atmos. Sci.*, **35**, 1861-1883.

Businger, J.A., 1982: Equations and concepts. Chapt. 1 in *Atmospheric Turbulence and Air Pollution Modelling*, Nieuwstadt and van Dop (Editors). Reidel. 358pp.

Carlson, M.A. and R.B. Stull, 1986: Subsidence in the nocturnal boundary layer. *J. Climate and Appl. Meteor.*, **25**, 1088-1099.

Clarke, R.H., A.J. Dyer, R.R. Brook, D.G. Reid, and A.J. Troup, 1971: *The Wangara Experiment: Boundary Layer Data*. Div. of Meteor. Physics Tech. Paper No. 19. Commonwealth Scientific and Industrial Research Organization (CSIRO), Australia. 341pp.

Crum, T.D., R.B. Stull, and E.W. Eloranta, 1987: Coincident lidar and aircraft observations of entrainment into thermals and mixed layers. *J. Climate and Appl. Meteor.*, **26**, 774-788.

Mahrt, L., 1986: On the shallow motion approximations. *J. Atmos. Sci.*, **43**, 1036-1044.

Stull, R.B. and E.W. Eloranta, 1984: Boundary layer experiment - 1983. *Bull. Am. Meteor. Soc.*, **65**, 450-456.

3.8 Exercises

1) Suppose that there is an air pollutant called gallacticum that is found within long, narrow spaceships. This pollutant decomposes faster in warmer air than in cooler air. Hence, the conservation equation for gallacticum is

$$\frac{dc}{dt} = - a c T$$

where c is the concentration of gallacticum, a is a constant, T is the absolute air temperature, and t is time.

    Derive the prognostic equation for the mean concentration of gallacticum that applies to turbulent flow within the space ship. You can scale your equations to the space ship by assuming that, within the region of interest, the only mean wind is the forced ventilation current, $\overline{U}$, down the length of the space ship (in the x-direction). There is, however, horizontal homogeneity of mean quantities in the x-direction only. Be sure to put the turbulence term(s) into flux form.

2) Why is an understanding of turbulence necessary for studying and modeling the boundary layer?

3) Expand the following term, and describe its physical meaning.

$$\delta_{ij} \, \overline{U}_k \, \frac{\partial \overline{u_i' u_j'}}{\partial x_k}$$

4) List the steps, assumptions, simplifications and substitutions (in their proper order) used to get the following equation from (3.2.3b). Do NOT do the whole derivation, just list the steps.

$$\frac{\partial \overline{U}}{\partial t} = - f_c (\overline{V}_g - \overline{V}) - \frac{\partial \overline{u'w'}}{\partial z}$$

5) Very briefly define the following, and comment or give examples of their use in micrometeorology:
a) kinematic heat flux
b) Reynolds stress
c) horizontal homogeneity
d) Boussinesq approximation

6) The forecast equation for mean wind in a turbulent flow is

$$\frac{\partial \overline{U}_i}{\partial t} + \overline{U}_j \frac{\partial \overline{U}_i}{\partial x_j} = - \delta_{i3} g - f_c \epsilon_{ij3} \overline{U}_j - \frac{1}{\rho} \frac{\partial \overline{p}}{\partial x_i} + \nu \frac{\partial^2 \overline{U}_i}{\partial x_j^2} - \frac{\partial \overline{u_i' u_j'}}{\partial x_j}$$

  A   B   C   D   E   F   G

a) Name each term, and give its physical interpretation.

b) Starting with the equation above, derive the equation for $\partial \overline{V}/\partial t$, assuming $\overline{U} = 0$.

7) Given the nighttime data of Fig 3.10, estimate the vertical profile of temperature change associated with the advective contribution between 2100 and 2230 CDT.

8) Suppose that the boundary layer warms by 10 °C during a 6 h period. If:

$$\frac{\partial \overline{w'\theta'}}{\partial z} = \frac{\overline{w'\theta'}_{top} - \overline{w'\theta'}_{bottom}}{z_{top} - z_{bottom}}$$

a) then what is the average value of turbulent heat flux at the earth's surface for a 1 km thick BL having no heat flux at its top?

b) If $u_* = 0.2$ m/s, then find $\theta_*{}^{SL}$.

9) Suppose that:

$$\overline{u'w'} = -(u_* + c\,z)^2, \quad \overline{v'w'} = 0 \text{ for all } z, \quad \overline{U}_g = 5 \text{ m/s at all heights}, \quad \overline{V}_g = 5 \text{ m/s at}$$

all heights, $f_c = 10^{-4}$ s^{-1}, $u_* = 0.3$ m/s, and $c = 0.001$ s^{-1}.

Find the acceleration of the air in the x-direction at a height of 100 m in the BL, assuming that initially $\overline{U} = 4$ m/s and $\overline{V} = 2$ m/s at that height.

10) Given the term $U\,\partial V^2/\partial x$, which represents the advection of mean horizontal v-component of kinetic energy. Expand the variables U and V into mean and turbulent parts, Reynolds average, and simplify as much as possible.

11) Show the steps necessary to put $\overline{u_j'\,\partial T'^2/\partial x_j}$ into flux form.

12) Expand the following Coriolis term $-2\,\varepsilon_{ijk}\,\Omega_j\,U_k$ for the case of i=1, in terms of latitude, velocity, and rotation rate of the earth. Assume that there is no subsidence.

13) Given the profile of momentum flux,

$\overline{u'w'}$, sketched here, sketch a mean wind profile between z=0 and z=2z$_i$ that could reasonably occur and be consistant with the momentum flux. Assume a slab-like mixed layer.

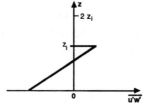

14) Let C be the concentration of hockipuculis bacteria in the air. This contagious bacteria, which sweeps across the northern states each winter, is known to increase as ice forms on the lakes. Researchers at the Institute for Sieve Studies have discovered the following conservation equation for hockipuculis in the air:

$$\frac{dC}{dt} = \frac{a\,C}{\theta}$$

where "a" is a constant. Find the conservation equation for $\overline{C}$ in a turbulent atmosphere. Assume horizontal homogeneity and no subsidence.

15) Evaluate $H = M N$, where $M = \varepsilon_{ijk} \dfrac{\partial \overline{U}_j}{\partial x_i}$ and $N = \delta_{2k} \overline{\theta'^2}$.

16) A virulent gas called cyclonide has recently been detected near weathermap display areas. Meteorologists who enhale this gas frequently become euphoric when hurricanes, tornadoes, low-pressure systems, and other cyclones are being displayed on the weather maps.

  In your efforts to eliminate this scourge of the meteorological community, you have discovered that cyclonide is neither created nor destroyed but is advected from place to place. If some of this gas escapes into the turbulent boundary layer, you will be asked to forecast its mean concentration. In anticipation of this request, derive the prognostic equation for mean cyclonide concentration. State any assumptions made.

17) Given a kinematic heat flux of 0.2 K m/s at the ground, and a flux of -0.1 K m/s at the top of a 1 km thick mixed layer, calculate the average warming rate of the mixed layer.

18) Expand the following, and eliminate all terms that are zero:

$$\frac{\partial(\varepsilon_{jkl} U_i U_m \delta_{ki} \delta_{jl})}{\partial x_m}$$

19) A consortium of personal computer manufacturers has contracted with a local genetic engineering firm to create a new virus call RFV. When humans breath this virus, it causes Ramchip Fever. Symptoms include: an insatiable urge to buy a computer, keyboard finger twitch, memory overflow, a love for mouses, and severe joystick spasms. Parents who breath RFV develop a guilt complex that their offspring will flunk out of school unless they buy a computer.

  The concentration, c, of RFV in the air is governed by the following conservation equation:

$$\frac{dc}{dt} = a c^2 T$$

where "a" is a constant and T is absolute temperature.

a) Derive the forecast equation for $\overline{C}$ in turbulent air. Put it into flux form.

b) Scale the answer by assuming horizontal homogeneity and no subsidence.

20) Starting with (3.2.4b), derive a forecast equation for $\overline{q}$ in turbulent flow. That is, derive an equation like (3.4.4b), except for water vapor only. State all assumptions and simplifications used.

21) If a volume of boundary layer air initially contains 2 g/kg of liquid water droplets, and these droplets completely evaporate during 15 minutes, then find $\partial \overline{q}/\partial t$ and $\partial \overline{\theta}/\partial t$ associated with this evaporation. What is the value (with its units) of E, in (3.2.4b) and (3.2.5)?

22) Given typical values for atmospheric air density near sea level, find:

   a) $\bar{\rho}g$ in units of (mb/m), and in units of (kPa/m) [useful for converting between pressure and height coordinates].

   b) $\bar{\rho}\, C_p$ in units of (mb/ K), and in units of (kPa/K). State all assumptions used.

23) Given typical mean air densities and virtual potential temperatures at sea level in the boundary layer:

   a) Find the density fluctuation, ρ', that corresponds to an air parcel with $\theta_v' = +2$ C.

   b) Find the vertical acceleration of that air parcel, neglecting pressure and viscous effects.

   c) Find the pressure fluctuation, p', if the parcel is restrained from accelerating (neglect viscous effects).

24) What magnitude of V-component geostrophic departure (deviation of the actual wind from its geostrophic value) is necessary to cause the U-component of wind to accelerate 5m/s in one hour?   State all assumptions.

25) Use current weather maps (analyses and/or forecasts) to evaluate terms I through IV in (3.3.4) for any one location of interest such as the town you are in. Do it for low level (BL) data for any one variable such as potential temperature or humidity. Compare and discuss the magnitudes of these terms.

26) Given $\overline{u'w'} = -0.3$ m^2 s^{-2}, find the value of the Reynolds stress in units of N·m^{-2}.

27) Look up the value for thermal diffusivity, ν_θ, for air at sea level. Given this value, what curvature in the mean temperature profile would be necessary to cause a warming rate of 5 K/hr?   Where, if anywhere, would such curvatures be expected to be found in the boundary layer?

28) Given equations (3.5.3), list all of the necessary initial and boundary conditions necessary to solve those equations for $\bar{\theta}$, $\bar{q}$, $\bar{U}$ and $\bar{V}$.

29) a) Determine the warming rate in the mixed layer, given the soundings of Fig 3.5.

   b) Using the heat flux data of Fig 3.2, what percentage of the warming rate from part (a) can be explained by the turbulent flux divergence term?

   c) Suggest physical mechanisms to explain the remaining percentages of warming.

30) Same as question (30), except for moisture using Figs 3.6 and 3.2.

4 Prognostic Equations for Turbulent Fluxes and Variances

In the previous chapter, we summarized the equations needed to forecast mean wind, temperature, humidity, and pollutants. The last term in each of equations (3.5.3c) through (3.5.3g) contains a covariance like $\overline{u_j'\theta'}$ or $\overline{u_j'c'}$. In order to use those previous equations, we can either evaluate the covariances experimentally, or we can derive additional equations to forecast the covariances.

In this chapter, prognostic equations are derived for variances and covariances. Variances give us information about turbulence energies and intensities, while covariances describe kinematic turbulent fluxes. While the previous chapter dealt primarily with the mean state, this chapter deals with the turbulent state of the atmosphere.

4.1 Prognostic Equations for the Turbulent Departures

Turbulent departures of variables are the deviations from their respective means; i.e., θ', u', v', w', q', and c'. In theory, prognostic equations for these departures could be used to forecast each individual gust, given accurate initial and boundary conditions for the gust. Unfortunately, the time span over which such a forecast is likely to be accurate is proportional to the lifetime of the eddy itself — on the order of a few seconds for the smallest eddy to about 15 minutes for the larger thermals. For most meteorological applications, such durations are too short to be of direct use. Instead, we will use the prognostic equations derived in this section as an intermediate step towards finding forecast equations for variances and covariances of the variables.

115

4.1.1 Momentum

Start with the expanded version of the momentum conservation equation (3.4.3a), rewritten here for convenience:

$$\frac{\partial \overline{U}_i}{\partial t} + \frac{\partial u_i{'}}{\partial t} + \overline{U}_j \frac{\partial \overline{U}_i}{\partial x_j} + \overline{U}_j \frac{\partial u_i{'}}{\partial x_j} + u_j{'} \frac{\partial \overline{U}_i}{\partial x_j} + u_j{'} \frac{\partial u_i{'}}{\partial x_j} =$$

$$- \delta_{i3} g + \delta_{i3} \left(\frac{\theta_v{'}}{\overline{\theta}_v} \right) g + f_c \varepsilon_{ij3} \overline{U}_j + f_c \varepsilon_{ij3} u_j{'} - \left(\frac{1}{\overline{\rho}} \right) \frac{\partial \overline{P}}{\partial x_i} - \left(\frac{1}{\overline{\rho}} \right) \frac{\partial p'}{\partial x_i} + \nu \frac{\partial^2 \overline{U}_i}{\partial x_j^2} + \nu \frac{\partial^2 u_i{'}}{\partial x_j^2}$$

Next, from this equation for the total wind ($\overline{U}_i + u_i{'}$), subtract the mean part (3.4.3c), also rewritten here:

$$\frac{\partial \overline{U}_i}{\partial t} + \overline{U}_j \frac{\partial \overline{U}_i}{\partial x_j} = - \delta_{i3} g + f_c \varepsilon_{ij3} \overline{U}_j - \left(\frac{1}{\overline{\rho}} \right) \frac{\partial \overline{P}}{\partial x_i} + \nu \frac{\partial^2 \overline{U}_i}{\partial x_j^2} - \frac{\partial \overline{(u_i{'} u_j{'})}}{\partial x_j}$$

This leaves a prognostic equation for just the turbulent gust, $u_i{'}$:

$$\frac{\partial u_i{'}}{\partial t} + \overline{U}_j \frac{\partial u_i{'}}{\partial x_j} + u_j{'} \frac{\partial \overline{U}_i}{\partial x_j} + u_j{'} \frac{\partial u_i{'}}{\partial x_j} =$$

$$+ \delta_{i3} \left(\frac{\theta_v{'}}{\overline{\theta}_v} \right) g + f_c \varepsilon_{ij3} u_j{'} - \left(\frac{1}{\overline{\rho}} \right) \frac{\partial p'}{\partial x_i} + \nu \frac{\partial^2 u_i{'}}{\partial x_j^2} + \frac{\partial \overline{(u_i{'} u_j{'})}}{\partial x_j} \qquad (4.1.1)$$

4.1.2 Moisture

To simplify future derivations, we will focus on just the vapor portion of the total humidity. For specific humidity of water vapor, start with (3.4.4a), except replace every occurrence of q_T by q, and include the phase change term E (see 3.2.4b). For simplicity, we will assume that body force terms S_q and E are mean terms only. The result is:

$$\frac{\partial \overline{q}}{\partial t} + \frac{\partial q'}{\partial t} + \overline{U}_j \frac{\partial \overline{q}}{\partial x_j} + \overline{U}_j \frac{\partial q'}{\partial x_j} + u_j{'} \frac{\partial \overline{q}}{\partial x_j} + u_j{'} \frac{\partial q'}{\partial x_j} =$$

$$v_q \frac{\partial^2 \overline{q}}{\partial x_j^2} + v_q \frac{\partial^2 q'}{\partial x_j^2} + \frac{(S_q + E)}{\overline{\rho}_{air}} \qquad (4.1.2a)$$

Next, subtract the equation for the mean (3.4.4b), again replacing every q_T with q:

$$\frac{\partial \overline{q}}{\partial t} + \overline{U}_j \frac{\partial \overline{q}}{\partial x_j} = v_q \frac{\partial^2 \overline{q}}{\partial x_j^2} + \frac{(S_q + E)}{\overline{\rho}_{air}} - \frac{\partial (\overline{u_j' q'})}{\partial x_j} \qquad (4.1.2b)$$

leaving a prognostic equation for the perturbation part, q':

$$\frac{\partial q'}{\partial t} + \overline{U}_j \frac{\partial q'}{\partial x_j} + u_j' \frac{\partial \overline{q}}{\partial x_j} + u_j' \frac{\partial q'}{\partial x_j} = v_q \frac{\partial^2 q'}{\partial x_j^2} + \frac{\partial (\overline{u_j' q'})}{\partial x_j} \qquad (4.1.2c)$$

The reader is invited to derive the equations for the case where S_q and E also have perturbation components.

4.1.3 Heat

Start with (3.4.5a) and subtract (3.4.5b) to leave

$$\frac{\partial \theta'}{\partial t} + \overline{U}_j \frac{\partial \theta'}{\partial x_j} + u_j' \frac{\partial \overline{\theta}}{\partial x_j} + u_j' \frac{\partial \theta'}{\partial x_j} = v_\theta \frac{\partial^2 \theta'}{\partial x_j^2} + \frac{\partial (\overline{u_j' \theta'})}{\partial x_j} - \frac{1}{\overline{\rho} \, C_p} \frac{\partial Q_j^{*'}}{\partial x_j} \qquad (4.1.3)$$

4.1.4 A Scalar Quantity

Start with (3.4.6a) and subtract (3.4.6b) to leave

$$\frac{\partial c'}{\partial t} + \overline{U}_j \frac{\partial c'}{\partial x_j} + u_j' \frac{\partial \overline{C}}{\partial x_j} + u_j' \frac{\partial c'}{\partial x_j} = v_c \frac{\partial^2 c'}{\partial x_j^2} + \frac{\partial (\overline{u_j' c'})}{\partial x_j} \qquad (4.1.4)$$

4.2 Free Convection Scaling Variables

Before deriving equations for variances and fluxes, we must detour a bit to learn how experimental data is scaled for presentation. We can then show case study examples of data that correspond to the equations we develop.

In Chapter 1, it was stated that turbulence can be produced by buoyant convective processes (i.e., thermals of warm air rising) and by mechanical processes (i.e., wind shear). Sometimes one process dominates. When buoyant convective processes dominate, the boundary layer is said to be in a state of *free convection*. When mechanical processes dominate, the boundary layer is in a state of *forced convection*. Free convection occurs over land on clear sunny days with light or calm winds. Forced convection occurs on overcast days with stronger winds. In this section, we will focus on free-convection scales; forced-convection scales have already been introduced in section 2.10.

For the free-convection case, strong solar heating at the surface creates a pronounced diurnal cycle in turbulence and ML depth. In chapter 3, profiles of heat and moisture flux were made nondimensional to remove these diurnal changes. The resulting profiles of heat flux, for example, presented height in terms of a fraction of the total ML depth, and presented flux values as a fraction of the surface flux values.

Such a scheme to remove nonstationary effects can be easily applied to other variables, and is quite useful for studying the relative contributions of the various terms in the variance and flux equations just presented. Some of the appropriate scaling variables for free convection conditions are presented here. Appendix A lists a more complete summary of scaling variables.

Length Scale: Thermals rise until they hit the stable layer capping the ML. As a result, the thermal size scales to z_i. Thermals are the dominant eddy in the convective boundary layer, and all smaller eddies feed on the thermals for energy. Thus, we would expect many turbulent processes to scale to z_i in convective situations.

Velocity Scale: The strong diurnal cycle in solar heating creates a strong heat flux into the air from the earth's surface. The buoyancy associated with this flux fuels the thermals. We can define a *buoyancy flux* as $(g/\overline{\theta_v}) \overline{w'\theta_v'}$.

Although the surface buoyancy flux could be used directly as a scaling variable, it is usually more convenient to generate a velocity scale instead, using the two variables we know to be important in free convection: buoyancy flux at the surface, and z_i. Combining these yields a velocity scale known as the *free convection scaling velocity*, w_*, also sometimes called the *convective velocity scale* for short:

$$ w_* = \left[\frac{g\, z_i}{\overline{\theta_v}} \left(\overline{w'\theta_v'} \right)_s \right]^{1/3} \tag{4.2a} $$

This scale appears to work quite well; for example, the magnitude of the vertical velocity fluctuations in thermals is on the same order as w_*. For deep MLs with vigorous heating at the ground, w_* can be on the order of 1 to 2 m/s. Fig 4.1 shows examples of the diurnal variation of w_*.

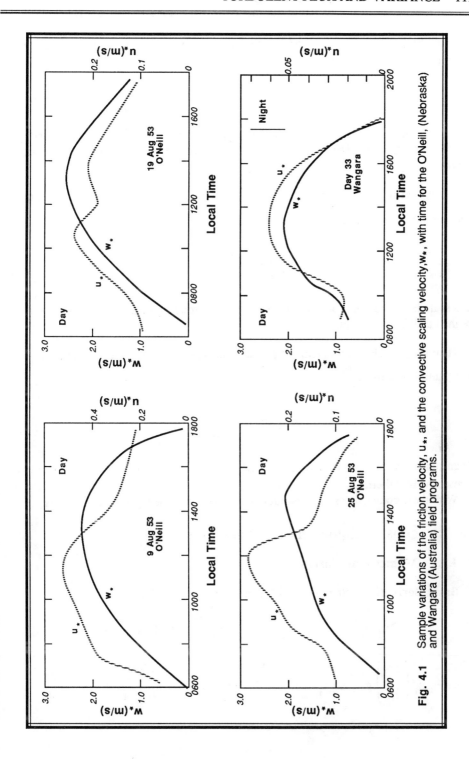

Fig. 4.1 Sample variations of the friction velocity, u_*, and the convective scaling velocity, w_*, with time for the O'Neill, (Nebraska) and Wangara (Australia) field programs.

Time Scale: The velocity and length scales can be combined to give the following free convection time scale, t_*:

$$t_* = \frac{z_i}{w_*} \tag{4.2b}$$

This time scale is on the order of 5 to 15 minutes for many MLs. Observations suggest that this is roughly the time it takes for air in a thermal to cycle once between the bottom and the top of the ML.

Temperature Scale: Using surface heat flux with w_*, we can define a temperature scale for the mixed layer, θ_*^{ML}, by:

$$\theta_*^{ML} = \frac{\left(\overline{w'\theta'}\right)_s}{w_*} \tag{4.2c}$$

This scale is on the order of 0.01 to 0.3 K, which is roughly how much warmer thermals are than their environment.

Humidity Scale: Surface moisture flux and w_* can be combined to define a mixed layer humidity scale, q_*^{ML}:

$$q_*^{ML} = \frac{\left(\overline{w'q'}\right)_s}{w_*} \tag{4.2d}$$

Magnitudes are on the order of 0.01 to 0.5 g_{water} $(kg_{air})^{-1}$ and scale well to moisture excesses within thermals.

With these convective scales in mind, we can return to the equation derivations.

4.3 Prognostic Equations for Variances

4.3.1 Momentum Variance

Basic Derivation. Start with (4.1.1) and multiply by $2u_i'$:

$$2u_i' \frac{\partial u_i'}{\partial t} + 2\overline{U}_j u_i' \frac{\partial u_i'}{\partial x_j} + 2u_i'u_j' \frac{\partial \overline{U}_i}{\partial x_j} + 2u_i'u_j' \frac{\partial u_i'}{\partial x_j} =$$

$$+ 2\delta_{i3}u_i'\left(\frac{\theta_v'}{\overline{\theta_v}}\right)g + 2f_c\varepsilon_{ij3}u_i'u_j' - 2\left(\frac{u_i'}{\overline{\rho}}\right)\frac{\partial p'}{\partial x_i} + 2v\,u_i' \frac{\partial^2 u_i'}{\partial x_j^2} + 2u_i' \frac{\partial \overline{(u_i'u_j')}}{\partial x_j}$$

Next, use the product rule of calculus to convert terms like $2u_i' \partial u_i'/\partial t$ into $\partial (u_i')^2/\partial t$:

$$\frac{\partial u_i'^2}{\partial t} + \overline{U}_j \frac{\partial u_i'^2}{\partial x_j} + 2u_i'u_j' \frac{\partial \overline{U}_i}{\partial x_j} + u_j' \frac{\partial u_i'^2}{\partial x_j} =$$

$$+ 2\, \delta_{i3} u_i' \left(\frac{\theta_v'}{\overline{\theta}_v} \right) g + 2 f_c \, \varepsilon_{ij3} u_i' u_j' - 2\left(\frac{u_i'}{\overline{\rho}} \right) \frac{\partial p'}{\partial x_i} + 2\nu u_i' \frac{\partial^2 u_i'}{\partial x_j^2} + 2u_i' \frac{\partial\, \overline{(u_i' u_j')}}{\partial x_j}$$

For step three, average the whole equation and apply Reynolds averaging rules:

$$\frac{\overline{\partial u_i'^2}}{\partial t} + \overline{U}_j \frac{\overline{\partial u_i'^2}}{\partial x_j} + 2\overline{u_i'u_j' \frac{\partial \overline{U}_i}{\partial x_j}} + \overline{u_j' \frac{\partial u_i'^2}{\partial x_j}} =$$

$$+ 2\, \delta_{i3} \overline{u_i' \left(\frac{\theta_v'}{\overline{\theta}_v} \right)} g + 2 f_c \, \overline{\varepsilon_{ij3} u_i' u_j'} - 2\overline{\left(\frac{u_i'}{\overline{\rho}} \right) \frac{\partial p'}{\partial x_i}} + 2\nu \overline{u_i' \frac{\partial^2 u_i'}{\partial x_j^2}} + 2\overline{u_i' \frac{\partial\, \overline{(u_i' u_j')}}{\partial x_j}}$$

where the last term is zero because $\overline{u_i'} = 0$.   If we multiply the turbulent continuity

equation by $u_i'^2$ and Reynolds average to get $\overline{u_i'^2 \, \partial u_j'/\partial x_j} = 0$, then we can add this equation to the equation above to put the last term before the equal sign into *flux form*: $\overline{\partial (u_j' \, u_i'^2)/\partial x_j}$. This leaves:

$$\frac{\overline{\partial u_i'^2}}{\partial t} + \overline{U}_j \frac{\overline{\partial u_i'^2}}{\partial x_j} + 2\overline{u_i'u_j'} \frac{\partial \overline{U}_i}{\partial x_j} + \frac{\partial\, \overline{(u_j' u_i'^2)}}{\partial x_j} =$$

$$+ 2\, \delta_{i3} \overline{u_i' \left(\frac{\theta_v'}{\overline{\theta}_v} \right)} g + 2 f_c \, \varepsilon_{ij3} \overline{u_i' u_j'} - 2\overline{\left(\frac{u_i'}{\overline{\rho}} \right) \frac{\partial p'}{\partial x_i}} + 2\nu \overline{u_i' \frac{\partial^2 u_i'}{\partial x_j^2}} \qquad (4.3.1a)$$

This general form of the prognostic equation for the variance of wind speed, $\overline{u_i'^2}$, is usually simplified further before being used for boundary layer flows.

Dissipation.   Consider a term of the form $\partial^2(\overline{u_i'^2})/\partial x_j^2$.   Using simple rules of calculus, we can rewrite it as:

$$\frac{\partial^2(\overline{u_i'^2})}{\partial x_j^2} = \frac{\partial}{\partial x_j}\left[\frac{\partial(\overline{u_i'^2})}{\partial x_j}\right] = \frac{\partial}{\partial x_j}\left[\overline{2u_i'\frac{\partial u_i'}{\partial x_j}}\right] = 2\overline{\frac{\partial u_i'}{\partial x_j}\frac{\partial u_i'}{\partial x_j}} + 2\overline{u_i'\frac{\partial^2 u_i'}{\partial x_j^2}} =$$

$$2\overline{\left(\frac{\partial u_i'}{\partial x_j}\right)^2} + 2\overline{u_i'\frac{\partial^2 u_i'}{\partial x_j^2}}$$

If we multiply the last term above by v, then it would be identical to the last term in (4.3.1a). Thus, we can write the last term in (4.3.1a) as

$$2v\overline{u_i'\frac{\partial^2 u_i'}{\partial x_j^2}} = v\frac{\partial^2(\overline{u_i'^2})}{\partial x_j^2} - 2v\overline{\left(\frac{\partial u_i'}{\partial x_j}\right)^2} \qquad (4.3.1b)$$

The first term on the right, which physically represents the molecular diffusion of velocity variance, contains the curvature of a variance. The variance changes fairly smoothly with distance within the boundary layer, its curvature being on the order of 10^{-6} s^{-2} in the ML to 10^{-2} s^{-2} in the SL. When multiplied by v, the first term ranges in magnitude between 10^{-11} and 10^{-7} $m^2\,s^{-3}$.

The last term on the right can be much larger. For example, if the eddy velocity changes by only 0.1 m/s across a very small size eddy (for example, 1 cm in diameter), then the instantaneous shear across that eddy is 10 s^{-1}. For smaller size eddies, the shear is larger. When this value is squared, averaged, and multiplied by $2v$, the magnitudes observed in the turbulent boundary layer range between about 10^{-6} and 10^{-2} $m^2\,s^{-3}$. Typical values in the ML are on the order of 10^{-4} to 10^{-3} $m^2\,s^{-3}$, while in the surface layer, values on the order of 10^{-2} $m^2\,s^{-3}$ can be found.   Thus, we can neglect the first term on the right and use:

$$2v\overline{u_i'\frac{\partial^2 u_i'}{\partial x_j^2}} \cong -2v\overline{\left(\frac{\partial u_i'}{\partial x_j}\right)^2} \qquad (4.3.1c)$$

The *viscous dissipation*, ε , is defined as:

$$\varepsilon = +\nu \overline{\left(\frac{\partial u_i{}'}{\partial x_j}\right)^2} \qquad\qquad (4.3.1d)$$

It is obvious that this term is always positive, because it is a squared quantity. Therefore, when used in (4.3.1a) with the negative sign as required by (4.3.1c), it is always causing a decrease in the variance with time. That is, *it is always a loss term*. In addition, it becomes larger in magnitude as the eddy size becomes smaller. For these small eddies, the eddy motions are rapidly damped by viscosity and irreversibly converted into heat. [This heating rate is so small, however, that it has been neglected in the heat conservation equation (3.4.5b).]

Pressure Perturbations. Using the product rule of calculus again, the pressure

term $-2 \, \overline{(u_i{}'/\overline{\rho})} \, \partial p'/\partial x_i$ in (4.3.1a) can be rewritten as

$$-2\overline{\left(\frac{u_i{}'}{\overline{\rho}}\right)\frac{\partial p'}{\partial x_i}} = -\left(\frac{2}{\overline{\rho}}\right)\frac{\partial\,\overline{(u_i{}'p')}}{\partial x_i} + 2\overline{\left(\frac{p'}{\overline{\rho}}\right)\left[\frac{\partial u_i{}'}{\partial x_i}\right]}$$

The last term is called the **pressure redistribution term**. The factor in square brackets consists of the sum of three terms: $\partial u'/\partial x$, $\partial v'/\partial y$, and $\partial w'/\partial z$. These terms sum to zero because of the turbulence continuity equation (3.4.2c); hence, the last term in the equation above does not change the total variance (by total variance we mean the sum of all three variance components). But it does tend to take energy out of the components having the most energy and put it into components with less energy. Thus it makes the turbulence more isotropic, and is also known as the **return-to-isotropy term**.
Terms like $\partial u'/\partial x$ are larger for the smaller size eddies. Thus, we would expect that smaller size eddies are more isotropic than larger ones. As we shall see later, this is indeed the case in the boundary layer.
The end result of this analysis is that:

$$-2\overline{\left(\frac{u_i{}'}{\overline{\rho}}\right)\frac{\partial p'}{\partial x_i}} \cong -\left(\frac{2}{\overline{\rho}}\right)\frac{\partial\,\overline{(u_i{}'p')}}{\partial x_i} \qquad\qquad (4.3.1e)$$

Coriolis Term. The Coriolis term $2f_c\varepsilon_{ij3} \, \overline{u_i{}'u_j{}'}$ is identically zero for velocity

variances, as can be seen by performing the sums implied by the repeated indices:

$$2f_c \varepsilon_{ij3} \overline{u_i'u_j'} = 2f_c \varepsilon_{213} \overline{u_2'u_1'} + 2f_c \varepsilon_{123} \overline{u_1'u_2'}$$

$$= -2f_c \overline{u_2'u_1'} + 2f_c \overline{u_1'u_2'}$$

$$= 0 \qquad\qquad (4.3.1f)$$

because $\overline{u_1'u_2'} = \overline{u_2'u_1'}$ (see section 2.9.2). Many of the terms in the above sum were not written out because the alternating unit tensor forced them to zero.

Physically, this means that Coriolis force can not generate turbulent kinetic energy. Kinetic energy enters the picture because the variance $\overline{u_i'^2}$ is nothing more than twice the turbulence kinetic energy per unit mass. The Coriolis term merely redistributes energy from one horizontal direction to another. Furthermore, the magnitude of the redistribution term $2f_c \overline{u_1'u_2'}$ is about three orders of magnitude smaller than the other terms in (4.3.1a). For that reason, the Coriolis terms are usually neglected in the turbulence variance and covariance equations, even for the cases where they are not identically zero.

Simplified Velocity Variance Budget Equations. Inserting the simplifications of the previous subsections in equation (4.3.1a) and rearranging the terms gives:

$$\frac{\partial \overline{u_i'^2}}{\partial t} + \overline{U_j}\frac{\partial \overline{u_i'^2}}{\partial x_j} = +2\,\delta_{i3}\,\frac{g\,(\overline{u_i'\theta_v'})}{\overline{\theta_v}} - 2\overline{u_i'u_j'}\frac{\partial \overline{U_i}}{\partial x_j} - \frac{\partial\,(\overline{u_j'u_i'^2})}{\partial x_j} - \frac{2}{\overline{\rho}}\frac{\partial\,(\overline{u_i'p'})}{\partial x_i} - 2\varepsilon$$

$$(4.3.1g)$$

| I | II | III | IV | V | VI | VII |

Term I   represents local storage of variance.

Term II   describes the advection of variance by the mean wind.

Term III   is a production or loss term, depending on whether the buoyancy flux

$\overline{w'\theta_v'}$ is positive (e.g., daytime over land) or negative (e.g., night over land).

Term IV   is a production term. The momentum flux $\overline{u_i'u_j'}$ is usually negative in the boundary layer because the momentum of the wind is lost downward to the ground; thus, it results in a positive contribution to variance when multiplied by a negative sign.

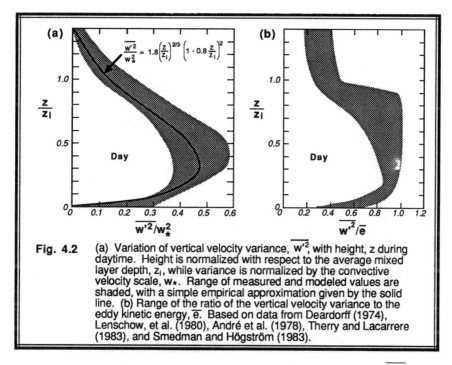

Fig. 4.2   (a) Variation of vertical velocity variance, $\overline{w'^2}$, with height, z during daytime. Height is normalized with respect to the average mixed layer depth, z_i, while variance is normalized by the convective velocity scale, w_*. Range of measured and modeled values are shaded, with a simple empirical approximation given by the solid line. (b) Range of the ratio of the vertical velocity variance to the eddy kinetic energy, $\bar{e}$. Based on data from Deardorff (1974), Lenschow, et al. (1980), André et al. (1978), Therry and Lacarrere (1983), and Smedman and Högström (1983).

Term V   is a turbulent transport term. It describes how variance $\overline{u_i'^2}$ is moved around by the turbulent eddies u_j'.

Term VI   describes how variance is redistributed by pressure perturbations. It is often associated with oscillations in the air (i.e., *buoyancy or gravity waves*).

Term VII   represents the viscous dissipation of velocity variance.

We can also examine the prognostic equations for each individual component of velocity variance if we relax slightly the summation requirement associated with repeated indices. For example, in the above equation, we could let i=2 to write the forecast equation for $\overline{v'^2}$. Any other repeated indices, such as j, continue to imply a sum. When we perform such a split, remembering to reinsert the return-to-isotropy terms (because for any one component, it is nonzero), we find:

$$\frac{\partial \overline{u'^2}}{\partial t} + \overline{U}_j \frac{\partial \overline{u'^2}}{\partial x_j} = \quad -2\overline{u'u_j'}\frac{\partial \overline{U}}{\partial x_j} - \frac{\partial\,\overline{(u_j'u'^2)}}{\partial x_j} - \frac{2}{\overline{\rho}}\frac{\partial\overline{(u'p')}}{\partial x} + \frac{\overline{2p'}}{\overline{\rho}}\frac{\partial u'}{\partial x} - 2\nu\overline{\left(\frac{\partial u'}{\partial x_j}\right)^2}$$

$$(4.3.1h)$$

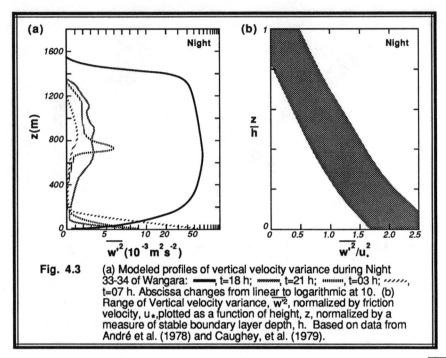

Fig. 4.3   (a) Modeled profiles of vertical velocity variance during Night 33-34 of Wangara: ━━━, t=18 h; ∞∞∞∞, t=21 h; ▪▪▪▪▪▪, t=03 h; ⁄⁄⁄⁄⁄, t=07 h. Abscissa changes from linear to logarithmic at 10. (b) Range of Vertical velocity variance, $\overline{w'^2}$, normalized by friction velocity, u_*, plotted as a function of height, z, normalized by a measure of stable boundary layer depth, h. Based on data from André et al. (1978) and Caughey, et al. (1979).

$$\frac{\partial \overline{v'^2}}{\partial t} + \overline{U_j}\frac{\partial \overline{v'^2}}{\partial x_j} = \qquad -2\overline{v'u_j'}\frac{\partial \overline{V}}{\partial x_j} - \frac{\partial\,(\overline{u_j'v'^2})}{\partial x_j} - \frac{2}{\rho}\frac{\partial(\overline{v'p'})}{\partial y} + \frac{\overline{2p'}}{\rho}\frac{\partial v'}{\partial y} - 2\nu\overline{\left(\frac{\partial v'}{\partial x_j}\right)^2}$$

$$(4.3.1i)$$

$$\frac{\partial \overline{w'^2}}{\partial t} + \overline{U_j}\frac{\partial \overline{w'^2}}{\partial x_j} = \frac{2g(\overline{w'\theta_v'})}{\overline{\theta_v}} - 2\overline{w'u_j'}\frac{\partial \overline{W}}{\partial x_j} - \frac{\partial(\overline{u_j'w'^2})}{\partial x_j} - \frac{2}{\rho}\frac{\partial(\overline{w'p'})}{\partial z} + \frac{\overline{2p'}}{\rho}\frac{\partial w'}{\partial z} - 2\nu\overline{\left(\frac{\partial w'}{\partial x_j}\right)^2}$$

$$(4.3.1.j)$$

| I | II | III | IV | V | VI | VIII | VII |

Terms I through VII have the same meaning as before. Term VIII represents pressure redistribution, which is associated with the return-to-isotropy term.

Case Study Examples. *Budget study* is the name given to an evaluation of the contributions of each term in prognostic equations such as the ones just derived. Some terms are very difficult to measure in field experiments, which is why computer simulation efforts are made. In the budget studies that follow, field data and numerical simulations are combined, and the range of values is indicated. In most cases, field measurements have significantly more scatter than the simulations.

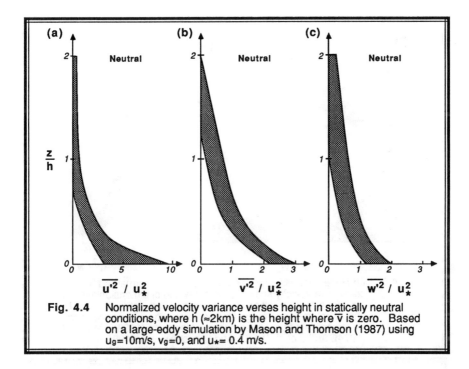

Fig. 4.4   Normalized velocity variance verses height in statically neutral conditions, where h ($\approx$2km) is the height where $\overline{v}$ is zero. Based on a large-eddy simulation by Mason and Thomson (1987) using u_g=10m/s, v_g=0, and u_*= 0.4 m/s.

Fig 4.2 shows that vertical velocity variance during the daytime is small near the surface, increases to a maximum about a third of the distance from the ground to the top of the ML, and then decreases with height. This is related to the vertical acceleration experienced by thermals during their initial rise, which is reduced by dilution with environmental air, by drag, and by the warming and stabilizing of the environment near the top of the ML.   In cloud-free conditions with light winds, glider pilots and birds would expect to find the maximum lift at $z/z_i = 0.3$.

At night, turbulence rapidly decreases over the residual layer, leaving a much thinner layer of turbulent air near the ground, as is shown in Fig 4.3. The depth of this turbulent SBL is often relatively small (h $\cong$ 200 m). In statically neutral conditions the variances also decrease with height from large values at the surface, as shown in Fig 4.4 (Mason and Thompson, 1987); however, the depth scale is much larger (h $\cong$ 2 km).

Fig 4.5 shows that the horizontal components are often largest near the ground during the day, associated with the strong wind shears in the surface layer. The horizontal variance is roughly constant throughout the ML, but decreases with height above the ML top.   At night, the horizontal variance decreases rapidly with height to near zero at the top of the SBL (Fig 4.6). This shape is similar to that of the vertical velocity variance.

The budget term discussion for velocity variance will be deferred to Chapter 5 because of the close association of velocity variance with turbulent kinetic energy.

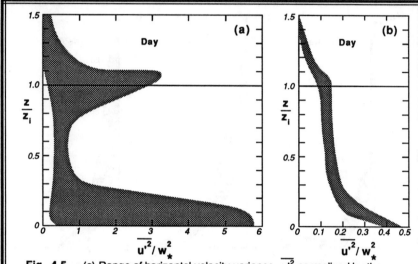

Fig. 4.5   (a) Range of horizontal velocity variance, $\overline{u'^2}$, normalized by the convective velocity scale, w_*^2, as a function of dimensionless height z/z_i, for typical conditions with combined convection and wind shear. (b) Idealized range for free convection with no mean shear. Based on data from Smedman and Hogstrom (1983), Deardorff (1974), André et al. (1978), and Lenschow et al. (1980).

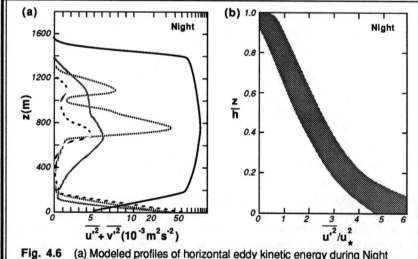

Fig. 4.6   (a) Modeled profiles of horizontal eddy kinetic energy during Night 33-34 of Wangara: ▬▬, t=18 h; ▬▬▬, t=21 h; ▪▪▪▪▪▪,t=03 h; ✦✦ , t=07 h. Abscissa changes from linear to logarithmic at 10. (b) Range of horizontal velocity variance, $\overline{u'^2}$, normalized by friction velocity, u_*, plotted as a function of height, z, normalized by a measure of stable boundary layer depth, h. Based on data from André, et al. (1978) and Caughey, et al. (1979).

4.3.2 Moisture Variance

Budget Equation. In the following development, only the vapor part of the specific humidity will be used, although similar derivations could be performed for the nonvapor part too. Start with (4.1.2c), multiply by 2q', and use the product rule of calculus to convert terms like $2q' \, \partial q'/\partial t$ into terms like $\partial(q'^2)/\partial t$, to yield:

$$\frac{\partial q'^2}{\partial t} + \overline{U}_j \frac{\partial q'^2}{\partial x_j} + 2q'u_j' \frac{\partial \overline{q}}{\partial x_j} + u_j' \frac{\partial q'^2}{\partial x_j} = 2q'v_q \frac{\partial^2 q'}{\partial x_j^2} + 2q' \frac{\partial \, \overline{(u_j'q')}}{\partial x_j}$$

Next, average and apply Reynolds averaging rules:

$$\frac{\partial \overline{q'^2}}{\partial t} + \overline{U}_j \frac{\partial \overline{q'^2}}{\partial x_j} + 2\overline{q'u_j'} \frac{\partial \overline{q}}{\partial x_j} + \overline{u_j' \frac{\partial q'^2}{\partial x_j}} = \overline{2q'v_q \frac{\partial^2 q'}{\partial x_j^2}}$$

To change this into flux form, add the averaged turbulent continuity equation multiplied by q'^2 (i.e., add $\overline{q'^2 \, \partial u_j'/\partial x_j} = 0$), and rearrange slightly:

$$\frac{\partial \overline{q'^2}}{\partial t} + \overline{U}_j \frac{\partial \overline{q'^2}}{\partial x_j} = -2\overline{q'u_j'} \frac{\partial \overline{q}}{\partial x_j} - \frac{\partial \, \overline{(u_j'q'^2)}}{\partial x_j} + 2v_q \overline{q' \frac{\partial^2 q'}{\partial x_j^2}}$$

As was done for momentum, the last term is split into two parts, one of which (the molecular diffusion of specific humidity variance) is small enough to be neglected. The

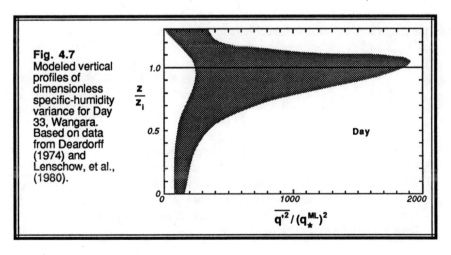

Fig. 4.7 Modeled vertical profiles of dimensionless specific-humidity variance for Day 33, Wangara. Based on data from Deardorff (1974) and Lenschow, et al., (1980).

$\overline{q'^2} / (q_*^{ML})^2$

Day

$\dfrac{z}{z_i}$

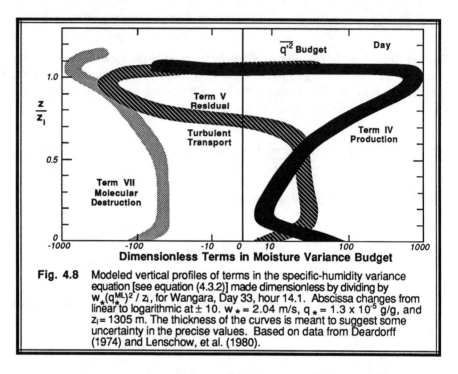

Fig. 4.8   Modeled vertical profiles of terms in the specific-humidity variance equation [see equation (4.3.2)] made dimensionless by dividing by $w_*(q_*^{ML})^2 / z_i$, for Wangara, Day 33, hour 14.1. Abscissa changes from linear to logarithmic at $\pm$ 10. $w_* = 2.04$ m/s, $q_* = 1.3 \times 10^{-5}$ g/g, and $z_i = 1305$ m. The thickness of the curves is meant to suggest some uncertainty in the precise values. Based on data from Deardorff (1974) and Lenschow, et al. (1980).

remaining part is defined as twice the molecular dissipation term, ε_q, by analogy with momentum:

$$\varepsilon_q = \nu_q \overline{\left(\frac{\partial q'}{\partial x_j}\right)^2}$$

Thus, the prognostic equation for specific humidity variance is

$$\underbrace{\frac{\partial \overline{q'^2}}{\partial t}}_{I} + \underbrace{\overline{U}_j \frac{\partial \overline{q'^2}}{\partial x_j}}_{II} = \underbrace{- 2\overline{q'u_j'}\,\frac{\partial \overline{q}}{\partial x_j}}_{IV} - \underbrace{\frac{\partial\,\overline{(u_j'q'^2)}}{\partial x_j}}_{V} - \underbrace{2\varepsilon_q}_{VII} \qquad (4.3.2)$$

Term I   represents local storage of humidity variance
Term II   describes the advection of humidity variance by the mean wind
Term IV   is a production term, associated with turbulent motions occurring within a mean moisture gradient
Term V   represents the turbulent transport of humidity variance
Term VII   is the molecular dissipation.

Case Study Examples. Fig 4.7 shows that humidity variance is small near the ground, because thermals have nearly the same humidity as their environment. At the top of the ML, however, drier air from aloft is being entrained down between the moist thermals, creating large humidity variances. Part of this variance might be associated with the excitation of gravity/buoyancy waves by the penetrative convection.

Fig 4.8 shows production terms balancing loss terms in the budget, assuming a steady state situation where storage and mean advection are neglected. Notice that the transport terms (found as a residual) are positive in the bottom half of the ML, but are negative in the top half. The integrated effects of these terms are zero. Such is the case for most transport terms — they merely move moisture variance from one part of the ML (where there is excess production) to another part (where there is excess dissipation), leaving zero net effect when averaged over the whole ML.

4.3.3 Heat (Potential Temperature Variance)

Budget Equations. As was done with the moisture equation, start with (4.1.3), multiply by $2\theta'$, use the product rule of calculus, Reynolds average, put into flux form, neglect molecular diffusion but retain the molecular dissipation, and rearrange to yield:

$$\frac{\partial \overline{\theta'^2}}{\partial t} + \overline{U}_j \frac{\partial \overline{\theta'^2}}{\partial x_j} = - 2\overline{\theta' u_j'} \frac{\partial \overline{\theta}}{\partial x_j} - \frac{\partial (\overline{u_j' \theta'^2})}{\partial x_j} - 2\varepsilon_\theta - \left(\frac{2}{\overline{\rho C_P}}\right) \overline{\theta' \frac{\partial Q_j^*{}'}{\partial x_j}}$$

$$(4.3.3)$$

$$\text{I} \qquad \text{II} \qquad \text{IV} \qquad \text{V} \qquad \text{VII} \qquad \text{VIII}$$

The terms above have physical representations analogous to those in (4.3.2). Term VIII is the radiation destruction term (sometimes given the symbol ε_R). It is difficult to

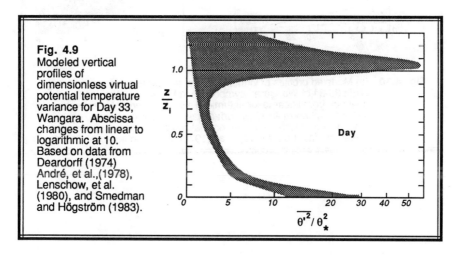

Fig. 4.9
Modeled vertical profiles of dimensionless virtual potential temperature variance for Day 33, Wangara. Abscissa changes from linear to logarithmic at 10. Based on data from Deardorff (1974) André, et al.,(1978), Lenschow, et al. (1980), and Smedman and Högström (1983).

measure this term directly, but sometimes it is modeled as $\varepsilon_R \cong (0.036 \text{ m/s}) \cdot \varepsilon \, \overline{\theta'^2} / \overline{e}^{\,3/2}$, where ε_R is about 1% to 10% of ε_θ (Coantic and Simonin, 1984).

Case Study Examples. The temperature variance at the top of the ML (Fig 4.9) is similar to humidity variance, because of the contrast between warmer entrained air and the cooler overshooting thermals. Gravity waves may also contribute to the variance. There is a greater difference near the bottom of the ML, however, because warm thermals in a cooler environment enhance the magnitude of the variance there. At night, Fig 4.10 shows that the largest temperature fluctuations are near the ground in the NBL, with weaker, sporadic turbulence in the RL aloft.

Fig 4.11 shows the contributions to the heat budget during daytime, again neglecting storage and advection. The radiation destruction term is small, but definitely nonzero. The dissipation is largest near the ground, as is the turbulent transport of temperature variance. Fig 4.12 shows the corresponding budget terms at night.

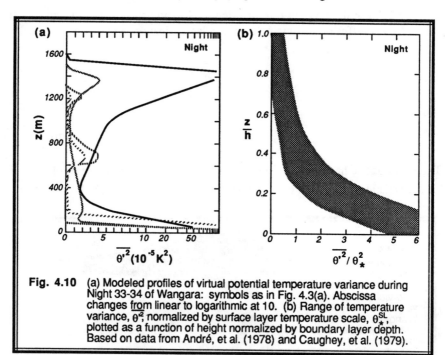

Fig. 4.10   (a) Modeled profiles of virtual potential temperature variance during Night 33-34 of Wangara: symbols as in Fig. 4.3(a). Abscissa changes from linear to logarithmic at 10. (b) Range of temperature variance, θ'^2, normalized by surface layer temperature scale, θ_*^{SL}, plotted as a function of height normalized by boundary layer depth. Based on data from André, et al. (1978) and Caughey, et al. (1979).

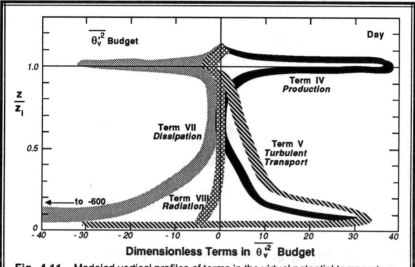

Fig. 4.11 Modeled vertical profiles of terms in the virtual potential temperature variance budget equation made dimensionless by dividing by $w_*(\theta_*^{ML})^2/z_i$. For Day 33, Wangara, hour 14.1. Thickness of lines denotes uncertainty or variability in the precisevalues.Based on data from Deardorff (1974), Lenschow et al. (1980), Andre et al. (1978), Caughey and Palmer (1979) and Zhou, et al. (1985).

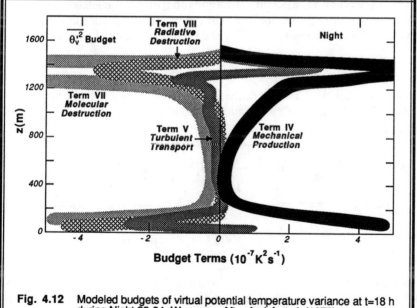

Fig. 4.12 Modeled budgets of virtual potential temperature variance at t=18 h during Night 33-34, Wangara. After André et al. (1978).

4.3.4 A Scalar Quantity (Tracer Concentration Variance)

Analogous with the moisture equation, start with (4.1.4), multiply by 2c', use the product rule of calculus, Reynolds average, put into flux form, neglect molecular diffusion but retain the molecular dissipation, and rearrange to yield:

$$\frac{\partial \overline{c'^2}}{\partial t} + \overline{U}_j \frac{\partial \overline{c'^2}}{\partial x_j} = -2\overline{c'u_j'} \frac{\partial \overline{C}}{\partial x_j} - \frac{\partial \overline{(u_j' c'^2)}}{\partial x_j} - 2\varepsilon_c \qquad (4.3.4)$$

$$\quad\; \text{I} \qquad\quad \text{II} \qquad\qquad \text{IV} \qquad\qquad \text{V} \qquad\quad \text{VII}$$

The terms above have physical representations analogous to those in (4.3.2).

No case study examples are shown for tracer variances because they vary so widely from constituent to constituent.

4.4 Prognostic Equations for Turbulent Fluxes

The equations of section 3.5.3 contain divergence terms of turbulent fluxes (e.g., $\overline{u_i'u_j'}$, $\overline{u_j'\theta'}$, $\overline{u_j'q'}$, and $\overline{u_j'c'}$). These fluxes are unknowns in equations (3.5.3). If prognostic equations for the fluxes can be found, one hopes that there will be as many equations as unknowns, allowing determination of the boundary layer wind and turbulence state. In this section, we will derive equations for the unknown fluxes; unfortunately, these new equations will contain additional new unknowns.

4.4.1 Momentum Flux

Budget Equations. Two perturbation equations are combined to produce flux equations. To obtain the first equation, start with (4.1.1), multiply it by u_k', and Reynolds average:

$$\overline{u_k'\frac{\partial u_i'}{\partial t}} + \overline{u_k'\overline{U}_j\frac{\partial u_i'}{\partial x_j}} + \overline{u_k'u_j'\frac{\partial \overline{U}_i}{\partial x_j}} + \overline{u_k'u_j'\frac{\partial u_i'}{\partial x_j}} =$$

$$+ \delta_{i3}\overline{u_k'\left(\frac{\theta_v'}{\overline{\theta}_v}\right)}g + f_c\varepsilon_{ij3}\overline{u_k'u_j'} - \overline{\left(\frac{u_k'}{\overline{\rho}}\right)\frac{\partial p'}{\partial x_i}} + \nu\overline{u_k'\frac{\partial^2 u_i'}{\partial x_j^2}}$$

For the second equation, interchange the i and k indices (i.e., replace each occurrence of i with k, and each occurrence of k with i). Such an interchange will not change the meaning of the equation, because summed terms will continue to be summed, and unsummed terms will continue to represent the three components. The result is:

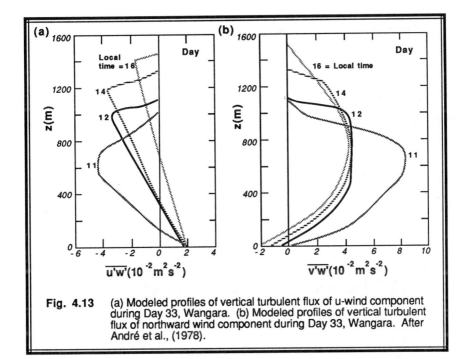

Fig. 4.13   (a) Modeled profiles of vertical turbulent flux of u-wind component during Day 33, Wangara. (b) Modeled profiles of vertical turbulent flux of northward wind component during Day 33, Wangara. After André et al., (1978).

$$\overline{u_i' \frac{\partial u_k}{\partial t}} + \overline{u_i' \overline{U}_j \frac{\partial u_k'}{\partial x_j}} + \overline{u_i' u_j' \frac{\partial \overline{U}_k}{\partial x_j}} + \overline{u_i' u_j' \frac{\partial u_k'}{\partial x_j}} =$$

$$+ \delta_{k3} \overline{u_i' \left(\frac{\theta_v'}{\overline{\theta}_v} \right)} + f_c \varepsilon_{kj3} \overline{u_i' u_j'} - \overline{\left(\frac{u_i'}{\overline{\rho}} \right) \frac{\partial p'}{\partial x_k}} + \nu \overline{u_i' \frac{\partial^2 u_k'}{\partial x_j^2}}$$

Next, add these two equations together, and use the product rule of calculus to produce combinations like $\overline{u_i' \partial u_k'/\partial t} + \overline{u_k' \partial u_i'/\partial t} = \partial(\overline{u_i' u_k'})/\partial t$:

$$\frac{\partial \overline{u_i' u_k'}}{\partial t} + \frac{\overline{U}_j \partial \overline{u_i' u_k'}}{\partial x_j} + \overline{u_i' u_j'} \frac{\partial \overline{U}_k}{\partial x_j} + \overline{u_k' u_j'} \frac{\partial \overline{U}_i}{\partial x_j} + \frac{\overline{u_j' \partial u_i' u_k'}}{\partial x_j} = \delta_{k3} \overline{u_i' \left(\frac{\theta_v'}{\overline{\theta}_v} \right)} g + \delta_{i3} \overline{u_k' \left(\frac{\theta_v'}{\overline{\theta}_v} \right)} g$$

$$+ f_c \varepsilon_{kj3} \overline{u_i' u_j'} + f_c \varepsilon_{ij3} \overline{u_k' u_j'} - \overline{\left(\frac{u_i'}{\overline{\rho}} \right) \frac{\partial p'}{\partial x_k}} - \overline{\left(\frac{u_k'}{\overline{\rho}} \right) \frac{\partial p'}{\partial x_i}} + \nu \overline{u_i' \frac{\partial^2 u_k'}{\partial x_j^2}} + \nu \overline{u_k' \frac{\partial^2 u_i'}{\partial x_j^2}}$$

The turbulent continuity equation multiplied by $u_i'u_k'$ and averaged $(\overline{u_i'u_k'\, \partial u_j'/\partial x_j} = 0)$ can be added to the last term before the equal sign to put it into flux form.

Each pressure term can be rewritten using the product rule of calculus, as shown in the following example: $\overline{u_k'\, \partial p'/\partial x_i} = \partial(\overline{p'u_k'})/\partial x_i - \overline{p'\, \partial u_k'/\partial x_i}$. Also, the viscosity terms can be rearranged using the product rule (left as an exercise to the reader). The final form for the momentum flux budget equation is:

$$\underbrace{\frac{\partial\left(\overline{u_i'u_k'}\right)}{\partial t}}_{\text{I}} + \underbrace{\frac{\overline{U}_j\partial\left(\overline{u_i'u_k'}\right)}{\partial x_j}}_{\text{II}} = \underbrace{-\left(\overline{u_i'u_j'}\right)\frac{\partial \overline{U}_k}{\partial x_j}}_{\text{III}} \underbrace{-\left(\overline{u_k'u_j'}\right)\frac{\partial \overline{U}_i}{\partial x_j}}_{\text{III}} \underbrace{-\frac{\partial\left(\overline{u_i'u_j'u_k'}\right)}{\partial x_j}}_{\text{IV}}$$

$$+ \underbrace{\left(\frac{g}{\overline{\theta}_v}\right)\left[\delta_{k3}\,\overline{u_i'\theta_v'} + \delta_{i3}\,\overline{u_k'\theta_v'}\right]}_{\text{V}\qquad\qquad\text{V}} + \underbrace{f_c\left[\varepsilon_{kj3}\,\overline{u_i'u_j'} + \varepsilon_{ij3}\,\overline{u_k'u_j'}\right]}_{\text{VI}\qquad\qquad\text{VI}}$$

$$\underbrace{-\frac{1}{\overline{\rho}}\left[\frac{\partial\left(\overline{p'u_k'}\right)}{\partial x_i}}_{\text{VII}} + \underbrace{\frac{\partial\left(\overline{p'u_i'}\right)}{\partial x_k}}_{\text{VII}} - \underbrace{\overline{p'\left(\frac{\partial u_i'}{\partial x_k} + \frac{\partial u_k'}{\partial x_i}\right)}\right]}_{\text{VIII}} + \underbrace{\frac{v\partial^2\left(\overline{u_i'u_k'}\right)}{\partial x_j^2}}_{\text{IX}} - \underbrace{\frac{2v\overline{\partial u_i'\,\partial u_k'}}{\partial x_j^2}}_{\text{X}}$$

$$(4.4.1a)$$

Term I Storage of momentum flux $\overline{u_i'u_k'}$.

Term II Advection of momentum flux by the mean wind.

Terms III Production of momentum flux by the mean wind shears.

Term IV Transport of momentum flux by turbulent motions (turbulent diffusion)

Term V Buoyant production or consumption.

Term VI Coriolis effects.

Term VII Transport by the pressure correlation term (pressure diffusion).

Term VIII Redistribution by the return-to-isotropy term (named by analogy with the velocity variance equation).

Term IX Molecular diffusion of turbulent momentum flux.

Term X Viscous dissipation term (named by analogy with the velocity variance equation). This term is often abbreviated by the symbol $2\varepsilon_{u_iu_k}$.

Scaling arguments, based on observations of the magnitudes of the terms in the above equation, suggest that terms VI, VII, and IX are smaller than the rest, leaving:

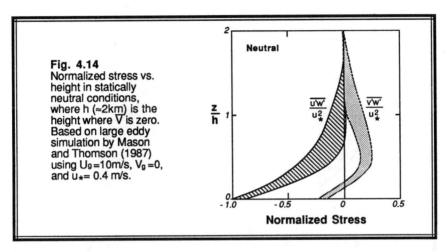

Fig. 4.14
Normalized stress vs. height in statically neutral conditions, where h ($\approx$2km) is the height where $\overline{V}$ is zero. Based on large eddy simulation by Mason and Thomson (1987) using U_g=10m/s, V_g=0, and u_*= 0.4 m/s.

$$\frac{\partial\left(\overline{u_i'u_k'}\right)}{\partial t} + \frac{\overline{U}_j\partial\left(\overline{u_i'u_k'}\right)}{\partial x_j} = -\left(\overline{u_i'u_j'}\right)\frac{\partial\overline{U}_k}{\partial x_j} - \left(\overline{u_k'u_j'}\right)\frac{\partial\overline{U}_i}{\partial x_j} - \frac{\partial\left(\overline{u_i'u_j'u_k'}\right)}{\partial x_j}$$

$$\text{I} \qquad \text{II} \qquad \text{III} \qquad \text{III} \qquad \text{IV}$$

$$+ \left(\frac{g}{\overline{\theta}_v}\right)\left[\delta_{k3}\,\overline{u_i'\theta_v'} + \delta_{i3}\,\overline{u_k'\theta_v'}\right] + \frac{\overline{p'}}{\rho}\left(\frac{\partial u_i'}{\partial x_k} + \frac{\partial u_k'}{\partial x_i}\right) - 2\varepsilon_{u_iu_k} \qquad (4.4.1b)$$

$$\text{V} \qquad\qquad \text{V} \qquad\qquad\qquad \text{VIII} \qquad\qquad \text{X}$$

Each term in the equation above contains unrepeated i and k indices. Remembering that i and k can each take on three values, that means (4.4.1a or b) represents 9 separate equations. Thus, the above equations can be used to forecast each of the nine terms in the Reynolds stress tensor, although as stated in Chapt. 2 the number of independent terms is reduced to 6 by symmetries.

As an example of an application of this equation for one term, choose a coordinate system aligned with the mean wind. Neglect subsidence and assume horizontal homogeneity. The $\overline{u'w'}$ component (i=1, k=3) of (4.4.1b) is thus:

$$\frac{\partial\left(\overline{u'w'}\right)}{\partial t} = -\frac{\overline{w'^2}\partial\overline{U}}{\partial z} - \frac{\partial\left(\overline{u'w'w'}\right)}{\partial z} + \frac{g\overline{u'\theta_v'}}{\overline{\theta}_v} + \frac{\overline{p'}}{\rho}\left(\frac{\partial u'}{\partial z} + \frac{\partial w'}{\partial x}\right) - 2\varepsilon_{uw}$$

$$(4.4.1c)$$

In general, the molecular (viscous) dissipation terms for the variance and covariance (flux) equations are abbreviated as $2\varepsilon_\xi$, where ξ represents the variance or covariance.

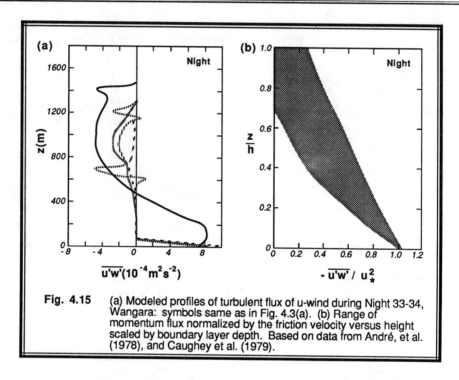

Fig. 4.15   (a) Modeled profiles of turbulent flux of u-wind during Night 33-34, Wangara: symbols same as in Fig. 4.3(a). (b) Range of momentum flux normalized by the friction velocity versus height scaled by boundary layer depth. Based on data from André, et al. (1978), and Caughey et al. (1979).

The only exception is the momentum variance equation (4.3.1g), where 2ε without a subscript is usually used.

Case Study Examples. Fig 4.13 shows vertical profiles of $\overline{u'w'}$ and $\overline{v'w'}$ for daytime cases, while Fig 4.14 shows a neutral example. Nighttime fluxes (Fig 4.15) are often much weaker than daytime fluxes, except near the ground where wind shear at night can maintain the turbulence intensity. The surface values of these fluxes are almost always of the sign appropriate for bringing momentum down from aloft. For mid-latitude situations with predominantly westerly flow that increases speed with height in the surface layer (i.e., $\overline{U}$ is positive), we find that the momentum flux ($\overline{u'w'}$) is negative near the ground. In Fig 4.15a, the surface flux is positive because the mean surface wind in from the east (i.e., $\overline{U}$ is negative).

Fig 4.16 shows the contribution of many of the terms in (4.4.1b) to the overall budgets of $\overline{u'w'}$ (eastward momentum flux budget) and $\overline{v'w'}$ (northward momentum flux budget). Again, steady state is assumed and mean advection is neglected. The resulting values of terms at any one height should thus sum to zero. Large values of many of the terms are observed both at the top and the bottom of the ML, where the strongest mean wind shears are found.

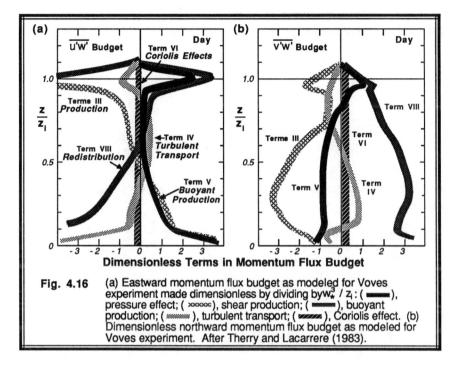

Fig. 4.16   (a) Eastward momentum flux budget as modeled for Voves experiment made dimensionless by dividing by w_*^3 / z_i: (▬▬), pressure effect; (∞∞∞), shear production; (▬▬), buoyant production; (∾∾∾), turbulent transport; (∾∾∾), Coriolis effect. (b) Dimensionless northward momentum flux budget as modeled for Voves experiment. After Therry and Lacarrere (1983).

4.4.2 Moisture Flux

Budget Equations.   As for momentum flux, the derivation combines two perturbation equations to produce a flux equation.   For the first equation start with the momentum perturbation equation (4.1.1), multiply it by the moisture perturbation q', and Reynolds average:

$$\overline{q'\frac{\partial u_i'}{\partial t}} + \overline{\bar{U}_j q'\frac{\partial u_i'}{\partial x_j}} + \overline{q'u_j'\frac{\partial \bar{U}_i}{\partial x_j}} + \overline{q'u_j'\frac{\partial u_i'}{\partial x_j}} =$$

$$+ \overline{q'\delta_{i3}\left(\frac{\theta_v'}{\bar{\theta}_v}\right)}g + f_c\varepsilon_{ij3}\overline{u_j'q'} - \overline{\left(\frac{q'}{\rho}\right)\frac{\partial p'}{\partial x_i}} + \nu\overline{q'\frac{\partial^2 u_i'}{\partial x_j^2}}$$

Similarly for the second equation, start with the moisture perturbation equation (4.1.2c) and multiply by u_i' and Reynolds average:

$$\overline{u_i'\frac{\partial q'}{\partial t}} + \overline{\bar{U}_j u_i'\frac{\partial q'}{\partial x_j}} + \overline{u_i'u_j'\frac{\partial \bar{q}}{\partial x_j}} + \overline{u_i'u_j'\frac{\partial q'}{\partial x_j}} = \nu_q\overline{u_i'\frac{\partial^2 q'}{\partial x_j^2}}$$

Next, add these two equations, put the turbulent flux divergence term into flux form using the turbulent continuity equation, and combine terms:

$$\frac{\partial\left(\overline{q'u_i'}\right)}{\partial t} + \overline{U}_j\frac{\partial\left(\overline{q'u_i'}\right)}{\partial x_j} + \overline{q'u_j'}\frac{\partial\overline{U}_i}{\partial x_j} + \overline{u_i'u_j'}\frac{\partial\overline{q}}{\partial x_j} + \frac{\partial\left(\overline{q'u_j'u_i'}\right)}{\partial x_j} =$$

$$+ \delta_{i3}\left(\frac{\overline{q'\theta_v'}}{\overline{\theta}_v}\right)g + f_c\varepsilon_{ij3}\overline{u_j'q'} - \overline{\left(\frac{q'}{\rho}\right)\frac{\partial p'}{\partial x_i}} + \nu\,\overline{q'\frac{\partial^2 u_i'}{\partial x_j^2}} + \nu_q\,\overline{u_i'\frac{\partial^2 q'}{\partial x_j^2}}$$

Then, split the pressure term into two parts, and assume $\nu \cong \nu_q$ to combine the molecular diffusion terms:

$$\underset{\text{I}}{\frac{\partial\left(\overline{q'u_i'}\right)}{\partial t}} + \underset{\text{II}}{\overline{U}_j\frac{\partial\left(\overline{q'u_i'}\right)}{\partial x_j}} = \underset{\text{III}}{-\overline{q'u_j'}\frac{\partial\overline{U}_i}{\partial x_j}} - \underset{\text{XI}}{\overline{u_i'u_j'}\frac{\partial\overline{q}}{\partial x_j}} - \underset{\text{IV}}{\frac{\partial\left(\overline{q'u_j'u_i'}\right)}{\partial x_j}}$$

$$+ \underset{\text{V}}{\delta_{i3}\left(\frac{\overline{q'\theta_v'}}{\overline{\theta}_v}\right)g} + \underset{\text{VI}}{f_c\varepsilon_{ij3}\left(\overline{u_j'q'}\right)} - \left(\frac{1}{\overline{\rho}}\right)\left[\underset{\text{VII}}{\frac{\partial\left(\overline{p'q'}\right)}{\partial x_i}} - \underset{\text{VIII}}{\overline{p'\frac{\partial q'}{\partial x_i}}}\right]$$

$$+ \underset{\text{IX}}{\frac{\nu\partial^2\left(\overline{q'u_i'}\right)}{\partial x_j^2}} - \underset{\text{X}}{2\nu\overline{\left(\frac{\partial u_i'}{\partial x_j}\right)\left(\frac{\partial q'}{\partial x_j}\right)}} \tag{4.4.2a}$$

The terms in this equation have meanings analogous to those in the momentum flux equation (4.4.1a), except for the additional term (XI), which is a production/loss term related to the mean moisture gradient. Remember that an additional term must be added if the body source is assumed to have perturbations too.

Substituting $2\varepsilon_{u_iq}$ for the last term, and neglecting the Coriolis term, the pressure diffusion term, and the molecular diffusion term leaves:

$$\underset{\text{I}}{\frac{\partial\left(\overline{q'u_i'}\right)}{\partial t}} + \underset{\text{II}}{\overline{U}_j\frac{\partial\left(\overline{q'u_i'}\right)}{\partial x_j}} = \underset{\text{III}}{-\overline{q'u_j'}\frac{\partial\overline{U}_i}{\partial x_j}} - \underset{\text{XI}}{\overline{u_i'u_j'}\frac{\partial\overline{q}}{\partial x_j}} - \underset{\text{IV}}{\frac{\partial\left(\overline{q'u_j'u_i'}\right)}{\partial x_j}}$$

$$+ \delta_{i3} \left(\frac{\overline{q'\theta_v'}}{\overline{\theta_v}} \right) g + \left(\frac{1}{\overline{\rho}} \right) \left[\overline{\frac{p'\partial q'}{\partial x_i}} \right] - 2\,\varepsilon_{u_i q} \qquad (4.4.2b)$$

$$\phantom{+ \delta_{i3}}\underset{V}{} \qquad\qquad \underset{VIII}{} \qquad\qquad \underset{X}{}$$

Terms I and II are the storage and advection, terms III, XI, and V relate to production/consumption; term IV is turbulent transport; term VIII is redistribution; and term X is the molecular destruction (dissipation) of turbulent moisture flux.

Physically, term V relates the correlation (covariance) between moisture and temperature to the production of moisture flux. One would expect that warmer air rises (3.3.3b); thus, if warmer air is also moister (i.e., a positive correlation), then the moist air would probably rise, thereby contributing to the moisture flux.

Physically, term XI suggests production of moisture flux when there is a momentum flux in a mean moisture gradient. The turbulent momentum flux implies a turbulent movement of air. If that movement occurs across a mean moisture gradient, then moisture fluctuation would be expected, as is suggested by analogy to Fig 2.13.

For the special case of vertical moisture flux (i=3) in a horizontally homogeneous setting with no subsidence, (4.4.2b) reduces to:

$$\frac{\partial(\overline{q'w'})}{\partial t} = -\overline{w'^2}\,\frac{\partial \overline{q}}{\partial z} - \frac{\partial(\overline{q'w'w'})}{\partial z} + \left(\overline{q'\theta_v'} \right) \frac{g}{\overline{\theta_v}} + \left(\frac{1}{\overline{\rho}} \right) \left[\overline{p'\frac{\partial q'}{\partial z}} \right] - 2\,\varepsilon_{wq}$$

$$\underset{I}{} \qquad\quad \underset{XI}{} \qquad\quad \underset{IV}{} \qquad\quad \underset{V}{} \qquad\quad \underset{VIII}{} \qquad\quad \underset{X}{}$$

Case Study Examples. The slope of the moisture flux curve in Fig 4.17 indicates that dry air is being entrained at the top of the ML fast enough to reduce the mean ML

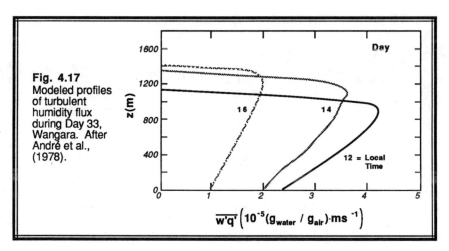

Fig. 4.17
Modeled profiles of turbulent humidity flux during Day 33, Wangara. After André et al., (1978).

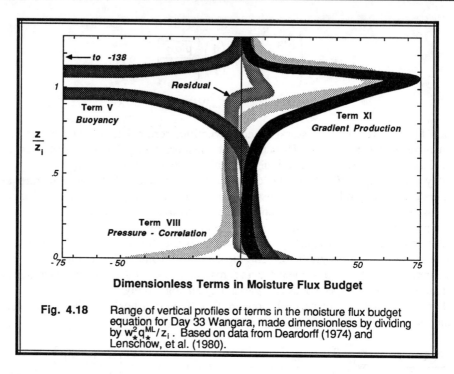

Dimensionless Terms in Moisture Flux Budget

Fig. 4.18   Range of vertical profiles of terms in the moisture flux budget equation for Day 33 Wangara, made dimensionless by dividing by $w_*^2 q_*^{ML}/z_i$. Based on data from Deardorff (1974) and Lenschow, et al. (1980).

humidity, in spite of evaporation from the surface.   Fig 4.18 shows a moisture flux budget, with the greatest production and loss values at the top of the ML for this case. Measurements of some of the individual terms are also shown in that figure.

4.4.3 Heat Flux

Budget Equations. The heat flux derivation is similar to that of the moisture flux.

$$\underbrace{\frac{\partial \left(\overline{\theta' u_i'} \right)}{\partial t}}_{\text{I}} + \underbrace{\overline{U}_j \frac{\partial \left(\overline{\theta' u_j'} \right)}{\partial x_j}}_{\text{II}} = \underbrace{- \overline{\theta' u_j'} \frac{\partial \overline{U}_i}{\partial x_j}}_{\text{III}} \underbrace{- \overline{u_i' u_j'} \frac{\partial \overline{\theta}}{\partial x_j}}_{\text{XI}} \underbrace{- \frac{\partial \left(\overline{\theta' u_j' u_i'} \right)}{\partial x_j}}_{\text{IV}}$$

$$\underbrace{+ \delta_{i3} \left(\frac{\overline{\theta' \theta_v'}}{\overline{\theta}_v} \right) g}_{\text{V}} + \underbrace{f_c \varepsilon_{ij3} \left(\overline{u_j' \theta'} \right)}_{\text{VI}} - \underbrace{\left(\frac{1}{\overline{\rho}} \right) \left[\frac{\partial \left(\overline{p' \theta'} \right)}{\partial x_i}}_{\text{VII}} - \underbrace{\overline{p' \frac{\partial \theta'}{\partial x_i}} \right]}_{\text{VIII}}$$

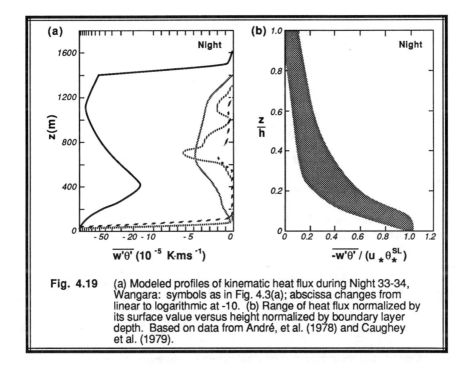

Fig. 4.19   (a) Modeled profiles of kinematic heat flux during Night 33-34, Wangara: symbols as in Fig. 4.3(a); abscissa changes from linear to logarithmic at -10. (b) Range of heat flux normalized by its surface value versus height normalized by boundary layer depth. Based on data from André, et al. (1978) and Caughey et al. (1979).

$$+ \; \frac{\nu \partial^2 \left(\overline{\theta' u_i'} \right)}{\partial x_j^2} \; - \; 2\nu \overline{\left(\frac{\partial u_i'}{\partial x_j} \right) \left(\frac{\partial \theta'}{\partial x_j} \right)} \; - \; \left(\frac{1}{\overline{\rho} \, \overline{C_p}} \right) \overline{u_i' \frac{\partial Q_j^{*'}}{\partial x_j}} \qquad (4.4.3a)$$

$$\text{IX} \qquad\qquad\qquad \text{X} \qquad\qquad\qquad \text{XII}$$

Namely, we started with the velocity perturbation equation multiplied by θ', and the temperature perturbation equation multiplied by u_i'. Both equations were Reynolds averaged and summed. The turbulent continuity equation was used to put the turbulence diffusion term into flux form.

The terms in this equation have analogous meanings as for the moisture flux equation (4.4.2a). Often, term V is approximated by $\delta_{i3} \, g \, (\overline{\theta_v'^2} / \overline{\theta_v})$. Term XII describes the correlation between velocity fluctuations and with radiation fluctuations.

Substituting $2 \, \varepsilon_{u_i\theta}$ for term X, and neglecting the Coriolis, pressure diffusion, radiation, and the molecular diffusion terms for simplicity leaves:

$$\frac{\partial\left(\overline{\theta'u_i'}\right)}{\partial t} + \overline{U}_j\frac{\partial\left(\overline{\theta'u_i'}\right)}{\partial x_j} = -\overline{\theta'u_j'}\frac{\partial\overline{U}_i}{\partial x_j} - \overline{u_i'u_j'}\frac{\partial\overline{\theta}}{\partial x_j} - \frac{\partial\left(\overline{\theta'u_j'u_i'}\right)}{\partial x_j}$$

$$\qquad\quad\text{I}\qquad\qquad\text{II}\qquad\qquad\qquad\text{III}\qquad\qquad\text{XI}\qquad\qquad\text{IV}$$

$$+ \delta_{i3}\left(\frac{\overline{\theta'\theta_v'}}{\overline{\theta}_v}\right)g + \left(\frac{1}{\overline{\rho}}\right)\left[\overline{\frac{p'\partial\theta'}{\partial x_i}}\right] - 2\,\varepsilon_{u_i\theta} \qquad (4.4.3b)$$

$$\qquad\qquad\text{V}\qquad\qquad\qquad\text{VIII}\qquad\qquad\text{X}$$

For the special case of vertical heat flux (i=3) in a horizontally homogeneous setting with no subsidence, (4.4.3b) reduces to:

$$\frac{\partial\left(\overline{\theta'w'}\right)}{\partial t} = -\overline{w'^2}\frac{\partial\overline{\theta}}{\partial z} - \frac{\partial\left(\overline{\theta'w'w'}\right)}{\partial z} + \left(\overline{\theta'\theta_v'}\right)\left(\frac{g}{\overline{\theta}_v}\right) + \left(\frac{1}{\overline{\rho}}\right)\left[\overline{p'\frac{\partial\theta'}{\partial z}}\right] - 2\varepsilon_{w\theta}$$

$$\quad\text{I}\qquad\qquad\text{XI}\qquad\qquad\text{IV}\qquad\qquad\qquad\text{V}\qquad\qquad\qquad\text{VIII}\qquad\qquad\text{X}$$

$$\qquad\qquad\qquad\qquad\qquad\qquad\qquad\qquad\qquad\qquad\qquad\qquad\qquad (4.4.3c)$$

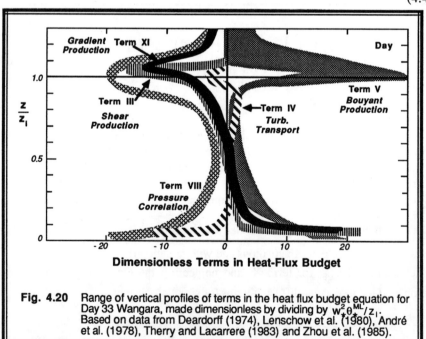

Fig. 4.20   Range of vertical profiles of terms in the heat flux budget equation for Day 33 Wangara, made dimensionless by dividing by $w_*^2\theta_*^{ML}/z_i$. Based on data from Deardorff (1974), Lenschow et al. (1980), André et al. (1978), Therry and Lacarrere (1983) and Zhou et al. (1985).

Case Study Examples. The linear decrease of heat flux with height has been discussed in chapter 3 for the daytime cases (Figs 3.1-3.3 and 3.7). Nighttime heat flux values are much weaker, as shown in Fig 4.19.

Fig 4.20 shows the steady-state, non-advective heat flux budget terms. Large values of terms occur at the surface because of the strong turbulence and mean temperature gradients there. In the middle of the ML, weak potential temperature gradients are associated with smaller magnitudes of the terms there. At the top of the ML the terms become large again, associated with the strong temperature contrast across the ML top.

4.4.4 Flux of a Scalar   (Pollutant or Tracer Flux)

Using the same procedures as before, the pollutant flux budget equation is

$$\underbrace{\frac{\partial\left(\overline{c'u_i'}\right)}{\partial t}}_{I} + \underbrace{\overline{U}_j\frac{\partial\left(\overline{c'u_i'}\right)}{\partial x_j}}_{II} = - \underbrace{\overline{c'u_j'}\frac{\partial\overline{U}_i}{\partial x_j}}_{III} - \underbrace{\overline{u_i'u_j'}\frac{\partial\overline{C}}{\partial x_j}}_{XI} - \underbrace{\frac{\partial\left(\overline{c'u_j'u_i'}\right)}{\partial x_j}}_{IV}$$

$$+ \underbrace{\delta_{i3}\left(\frac{\overline{c'\theta_v'}}{\overline{\theta}_v}\right)g}_{V} + \underbrace{f_c\varepsilon_{ij3}\left(\overline{u_j'c'}\right)}_{VI} - \underbrace{\left(\frac{1}{\overline{\rho}}\right)\left[\frac{\partial\left(\overline{p'c'}\right)}{\partial x_i}\right.}_{VII} \underbrace{\left.- \overline{p'\frac{\partial c'}{\partial x_i}}\right]}_{VII} + \underbrace{v\overline{\partial^2\left(\overline{c'u_i'}\right)}{\partial x_j^2}}_{IX} \underbrace{- 2v\overline{\left(\frac{\partial u_i'}{\partial x_j}\right)\left(\frac{\partial c'}{\partial x_j}\right)}}_{X}$$

$$(4.4.4a)$$

The terms in this equation have meanings analogous to those of the moisture flux equation (4.4.2a). Substituting $2\,\varepsilon_{u_ic}$ for the last term, and neglecting the Coriolis term, the pressure diffusion term, and the molecular diffusion term leaves:

$$\underbrace{\frac{\partial\left(\overline{c'u_i'}\right)}{\partial t}}_{I} + \underbrace{\overline{U}_j\frac{\partial\left(\overline{c'u_i'}\right)}{\partial x_j}}_{II} = - \underbrace{\overline{c'u_j'}\frac{\partial\overline{U}_i}{\partial x_j}}_{III} - \underbrace{\overline{u_i'u_j'}\frac{\partial\overline{C}}{\partial x_j}}_{XI} - \underbrace{\frac{\partial\left(\overline{c'u_j'u_i'}\right)}{\partial x_j}}_{IV}$$

$$+ \underbrace{\delta_{i3}\left(\frac{\overline{c'\theta_v'}}{\overline{\theta}_v}\right)g}_{V} + \underbrace{\left(\frac{1}{\overline{\rho}}\right)\left[\overline{p'\frac{\partial c'}{\partial x_i}}\right]}_{VIII} \underbrace{- 2\,\varepsilon_{u_ic}}_{X} \qquad (4.4.4b)$$

For the special case of vertical pollutant flux (i=3) in a horizontally homogeneous setting with no subsidence, (4.4.4b) reduces to:

$$\frac{\partial \left(\overline{c'w'}\right)}{\partial t} = -\overline{w'^2}\frac{\partial \overline{C}}{\partial z} - \frac{\partial \left(\overline{c'w'w'}\right)}{\partial z} + \overline{c'\theta_v'}\left(\frac{g}{\overline{\theta_v}}\right) + \left(\frac{1}{\overline{\rho}}\right)\left[\overline{p\frac{\partial c'}{\partial z}}\right] - 2\,\varepsilon_{wc} \quad (4.4.4c)$$

4.4.5 Buoyancy Flux

Both the definition of w_* and the buoyant production term in equation (4.3.1j) contain

a buoyancy flux defined by $(g/\overline{\theta_v})\,(\overline{w'\theta_v'})_s$. The flux of virtual potential temperature

$\overline{w'\theta_v'}$ is different than the heat flux $\overline{w'\theta'}$ — the two must not be interchanged. We

can, however, use the definition of virtual potential temperature to derive a diagnostic

equation for $\overline{w'\theta_v'}$ in terms of $\overline{w'\theta'}$.

Start with

$$\theta_v \cong \theta\left[1 + 0.61r - r_L\right] \qquad (4.4.5a)$$

from section 1.5 (or Appendix D), where it is understood that r equals the saturation value
whenever r_L is nonzero. Expand the dependent variables into mean and turbulent parts:

$$\overline{\theta}_v + \theta_v' = \left(\overline{\theta} + \theta'\right)\left[1 + 0.61\left(\overline{r} + r'\right) - \left(\overline{r_L} + r_L'\right)\right]$$

$$= \overline{\theta}\left[1 + 0.61\overline{r} - \overline{r_L}\right] + \overline{\theta}\left[0.61r' - r_L'\right] + \theta'\left[1 + 0.61\overline{r} - \overline{r_L}\right] + \theta'\left[0.61r' - r_L'\right]$$

Multiply this equation by w' and Reynolds average

$$\overline{w'\theta_v'} = \overline{\theta}\left[0.61\left(\overline{w'r'}\right) - \left(\overline{w'r_L'}\right)\right] + \left(\overline{w'\theta'}\right)\left[1 + 0.61\overline{r} - \overline{r_L}\right] + \overline{w'\theta'\left[0.61r' - r_L'\right]}$$
$$(4.4.5b)$$

The last terms are triple correlations ($\overline{w'\theta'r'}$ and $\overline{w'\theta'r_L'}$). Observations in the

atmosphere suggest that they are small enough compared to the other terms to be
neglected, although these estimates are difficult to measure and fraught with error. Thus,
the usual form for the virtual heat flux is:

$$\overline{w'\theta_v'} \cong \overline{\theta}\left[0.61\left(\overline{w'r'}\right) - \left(\overline{w'r_L'}\right)\right] + \left(\overline{w'\theta'}\right)\left[1 + 0.61\overline{r} - \overline{r_L}\right] \quad (4.4.5c)$$

For the special case of no liquid water in the air, this reduces to:

$$\overline{w'\theta_v'} \cong \left(\overline{w'\theta'}\right)\left[1 + 0.61\bar{r}\right] + 0.61\bar{\theta}\left(\overline{w'r'}\right) \qquad (4.4.5d)$$

Modelers and theorists often have no recourse but to use this equation whenever buoyancy flux is needed. Experimentalists, on the other hand, often have direct observations of instantaneous values of θ and r, enabling them to calculate θ_v using

(4.4.5a). Knowing θ_v, it is easy to then find $\overline{\theta}_v$ and θ_v', as is done with any other variable. The resulting flux found by forming the product of w' and θ_v' yields a more accurate virtual potential temperature flux than (4.4.5c or d).

Case study examples for the buoyancy flux were shown in Figs 3.1-3.3.

4.5 References

André, J.-C.,G. De Moor, P. Lacarrère, G. Therry, &R. du Vachat, 1978: Modeling the 24-hour evolution of the mean and turbulent structures of the planetary boundary layer. *J. Atmos. Sci.*, **35**, 1861-1883.

Caughey, S.J., J.C.Wyngaard and J.C. Kaimal, 1979: Turbulence in the evolving stable boundary layer. *J. Atmos. Sci.*, **36**, 1041-1052.

Coantic, M. and O. Simonin, 1984: Radiative effects on turbulent temperature spectra and budgets in the planetary boundary layer. *J. Atmos. Sci.*, **41**, 2629-2651.

Deardorff, J.W., 1974: Three-dimensional numerical study of turbulence in an entraining mixed layer. *Bound.-Layer Meteor.*, **7**, 199-226.

Lenschow, D.H., J.C. Wyngaard, & W.T. Pennell, 1980: Mean field and second moment budgets in a baroclinic, convective boundary layer. *J. Atmos. Sci.*, **37**, 1313-1326.

Mason, P.J. and D.J. Thompson, 1987: Large eddy simulations of the neutral-static-stability planetary boundary layer. *Quart. J. Roy. Meteor. Soc.*, **113**, 413-443.

Smedman, A.-S. and U. Högström, 1983: Turbulent characteristics of a shallow convective internal boundary layer. *Bound.-Layer Meteor.*, **25**, 271-287.

Therry, G. and P. Lacarrère, 1983: Improving the eddy kinetic energy model for blanetary boundary layer description. *Bound.-Layer Meteor.*, **25**, 63-88.

Zhou, M.Y., D.H. Lenschow, B.B. Stankov, J.C. Kaimal, and J.E. Gaynor, 1985: Wave and turbulent structure in a shallow baroclinic convective boundary layer and overalying inversion. *J. Atmos. Sci.*, **42**, 47-57.

4.6 Exercises

1) Confirm that $-2\overline{u_j'q'}\,\partial\bar{q}/\partial x_j$ is a production term and not a loss term for equation (4.3.2).   Hint, review section 2.7 and Fig 2.13.

2) Given values for the viscous dissipation rate of velocity variance (see section 4.3.1), express that rate as a heating rate $\partial\bar{\theta}/\partial t$ for air, and compare its magnitude with the magnitudes of other terms in (3.4.5b).   Hint, remember that viscosity dissipates turbulent motions into heat.

3) Given the general form for the momentum flux equation (4.4.1b), write out the equations for the following components:

   a) $\overline{u'w'}$ b) $\overline{v'w'}$ c) $\overline{u'v'}$

4) Given the momentum flux equation (4.4.1b), show how to transform that equation into an equation for velocity variance $\overline{u_i'^2}$.

5) Show how the two viscosity terms in the equation just before (4.4.1a) can be manipulated into the form shown in (4.4.1a).   Hint, start with $\partial^2(\overline{u_i'u_k'})/\partial x_j^2$.

6) a) Given the data from Figs 3.1-3.6 of chapter 3, calculate w_* for each of the flights.

   b) Also calculate t_*, θ_*^{ML}, and q_*^{ML}.

7) In Fig 3.1a of chapter 3 are plotted two data points at each height. One data point represents heat flux and one represents moisture flux. Using the values from this figure, calculate $\overline{w'\theta_v'}$ for each of those heights, and plot the result.   Do NOT normalize your results by the surface value.

8) Given Fig 4.13, calculate the value of terms VI and IX of equation (4.4.1a). By comparing these values with the magnitudes of the other terms, are we justified in neglecting them to derive (4.4.1b)?

9) Some of the terms in (4.4.1a) involve correlations with pressure perturbation. Discuss how you would design an instrument for measuring p' , and what some of the difficulties might be.

10) Many of the prognostic equations in this chapter include triple-correlation terms such as $\overline{w'w'\theta'}$. Discuss the steps (but do not do the complete derivation) you would take to derive a prognostic equation for $\partial(\overline{w'w'\theta'})/\partial t$. Hint, review the general methodology used to derive equations for $\partial(\overline{w'\theta'})/\partial t$.

11) Suppose that your best friend sneezes into still air, creating a SEG cloud of concentration c, where SEG = Someone Else's Germs. These germs either multiply or die depending on the temperature. If the following conservation equation describes SEG: $dc/dt = b c (T-T_o)$,   where b and T_o are constants and T is temperature, then derive a prognostic equation for $\overline{w'c'}$.

5 Turbulence Kinetic Energy, Stability and Scaling

Turbulence kinetic energy (TKE) is one of the most important variables in micrometeorology, because it is a measure of the intensity of turbulence. It is directly related to the momentum, heat, and moisture transport through the boundary layer. Turbulence kinetic energy is also sometimes used as a starting point for approximations of turbulent diffusion.

The individual terms in the TKE budget equation describe physical processes that generate turbulence. The relative balance of these processes determines the ability of the flow to maintain turbulence or become turbulent, and thus indicates flow stability. Some important dimensionless groups and scaling parameters are also based on terms in the TKE equation. For these reasons, our study of turbulence kinetic energy will begin with the TKE budget equation, and end in a general discussion of stability and scaling.

5.1 The TKE Budget Derivation

The definition of TKE presented in section 2.5 is $\mathrm{TKE}/m = \bar{e} = 0.5\,(\overline{u'^2} + \overline{v'^2} + \overline{w'^2})$.

Using summation notation, it is easy to rewrite this as $\bar{e} = 0.5\,\overline{u_i'^2}$. We recognize immediately that TKE/m is nothing more than the summed velocity variances divided by two. Therefore, starting with the prognostic equation for the sum of velocity variances (4.3.1g) and dividing by two easily gives us the TKE budget equation:

$$\frac{\partial \overline{e}}{\partial t} + \overline{U}_j \frac{\partial \overline{e}}{\partial x_j} = + \delta_{i3} \frac{g}{\overline{\theta}_v}\left(\overline{u_i'\theta_v'}\right) - \overline{u_i'u_j'} \frac{\partial \overline{U}_i}{\partial x_j} - \frac{\partial\left(\overline{u_j'e}\right)}{\partial x_j} - \frac{1}{\overline{\rho}} \frac{\partial\left(\overline{u_i'p'}\right)}{\partial x_i} - \varepsilon$$

$$(5.1a)$$

I   II   III   IV   V   VI   VII

Term I   represents local *storage* or tendency of TKE.
Term II   describes the *advection* of TKE by the mean wind.
Term III   is the *buoyant production or consumption term*. It is a

production or loss term depending on whether the heat flux $\overline{u_i'\theta_v'}$ is

positive (during daytime over land) or negative (at night over land).
Term IV   is a *mechanical or shear production/loss term.* The momentum

flux $\overline{u_i'u_j'}$ is usually of opposite sign from the mean wind shear,

because the momentum of the wind is usually lost downward to the
ground. Thus, Term IV results in a positive contribution to TKE when
multiplied by a negative sign.
Term V   represents the *turbulent transport* of TKE. It describes how TKE
is moved around by the turbulent eddies u_j'.
Term VI   is a *pressure correlation term* that describes how TKE is
redistributed by pressure perturbations. It is often associated with
oscillations in the air (*buoyancy or gravity waves*).
Term VII   represents the viscous *dissipation* of TKE; i.e., the conversion of
TKE into heat.

If we choose a coordinate system aligned with the mean wind, assume horizontal
homogeneity, and neglect subsidence, then a special form of the TKE budget equation can
be written

$$\frac{\partial \overline{e}}{\partial t} = \frac{g}{\overline{\theta}_v}\left(\overline{w'\theta_v'}\right) - \overline{u'w'} \frac{\partial \overline{U}}{\partial z} - \frac{\partial\left(\overline{w'e}\right)}{\partial z} - \frac{1}{\overline{\rho}} \frac{\partial\left(\overline{w'p'}\right)}{\partial z} - \varepsilon \qquad (5.1b)$$

I   III   IV   V   VI   VII

Turbulence is *dissipative*. Term VII is a loss term that always exists whenever
TKE is nonzero. Physically, this means that turbulence will tend to decrease and
disappear with time, unless it can be generated locally or transported in by mean,
turbulent, or pressure processes. Thus, TKE is not a conserved quantity. The boundary
layer can be turbulent only if there are specific physical processes generating the
turbulence. In the next subsections, the role of each of the terms is examined in more
detail.

5.2 Contributions to the TKE Budget

5.2.1 Term 1:   Storage

Fig 2.10 shows that there can be substantial variation in the magnitude of TKE with time at any one height. Fig 5.1 shows a simulation of TKE over a two day period, where a dramatic increase and decrease of TKE occurs within each diurnal cycle. An increase in TKE from a small early morning value to a larger early afternoon value represents a net storage of TKE in the air.   In particular, nonturbulent FA air just above the ML top must be *spun up* (i.e., its turbulence intensity must increase from near zero to the current ML value) as entrainment incorporates it into the ML.

Over a land surface experiencing a strong diurnal cycle, typical order of magnitudes for this term range from about 5×10^{-5} m^2 s^{-3} for surface-layer air over a 6 h interval, to about 5×10^{-3} m^2 s^{-3} for FA air that is spun up over 15 min (i.e., over a time interval corresponding to t_*).   Fig 5.2 shows sample observations of TKE made in the surface layer, where TKE varies by about two orders of magnitude.

During the later afternoon and evening, a corresponding *spin down* (i.e., decrease of TKE with time) occurs where dissipation and other losses exceed the production of turbulence. The storage term is thus negative during this transition phase.

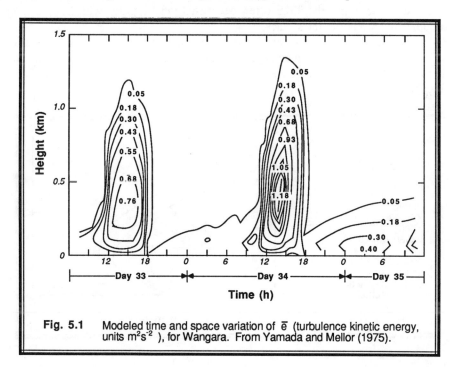

Fig. 5.1   Modeled time and space variation of $\bar{e}$ (turbulence kinetic energy, units m^{2}s^{-2}), for Wangara. From Yamada and Mellor (1975).

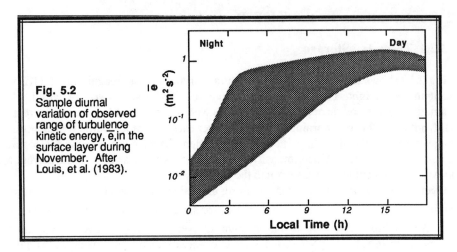

Fig. 5.2
Sample diurnal variation of observed range of turbulence kinetic energy, $\bar{e}$, in the surface layer during November. After Louis, et al. (1983).

Fig 5.3 indicates that the vertical profile of TKE can sometimes increase to a maximum at a height of about $z/z_i \cong 0.3$ when free convection dominates, as modeled for the Wangara experiment. When strong winds are present, the TKE might be nearly constant with height within the BL, or might decrease slightly with height as shown in Fig 5.3 for BLX83 data. At night, the TKE often decreases very rapidly with height, from a maximum value just above the surface.

Over surfaces such as oceans that do not experience a large diurnal cycle, the storage term is often so small that it can be neglected (i.e., steady state can be assumed). This is not to say that there is no turbulence, just that the intensity of turbulence is not changing significantly with time.

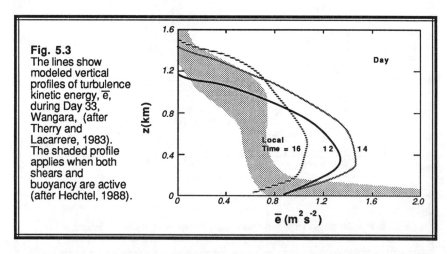

Fig. 5.3
The lines show modeled vertical profiles of turbulence kinetic energy, $\bar{e}$, during Day 33, Wangara, (after Therry and Lacarrere, 1983). The shaded profile applies when both shears and buoyancy are active (after Hechtel, 1988).

5.2.2 Term II: Advection

Little is known about this term. When averaged over a horizontal area larger than about 10 km by 10 km, it is often assumed that there is little horizontal variation in TKE, thereby making the advection term negligible. This is probably a good assumption over most land surfaces.

On a smaller scale, however, it is clear that this term must be important. For example, picture a reservoir of water cooler than the surrounding land. The lack of heating over the reservoir would allow turbulence to decay in the overlying air, while air over the adjacent land surfaces could be in a state of active convection. A mean wind advecting air across the shores of this reservoir would thus cause significant change in the TKE budget. Over ocean surfaces, the advection term would probably be negligible even on the small scales.

5.2.3 Term III: Buoyant Production/Consumption

Production. Fig 5.4 shows the variation of a number of TKE budget terms with height within a fair-weather convective ML. The most important part of the buoyancy

term is the flux of virtual potential temperature, $\overline{w'\theta_v'}$. As we have already studied in

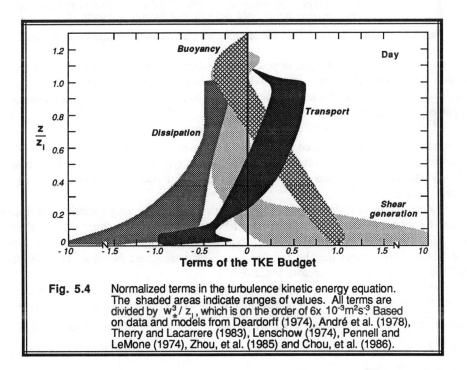

Fig. 5.4   Normalized terms in the turbulence kinetic energy equation. The shaded areas indicate ranges of values. All terms are divided by w_*^3 / z_i, which is on the order of 6x $10^{-3}m^2s^{-3}$ Based on data and models from Deardorff (1974), André et al. (1978), Therry and Lacarrere (1983), Lenschow (1974), Pennell and LeMone (1974), Zhou, et al. (1985) and Chou, et al. (1986).

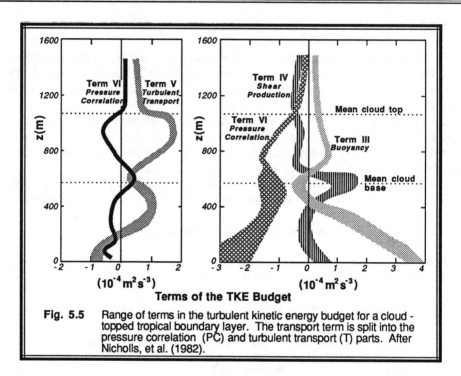

Terms of the TKE Budget

Fig. 5.5   Range of terms in the turbulent kinetic energy budget for a cloud - topped tropical boundary layer. The transport term is split into the pressure correlation (PC) and turbulent transport (T) parts. After Nicholls, et al. (1982).

the previous chapters, this flux is positive and decreases roughly linearly with height within the bottom 2/3 of the convective ML.   Near the ground, term III is large and positive, corresponding to a large generation rate of turbulence whenever the underlying surface is warmer than the air.

When positive, this term represents the effects of *thermals* in the ML.   Active thermal convection is associated with large values of this term, as large as $1 \times 10^{-2} \, m^2 \, s^{-3}$ near the ground.   Thus, we often associate this term with sunny days over land, or cold air advection over a warmer underlying surface.   For cloudy days over land, it can be much smaller.

In convective boundary layers capped with actively growing cumulus clouds, the positive buoyancy within the cloud can contribute to the production (term III) of TKE (see Fig 5.5).   Between this cloud layer contribution and the contribution near the bottom of the subcloud layer, there may be a region near cloud base where the air is statically stable and the buoyancy term is therefore negative.

Because Term III is so important on days of free convection, it is often used to normalize all the other terms.   For example, using the definitions of w_* and z_i presented earlier, it is easy to show that Term III = $(w_*)^3 / z_i$ at the surface. Dividing (5.1b) by $(w_*)^3 / z_i$ gives a dimensionless form of the TKE budget equation that is useful for free convection situations:

$$\frac{z_i}{w_*^3}\frac{\partial \bar{e}}{\partial t} = \frac{g\,z_i\left(\overline{w'\theta_v'}\right)}{w_*^3\,\overline{\theta_v}} - \frac{z_i\,\overline{u'w'}}{w_*^3}\frac{\partial \overline{U}}{\partial z} - \frac{z_i}{w_*^3}\frac{\partial\left(\overline{w'e}\right)}{\partial z} - \frac{z_i}{w_*^3}\,\bar{\rho}\,\frac{\partial\left(\overline{w'p'}\right)}{\partial z} - \frac{z_i\,\varepsilon}{w_*^3}$$

(5.2.3)

    I   III   IV   V   VI   VII

By definition, the dimensionless Term III is unity at the surface. Equations that are made dimensionless by dividing by scaling parameters are said to be *normalized*. The normalization scheme expressed by (5.2.3) is used in most of the figures in this section, and indeed has been used in the previous chapter too.

As is evident in (4.3.1j), the buoyancy term acts only on the vertical component of TKE. Hence, this production term is *anisotropic* (i.e., not isotropic). The return-to-isotropy terms of (4.3.1h-j) are responsible for moving some of the vertical kinetic energy into the horizontal directions. Again, the anisotropic nature of Term III confirms our picture of strong up and downdrafts within thermals.

Consumption. In statically stable conditions, an air parcel displaced vertically by turbulence would experience a buoyancy force pushing it back towards its starting height. *Static stability thereby tends to suppress, or consume, TKE*, and is

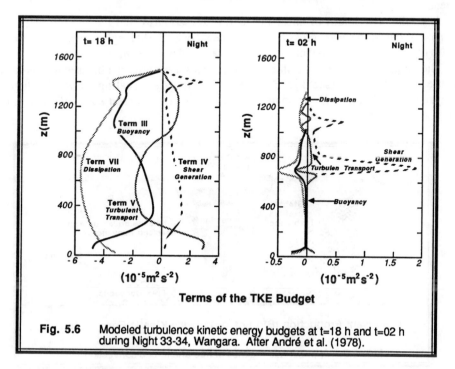

Fig. 5.6   Modeled turbulence kinetic energy budgets at t=18 h and t=02 h during Night 33-34, Wangara. After André et al. (1978).

associated with negative values of term III. Such conditions are present in the SBL at night over land, or anytime the surface is colder than the overlying air. An example of the decay of turbulence in negatively buoyant conditions just after sunset is shown in the budget profiles of Fig 5.6.

This same type of consumption can occur at the top of a ML, where warmer air entrained downward by turbulence opposes the descent because of its buoyancy (Stage and Businger, 1981). This is related to the negative values of the buoyancy term near the top of the ML in Fig 5.4.

5.2.4 Term IV: Mechanical (Shear) Production

When there is a turbulent momentum flux in the presence of a mean wind shear, the interaction between the two tends to generate more turbulence. Even though a negative sign precedes Term IV, the momentum flux is usually of opposite sign from the mean shear, resulting in production, not loss, of turbulence.

Fig. 5.4 shows case studies of the contribution of shear production to the TKE budget for convective situations. The greatest wind shear magnitude occurs at the surface. Not surprisingly, the maximum shear production rate also occurs there.   As shown in Chapters 1 and 3, the wind speed frequently varies little with height in the ML above the surface layer, resulting in near zero shear and near zero shear production of turbulence. Shear production is often associated with the surface layer because of its limited vertical extent.

A smaller maximum of shear production sometimes occurs at the top of the ML because of the wind shear across the entrainment zone. In that region, the subgeostrophic winds of the ML recover to their geostrophic values above the ML.

The relative contributions of the buoyancy and shear terms can be used to classify the nature of convection (see Fig 5.7) *Free convection* scaling is valid when the buoyancy term is much larger than the mechanical term, *forced convection* scaling is valid when the opposite is true.

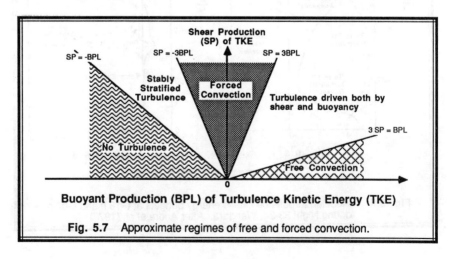

Fig. 5.7   Approximate regimes of free and forced convection.

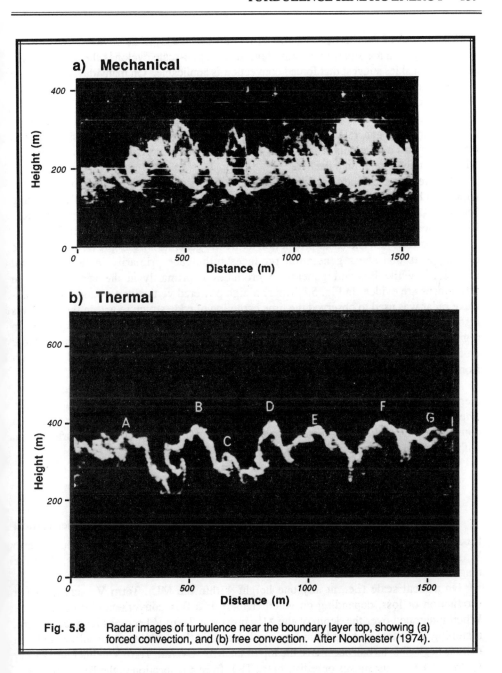

Fig. 5.8 Radar images of turbulence near the boundary layer top, showing (a) forced convection, and (b) free convection. After Noonkester (1974).

Magnitudes of the shear production term in the surface layer are obviously greatest on a windy day, and are small on a calm day. In synoptic-scale cyclones the strong winds and overcast skies suggest that forced convection is applicable. On many days, turbulence is neither in a state of free nor forced convection because both the shear and buoyancy terms are contributing to the production of turbulence.

At night over land, or anytime the ground is colder than the air, the shear term is often the only term that generates turbulence. We have seen from Fig 5.4 that the shear term is active over just a relatively small depth of air, so it is not surprising that, over land, the NBL is usually thinner than the ML.

The greatest shears are associated with the change of U and V components of mean wind with height. Except in thunderstorms, shear of W is negligible in the BL. Looking back on (4.3.1h-j), the shear production is greatest into the x and y components of TKE. Hence, shear production is also an anisotropic forcing — strongest in the horizontal.

Both the buoyant and shear production terms can generate anisotropic turbulence. The difference is that shear generation produces turbulence primarily in the horizontal directions, while buoyant generation produces it primarily in the vertical. These differences are evident in Fig. 5.8, where a high powered vertically pointing continuous-wave radar was used to observed the time evolution of eddy structure within the BL. This instrument senses moisture contrasts between dry and moist air. The boundary between regions of different moisture appear white in the photographs, while regions of more uniform high or low humidity appear black. Taylor's hypothesis has been used to convert from time-height graphs to vertical cross sections.

In the one photograph made in free convection, the "inverted U-shaped" tops of thermals shows up as white because they separate the dry FA air from the moister ML air. These structures are predominantly vertical. In the other photograph made in forced convection, the eddies are sheared into a much more horizontal or slanting orientation, with a much more chaotic appearance.

5.2.5 Term V: Turbulent Transport

The quantity $\overline{w'e}$ represents the vertical turbulent flux of TKE. As for other vertical fluxes, the change in flux with height is more important than the magnitude of flux. Term V is a flux divergence term; if there is more flux into a layer than leaves, then the magnitude of TKE increases.

On a local scale (i.e., at any one height within the ML), Term V acts as either production or loss, depending on whether there is a flux convergence or divergence. When integrated over the depth of the ML, however, Term V becomes identically zero, assuming as bottom and top boundary conditions that the earth is not turbulent, and that there is negligible turbulence above the top of the ML. Overall, Term V neither creates nor destroys TKE, it just moves or redistributes TKE from one location in the BL to another.

Fig 5.9 shows vertical profiles of $\overline{w'e}$ for daytime, convective cases. Most of these

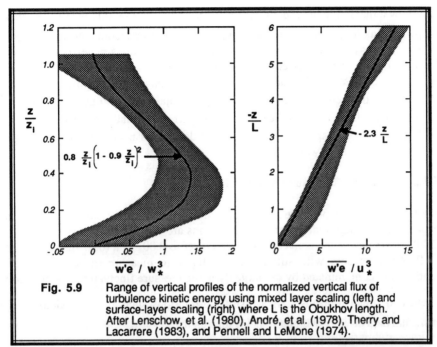

Fig. 5.9   Range of vertical profiles of the normalized vertical flux of turbulence kinetic energy using mixed layer scaling (left) and surface-layer scaling (right) where L is the Obukhov length. After Lenschow, et al. (1980), André, et al. (1978), Therry and Lacarrere (1983), and Pennell and LeMone (1974).

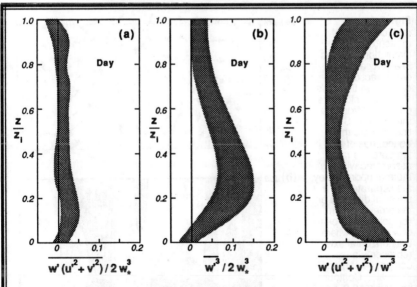

Fig. 5.10   (a) Range of vertical profiles of the normalized vertical flux of horizontal variance; (b), the vertical flux of vertical variance; (c), and the ratio of the two during daytime. After Lenschow, et al. (1980).

profiles show a maximum of $\overline{w'e}$ at $z/z_i = 0.3$ to 0.5. Below this maximum, there is more upward flux leaving the top of any one layer than enters from below, making a net divergence or loss of TKE. Above the maximum, there is a net convergence or production of TKE. The net effect is that some of the TKE produced near the ground is transported up to the top half of the ML before it is dissipated, as confirmed in Fig 5.4.

If one splits the vertical turbulent transport of total TKE into transport of w'^2 and $(u'^2+v'^2)$, then one finds that it is the vertical transport of w'^2 that dominates in the middle of the ML, and the transport of $(u'^2+v'^2)$ that dominates near the surface. Fig 5.10 shows these transports, as well as their ratio.

5.2.6 Term VI: Pressure Correlation

Turbulence. Static pressure fluctuations are exceedingly difficult to measure in the atmosphere. The magnitudes of these fluctuations are very small, being on the order of 0.005 kPa (0.05 mb) in the convective surface layer to 0.001 kPa (0.01 mb) or less in the ML. Pressure sensors with sufficient sensitivity to measure these static pressure fluctuations are contaminated by the large dynamic pressure fluctuations associated with

turbulent and mean motions. As a result, correlations such as $\overline{w'p'}$ calculated from experimental data often contain more noise than signal.

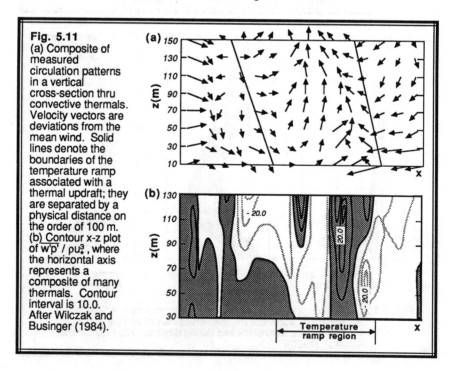

Fig. 5.11 (a) Composite of measured circulation patterns in a vertical cross-section thru convective thermals. Velocity vectors are deviations from the mean wind. Solid lines denote the boundaries of the temperature ramp associated with a thermal updraft; they are separated by a physical distance on the order of 100 m. (b) Contour x-z plot of $\overline{w'p'} / \rho u_*^3$, where the horizontal axis represents a composite of many thermals. Contour interval is 10.0. After Wilczak and Businger (1984).

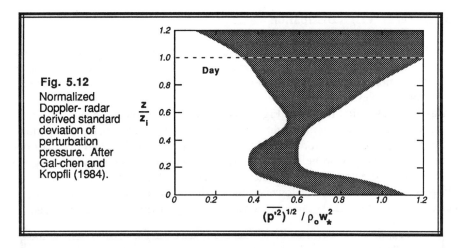

Fig. 5.12
Normalized
Doppler- radar
derived standard
deviation of
perturbation
pressure. After
Gal-chen and
Kropfli (1984).

What little is known about the behavior of pressure correlation terms is estimated as a residual in the budget equations discussed previously. Namely, if all of the other terms in a budget equation are measured or parameterized, then the residual necessary to make the equation balance includes an estimate of the unknown term(s) together with the accumulated errors. An obvious hazard of this approach is that the accumulated errors from all of the other terms can be quite large.

Estimates of $\overline{w'p'}$ in the surface layer are shown in Fig 5.11 using this method, composited with respect to a large number of convective plume structures. We see quite a variation both in the vertical and horizontal. Here, the plume is defined by its temperature ramp signal. Fig 5.12 shows estimates of pressure variance based on Doppler radar measurements of motion within the ML.

Waves. Recall from chapters 1 and 2 that perturbations from a mean can describe waves as well as turbulence. Given measured values of $\overline{w'p'}$, it is impossible to separate the wave and turbulence contributions without additional information.

Work in linear gravity wave theory shows that $\overline{w'p'}$ is equal to the upward flux of wave energy for a vertically propagating internal gravity wave within a statically stable environment. This suggests that turbulence energy can be lost from the ML top in the form of internal gravity waves being excited by thermals penetrating the stable layer at the top of the ML. The amount of energy lost may be on the order of less than 10% of the total rate of TKE dissipation, but the resulting waves can sometimes enhance or trigger clouds.

Turbulence within stable NBLs can also be lost in the form of waves. One concludes that the pressure correlation term not only acts to redistribute TKE within the BL, but it can also drain energy out of the BL.

5.2.7 Term VII: Dissipation

As discussed in section 4.3.1, molecular destruction of turbulent motions is greatest for the smallest size eddies. The more intense this small-scale turbulence, the greater the rate of dissipation. Small-scale turbulence is, in turn, driven by the cascade of energy from the larger scales.

Daytime dissipation rates (see Fig 5.13) are often largest near the surface, and then become relatively constant with height in the ML. Above the ML top, the dissipation rate rapidly decreases to near zero. At night (see Fig 5.14), both TKE and dissipation rate decrease very rapidly with height. Because turbulence is not conserved, the greatest TKEs, and hence greatest dissipation rates, are frequently found where TKE production is the largest — near the surface. However, the dissipation rate is not expected to perfectly balance the production rate because of the various transport terms in the TKE budget.

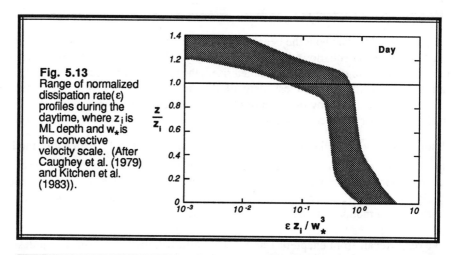

Fig. 5.13
Range of normalized dissipation rate(ϵ) profiles during the daytime, where z_i is ML depth and w_* is the convective velocity scale. (After Caughey et al. (1979) and Kitchen et al. (1983)).

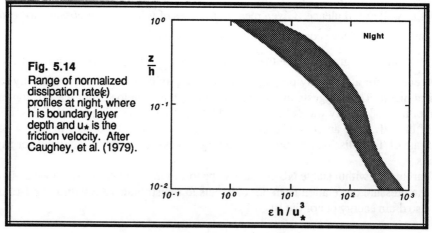

Fig. 5.14
Range of normalized dissipation rate(ϵ) profiles at night, where h is boundary layer depth and u_* is the friction velocity. After Caughey, et al. (1979).

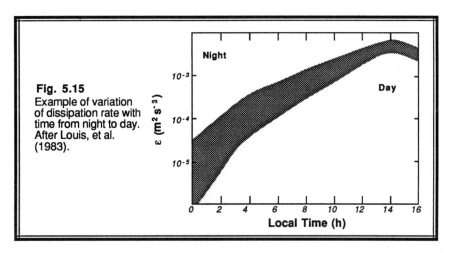

Fig. 5.15
Example of variation
of dissipation rate with
time from night to day.
After Louis, et al.
(1983).

The close relationship between TKE production rate, intensity of turbulence, and dissipation rate is shown in Fig 5.15. At night where only shear can produce turbulence, the dissipation rate is small because the associated TKE is small (refer back to Fig 5.2). After sunrise, buoyant production greatly increases the turbulence intensity, resulting in the associated increase in dissipation seen in Fig 5.15.

5.2.8 Example

Problem: At a height of $z = 300$ m in a 1000 m thick mixed layer the following conditions were observed: $\partial \overline{U}/\partial z = 0.01$ s^{-1}, $\overline{\theta}_v = 25°C$, $\overline{w'\theta_v'} = 0.15$ K m/s, and $\overline{u'w'} = -0.03$ m^{2}s^{-2}. Also, the surface virtual heat flux is 0.24 K m/s. If the pressure and turbulent transports are neglected, then (a) what dissipation rate is required to maintain a locally steady state at $z = 300$ m; and (b) what are the values of the normalized TKE terms?

Solution: (a) Since no information was given about the V-component of velocity or stress, let's assume that the x-axis has been chosen to be aligned with the mean wind. Looking at the TKE budget (5.1b), we know that term I must be zero for steady state, and terms V and VI are zero as specified in the statement of the problem. Thus, the remaining terms can be manipulated to solve for ε:

$$\varepsilon = \frac{g}{\overline{\theta}_v} \overline{w'\theta_v'} - \overline{u'w'} \frac{\partial \overline{U}}{\partial z}$$

Plugging in the values given above yields:

$$\varepsilon = \{(9.8 \text{ m·s}^{-2}) / [(273.15+25)\text{K}]\} \cdot (0.15 \text{ K·m·s}^{-1}) \quad - \quad (-0.03 \text{ m}^2\text{s}^{-2}) \cdot (0.01 \text{ s}^{-1})$$

$$\varepsilon = 4.93 \times 10^{-3} + 3 \times 10^{-4} \ (\text{m}^2\text{s}^{-3})$$

$$\varepsilon = 5.23 \times 10^{-3} \ (\text{m}^2\text{s}^{-3})$$

(b) To normalize the equations as in (5.2.3), we first use (4.2a) to give $w_*^3 / z_i =$ $(g/\overline{\theta}_v) \cdot \overline{w'\theta_v'}$, which for our case equals 7.89×10^{-3} $(\text{m}^2\text{s}^{-3})$. Dividing our terms by this value, and rewriting in the same order as (5.2.3) yields:

	0	=	0.625	+	0.038	-	0	-	0	-	0.663
Term:	I		III		IV		V		VI		VII

Discussion: This buoyant production term is about an order of magnitude larger than the mechanical production term, meaning that the turbulence is in a state of free convection. In regions of strong turbulence production, the transport term usually removes some of the TKE and deposits it where there is a net loss of TKE, such as in the entrainment zone. Thus, we might expect that the local dissipation rate at z = 300 m is smaller than the value calculated above.

5.3 TKE Budget Contributions as a Function of Eddy Size

As will be shown in chapter 8, the TKE budget equation can be written in a spectral form where the the contributions of each term in (5.1) can be examined as a function of wavelength or eddy size. Fig 5.16b shows the following terms as a function of wavenumber: buoyant production (Term III), shear production (Term IV), and dissipation (Term VII), all measured at one height in the BL. The turbulent transport and pressure redistribution term calculations were inaccurate, and hence left out of these figures.

One additional term appears in the spectral form of the TKE equation: the transfer of energy across the spectrum. In this case, as in most atmospheric cases, the transfer is from large size eddies (low wavenumbers) to small sizes (high wavenumbers). The concept behind this cascade of energy was introduced in Chapter 2. The rate of flow of this energy, shown in Fig 5.16a, is greatest for middle size eddies. Not only is it largest there, but it is also relatively constant with wavenumber. Hence, there is no net divergence or convergence of energy in the middle of the spectral domain, but there is a large amount of energy flowing through that domain. The slope of the curves in Fig 5.16a determines the magnitude of the transport term in curves b&d.

Large size eddies are presented on the left side of these figures, and small on the right. We see in Fig 5.16b that there is little energy at the very largest sizes, corresponding to the spectral gap. Once we get down to a normalized wavenumber of 0.01, we see large magnitudes of the shear and buoyant production terms. The production is not dissipated at these sizes however. Instead, there is the cascade or transport of energy away from the large size eddies towards the smaller sizes where it is deposited. At the small-eddy end of the spectrum (large wavenumber of 100-1000), the production terms are near zero. Instead, dissipation is large.

One measure of the smallest scales of turbulence is the Kolmogorov microscale, η, given by: $\eta = (v^3/\varepsilon)^{1/4}$. This scaling assumes that the smallest eddies see only turbulent energy cascading down the spectrum at rate ε, and feel only the viscous damping of v. For the example of Fig 5.16, $\eta \cong 1$ mm, which occurs at a normalized frequency of about 3000.

The nature of the atmospheric turbulence spectrum is directly related to the fact that production and dissipation are not happening at the same scales. Production is feeding only the larger size eddies (anisotropically, as we learned earlier), but dissipation is acting only on the smaller sizes. Thus, the rate of transport across the middle part of the spectrum is equal to the rate of dissipation, ε, at the small-eddy end. Such transfer can be thought of as happening inertially — larger eddies creating or bumping into smaller ones, and transferring some of their inertia in the process. This middle portion of the spectrum is called the ***inertial subrange***.

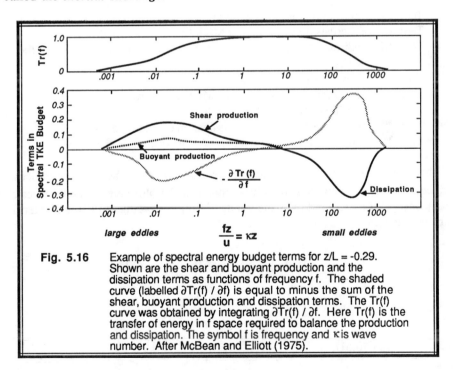

Fig. 5.16   Example of spectral energy budget terms for z/L = -0.29. Shown are the shear and buoyant production and the dissipation terms as functions of frequency f. The shaded curve (labelled $\partial Tr(f) / \partial f$) is equal to minus the sum of the shear, buoyant production and dissipation terms. The Tr(f) curve was obtained by integrating $\partial Tr(f) / \partial f$. Here Tr(f) is the transfer of energy in f space required to balance the production and dissipation. The symbol f is frequency and κ is wave number. After McBean and Elliott (1975).

5.4 Mean Kinetic Energy and Its Interaction with Turbulence

Term IV in the TKE budget (5.1) involves the production of TKE by interaction of turbulence with the mean wind. One might expect that the production of TKE is accompanied by a corresponding loss of kinetic energy from the mean flow.

To study that possibility, start with the prognostic equation for mean wind in turbulent flow (3.4.3c), multiply by $\overline{U}_i$, and use the chain rule to derive the following equation for mean kinetic energy per unit mass [MKE/m $= 0.5(\overline{U}^2 + \overline{V}^2 + \overline{W}^2) = 0.5\,\overline{U}_i{}^2$]:

$$\frac{\partial\left(0.5\overline{U}_i{}^2\right)}{\partial t} + \overline{U}_j\,\frac{\partial\left(0.5\,\overline{U}_i{}^2\right)}{\partial x_j} = -g\delta_{i3}\overline{U}_i + f_c\varepsilon_{ij3}\overline{U}_i\,\overline{U}_j - \frac{\overline{U}_i}{\overline{\rho}}\,\frac{\partial\overline{P}}{\partial x_i} + \nu\overline{U}_i\,\frac{\partial^2\overline{U}_i}{\partial x_j^2} - \overline{U}_i\,\frac{\partial\left(\overline{u_i'u_j'}\right)}{\partial x_j}$$

$$\text{(5.4a)}$$

| I | II | III | IV | V | VI | X |

Term I   represents storage of MKE.
Term II   describes the advection of MKE by the mean wind.
Term III   indicates that gravitational acceleration of vertical motions alter the MKE.
Term IV   shows the effects of the Coriolis force.
Term V   represents the production of MKE when pressure gradients accelerate the mean flow.
Term VI   represents the molecular dissipation of mean motions.
Term X   indicates the interaction between the mean flow and turbulence.

When the Coriolis term (IV) is summed over all values of the repeated indices, the result equals zero. This confirms our observation that Coriolis force can neither create nor destroy energy; it merely redirects the winds. Using the product rule, the last term (X) can be rewritten as

$$-\overline{U}_i\,\frac{\partial\left(\overline{u_i'u_j'}\right)}{\partial x_j} = \overline{u_i'u_j'}\,\frac{\partial\overline{U}_i}{\partial x_j} - \frac{\partial\left(\overline{u_i'u_j'}\,\overline{U}_i\right)}{\partial x_j}$$

This leaves

$$\frac{\partial\left(0.5\overline{U}_i{}^2\right)}{\partial t} + \overline{U}_j\,\frac{\partial\left(0.5\overline{U}_i{}^2\right)}{\partial x_j} = -g\overline{W} - \frac{\overline{U}_i}{\overline{\rho}}\,\frac{\partial\overline{P}}{\partial x_i} + \nu\overline{U}_i\,\frac{\partial^2\overline{U}_i}{\partial x_j^2} + \overline{u_i'u_j'}\,\frac{\partial\overline{U}_i}{\partial x_j} - \frac{\partial\left(\overline{u_i'u_j'}\,\overline{U}_i\right)}{\partial x_j}$$

$$\text{(5.4b)}$$

If we compare the TKE equation (5.1) with the MKE equation (5.4b):

$$\frac{\partial(\text{TKE/m})}{\partial t} = \dots - \overline{u_i'u_j'}\frac{\partial \overline{U_i}}{\partial x_j}$$

$$\frac{\partial(\text{MKE/m})}{\partial t} = \dots + \overline{u_i'u_j'}\frac{\partial \overline{U_i}}{\partial x_j}$$

we see that they both contain a term describing the interaction between the mean flow and turbulence. The sign of these terms differ. *Thus, the energy that is mechanically produced as turbulence is lost from the mean flow*, and vice versa.

5.5 Stability Concepts

Unstable flows become or remain turbulent. Stable flows become or remain laminar. There are many factors that can cause laminar flow to become turbulent, and other factors that tend to stabilize flows. If the the net effect of all the destabilizing factors exceeds the net effect of the stabilizing factors, then turbulence will occur. In many cases, these factors can be interpreted as terms in the TKE budget equation.

To simplify the problem, investigators have historically paired one destabilizing factor with one stabilizing factor, and expressed these factors as a dimensionless ratio. Examples of these ratios are the Reynolds number, Richardson number, Rossby number, Froude number, and Rayleigh number. Some other stability parameters such as static stability, however, are not expressed in dimensionless form.

5.5.1 Static Stability and Convection

Static stability is a measure of the capability for buoyant convection. The word "static" means "having no motion"; hence this type of stability does not depend on wind. Air is statically unstable when less-dense air (warmer and/or moister) underlies more-dense air. The flow responds to this instability by supporting convective circulations such as thermals that allow buoyant air to rise to the top of the unstable layer, thereby stabilizing the fluid. Thermals also need some trigger mechanism to get them started. In the real boundary layer, there are so many triggers (hills, buildings, trees, dark fields, or other perturbations to the mean flow) that convection is usually insured, given the static instability.

Local Definitions. The traditional definition taught in basic meteorology classes is local in nature; namely, the static stability is determined by the local lapse rate. The local definition frequently fails in convective MLs, because the rise of thermals from near the surface or their descent from cloud top depends on their excess buoyancy and not on the ambient lapse rate.

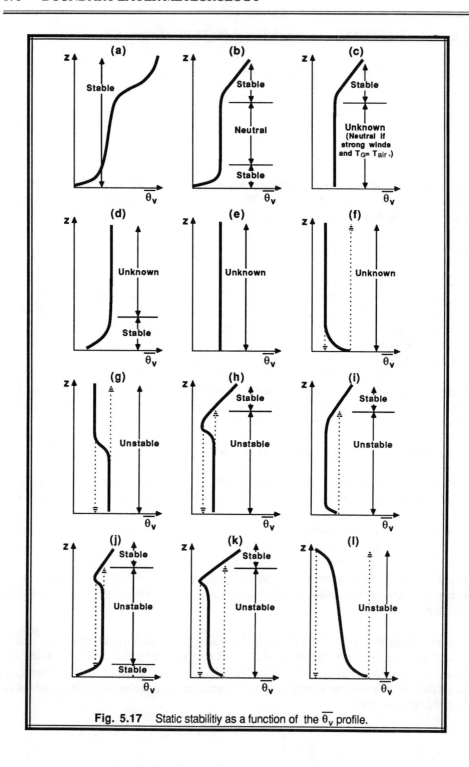

Fig. 5.17   Static stabilitiy as a function of the $\overline{\theta_v}$ profile.

As an example, in the middle 50% of the convective ML the lapse rate is nearly adiabatic, causing an incorrect classification of neutral stability if the traditional local definition is used. We must make a clear distinction between the phrases "adiabatic lapse rate" and "neutral stability".   An *adiabatic lapse rate* (in the virtual potential temperature sense) may be statically stable, neutral, or unstable, depending on convection and the buoyancy flux. *Neutral stability* implies a very specific situation: adiabatic lapse rate AND no convection.   The two phrases should NOT be used interchangeably, and the phrase "neutral lapse rate" should be avoided altogether.

We   conclude   that *measurement of the local lapse rate alone is INSUFFICIENT to determine the static stability*.   Either knowledge of the whole $\overline{\theta}_v$ profile is needed (described next), or measurement of the turbulent buoyancy flux must be made.

Nonlocal Definitions.   It is better to examine the stability of the whole layer, and make a layer determination of stability such as was done in section 1.6.4. For example, if $\overline{w'\theta_v'}$ at the earth's surface is positive, or if displaced air parcels will rise from the ground or sink from cloud top as thermals traveling across a BL, then the whole BL is said to be *unstable* or *convective*. If $\overline{w'\theta_v'}$ is negative at the surface, or if displaced air parcels return to their starting point, then the BL is said to be *stable*.

If, when integrated over the depth of the boundary layer, the mechanical production term in the TKE equation (5.1) is much larger than the buoyancy term, or if the buoyancy term is near zero, then the boundary layer is said to be *neutral*. In some of the older literature, the boundary layer of this latter case is also sometimes referred to as an *Ekman boundary layer*. During fair weather conditions over land, the BL touching the ground is rarely neutral. Neutral conditions are frequently found in the RL aloft. In overcast conditions with strong winds but little temperature difference between the air and the surface, the BL is often close to neutral stability.

In the absence of knowledge of convection or measurements of buoyancy flux, an alternate determination of static stability is possible if the $\overline{\theta}_v$ profile over the whole BL is known, as sketched in Fig 5.17. As is indicated in the figure, if only portions of the profile are known, then the stability might be indeterminate. Also, it is clear that there are many situations where the traditional local definition fails.

5.5.2 Example

Problem. Given the sounding at right, identify the static stability of the air at z = 600 m.

z (m)	$\overline{\theta}_v$(K)
1000	298
800	299
600	299
400	299
200	298
0	295

Solution. Using a local definition in the absence of heat fluxes, if we look downward from 600 m until a diabatic layer is encountered, we find a stable layer with cooler temperatures at 200 m. Before we reach any hasty conclusions, however, we must look up from 600 m. Doing so we find cooler unstable air at 1000 m. Thus, the static stability is <u>unstable</u> at 600 m.

Discussion. The whole adiabatic layer is unstable, considering the nonlocal approach of a cool parcel sinking from above. This sounding is characteristic of stratocumulus.

5.5.3 Dynamic Stability and Kelvin-Helmholtz Waves

The word "dynamic" refers to motion; hence, dynamic stability depends in part on the winds. Even if the air is statically stable, wind shears may be able to generate turbulence dynamically.

Some laboratory experiments have been performed (Thorpe, 1969, 1973; Woods 1969) using denser fluids underlying less-dense fluids with a velocity shear between the layers to simulate the stable stratification and shears of the atmosphere. Fig 5.18 is a sketch of the resulting flow behavior. The typical sequence of events is:

(1) A shear exists across a density interface. Initially, the flow is laminar.

(2) If a critical value of shear is reached (see section 5.6), then the flow becomes dynamically unstable, and gentle waves begin to form on the interface. The crests of these waves are normal to the shear direction

(3) These waves continue to grow in amplitude, eventually reaching a point where each wave begins to "roll up" or "break". This "breaking" wave is called a *Kelvin-Helmholtz (KH) wave*, and is based on different physics than surface waves that "break" on an ocean beach.

(4) Within each wave, there exists some lighter fluid that has been rolled under denser fluid, resulting in patches of static instability. On radar, these features appear as braided ropelike patterns, "cat's eye" patterns or breaking wave patterns.

(5) The static instability, combined with the continued dynamic instability, causes each wave to become turbulent.

(6) The turbulence then spreads throughout the layer, causing a diffusion or mixing of the different fluids. During this diffusion process, some momentum is transferred between the fluids, reducing the shear between the layers. What was formerly a sharp, well-defined, interface becomes a broader, more diffuse shear layer with weaker shear and static stability.

(7) This mixing can reduce the shear below a critical value and eliminate the dynamic instability.

(8) In the absence of continued forcing to restore the shears, turbulence decays in the interface region, and the flow becomes laminar again.

This sequence of events is suspected to occur during the onset of *clear air turbulence (CAT)*. These often occur above and below strong wind jets, such as the nocturnal jet and the planetary-scale jet stream. In these situations, however, continued dynamic forcings can allow turbulence to continue for hours to days. These regions of CAT have large horizontal extent (hundreds of kilometers in some cases), but usually limited vertical extent (tens to hundred of meters). They can be visualized as large pancake-shaped regions of turbulence. Aircraft encountering CAT can often climb or descend into smoother air.

Although KH waves are probably a frequent occurrence within statically stable shear layers, they are only rarely observed with the naked eye. Occasionally, there is sufficient moisture in the atmosphere to allow cloud droplets to act as visible tracers. Clouds that form in the rising portions of the waves often form parallel bands called *billow clouds*. The orientation of these bands is perpendicular to the shear vector. One must remember that the wind SHEAR vector need not necessarily point in the same direction as the mean wind vector.

For both static and dynamic instabilities, and many other instabilities for that matter, it is interesting to note that the fluid reacts in a manner to undo the cause of the instability. This process is strikingly similar to *LeChatelier's principle* of chemistry, which states that "if some stress is brought to bear upon a system in equilibrium, a change occurs such that the equilibrium is displaced in a direction which tends to undo the effect of the stress". Thus, turbulence is a mechanism whereby fluid flows tend to undo the cause of the instability. In the case of static instabilities, convection occurs that tends to move more buoyant fluid upward, thereby stabilizing the system. For dynamic instability, turbulence tends to reduce the wind shears, also stabilizing the system.

With this in mind, it is apparent that turbulence acts to eliminate itself. After the unstable system has been stabilized, turbulence tends to decay. Given observations of turbulence occurring for long periods of time within the boundary layer, it is logical to surmise that there must be external forcings tending to destabilize the BL over long time periods. In the case of static instability, the solar heating of the ground by the sun is that external forcing. In the case of dynamic instabilities, pressure gradients imposed by synoptic-scale features drive the winds against the dissipative effects of turbulence.

By comparing the relative magnitudes of the shear production and buoyant consumption terms of the TKE equation, we can hope to estimate when the flow might become dynamically unstable. The Richardson number, Ri, described in the next subsection, can be used as just such an indicator.

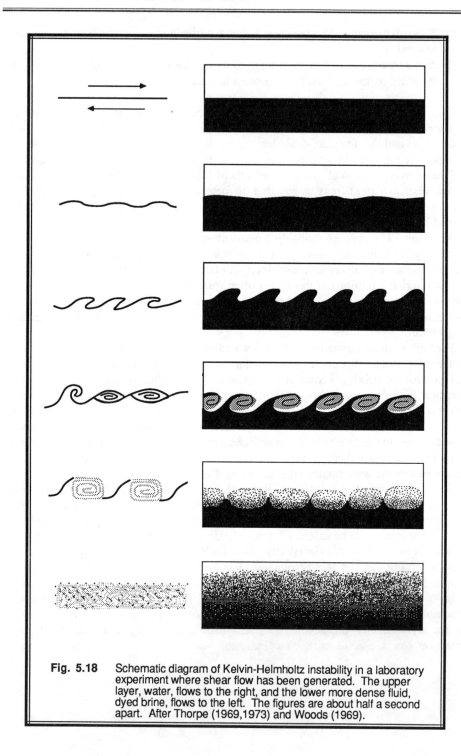

Fig. 5.18   Schematic diagram of Kelvin-Helmholtz instability in a laboratory experiment where shear flow has been generated. The upper layer, water, flows to the right, and the lower more dense fluid, dyed brine, flows to the left. The figures are about half a second apart. After Thorpe (1969,1973) and Woods (1969).

5.6 The Richardson Number

5.6.1 Flux Richardson Number

In a statically stable environment, turbulent vertical motions are acting against the restoring force of gravity. Thus, buoyancy tends to suppress turbulence, while wind shears tend to generate turbulence mechanically. The buoyant production term (Term III) of the TKE budget equation (5.1b) is negative in this situation, while the mechanical production term (Term IV) is positive. Although the other terms in the TKE budget are certainly important, a simplified but nevertheless useful approximation to the physics is possible by examining the ratio of Term III to Term IV.   This ratio, called the *flux Richardson number*, R_f, is given by

$$R_f = \frac{\left(\dfrac{g}{\overline{\theta}_v}\right) \overline{(w'\theta_v')}}{\overline{(u_i'u_j')} \dfrac{\partial \overline{U}_i}{\partial x_j}} \tag{5.6.1a}$$

where the negative sign on Term IV is dropped by convention. The Richardson number is dimensionless. The denominator consists of 9 terms, as implied by the summation notation.

If we assume horizontal homogeneity and neglect subsidence, then the above equation reduces to the more common form of the flux Richardson number:

$$R_f = \frac{\left(\dfrac{g}{\overline{\theta}_v}\right) \overline{(w'\theta_v')}}{\overline{(u'w')} \dfrac{\partial \overline{U}}{\partial z} + \overline{(v'w')} \dfrac{\partial \overline{V}}{\partial z}} \tag{5.6.1b}$$

For statically unstable flows, R_f is usually negative (remember that the denominator is usually negative). For neutral flows, it is zero. For statically stable flows, R_f is positive.

Richardson proposed that $R_f = +1$ is a critical value, because the mechanical production rate balances the buoyant consumption of TKE. At any value of R_f less than +1, static stability is insufficiently strong to prevent the mechanical generation of turbulence. For negative values of R_f, the numerator even contributes to the generation of turbulence. Therefore, he expected that

Flow IS turbulent (dynamically unstable) when $R_f < +1$
Flow BECOMES laminar (dynamically stable) when $R_f > +1$
We recognize that statically unstable flow is, by definition, always dynamically unstable.

5.6.2 Gradient Richardson Number

A peculiar problem arises in the use of R_f ; namely, we can calculate its value only for turbulent flow because it contains factors involving turbulent correlations like $\overline{w'\theta_v'}$. In other words, we can use it to determine whether turbulent flow will become laminar, but not whether laminar flow will become turbulent.

Using the reasoning of section 2.7 and Fig 2.13, it is logical to suggest that the value of the turbulent correlation $-\overline{w'\theta_v'}$ might be proportional to the lapse rate $\partial\overline{\theta_v}/\partial z$.

Similarly, we might suggest that $-\overline{u'w'}$ is proportional to $\partial\overline{U}/\partial z$, and that $-\overline{v'w'}$ is proportional to $\partial\overline{V}/\partial z$. These arguments form the basis of a theory known as K-theory or eddy diffusivity theory, which will be discussed in much more detail in chapter 6. For now, we will just assume that the proportionalities are possible, and substitute those in (5.6.1b) to give a new ratio called the *gradient Richardson number*, Ri :

$$Ri = \frac{\dfrac{g}{\overline{\theta_v}}\dfrac{\partial\overline{\theta_v}}{\partial z}}{\left[\left(\dfrac{\partial\overline{U}}{\partial z}\right)^2 + \left(\dfrac{\partial\overline{V}}{\partial z}\right)^2\right]} \qquad (5.6.2)$$

When investigators refer to a "Richardson number" without specifying which one, they usually mean the gradient Richardson number.

Theoretical and laboratory research suggest that laminar flow becomes unstable to KH-wave formation and the ONSET of turbulence when Ri is smaller than the *critical Richardson number*, R_c . Another value, R_T, indicates the termination of turbulence. The dynamic stability criteria can be stated as follows:

Laminar flow becomes turbulent when $Ri < R_c$.

Turbulent flow becomes laminar when $Ri > R_T$.

Although there is still some debate on the correct values of R_c and R_T, it appears that $R_c = 0.21$ to 0.25 and $R_T = 1.0$ work fairly well. Thus, there appears to be a *hysteresis* effect because R_T is greater than R_c.

One hypothesis for the apparent hysteresis is as follows. Two conditions are needed for turbulence: instability, and some trigger mechanism. Suppose that dynamic instability occurs whenever $Ri < R_T$. If one trigger mechanism is existing turbulence in or adjacent to the unstable fluid, then turbulence can continue as long as $Ri < R_T$ because of the presence of both the instability and the trigger. If KH waves are another trigger mechanism, then in the absence of existing turbulence one finds that Ri must get well below R_T before KH waves can form. Laboratory and theoretical work have shown that the criterion for KH wave formation is Ri < R_c. This leads to the apparent hysteresis, because the Richardson number of nonturbulent flow must be lowered to R_c before turbulence will start, but once turbulent, the turbulence can continue until the Richardson number is raised above R_T.

5.6.3 Bulk Richardson Number

The theoretical work yielding $R_c \cong 0.25$ is based on local measurements of the wind shear and temperature gradient. Meteorologists rarely know the actual local gradients, but can approximate the gradients using observations made at a series of discrete height intervals. If we approximate $\partial \overline{\theta}_v / \partial z$ by $\Delta \overline{\theta}_v / \Delta z$, and approximate $\partial \overline{U} / \partial z$ and $\partial \overline{V} / \partial z$ by $\Delta \overline{U} / \Delta z$ and $\Delta \overline{V} / \Delta z$ respectively, then we can define a new ratio known as the *bulk Richardson number*, R_B :

$$R_B = \frac{g \; \Delta \overline{\theta}_v \; \Delta z}{\overline{\theta}_v \; [(\Delta \overline{U})^2 + (\Delta \overline{V})^2]} \qquad (5.6.3)$$

It is this form of the Richardson number that is used most frequently in meteorology, because rawinsonde data and numerical weather forecasts supply wind and temperature measurements at discrete points in space. The sign of the finite differences are defined, for example, by $\Delta \overline{U} = \overline{U}(z_{top}) - \overline{U}(z_{bottom})$.

Unfortunately, the critical value of 0.25 applies only for local gradients, not for finite differences across thick layers. In fact, the thicker the layer is, the more likely we are to average out large gradients that occur within small subregions of the layer of interest. The net result is (1) we introduce uncertainty into our prediction of the occurrence of turbulence, and (2) we must use an artificially large (theoretically unjustified) value of the critical Richardson number that gives reasonable results using our smoothed gradients. The thinner the layer, the closer the critical Richardson number will likely be to 0.25. Since data points in soundings are sometimes spaced far apart in the vertical, approximations such as shown in the graph and table in Fig 5.19 can be used to estimate the probability and intensity of turbulence (Lee, et al., 1979).

Table 5-1 shows a portion of a rawinsonde sounding, together with the corresponding values of bulk Richardson number. The resulting turbulence diagnosis is given in the rightmost column of Table 5-1. Note that the Richardson number itself says nothing about the intensity of turbulence, only about the yes/no presence of turbulence.

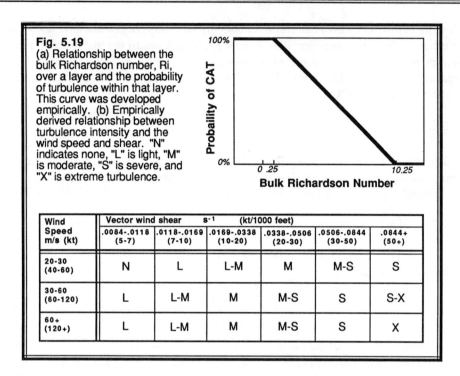

Fig. 5.19 (a) Relationship between the bulk Richardson number, Ri, over a layer and the probability of turbulence within that layer. This curve was developed empirically. (b) Empirically derived relationship between turbulence intensity and the wind speed and shear. "N" indicates none, "L" is light, "M" is moderate, "S" is severe, and "X" is extreme turbulence.

Wind Speed m/s (kt)	Vector wind shear s⁻¹ (kt/1000 feet)					
	.0084-.0118 (5-7)	.0118-.0169 (7-10)	.0169-.0338 (10-20)	.0338-.0506 (20-30)	.0506-.0844 (30-50)	.0844+ (50+)
20-30 (40-60)	N	L	L-M	M	M-S	S
30-60 (60-120)	L	L-M	M	M-S	S	S-X
60+ (120+)	L	L-M	M	M-S	S	X

Fig 5.20 show examples of the evolution of the Richardson number during some nighttime case studies. Regions where the Richardson number is small are sometimes used as an indicator of the depth of the turbulent SBL. Here we see low Richardson numbers close to the ground, in addition to patches of low Richardson number aloft.

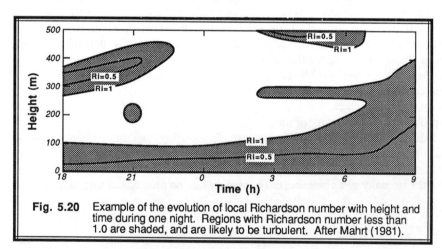

Fig. 5.20 Example of the evolution of local Richardson number with height and time during one night. Regions with Richardson number less than 1.0 are shaded, and are likely to be turbulent. After Mahrt (1981).

Table 5-1.   Example of a nighttime rawinsonde sounding analyzed to give stability, shear, Richardson number, and the probability and intensity of turbulence. Probabilities are expressed as a percent, and intensities are abbreviated by:

   N = no turbulence,   L = light (0.5 G),   M = moderate (1 G),   S=severe (2 G)

These intensity levels correspond to the turbulence reporting recommendations used in aviation, where the vertical acceleration measured in Gs (number of times the pull of gravity) is relative to the center of gravity of the aircraft. For practical purposes, a probability greater than 50% AND an intensity greater than L were required before a CAT forecast would be issued.

z (m)	Wind Dir (°)	Speed (m/s)	T (K)	θ (K)	Lapse (K/m)	Shear (s^{-1})	R_B	CAT Prob(%)	CAT Inten.
1591	154	9.8	281	294.4	0.0021	0.0034	6.19	41	N
1219	150	10.7	-	-	0.0021	0.0045	3.43	68	N
914	144	9.7	-	-	0.0021	0.0091	0.86	94	N-L
702	-	-	287.8	292.5	0.0020	0.0091	0.81	94	N-L
610	134	7.4	-	-	0.0020	0.0170	0.23	100	L-M
393	-	-	290.2	291.9	0.0204	0.0170	2.37	79	L-M
305	95	3.5	-	-	0.0204	0.0137	3.64	66	N
222	79	2.7	288.4	288.4	0.0133	0.0071	8.92	13	N
4	45	2.5	287.6	285.5	-	-	-	-	-

5.6.4 Examples

Problem A: Given the same data from problem 5.2.8, calculate the flux Richardson number and comment on the dynamic stability.

Solution. Since the flux Richardson number is defined as the ratio of the buoyancy term to the negative of the shear term, we can use the values for these terms already calculated in example 5.2.8:

$$R_f = \frac{\text{buoyancy term}}{-\text{ shear term}} = \frac{0.00493}{-0.0003} = -16.4$$

Discussion. A negative Richardson number is without question less than +1, and thus indicates dynamic instability and turbulence. This is a trivial conclusion, because any flow that is statically unstable is also dynamically unstable by definition.

Problem B: Given a fictitious SBL where $(g/\overline{\theta_v}) = 0.033$, $\partial \overline{U}/\partial z = u_* / (kz)$, $u_* = 0.4$, and where the lapse rate, c_1, is constant with height such that there is 6°C $\overline{\theta_v}$ increase with each 200 m of altitude gained. How deep is the turbulence?

Solution. We can use the gradient Richardson number as an indicator of dynamic stability and turbulence. Using the prescribed gradients, we find that:

$$Ri = \frac{\dfrac{g}{\overline{\theta_v}} \dfrac{\partial \overline{\theta_v}}{\partial z}}{\left(\dfrac{\partial \overline{U}}{\partial z}\right)^2} = \frac{\dfrac{g}{\overline{\theta_v}} c_1}{\left(\dfrac{u_*}{kz}\right)^2} = \frac{(0.033)\cdot(0.03)}{(0.4/0.4)^2} z^2 = (0.00099 \text{ m}^{-2}) z^2$$

If we use $R_c = 0.25$, then we can use this critical value in place of Ri above and solve for z at the critical height above which there is no turbulence:

$$z = \sqrt{(1010 \text{ m}^2) R_c} = \sqrt{252.5 \text{ m}^2} = 15.9 \text{ m}$$

Discussion. If we has used a critical termination value of $R_T = 1.0$, then we would have found a critical height of 31.8 m. Thus, below 15.9 m we expect turbulence, while above 31.8 m we expect laminar flow. Between these heights the turbulent state depends on the past history of the flow at that height. If previously turbulent, it is turbulent now.

5.7 The Obukhov Length

The Obukhov length (L) is a scaling parameter that is useful in the surface layer. To show how this parameter is related to the TKE equation, first recall that one definition of the surface layer is that region where turbulent fluxes vary by less that 10% of their magnitude with height. By making the constant flux (with height) approximation, one can use surface values of heat and momentum flux to define turbulence scales and nondimensionalize the TKE equation.

Start with the TKE equation (5.1a), multiply the whole equation by $(-k z/u_*^3)$, assume all turbulent fluxes equal their respective surface values, and focus on just terms III, IV, and VII:

$$\ldots = -\frac{k z g \, \overline{(w'\theta_v')}_s}{\overline{\theta_v} u_*^3} + \frac{k z \, \overline{(u_i'u_j')}_s}{u_*^3} + \ldots - \frac{k z \, \varepsilon|_s}{u_*^3} \qquad (5.7a)$$

$$\underset{\text{III}}{} \qquad\qquad \underset{\text{IV}}{} \qquad\qquad \underset{\text{VII}}{}$$

Each of these terms is now dimensionless.   The last term, a dimensionless dissipation rate, will not be pursued further here.

The **von Karman constant**, k, is a dimensionless number included by tradition. Its importance in the log wind profile in the surface layer is discussed in the next section. Investigators have yet to pin down its precise value, although preliminary experiments suggest that it is between about 0.35 and 0.42 . We will use a value of 0.4 in most of this book, although some of the figures adopted from the literature are based on k=0.35.

Term III is usually assigned the symbol, ζ, and is further defined as $\zeta \equiv z/L$, where L is the **Obukhov length**. Thus,

$$\zeta = \frac{z}{L} = \frac{-k\,z\,g\,\overline{(w'\theta_v')}_s}{\overline{\theta_v}\,u_*^3} \tag{5.7b}$$

The Obukhov length is given by:

$$L = \frac{-\,\overline{\theta_v}\,u_*^3}{k\,g\,\overline{(w'\theta_v')}_s} \tag{5.7c}$$

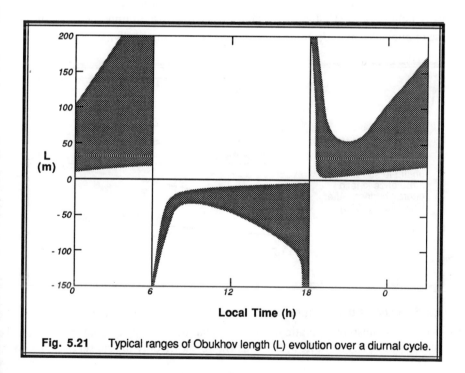

Fig. 5.21   Typical ranges of Obukhov length (L) evolution over a diurnal cycle.

One physical interpretation of the Obukhov length is that it is proportional to the height above the surface at which buoyant factors first dominate over mechanical (shear) production of turbulence. For convective situations, buoyant and shear production terms are approximately equal at $z = -0.5\,L$. Fig 5.21 shows the typical range of variations of the Obukhov length in fair weather conditions over land. Additional tables of values of these scaling parameters, extracted from field experiment data books, are presented in Appendix E.

The parameter ζ turns out to be very important for scaling and similarity arguments of the surface layer, as will be discussed in more detail in a later chapter. It is sometimes called a stability parameter, although its magnitude is not directly related to static nor dynamic stability. Only its sign relates to static stability: negative implies unstable, positive implies statically stable. A better description of ζ is "a surface-layer scaling parameter".

We can write an alternative form for ζ by employing the definition of w_*:

$$\zeta = \frac{z}{L} = -\frac{k\,z\,w_*^3}{z_i\,u_*^3} \qquad (5.7d)$$

Fig. 5.22 shows the variation of TKE budget terms with ζ, as ζ varies between 0 (statically neutral) and -1 (slightly unstable). The decrease in importance of shear and increase of buoyancy as ζ decreases from 0 to -1 is particularly obvious.

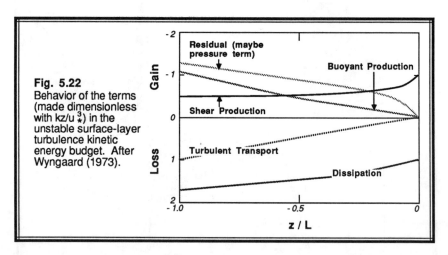

Fig. 5.22 Behavior of the terms (made dimensionless with kz/u_*^3) in the unstable surface-layer turbulence kinetic energy budget. After Wyngaard (1973).

Figs. 5.23 shows the variation of Ri with ζ from slightly unstable to slightly stable conditions. For unstable situations, $Ri \cong \zeta$. One must keep in mind that ζ can be calculated only for turbulent flow, thus this figure shows only the subset of all data that was turbulent. Nonturbulent flow can occur in stable situations, but it does not appear in this figure.

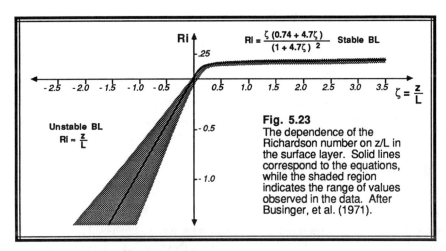

Fig. 5.23
The dependence of the Richardson number on z/L in the surface layer. Solid lines correspond to the equations, while the shaded region indicates the range of values observed in the data. After Businger, et al. (1971).

5.8 Dimensionless Gradients

We can simplify term IV of the dimensionless TKE equation (5.7a) by choosing a coordinate system aligned with the mean wind, assuming horizontal homogeneity, neglecting subsidence, and using the definition that $u_*^2 = -\overline{(u'w')}_s$:

$$\text{Term IV} = \frac{-k\,z}{u_*}\frac{\partial \overline{U}}{\partial z}$$

Based on this dimensionless term, we can define a *dimensionless wind shear*, ϕ_M, by

$$\phi_M = \frac{k\,z}{u_*}\frac{\partial \overline{U}}{\partial z} \tag{5.8a}$$

This parameter is primarily useful for studies of surface-layer wind profiles and momentum fluxes. In chapter 9 we will use ϕ_M in similarity theory to estimate momentum flux (as given by u_*) from the local mean wind shear. This is particularly valuable because it is easy to measure mean wind speeds at a variety of heights in the surface layer, but much more difficult and expensive to measure the eddy correlations such as $\overline{u'w'}$.

By analogy, a *dimensionless lapse rate*, ϕ_H, and a *dimensionless humidity gradient*, ϕ_E , can be defined:

$$\phi_H = \frac{k\,z}{\theta_*^{SL}}\frac{\partial\overline{\theta}}{\partial z} \tag{5.8b}$$

$$\phi_E = \frac{k\,z}{q_*^{SL}}\frac{\partial\overline{q}}{\partial z} \tag{5.8c}$$

These dimensionless gradients are equally as valuable as the dimensionless shear, because using similarity theory we can estimate the surface layer heat flux and moisture flux from simple measurements of lapse rate and moisture gradient, respectively.

5.9 Miscellaneous Scaling Parameters

5.9.1 Definitions

A few additional dimensionless scaling groups have been suggested in the literature to help explain boundary layer characteristics. Again, these are often inappropriately called stability parameters. One parameter that is useful in the surface layer is:

$$\mu^{SL} = \frac{k\,u_*}{f_c\,L} \tag{5.9.1a}$$

$$= \frac{g\,k^2\overline{(w'\theta_v')}_s}{\overline{\theta_v}\,f_c\overline{(u'w')}_s} \tag{5.9.1b}$$

$$= \frac{g\,k^2\theta_*^{SL}}{\overline{\theta_v}\,f_c\,u_*} \tag{5.9.1c}$$

Another scaling parameter that is useful in the ML is

$$\mu^{ML} = k\,\frac{z_i}{L} \tag{5.9.1d}$$

$$= \frac{-k^2\left(\dfrac{g}{\overline{\theta_v}}\right)\overline{(w'\theta_v')}_s}{u_*^3} \tag{5.9.1e}$$

$$= -k^2 \frac{w_*^3}{u_*^3} \qquad (5.9.1f)$$

It's important not to confuse either of these two parameters with the dynamic viscosity, which traditionally uses the same symbol.

Another parameter occasionally used is:

$$S_G = \frac{\overline{\theta}_s - \overline{\theta}_{air}}{\overline{M}^2 \left[1 + \log\left(\frac{10}{z} \right) \right]^2} \qquad (5.9.1g)$$

which looks like a modified Richardson number. Additional scaling parameters and dimensionless groups will be introduced in later chapters where appropriate.

5.9.2 Example

Problem: Given surface measurements: $u_* = 0.2$ m·s^{-1}, $g/\overline{\theta}_v = 0.0333$ m·s^{-1}K^{-1},

and $\overline{w'\theta_v'} = -0.05$ K·m·s^{-1};   and at 10 m: $\partial\overline{U}/\partial z = 20$ m·s^{-1}/100m, and $\partial\overline{\theta}_v/\partial z = 20$ °C/100m. Find scaling parameters L, ζ, θ_*^{SL}, ϕ_M, ϕ_H, and μ^{SL} at z = 10 m, at a latitude where $f_c = 10^{-4}$.

Solution: $L = \dfrac{-u_*^3}{k \, (g/\overline{\theta}_v) \, \overline{w'\theta_v'}} = \dfrac{(0.2)^3}{(0.4)(0.0333)(0.05)} = 12.0$ m

$$\zeta = \frac{z}{L} = \frac{10}{12} = 0.83$$

$$\phi_M = \frac{k \, z}{u_*} \frac{\partial\overline{U}}{\partial z} = \frac{(0.4)(10)}{(0.2)}(0.2) = 4.0$$

$$\theta_*^{SL} = \frac{-\overline{w'\theta_v'}_s}{u_*} = \frac{0.05}{0.2} = 0.25 \text{ K}$$

$$\phi_H = \frac{k\,z}{\theta_*^{SL}}\frac{\partial\overline{\theta_v}}{\partial z} = \frac{(0.4)\,(10)}{(0.25)}\,(0.2) = 3.2$$

$$\mu^{SL} = \frac{k\,u_*}{f_c\,L} = \frac{(0.4)\,(0.2)}{(10^{-4})\,(12)} = 66.7$$

5.10 Combined Stability Tables

Static and dynamic stability concepts are intertwined, as sketched in Fig 5.24a. Negative Richardson numbers always correspond to statically and dynamically unstable flow. This flow will definitely become turbulent. Positive Richardson numbers are always statically stable, but there is the small range of $0 < Ri < 1$ where positive Richardson numbers are dynamically unstable, and may be turbulent depending on the past history of the flow. Namely, nonturbulent flow will become turbulent at about $Ri = 0.25$, while flow that is presently turbulent will stay turbulent if $Ri < 1$.

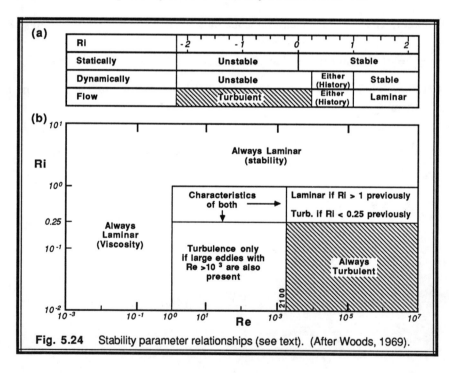

Fig. 5.24   Stability parameter relationships (see text). (After Woods, 1969).

The effects of viscosity and stability in suppressing turbulence are also intertwined, as sketched in Fig 5.24b. In Section 3.5.1 we defined the Reynolds number as the ratio of inertial to viscous forces, with no mention about buoyancy. In section 5.5.3, we defined a Richardson number as the ratio of buoyant to inertial or mechanical forces, with no mention of viscosity. In the atmosphere, the Reynolds number is usually so large that it corresponds to the rightmost edge of Fig 5.24b. Thus, we can essentially ignore viscous effects on stability in the atmosphere, and focus on the static and dynamic stability instead.

In conclusion, we see that the TKE equation is critical for determining the nature of flow in the BL. The relative contributions of various turbulence production and loss terms can be compared when rewritten as dimensionless scaling parameters. These parameters can be used to define layers within the BL where the physics is simplified, and where a variety of similarity scaling arguments can be made (see chapter 9 for details of similarity theory).

5.11 References

André, J.-C., G. De Moor, P. Lacarrère, G. Therry, and R. du Vachat, 1978: Modeling the 24-hour evolution of the mean and turbulent structures of the planetary boundary layer. *J. Atmos. Sci.*, **35**, 1861-1883.

Businger, J.A., J.C. Wyngaard, Y. Izumi and E.F. Bradley, 1971: Flux profile relationships in the atmospheric surface layer. *J. Atmos. Sci.*, **28**, 181-189.

Caughey, S.J., J.C. Wyngaard and J.C. Kaimal, 1979: Turbulence in the evolving stable boundary layer. *J. Atmos. Sci.*, **36**, 1041-1052.

Chou, S.-H., D. Atlas, and E.-N. Yeh, 1986: Turbulence in a convective marine atmospheric boundary layer. *J. Atmos. Sci.*, **43**, 547-564.

Deardorff, J.W., 1974: Three-dimensional numerical study of turbulence in an entraining mixed layer. *Bound.-Layer Meteor.*, **7**, 199-226.

Gal-Chen, T. and R.A. Kropfli, 1984: Buoyancy and pressure perturbations derived from dual-Doppler radar observations of the planetary boundary layer: applications for matching models with observations. *J. Atmos. Sci.*, **41**, 3007-3020.

Hechtel, L.M., 1988: The effects of nonhomogeneous surface heat and moisture fluxes on the convective boundary layer. *Preprints of the Am. Meteor. Soc. 8th Symposium on Turbulence and Diffusion in San Diego, April 1988*. 4pp.

Holtslag, A.A.M. and F.T.M. Nieuwstadt, 1986: Scaling the atmospheric boundary layer. *Bound.-Layer Meteor.*, **36**, 201-209.

Kitchen, M., J.R. Leighton and S.J. Caughey, 1983: Three case studies of shallow convection using a tethered balloon. *Bound.-Layer Meteor.*, **27**, 281-308.

Lee, D.R., R. B. Stull, and W.S. Irvine, 1979: *Clear Air Turbulence Forecasting Techniques*. AFGWC/TN-79/001. Air Force Global Weather Central, Offutt AFB, NE 68113. 73pp.

Lenschow, D.H., 1974: Model of the height variation of the turbulence kinetic energy budget in the unstable planetary boundary layer. *J. Atmos. Sci.*, **31**, 465-474.

Lenschow, D.H., J.C. Wyngaard and W.T Pennell, 1980: Mean field and second moment budgets in a baroclinic, convective boundary layer. *J. Atmos. Sci.*, **37**, 1313-1326.

Louis, J.F., A. Weill and D. Vidal-Madjar, 1983: Dissipation length in stable layers. *Bound.-Layer Meteor.*, **25**, 229-243.

Mahrt, L., 1981: Modelling the depth of the stable boundary layer. *Bound.-Layer Meteor.*, **21**, 3-19.

McBean, G.A. and J.A. Elliott, 1975: The vertical transports of kinetic energy by turbulence and pressure in the boundary layer. *J.Atmos. Sci.*, **32**, 753-766.

Nicholls, S., M.A. LeMone and G. Sommeria, 1982: The simulation of a fair weather marine boundary layer in GATE using a three dimensional model. *Quart. J. Roy. Meteor. Soc.*, **108**, 167-190.

Nicholls, S. and C.J. Readings, 1979: Aircraft observations of the structure of the lower boundary layer over the sea. *Quart. J. Roy. Meteor. Soc.*, **105**, 785-802.

Noonkester, V.R., 1974: Convective activity observed by FM-CW radar. Naval Electronics Lab. Center, NELC/TR 1919. San Diego, CA 92152. 70pp.

Pennell, W.T. and M.A. LeMone, 1974: An experimental study of turbulence structure in the fair-weather trade wind boundary layer. *J. Atmos. Sci.*, **31**, 1308-1323.

Stage, S.A. and J.A. Businger, 1981: A model for entrainment into a cloud-topped marine boundary layer. Part I. Model description and application to a cold air outbreak episode. *J. Atmos. Sci.*, **38**, 2213-2229.

Therry, G. and P. Lacarrère, 1983: Improving the eddy kinetic energy model for planetary boundary layer description. *Bound.-Layer Meteor.*, **25**, 63-88.

Thorpe, S.A., 1969: Experiments on the stability of stratified shear flows. *Radio Science,* **4**, 1327-1331.

Thorpe, S.A., 1973: CAT in the lab. *Weather,* **28**, 471-475.

Wilczak, J.M. and J.A. Businger, 1984: Large-scale eddies in the unstably stratified atmospheric surface layer. Part II. Turbulent pressure fluctuations and the budgets of heat flux, stress and turbulent kinetic energy. *J. Atmos. Sci.*, **41**, 3551-3567.

Woods, J.D., 1969: On Richardson's number as a criterion for laminar-turbulent-laminar transition in the ocean and atmosphere. *Radio Science,* **4**, 1289-1298.

Wyngaard, J.C., 1973: On surface layer turbulence. *Workshop on Micrometeorology*, D.A. Haugen (Ed.), Amer. Meteor. Soc., Boston. 101-149.

Yamada, T. and G. Mellor, 1975: A simulation of the Wangara atmospheric boundary layer data. *J. Atmos. Sci.*, **32**, 2309-2329.

Zhou, M.Y., D.H. Lenschow, B.B. Stankov, J.C. Kaimal, and J.E. Gaynor, 1985: Wave and turbulence structure in a shallow baroclinic convective boundary layer and overlying inversion. *J. Atmos. Sci.*, **42**, 47-57.

5.12 Exercises

1) Why doesn't turbulent energy cascade from small to large eddies (or wavelengths) in the boundary layer?
2) Refer to the TKE equation. Which term(s), if any, represent the production of turbulence during a day when there are light winds and strong solar heating of the boundary layer?
3) Given the following wind speeds measured at various heights in the boundary layer:

z (m)	U (m/s)
2000	10.0
1000	10.0
500	9.5
300	9.0
100	8.0
50	7.4
20	6.5
10	5.8
4	5.0
1	3.7

Assume that the potential temperature increases with height at the constant rate of 6 K/km. Calculate the bulk Richardson number for each layer and indicate the static and dynamic stability of each layer. Also, show what part of the atmosphere is expected to be turbulent in these conditions.

4) Derive an expression for the kinematic heat flux $\overline{w'\theta'}$ in terms of the dimensionless wind shear ϕ_M and dimensionless lapse rate ϕ_H.

5) Given the following TKE equation:

$$\frac{\partial \bar{e}}{\partial t} + \overline{U}_j \frac{\partial \bar{e}}{\partial x_j} = -\overline{u'w'}\frac{\partial \overline{U}}{\partial z} + \frac{g}{\overline{\theta}_v}\left(\overline{w'\theta_v'}\right) - \frac{\partial\left(\overline{w'e}\right)}{\partial z} - \frac{1}{\rho}\frac{\partial\left(\overline{w'p'}\right)}{\partial z} - \varepsilon$$

A   B   C   D   E   F   G

a. Which terms are always loss terms?
b. Which terms neither create nor destroy TKE?
c. Which terms can be either production or loss?
d. Which terms are due to molecular effects?
e. Which production terms are largest on a cloudy, windy day?
f. Which production terms are largest on a calm sunny day over land?
g. Which terms tend to make turbulence more homogeneous?
h. Which terms tend to make turbulence less isotropic?
i. Which terms describe the stationarity of the turbulence?
j. Which terms describe the kinetic energy lost from the mean wind?

6) Very briefly define the following, and comment or give examples of their use in micrometeorology.

a. inertial subrange

b. friction velocity

c. Obukhov length

d. return-to-isotropy term

e. static stability

f. convective velocity scale

g. Reynold's stress

h. turbulence closure problem

i. Richardson number

j. TKE

7) Fill in the table based on the regions A-H labeled on the attached diagram.

Property:	Lapse Rate	Heat Flux	Static Stability	Turbulent?	Name
Choices:	Subadiab.	Up	Stable	Yes	Noct. inversion
	Adiabatic	Zero	Neutral	Unknown	Cloud layer
	Superad.	Down	Unstable	No	Mixed layer
				Sporadic	Entrainment Zone
					Capping inversion
Region					Free atmosphere
A					Surface layer
B					
C	Subadiab.				
D					
E					Residual layer
F		Zero			
G				Unknown	
H			Stable		

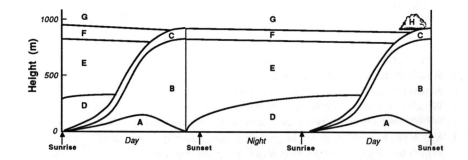

8) It has been suggested that in regions of strong static stability, the lower end of the inertial subrange (long wavelength, small wavenumber) occurs at a wavenumber, κ_b, given by: $\kappa_b \cong N_{BV}^{3/2} \, \varepsilon^{-1/2}$, where N_{BV} is the Brunt-Väisäla frequency, and ε is the TKE dissipation rate. Within the buoyancy subrange sketched below, would you expect turbulence to be isotropic? (Hint, buoyancy effects are important in a statically stable environment.)

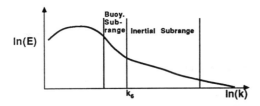

9) a) Rewrite the conservation equation for mean kinetic energy in terms of the geostrophic wind.

b) Suppose that $\overline{u'w'} = -0.05 \text{ m}^2 \text{ s}^{-2}$ and $\partial \overline{U}/\partial z = 5 \text{ s}^{-1}$ and $\overline{V} = 0$ within the surface layer. If there are no pressure gradients, then what is the value of the rate of change of mean kinetic energy, and what does it mean concerning the change in mean wind speed during a 1 minute period?

10) On the planet Krypton suppose that turbulent motions are affected by a strange form of viscosity that dissipates only the vertical motions. How would the TKE be affected?

11) What is the Reynolds stress? Why is it called a stress? How does it relate to u_*?

12) Define the following types of convection.   Under what weather conditions is each type of convection most likely? What term in the TKE equation is small under each condition?
a) Free convection
b) Forced convection.

13) Given the term: $U_j \, \partial(\frac{1}{2} V^2)/\partial x_j$, which represent the advection of total horizontal v-component of kinetic energy. Expand the variables U_j and V into mean and turbulent parts, Reynolds average, and simplify as much as possible.

14) Observations:

z(m):	12	8	2	$0.1 = z_o$
$\overline{\theta}$ (K):	300	301	303	308
$\overline{U}$(m/s)	5.4	5.0	3.4	0

Situation: Daytime boundary layer over land.
a) Find R_B at 2, 4, and 10m.
b) Comment on the static and dynamic stability of the air. Is the flow turbulent?

15) If the TKE at 10m is at steady state, and if $\varepsilon = 0.01$ m^2 s^{-3}, then is the transport term supplying or removing TKE from the air at z = 10m?

16) What is the static stability
of each of the layers in the
diagram at right?

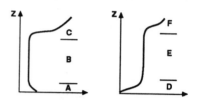

17) This problem is best saved until after the log-wind profile has been introduced. Given the following data:

$\overline{w'\theta'} = 0.2$ K m/s $u_* = 0.2$ m/s

z_i = 500 m $k = 0.4$

$g/\overline{\theta} = 0.0333$ m s^{-2} K^{-1} $z = 6$ m
z_0 = 0.01 m = roughness length   no moisture

Find:
a) L   f) R_f at 6m (make assumptions to find this)
b) z/L   g) Ri at 6m (make assumptions to find this)
c) w*   h) dynamic stability
d) θ*   i) flow state (turbulent or not)
e) static stability

18) Given the following sounding in the morning boundary layer. Determine whether each layer is stable or unstable (in both the static and dynamic sense), and state if the flow is turbulent. Indicate your results in the table to the right of the figures.

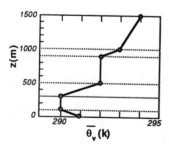

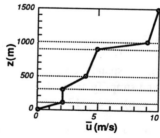

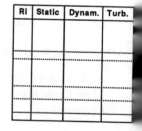

RI	Static	Dynam.	Turb.

19) Given the following turbulence statistics.

Where:	Location A		Location B	
When (UTC):	1000	1100	1000	1100
Statistic				
$\overline{u'^2}$ (m² s⁻²)	0.50	0.50	0.70	0.50
$\overline{v'^2}$ (m² s⁻²)	0.25	0.50	0.25	0.25
$\overline{w'^2}$ (m² s⁻²)	0.70	0.50	0.70	0.25

Where and when is the turbulence
a) Stationary   b) Homogeneous   c) Isotropic?

20) What boundary layer flow phenomena or characteristics have scale sizes on the order
of: a) 1 mm,   b) 10 m,   c) 1 km ?

21) Fill in the blanks in the table below:

Characteristic:	Phenomenon:	Convective turbulence in the BL	Mechanical turbulence in the BL
• Name of (or symbol for) a characteristic depth scale:			
• Name of (or symbol for) a characteristic velocity scale:			
• Name of the type of convection associated with this phenomenon:			
• Dominant production term in the TKE equation:			
• Sign of the gradient Richardson number:			
• Direction (horizontal or vertical) of the dominant anisotropic component of turbulence:			

22) a) Is θ_v conserved during adiabatic ascent of an unsaturated air parcel? If not, does it
increase or decrease with height?
b) Same question, but in saturated (cloudy) air.
c) If a saturated air parcel at 80 kPa (800 mb) has T = 4 °C (thus, r_s = 6.5 g/kg) and
has a total water mixing ratio of r_T = 8.0 g/kg, then calculate the virtual potential
temperature at that altitude.

23) Given the TKE equation with terms labeled A to E below:

$$\frac{\partial \bar{e}}{\partial t} = -\overline{u'w'}\frac{\partial \bar{U}}{\partial z} + \frac{g}{\theta_v}\overline{w'\theta_v'} - \frac{\partial}{\partial z}\overline{w'\left(\frac{p'}{\rho}+e\right)} - \varepsilon$$

   A   B   C   D   E

and given 4 regions of the stable boundary layer, labeled I to IV in the figure below, determine the sign (+ , - , or near zero) of each term in each region. (Assume: that term A is always zero; i.e., steady state.)

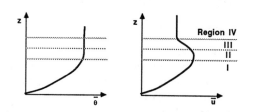

24) The dissipation rate of TKE is sometimes approximated by $e = \bar{e}^{-3/2}/l$, where l is the **dissipation length scale**. It is often assumed that $l = 5\ z$ in statically neutral conditions (Louis, et al., 1983). If the TKE shown in Fig 2.9b is assumed as an initial condition, there is no shear or buoyancy production or loss, and no redistribution nor turbulent transport, then at z = 100m:
a) What is the initial value of the dissipation rate?
b) How long will it take the TKE to decay to 10% of its initial value?

25) Given $\overline{w'\theta_v'} = 0.3$ K m/s, $\overline{u'w'} = -0.25$ m^2 s^{-2}, and $z_i = 1$ km, find:

a) u_* e) θ_*^{SL}

b) w_* f) R_f (assume $\partial U/\partial z = 0.1$ s^{-1})

c) t_*^{ML} g) Obukhov length (L)

d) θ_*^{ML}

26) Given the following sounding, indicate for each layer the
a) static stability
b) dynamic stability
c) existence of turbulence (assuming a laminar past history).

z (m)	$\overline{\theta}_v$ (K)	$\overline{U}$ (m/s)
80	305	18
70	305	17
60	301	15
50	300	14
40	298	10
30	294	8
20	292	7
10	292	7
0	293	2

27) Which Richardson number (flux, gradient, bulk) would you use for the following application? (Give the one best answer for each question).
a) Diagnose the possible existence of clear air turbulence using rawinsonde data.
b) Determine whether turbulent flow will become laminar.
c) Determine whether laminar flow will become turbulent in the boundary layer.

28) What is the difference between free and forced convection?

29) What is the Reynolds number? Of what importance is it to boundary layer flows?

30) What is the closure problem?

31) Given isotropic turbulence with $u_* = 0.5$ m/s and TKE/m $= 0.9$ m^2 s^{-2}, find the correlation coefficient, r, between w and u.

32) Given the TKE equation, name each term and describe how you could determine the value of each term.

33) Indicate the nature of the flow (laminar or turbulent) for each cell in the table below:

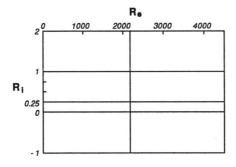

6
Turbulence Closure Techniques

At first glance, the large number of equations developed in Chapters 3-5 would suggest that we have a fairly complete description of turbulent flow. Unfortunately, a closer examination reveals that there are a large number of unknowns remaining in those equations. These unknowns must be dealt with in order end up with a useful description of turbulence that can be applied to real situations. In this Chapter, the unknowns are identified, and methods to parameterize them are reviewed. Simulation techniques such as large-eddy simulation are discussed in Chapter 10.

6.1 The Closure Problem

As will be demonstrated below, *the number of unknowns in the set of equations for turbulent flow is larger than the number of equations*. A variable is considered to be unknown if one doesn't have a prognostic or diagnostic equation defining it. When equations are included for these unknowns (changing them to known variables), one discovers even more new unknowns. Thus, for any finite set of those equations the description of turbulence is not *closed*. Alternately, the total statistical description of turbulence requires an infinite set of equations. This unfortunate conclusion is called the *closure problem*. It was first recognized in 1924 by Keller and Friedmann, and was associated with the non-linear characteristic of turbulence. It has remained one of the unsolved problems of classical physics.

To demonstrate the closure problem, recall from equations 3.5.3 in Chapter 3 that the forecast equation for a mean variable such as potential temperature has at least one turbulence term in it, such as $\partial(\overline{u_j'\theta'})/\partial x_j$. A quantity like $\overline{u_j'\theta'}$ is called a *double correlation*, or a *second statistical moment*. To eliminate this as an unknown we derived a forecast equation for it in Chapter 4 (equation 4.4.3). Unfortunately, this equation contained additional *triple correlation (third moment)* terms such as $\overline{\theta'u_i'u_j'}$. As you might expect, if we were to write an equation for this third moment, it would contain a fourth-moment quantity.

The matter is even worse than highlighted above, because $\overline{\theta'u_i'u_j'}$ really represents 9 terms, one for each value of i and j. Of these 9 terms, 6 remain as unknowns because of symmetries in the tensor matrix (e.g., $\overline{\theta'u_1'u_2'} = \overline{\theta'u_2'u_1'}$). Similar problems occur for the turbulence equations for momentum, as is shown in Table 6-1. There is an easy way to anticipate which unknowns remain at any level of closure after symmetries are considered, as is shown in Table 6-2 for momentum correlations. In the full equations of motion there are additional unknowns such as pressure correlations and terms involving viscosity.

Table 6-1. Simplified example showing a tally of equations and unknowns for various statistical moments of momentum, demonstrating the closure problem for turbulent flow. The full set of equations includes even more unknowns.

Prognostic Eq. for:	Moment	Equation	Number of Eqs.	Number of Unknowns
$\overline{U_i}$	First	$\dfrac{\partial \overline{U_i}}{\partial t} = \dots - \dfrac{\partial \overline{u_i'u_j'}}{\partial x_j}$	3	6
$\overline{u_i'u_j'}$	Second	$\dfrac{\partial \overline{u_i'u_j'}}{\partial t} = \dots - \dfrac{\partial \overline{u_i'u_j'u_k'}}{\partial x_k}$	6	10
$\overline{u_i'u_j'u_k'}$	Third	$\dfrac{\partial \overline{u_i'u_j'u_k'}}{\partial t} = \dots - \dfrac{\partial \overline{u_i'u_j'u_k'u_m'}}{\partial x_m}$	10	15

Table 6-2. Correlation triangles indicating the unknowns for various levels of turbulence closure, for the momentum equations only. Notice the pattern in these triangles, with the u, v, and w statistics at their respective vertices, and the cross correlations in between.

Order of Closure   **Correlation Triangle of Unknowns**

Zero
$$\overline{U}$$
$$\overline{V} \quad \overline{W}$$

First
$$\overline{u'^2}$$
$$\overline{u'v'} \qquad \overline{u'w'}$$
$$\overline{v'^2} \qquad \overline{v'w'} \qquad \overline{w'^2}$$

Second
$$\overline{u'^3}$$
$$\overline{u'^2v'} \qquad \overline{u'^2w'}$$
$$\overline{u'v'^2} \qquad \overline{u'v'w'} \qquad \overline{u'w'^2}$$
$$\overline{v'^3} \qquad \overline{v'^2w'} \qquad \overline{v'w'^2} \qquad \overline{w'^3}$$

To make the mathematical/statistical description of turbulence tractable, one approach is to use only a finite number of equations, and then approximate the remaining unknowns in terms of known quantities.   Such *closure approximations* or *closure assumptions* are named by the highest order prognostic equations that are retained. Using the equations in Table 6-1 as an example, for *first-order closure* the first equation is retained and the second moments are approximated.   Similarly, *second-order closure* retains the first two equations, and approximates terms involving third moments.

Some closure assumptions utilize only a portion of the equations available within a particular moment category. For example, if equations for the turbulence kinetic energy and temperature and moisture variance are used along with the first-moment equations of Table 6-1, the result can be classified as ***one-and-a-half order closure***. It clearly would not be full second-order closure because not all of the prognostic equations for the second moments (i.e., for the fluxes) are retained, yet it is higher order than first-order closure. One can similarly define ***zero-order closure*** and ***half-order closure*** methods.

Two major schools of thought of turbulence closure have appeared in the literature: *local* and ***nonlocal*** closure. Neither local nor nonlocal methods are exact, but both appear to work well for the physical situations for which the parameterizations are designed.

For ***local closure***, an unknown quantity at any point in space is parameterized by values and/or gradients of known quantities at the ***same*** point. Local closure thus assumes that turbulence is analogous to molecular diffusion. The Donaldson example in the next section demonstrates a local second-order closure. In the meteorological literature, local closure has been used at all orders up through third order.

For ***nonlocal closure***, the unknown quantity at one point is parameterized by values of known quantities at ***many points*** in space. This assumes that turbulence is a superposition of eddies, each of which transports fluid like an advection process. Nonlocal methods have been used mostly with first-order closure. Table 6-3 summarizes the myriad of closure methods which have often appeared in the meteorological literature. Generally, the higher-order local closures and the nonlocal closures yield more accurate solutions than lower order, but they do so at added expense and complexity.

Table 6-3. Classification of closure techniques that have been frequently reported in the literature. Bulk and similarity methods are discussed in more detail in Chapters 9, 11 and 12.

Order	Local	Nonlocal	Other (bulk or similarity methods)
Zero			X
Half	X	X	X
First	X	X	
One-and-a-half	X		
Second	X		
Third	X		

6.2 Parameterization Rules

Regardless of which order closure is used, there are ***unknown*** turbulence terms which must be parameterized as a function of ***known*** quantities and ***parameters***. A *known* quantity is any quantity for which a prognostic or diagnostic equation is retained. Using the equations in Table 6-1 for example, if we decide to use second-order closure, the unknown quantity $\overline{u_i'u_j'u_k'}$ can be parameterized as a function of $\overline{U_i}$ and $\overline{u_i'u_j'}$, because we have prognostic equations for these quantities. One must remember that the equations in Table 6-1 are only a subset of the full set of equations, therefore second-order closure can also employ other known first and second moments such as $\overline{\theta}$, $\partial\overline{\theta}/\partial z$, or $\overline{u_j'q'}$ in the parameterization. A *parameter* is usually a constant, the value of which is determined empirically. For example, the parameter can be a separate term, a multiplicative constant, or the value of a power or exponent.

By definition, a *parameterization* is an approximation to nature. In other words, we are replacing the true (natural) equation describing a value with some artificially constructed approximation. Sometimes parameterizations are employed because the true physics has yet to be discovered. Other times, the known physics are too complicated to use for a particular application, given cost or computer limitations. Parameterization will rarely be perfect. The hope is that it will be adequate.

Parameterization involves human interpretation and creativity, which means that different investigators can propose different parameterizations for the same unknown. In fact, Donaldson (1973) noted that "there are more models for closure of the equations of motion at the second-order correlation level than there are principal investigators working on the problem". Although there is likely to be an infinite set of possible parameterizations for any quantity, all acceptable parameterizations must follow certain common-sense rules.

Most importantly, the parameterization for an unknown quantity should be physically reasonable. In addition, the parameterization must:
• have the same dimensions as the unknown.
• have the same tensor properties.
• have the same symmetries.
• be invariant under an arbitrary transformation of coordinate systems.
• be invariant under a Galilean (i.e., inertial or Newtonian) transformation.
• satisfy the same budget equations and constraints.
These rules apply to all orders of closure.

As an example, Donaldson (1973) has proposed that the **unknown** $\overline{u_i'u_j'u_k'}$ be parameterized by:

$$ - \Lambda\, e^{1/2} \left[\frac{\partial\,\overline{(u_j'u_k')}}{\partial x_i} + \frac{\partial\,\overline{(u_i'u_k')}}{\partial x_j} + \frac{\partial\,\overline{(u_i'u_j')}}{\partial x_k} \right] $$

where Λ is a *parameter* having the dimension of length (m), and the **knowns** are e (turbulence kinetic energy per unit mass, $m^2 s^{-2}$) and $\overline{u_i'u_j'}$ (momentum flux, $m^2 s^{-2}$).

This parameterization has the same dimensions ($m^3 s^{-3}$) and the same tensor properties (unsummed i, j & k) as the original unknown. The symmetry of the original unknown is such that the order of the indices i, j, k is not significant. The same symmetry is achieved in the parameterization by having the sum of the three terms in square brackets. If only one term had been used instead of the sum, then a change in the order of the indices would have produced a different numerical result (because $\partial \overline{u'v'}/\partial z$ is not necessarily equal to $\partial \overline{u'w'}/\partial y$). Since the gradient of the momentum flux is taken in all three Cartesian directions in the square brackets, any rotation or displacement of the coordinate system will not change the result. Also, movement of the coordinate system at constant velocity c_i (a Galilean transformation) does not change the parameterization, as can be seen by setting $x_i = X_i + c_i t$.

The final rule is difficult to demonstrate here without a long explanation of all the constraints on the system, but as a sample we can look at one constraint. The original unknown appears in budget equation (4.4.1b) as $\partial \overline{u_i'u_j'u_k'}/\partial x_j$, which represents the turbulent transport of $\overline{u_i'u_k'}$. When the vertical component of this term ($\partial \overline{u_i'u_j'u_k'}/\partial z$) is integrated over the depth of the boundary layer it should equal zero, because it represents a movement of existing momentum flux from one height to another. The momentum flux drained by this term from one location within the turbulent domain should be deposited in a different part, yielding no net increase or decrease in total integrated momentum flux. This budget constraint is indeed satisfied by the parameterization above, because each of the terms in square brackets becomes zero when integrated over the whole BL depth.

The remainder of this Chapter reviews some of the parameterizations that have been presented in the literature. This review is by no means comprehensive — it is meant only to demonstrate the various types of closure and their features. Regardless of the type of parameterization used, the result closes the equations of motion for turbulent flow and allows them to be solved for various forecasting, diagnostic, and other practical applications.

6.3 Local Closure — Zero and Half Order

6.3.1 Zero Order

Zero-order closure implies that no prognostic equations are retained, not even the equations for mean quantities. In other words, the mean wind, temperature, humidity, and other mean quantities are parameterized directly as a function of space and time. Obviously, this is neither local or nonlocal closure because it avoids the parameterization of turbulence altogether. For this reason, we will not dwell on zero-order closure here, but will come back to it later in Chapter 9 under the topic of *similarity theory*.

6.3.2 Half-order

Half-order closure uses a subset of the first moment equations (3.5.3). A variation of this approach is called the **bulk method**, where a profile shape for wind or temperature is assumed, but where the resulting wind or temperature curve can be shifted depending on the bulk-average background wind or temperature within the whole layer.

For example, a boundary-layer (bulk) average $<\overline{\theta}(t)>$ is forecast using equations like (3.5.3), a profile shape $[\Delta\overline{\theta}(z)]$ is assumed, and then the final values of $\overline{\theta}(z,t)$ are found from $\overline{\theta}(z,t) = <\overline{\theta}(t)> + \Delta\overline{\theta}(z)$. Such schemes are used for bulk or **slab** mixed-layer models with $\Delta\overline{\theta}(z) = 0$ at all heights (see Chapter 11); for cloud models with $\Delta\overline{\theta}(z)$ modeled as linear functions of height within separate cloud and subcloud layers (see Chapter 13); and for stable boundary layers with $\Delta\overline{\theta}(z)$ approximated with either linear, polynomial, or exponential profile shapes (see Chapter 12).

6.4 Local Closure — First Order

6.4.1   Definition

First-order closure retains the prognostic equations for only the zero-order mean variables such as wind, temperature, and humidity. As an example, consider the idealized scenario of a dry environment, horizontally homogeneous, with no subsidence. The geostrophic wind is assumed to be known as a prescribed boundary condition. The governing prognostic equations (3.5.3) for the zero-order variables then reduce to:

$$\frac{\partial \overline{U}}{\partial t} = f_c(\overline{V} - \overline{V}_g) - \frac{\partial\,(\overline{u'w'})}{\partial z} \tag{6.4.1a}$$

$$\frac{\partial \overline{V}}{\partial t} = -f_c(\overline{U} - \overline{U}_g) - \frac{\partial\,(\overline{v'w'})}{\partial z}$$

$$\frac{\partial \overline{\theta}}{\partial t} = -\frac{\partial\,(\overline{w'\theta'})}{\partial z}$$

The unknowns in this set of equations are second moments: $\overline{u'w'}$, $\overline{v'w'}$ and $\overline{w'\theta'}$.

To close the above set of equations, we must parameterize the turbulent fluxes. If we let ξ be any variable, then one possible first-order closure approximation for flux $\overline{u_j'\xi'}$ is:

$$\overline{u_j'\xi'} = -K \frac{\partial \overline{\xi}}{\partial x_j} \qquad (6.4.1b)$$

where the parameter K is a scalar with units $m^2 s^{-1}$. For positive K, (6.4.1b) implies that the flux $\overline{u_j'\xi'}$ flows down the local gradient of $\overline{\xi}$. This closure approximation is often called **gradient transport theory** or **K-theory**. Although it is one of the simplest parameterizations, it frequently fails when larger-size eddies are present in the flow. Hence, we can catalog (6.4.1b) as a **small-eddy closure** technique.

K is known by a variety of names:
- eddy viscosity
- eddy diffusivity
- eddy-transfer coefficient
- turbulent-transfer coefficient
- gradient-transfer coefficient

because it relates the turbulent flux to the gradient of the associated mean variable. Sometimes, different K values are associated with different variables. A subscript "m" is used for momentum, resulting in K_m as the eddy viscosity. For heat and moisture, we will use K_H and K_E for the respective eddy diffusivities. There is some experimental evidence to suggest that for statically neutral conditions:

$$K_H = K_E = 1.35 \, K_m \qquad (6.4.1c)$$

It is not clear why K_m should be smaller than the other K values. Perhaps pressure-correlation effects contaminated the measurements upon which (6.4.1c) was based.

6.4.2   Examples

Problem A: Given $K_H = 5 \, m^2 s^{-1}$ for turbulence within a background stable environment, where the local lapse rate is $\partial \overline{\theta}/\partial z = 0.01$ K/m. Find $\overline{w'\theta'}$, the kinematic heat flux.

Solution: Use $\overline{u_j'\xi'} = -K \, \partial \overline{\xi}/\partial x_j$. Let $\overline{\xi}$ represent $\overline{\theta}$, and set j =3. This gives:

$$\overline{w'\theta'} = -K_H \frac{\partial \overline{\theta}}{\partial z} = -\left(5 \, m^2 s^{-1}\right)\left(0.01 \text{ K/m}\right) = -0.05 \text{ K m/s}$$

Discussion: A negative heat flux would normally be expected in a statically stable environment, assuming only small eddies were present. In other words, in an environment with warm air above colder air, turbulence moves warm air down the gradient to cooler air, which in this case is a downward (or negative) flux.

Problem B: Suggest parameterizations to close the set of equations (6.4.1a).

Solution:

$$\overline{w'\theta'} = -K_H \frac{\partial \overline{\theta}}{\partial z}$$

$$\overline{u'w'} = -K_m \frac{\partial \overline{U}}{\partial z}$$

$$\overline{v'w'} = -K_m \frac{\partial \overline{V}}{\partial z}$$

Discussion: If these equations are plugged back into (6.4.1a), then there are 3 equations for the 3 unknowns $\overline{\theta}$, $\overline{U}$, and $\overline{V}$.   This is a closed set which can be solved numerically if the K values are known.   Although the assumption that K=constant is an easy assumption, it is the least realistic.   It would be better to parameterize K as a function of the knowns: $\overline{\theta}$, $\overline{U}$, and $\overline{V}$, or of gradients of those knowns.

Problem C: Given $K_H = 5$ m^{2}s^{-1} for turbulence within a background horizontally homogeneous environment, find $\overline{u'\theta'}$.

Solution:

$$\overline{u'\theta'} = -K_H \frac{\partial \overline{\theta}}{\partial x}$$

But $\partial\overline{\theta}/\partial x = 0$ for a horizontally homogeneous environment. Thus $\overline{u'\theta'} = 0$.

Discussion: It makes no difference whether K_H is positive, negative or exceptionally large. K-theory will always yield zero flux in a uniform environment, regardless of the true flux.

6.4.3   Analogy with Viscosity

As we saw in Section 2.9.3, the molecular stress τ_{mol} can be approximated by $\tau_{mol} = \rho \, v \, \partial\overline{U}/\partial z$ for a Newtonian fluid.   By analogy, one might expect that the turbulent Reynold's stress can be also expressed in terms of the shear, with a corresponding change from molecular viscosity v to eddy viscosity K_m, yielding: $\tau_{Reynolds} = \rho \, K_m \, \partial\overline{U}/\partial z$. Dividing this latter expression by ρ gives the kinematic form (6.4.1b).   The product $\rho \cdot K_m$ is sometimes called the *Austausch coefficient*.

Since turbulence is much more effective than viscosity at causing mixing, one would expect $K_m > \nu$. Values of K_m reported in the literature vary from $0.1 \ m^2s^{-1}$ to $2000 \ m^2s^{-1}$, with typical values on the order of 1 to 10 m^2s^{-1}. Values of ν are much smaller, on the order of $1.5 \times 10^{-5} \ m^2s^{-1}$.

Magnitude is not the only difference between the molecular and eddy viscosities. A significant difference is that ν is a function of the fluid, while K_m is a function of the flow. Thus, while ν is uniquely determined by the chemical composition of the fluid and its state (temperature and pressure, etc.), K_m varies as the turbulence varies. Thus, one must parameterize K_m as a function of other variables such as z/L, Richardson number or the stability $\partial\overline{\theta}_v/\partial z$, as outlined in Section 6.3.2.4.

6.4.4 Mixing-Length Theory

The following development is patterned after the mixing-length arguments proposed by Prandtl in 1925. Assume that there is turbulence in a statically neutral environment, with a linear mean humidity gradient in the vertical as sketched in Fig 6.1a. If a turbulent eddy moves a parcel of air upward by amount z' towards some reference level Z during which there is no mixing nor other changes in the value of q within the parcel, then the humidity of that parcel will differ from the surrounding environment by amount q', where:

$$q' = -\left(\frac{\partial\overline{q}}{\partial z}\right)z' \qquad (6.4.4a)$$

If the background mean wind profile is also linear, then a similar expression can be written for u':

$$u' = -\left(\frac{\partial\overline{U}}{\partial z}\right)z' \qquad (6.4.4b)$$

In order for the parcel to move upward a distance z', it must have had some vertical velocity w'. If the nature of turbulence is such that w' is proportional to u', then we might expect that w' = - c u' for the the wind shear sketched in Fig 6.1b (i.e., for $\partial\overline{U}/\partial z > 0$), and w' = c u' for $\partial\overline{U}/\partial z < 0$, where c is a constant of proportionality. Substituting (6.4.4b) for u' in the above expression for w' yields

$$w' = c\left|\frac{\partial\overline{U}}{\partial z}\right|z' \qquad (6.4.4c)$$

where we find that the magnitude of the shear is important.

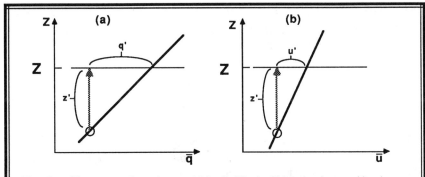

Fig. 6.1 Movement of an air parcel (shaded line) within a background having linear moisture and wind gradients (heavy line). The superposition of many such parcels, starting at different levels but all ending at level Z, forms the conceptual basis for "mixing length theory."

In Sections 2.6 and 2.7 it was shown that the kinematic eddy flux of moisture is $R = \overline{w'q'}$. Since we know q' from (6.4.4a) and w' from (6.4.4c), we need only multiply the two together, and then average over the spectrum of different size eddies z' to get the average flux R:

$$R = -c\,\overline{(z')^2}\,\left|\frac{\partial \overline{U}}{\partial z}\right| \cdot \left(\frac{\partial \overline{q}}{\partial z}\right) \qquad (6.4.4d)$$

We recognize $\overline{z'^2}$ as the variance of parcel displacement distance. The square root of it is a measure of the average distance a parcel moves in the mixing process that generated flux R. In this way, we can define a *mixing length*, l, by $l^2 = c\,\overline{z'^2}$. Thus, the final expression for moisture flux is

$$R = -l^2\,\left|\frac{\partial \overline{U}}{\partial z}\right| \cdot \left(\frac{\partial \overline{q}}{\partial z}\right) \qquad (6.4.4e)$$

This is directly analogous to K-theory if

$$K_E = l^2\,\left|\frac{\partial \overline{U}}{\partial z}\right| \qquad (6.4.4f)$$

because that leaves us with $R = -K_E\,(\partial \overline{q}/\partial z)$. In fact, mixing-length theory tells us via (6.4.4f) that the magnitude of K_E should increase as the shear increases (i.e., a measure of the intensity of turbulence) and as the mixing length increases (i.e., a measure of the ability of turbulence to cause mixing).

In the surface layer, the size of the turbulent eddies is limited by the presence of the earth's surface. Thus, it is sometimes assumed that $l^2 = k^2 z^2$, where k is the von Karman constant. The resulting expression for eddy viscosity in the surface layer is

$$K_E = k^2 z^2 \left| \frac{\partial \overline{U}}{\partial z} \right| \qquad (6.4.4g)$$

For SBLs, Delage (1974) proposed the following parameterization for mixing length that has since been used as a starting point for other parameterizations (Estournel and Guedalia, 1987; and Lasser and Arya, 1986):

$$\frac{1}{l} = \frac{1}{kz} + \frac{1}{0.0004 \, G \, f_c^{-1}} + \frac{\beta}{k \, L_L} \qquad (6.4.4h)$$

where L_L is a local Obukhov length (see Appendix A) based on local values of stress and heat flux above the surface, G is geostrophic wind speed, and β is an empirical constant.

Before leaving this section, we should examine the *limitations* of mixing-length theory. The relationship between w' and z' given by (6.4.4c) is only valid when turbulence is generated mechanically. Hence, mixing-length derivation is valid only for *statically neutral situations*, even though K-theory has been applied to statically stable conditions. Also, linear gradients of wind and moisture were assumed in (6.4.4a & b). In the real atmosphere, gradients are approximately linear only over small distances (i.e., the first-order term of a Taylor's series expansion). Hence, we see that mixing-length theory is a *small-eddy theory*.

6.4.5 Sample Parameterizations of K

The eddy viscosity is best not kept constant, as mentioned earlier, but should be parameterized as a function of the flow. The parameterizations for K should satisfy the following constraints: • K = 0 where there is no turbulence.
  • K = 0 at the ground (z=0).
  • K increases as TKE increases.
  • K varies with static stability (in fact, one might expect that a different value of K should be used in each of the coordinate directions for anisotropic turbulence).
  • K is non-negative (if one uses the analogy with viscosity).

This latter constraint has occasionally been ignored. The normal concept of an eddy viscosity or a small-eddy theory is that a turbulent flux flows down the gradient. Such a *down-gradient transport* means heat flows from hot to cold, moisture flows from moist to dry, and so forth. Such down-gradient transport is associated with positive values for K, and is indeed consistent with the analogy with molecular viscosity.

Table 6-4. Examples of parameterizations for the eddy viscosity, K, in the boundary layer.

Neutral Surface Layer:

$K = \text{constant}$	not the best parameterization
$K = u_*^2 \, T_0$	where u_* is the friction velocity
$K = U^2 \, T_0$	where T_0 is a timescale
$K = k \, z \, u_*$	where k is von Karman's constant
$K = k^2 z^2 \, [(\partial \overline{U}/\partial z)^2 + (\partial \overline{V}/\partial z)^2]^{1/2}$	from mixing-length theory
$K = l^2 \, (\partial \overline{U}/\partial z)^2$	where $l = k(z+z_0)/\{1+[k(z+z_0)/\Lambda]\}$, Λ=length scale

Diabatic Surface Layer   (generally, $K_{\text{statically unstable}} > K_{\text{neutral}} > K_{\text{statically stable}}$)

$K = k \, z \, u_* / \phi_M (z/L)$ where ϕ_M a dimensionless shear (see appendix A), and L is the Obukhov length (appendix A)

$K = k^2 z^2 \, [(\partial \overline{U}/\partial z) + \{(g/\overline{\theta}_v) \cdot |\partial \overline{\theta}_v/\partial z|\}^{1/2}]$ for statically unstable conditions

$K = k^2 z^2 \, [(\partial \overline{U}/\partial z) - (L_*/z)^{1/6} \{(15g/\overline{\theta}_v) \cdot |\partial \overline{\theta}_v/\partial z|\}^{1/2}]$ for statically stable conditions, where

$$L_* = -\theta \, u_*^2/(15 \, k \, g \, \theta_*)$$

Neutral or Stable Boundary Layer

$K = \text{constant}$ see Ekman Spiral derivation in next subsection

$K = K(h) + [(h-z)/(h-z_{SL})]^2 \{K(z_{SL}) - K(h) + (z-z_{SL})[\partial K/\partial z|_{z_{SL}} + 2(K(z_{SL})-K(h))/(h-z_{SL})]\}$

this is known as the O'Brien cubic polynomial approximation (O'Brien, 1970), see Fig 6-2, where z_{SL} represents the surface layer depth.

Unstable (Convective) Boundary Layer:

$K = 1.1 \, [(R_c - Ri) \, l^2 / Ri] \, |\partial \overline{U}/\partial z|$ for $\partial \overline{\theta}_v/\partial z > 0$ where $l = kz$ for $z < 200$ m and

$K = (1 - 18 \, Ri)^{-1/2} \, l^2 \, |\partial \overline{U}/\partial z|$ for $\partial \overline{\theta}_v/\partial z < 0$ $l = 70$ m for $z > 200$ m.

Numerical Model Approximation for Anelastic 3-D Flow:

$K = (0.25 \, \Delta)^2 \cdot |0.5 \, \Sigma_i \, \Sigma_j \, [\partial \overline{U}_i/\partial x_j + \partial \overline{U}_j/\partial x_i - (2/3)\delta_{ij}\Sigma_k(\partial \overline{U}_k/\partial x_k)]^2 \, |^{1/2}$ where Δ=grid size

In the real atmosphere, however, there are occasions where transport appears to flow UP the gradient (i.e., *counter-gradient*). As it turns out, this is physically explained by the fact that there are large eddies associated with rise of warm air parcels that transport heat from hot to cold, regardless of the local gradient of the background environment. Thus, in an attempt to make small-eddy K-theory work in large-eddy convective boundary layers, one must resort to negative values of K. Since this results in heat flowing from cold to hot, it is counter to our common-sense concept of diffusion.   Thus, *K-theory is not recommended for use in convective mixed layers.*

There has been no lack of creativity by investigators in designing parameterizations for K. Table 6-4 lists some of the parameterizations for K that have appeared in the literature (Bhumralkar, 1975), and Fig 6.2 show a typical variation of K with height in the boundary layer. Variations of K in the horizontal have also been suggested to explain phenomena such as mesoscale cellular convection (Ray, 1986).

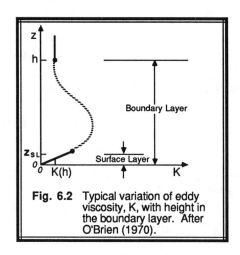

Fig. 6.2 Typical variation of eddy viscosity, K, with height in the boundary layer. After O'Brien (1970).

6.4.6   The Ekman Spiral in Atmospheres and Oceans

Even with first-order closure, the equations of motion (3.5.3) are often too difficult to solve analytically. However, for the special case of a steady state $[\partial \overline{(\ )}/\partial t = 0]$, horizontally homogeneous $[\partial \overline{(\ )}/\partial x = 0,\ \partial \overline{(\ )}/\partial y = 0]$, statically neutral $[\partial \overline{\theta}_v/\partial z = 0]$, barotropic atmosphere $[\overline{U}_g\ \&\ \overline{V}_g$ constant with height$]$ with no subsidence $[\overline{W}=0]$, the equations of motion can be reduced to:

$$\begin{cases} 0 = -f_c\left(\overline{V}_g - \overline{V}\right) - \dfrac{\partial\,(\overline{u'w'})}{\partial z} \\[3mm] 0 = +f_c\left(\overline{U}_g - \overline{U}\right) - \dfrac{\partial\,(\overline{v'w'})}{\partial z} \end{cases} \qquad (6.4.6a)$$

An analytic solution of these equations for the ocean was presented by Ekman in 1905, and was soon modified for the atmosphere.

Atmosphere: The following derivations are based on the approach of Businger (1982). Define the magnitude of the geostrophic wind, $\overline{G}$, by $\overline{G} = [\,\overline{U}_g^2 + \overline{V}_g^2\,]^{1/2}$. Pick an x-axis aligned with the geostrophic wind; thus, $\overline{V}_g = 0$ and $\overline{U}_g = \overline{G}$. Use first-order local closure K-theory, with constant K_m. Hence, $\overline{u'w'} = -K_m\,\partial\overline{U}/\partial z$ and $\overline{v'w'} = -K_m$ $\partial\overline{V}/\partial z$. Inserting these into (6.4.6a) leaves the following set of coupled second-order differential equations:

$$\begin{cases} f_c\,\overline{V} & = -\,K_m\,\dfrac{\partial^2\overline{U}}{\partial z^2} \\[3mm] f_c\,(\overline{U} - \overline{G}) & = +K_m\,\dfrac{\partial^2\overline{V}}{\partial z^2} \end{cases} \qquad (6.4.6b)$$

The four boundary conditions are $\overline{U} = 0$ at $z = 0$, $\overline{V} = 0$ at $z = 0$, $\overline{U} \to \overline{G}$ as $z \to \infty$, and $\overline{V} = 0$ as $z \to \infty$. It is assumed that the winds become geostrophic away from the surface.

The solution to this set of equations for the atmosphere is

$$\overline{U} = \overline{G}\left[\,1 - e^{-\gamma_E z}\cos(\gamma_E\,z)\right]$$

$$\overline{V} = \overline{G}\left[\,e^{-\gamma_E z}\sin(\gamma_E\,z)\right] \qquad (6.4.6c)$$

where $\gamma_E = [f_c/(2K_m)]^{1/2}$. The velocity vectors for this solution as a function of height are plotted in Fig 6.3a. The tip of the vectors trace out a spiral — hence the name *Ekman Spiral*.

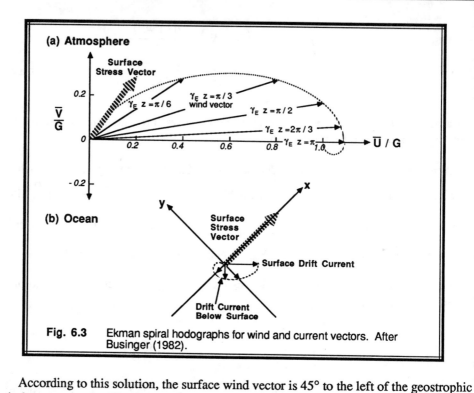

Fig. 6.3   Ekman spiral hodographs for wind and current vectors. After Businger (1982).

According to this solution, the surface wind vector is 45° to the left of the geostrophic wind vector in the Northern Hemisphere. Hence, the surface stress is also in this direction, because it is caused by the drag of the surface wind against the surface. Use u_*^2 as a measure of the surface stress magnitude, where $u_*^2 = [(\overline{u'w'})^2 + (\overline{v'w'})^2]^{1/2} = [(K_m \partial \overline{U}/\partial z)^2 + (K_m \partial \overline{V}/\partial z)^2]^{1/2}$ evaluated at z=0. Inserting (6.4.6c) into this expression yields:

$$u_*^2 = \overline{G} \left(K_m f_c \right)^{1/2} \tag{6.4.6d}$$

The wind speed is supergeostrophic at height $z = \pi/\gamma_E$, which is also the lowest height where the wind is parallel to geostrophic. Sometimes this height is used as an estimate of the **depth of the neutral boundary layer**. Hence, the **Ekman layer depth**, h_E, is defined as $h_E = \pi/\gamma_E$. If K_m is assumed to equal $u_* \cdot h_E$, then the Ekman layer depth for a neutral boundary layer becomes:

$$h_E = 2\pi^2 \left(\frac{u_*}{f_c} \right) \tag{6.4.6e}$$

The major conclusion from the Ekman solution is that friction reduces the boundary layer wind speed below geostrophic, and causes it to cross the isobars from high towards low pressure. In a synoptic situation where the isobars are curved, such as a low or high pressure system, the *cross-isobaric component of flow* near the surface causes convergence or divergence, respectively. Hence, mass continuity requires that there be rising air in low pressure systems, and descending air in highs. The process of inducing vertical motions by boundary layer friction is called *Ekman pumping*.

Ocean. The ocean drift current is driven by the wind stress at the surface, neglecting pressure gradients in the ocean.   Hence, the equations of motion reduce to:

$$f_c \overline{V} = - K_m \frac{\partial^2 \overline{U}}{\partial z^2}$$

$$f_c \overline{U} = + K_m \frac{\partial^2 \overline{V}}{\partial z^2} \qquad (6.4.6f)$$

This time, let us choose a coordinate system with the x-axis aligned with the surface stress, and z positive up. Thus, the four boundary conditions become: $K_m \partial \overline{U}/\partial z = u_*^2$ at z = 0, $\partial \overline{V}/\partial z = 0$ at z = 0, $\overline{U} \rightarrow 0$ as z $\rightarrow$ -∞, and $\overline{V} \rightarrow 0$ as z $\rightarrow$ -∞. Thus, the current is assumed to go to zero deep in the ocean. In the equations above, K_m and u_* refer to their ocean values, where $\rho_{water} \cdot u_{*water}^2$ = surface stress = $\rho_{air} \cdot u_{*air}^2$.
  The solution is:

$$\overline{U} = \left[\frac{u_*^2}{\left(K_m f_c \right)^{1/2}} \right] \left[e^{\gamma_E z} \cos\left(\gamma_E z - \frac{\pi}{4} \right) \right]$$

$$\overline{V} = \left[\frac{u_*^2}{\left(K_m f_c \right)^{1/2}} \right] \left[e^{\gamma_E z} \sin\left(\gamma_E z - \frac{\pi}{4} \right) \right] \qquad (6.4.6g)$$

where K_m and γ_E now apply to ocean values. This solution gives a surface current that is 45° to the right of the surface stress, making it parallel to the geostrophic wind in the atmosphere. Based on typical values of eddy viscosity in the air and ocean, the magnitude of the surface drift current is roughly G/30. Deeper in the ocean the current reduces in speed, and turns to the right as shown in Fig 6.3b. This causes horizontal convergence in the ocean under atmospheric regions of horizontal divergence, and vice versa. Hence, we expect *downwelling* water movement under synoptic high pressure systems, and *upwelling* under lows.

Discussion.   Although the Ekman solution is analytic and has been around for a long time, the conditions under which it was derived rarely happen in nature in the atmosphere.   At best, it gives an approximate quantitative solution for statically neutral boundary layers (i.e., mechanical turbulence production characteristic of strong winds, with no buoyancy effects).   *For convective mixed layers, the Ekman profile shape is not observed*, although it qualitatively agrees with the observed winds, which are subgeostrophic and cross-isobaric.   *Observed stable boundary layers can have supergeostrophic winds at low altitudes, making the Ekman solution even qualitatively incorrect.*

6.5 Local Closure — One-and-a-half Order

One-and-a-half-order closure retains the prognostic equations for the zero-order statistics such as mean wind, temperature, and humidity, and also retains equations for the variances of those variables.   The TKE equation is usually used in place of the velocity variance equations.   The following development is based on the work of Yamada & Mellor (1975).

As an example, consider the idealized scenario of a dry environment, horizontally homogeneous, with no subsidence.   The governing prognostic equations are (3.5.3) for the zero-order variables, (4.3.3) for the temperature variance, and (5.1b) for TKE:

$$\frac{\partial \overline{U}}{\partial t} = f_c\left(\overline{V} - \overline{V}_g\right) - \frac{\partial \,(\overline{u'w'})}{\partial z} \tag{6.5a}$$

$$\frac{\partial \overline{V}}{\partial t} = -f_c\left(\overline{U} - \overline{U}_g\right) - \frac{\partial \,(\overline{v'w'})}{\partial z}$$

$$\frac{\partial \overline{\theta}}{\partial t} = \qquad\qquad - \frac{\partial \,(\overline{w'\theta'})}{\partial z}$$

$$\frac{\partial \overline{e}}{\partial t} = -\overline{u'w'}\frac{\partial \overline{U}}{\partial z} - \overline{v'w'}\frac{\partial \overline{V}}{\partial z} + \left(\frac{g}{\overline{\theta}}\right)\overline{w'\theta'} - \frac{\partial \left[\overline{w'\,((p'/\rho) + e)}\right]}{\partial z} - \varepsilon$$

$$\frac{\partial \left(\overline{\theta'^2}\right)}{\partial t} = -2\,\overline{w'\theta'}\frac{\partial \overline{\theta}}{\partial z} - \frac{\partial \,(\overline{w'\theta'^2})}{\partial z} - 2\,\varepsilon_\theta - \varepsilon_R$$

The unknowns in this set of equations include second moments (fluxes): $\overline{u'w'}$, $\overline{v'w'}$, $\overline{w'\theta'}$, $\overline{w'p'}/\overline{\rho}$; third moments: $\overline{w'e}$, $\overline{w'\theta'^2}$;   and dissipations: ε, ε_θ, and ε_R.

At first glance, the addition of the variance equations appears to have hurt us rather than helped us. With first-order closure we had the 3 unknown fluxes: $\overline{u'w'}$, $\overline{v'w'}$, $\overline{w'\theta'}$. The addition of the variance equations did not eliminate these fluxes as unknowns, and in fact added 6 more unknowns. So why did we do it?

The reason is that knowledge of the TKE and temperature variance gives us a measure of the intensity and effectiveness of turbulence. Hence, we can use this information within improved parameterizations for eddy diffusivity, $K_m(\bar{e}, \overline{\theta'^2})$. One suggested set of parameterizations for the unknowns is:

$$\overline{u'w'} = -K_m\left(\bar{e}, \overline{\theta'^2}\right)\frac{\partial \overline{U}}{\partial z} \tag{6.5b}$$

$$\overline{v'w'} = -K_m\left(\bar{e}, \overline{\theta'^2}\right)\frac{\partial \overline{V}}{\partial z}$$

$$\overline{w'\theta'} = -K_H\left(\bar{e}, \overline{\theta'^2}\right)\frac{\partial \overline{\theta}}{\partial z} - \gamma_c\left(\bar{e}, \overline{\theta'^2}\right)$$

$$\overline{w'\left[(p'/\bar{\rho}) + e\right]} = \left(\frac{5}{3}\right)\Lambda_4 e^{-1/2}\frac{\partial \bar{e}}{\partial z}$$

$$\overline{w'\theta'^2} = \Lambda_3 e^{-1/2}\frac{\partial \overline{\theta'^2}}{\partial z}$$

$$\varepsilon_R = 0 \qquad \varepsilon = \frac{\bar{e}^{3/2}}{\Lambda_1} \qquad \varepsilon_\theta = \frac{\bar{e}^{1/2}\overline{\theta'^2}}{\Lambda_2}$$

where the Λ factors are empirical length-scale parameters. These length scales are often chosen by trial and error, in an attempt to make the simulated flow field match observed laboratory or field cases. One problem with the closure of (6.5b) is that the length scales are rather arbitrary.

The expressions for K are too complex to reproduce here, but can approximately be represented by:

$$K = \Lambda \bar{e}^{1/2} \tag{6.5c}$$

where Λ represents one of the empirical length scales.

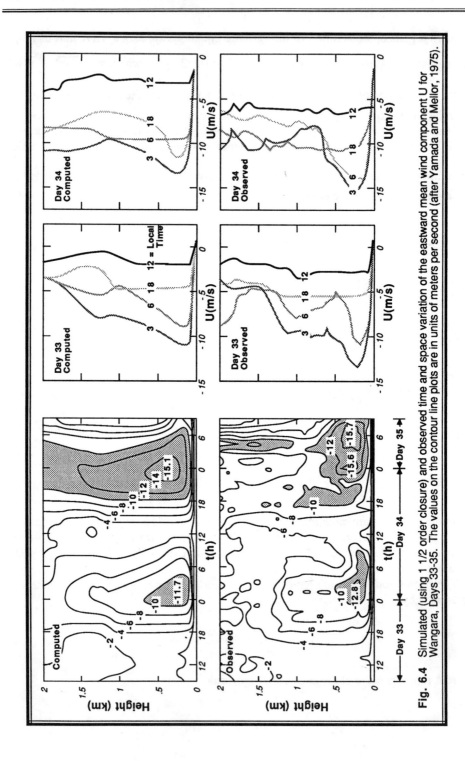

Fig. 6.4 Simulated (using 1 1/2 order closure) and observed time and space variation of the eastward mean wind component U for Wangara, Days 33-35. The values on the contour line plots are in units of meters per second (after Yamada and Mellor, 1975).

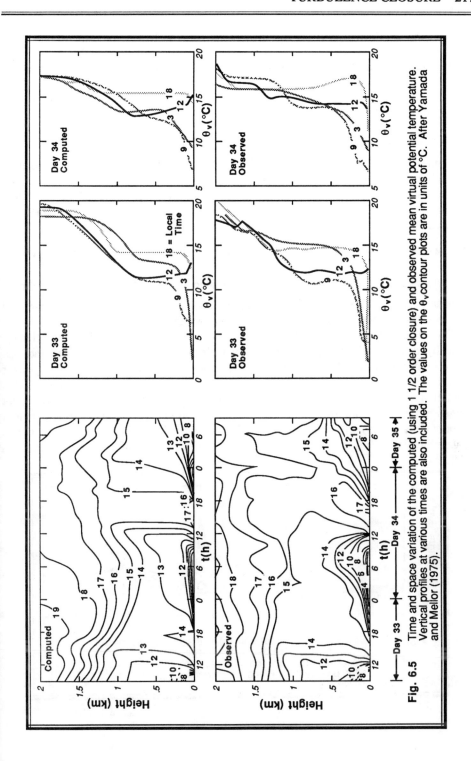

Fig. 6.5   Time and space variation of the computed (using 1 1/2 order closure) and observed mean virtual potential temperature. Vertical profiles at various times are also included. The values on the θ_v contour plots are in units of °C. After Yamada and Mellor (1975).

The second correlation terms are approximated as functions of gradients of mean values (e.g., the vertical flux of temperature, $\overline{w'\theta'}$, flows down the local vertical gradient of temperature $\partial\overline{\theta}/\partial z$). Similarly, the triple correlation terms are approximated as functions of gradients of second correlations (e.g., the vertical flux of temperature variance $\overline{w'\theta'^2}$ flows down the local vertical gradient of temperature variance $\partial\overline{\theta'^2}/\partial z$). The higher-order closure is thus very similar to the first-order closure — both depend on the local gradients and values of knowns.

The viscous dissipation terms of TKE and temperature variance are modeled as being proportional to their respective variables. For this reason, dissipation rates are sometimes used as a measure of the intensity of turbulence. More intense turbulence dissipates faster than weaker turbulence. The modeled dissipation rate has a timescale of $\Lambda/\overline{e}^{1/2}$. The γ_c parameter is added to the parameterization of $\overline{w'\theta'}$ to allow heat flux even when there is no mean gradient. This allows better representations of mixed layers. The pressure and turbulent transport terms assume that transport is down the mean TKE gradient.

The set of equations given by (6.5a & b) is too complex to solve analytically. Typically, these equations are approximated by their finite difference equivalents, and then solved numerically on a digital computer. Figs 6.4 through 6.6 shows a boundary-layer numerical simulation of a two day period from the Wangara field experiment. This simulation, produced by Yamada and Mellor using their one-and-a-half-order closure, not only shows the evolution of mean quantities such as wind (Fig 6.4) and virtual potential temperature (Fig 6.5), but it also shows the evolution of TKE (refer back to Fig 5.1) and temperature variance (Fig 6.6).

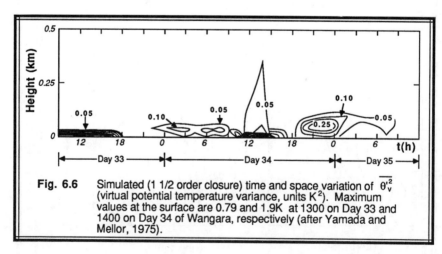

Fig. 6.6   Simulated (1 1/2 order closure) time and space variation of $\overline{\theta_v'^2}$ (virtual potential temperature variance, units K^2). Maximum values at the surface are 0.79 and 1.9K at 1300 on Day 33 and 1400 on Day 34 of Wangara, respectively (after Yamada and Mellor, 1975).

An alternative approach, called $\bar{e}$-ε *closure* (or *k*-ε *closure* in the engineering literature) avoids the Λ uncertainty by including a highly-parameterized prognostic equation for the dissipation rate in addition to the equation for TKE (Beljaars, et al., 1987; Kitada, 1987; Detering and Etling, 1985). The dissipation equation, which should be included with (6.5a), is sometimes written as:

$$\frac{\partial \varepsilon}{\partial t} = c_{\varepsilon 1} \frac{\varepsilon}{\bar{e}} \left[-\overline{u'w'} \frac{\partial \overline{U}}{\partial z} - \overline{v'w'} \frac{\partial \overline{V}}{\partial z} \right] + c_{\varepsilon 2} \frac{\varepsilon}{\bar{e}} \left(\frac{g}{\theta} \right) \overline{w'\theta'} - \frac{\partial \overline{w'\varepsilon}}{\partial z} - c_{\varepsilon 3} \frac{\varepsilon^2}{\bar{e}} \qquad (6.5d)$$

where the parameters are $c_{\varepsilon 1} = 1.44$, $c_{\varepsilon 2} = 1.0$, and $c_{\varepsilon 3} = 1.92$.

The following additional closure assumption should also be added to (6.5b):

$$\overline{w'\varepsilon'} = -\frac{K}{c_{\varepsilon 4}} \frac{\partial \varepsilon}{\partial z} \qquad (6.5e)$$

where $c_{\varepsilon 4} = 1.3$. In place of (6.5c), the eddy diffusivity can now be parameterized as:

$$K = \frac{(c_{\varepsilon 5} \bar{e})^2}{\varepsilon} \qquad (6.5f)$$

where $c_{\varepsilon 5} = 0.3$. Similarly, the remaining length scales in (6.5b) are hopefully more accurate because they are a function of prognostic variables:

$$\Lambda = \frac{\bar{e}^{3/2}}{\varepsilon} \qquad (6.5g)$$

The k-ε version of one-and-a-half-order closure has been used to simulate boundary layer evolution, flow over changes in roughness and topography, and sea-breeze fronts.

By studying this numerical simulation, we can learn some of the advantages of higher-order closure. (1) The higher-order scheme creates nearly well-mixed layers during the daytime that increase in depth with time. (2) At night, there is evidence of nocturnal jet formation along with the development of a statically stable layer near the ground. (3) Turbulence intensity increases to large values during the day, but maintains smaller values at night in the nocturnal boundary layer. First-order closure, on the other hand, gives no information on turbulence intensity or temperature variance. Furthermore, it has difficulty with well mixed layers that have zero gradients of mean variables. However, the benefits of higher-order closure do not come cheaply; they are gained at the expense of increased computer time and cost compared to first-order closure.

6.6   Local Closure — Second Order

The development of *higher-order closure* (usually meaning anything higher than first-order closure) was closely tied to the evolution of digital computer power. Although the use of higher-moment equations for turbulence forecasting was suggested in the early 1940's, the large number of unknown variables remained a stumbling block. Then, around 1950, Rotta and Chou and others suggested parameterizations for some of the unknowns. By the late 1960's, computer power improved to the point where second-order closure forecasts for clear air turbulence and shear flows were first made. In the early 1970's, the United States Environmental Protection Agency began funding some second-order closure pollution dispersion models, and by the mid 1970's a number of investigators were using such models. In fact, second-order closure appears to have started before one-and-a-half-order closure. In the late 1970's, some third-order closure models also started to appear in the literature, with many more third-order simulations published in the 1980's.

The set of second-order turbulence equations includes not only those from one-and-a-half-order, but it includes second moment terms as well (Wichmann and Schaller, 1986). Using the same idealized example as above, consider a dry environment, horizontally homogeneous, with no subsidence. The additional governing prognostic equations are

(4.4.1b) for $\overline{u_i'u_j'}$, and (4.4.3c) for $\overline{w'\theta'}$. The resulting set of coupled equations is:

$$\frac{\partial \overline{U_i}}{\partial t} = - f_c \varepsilon_{ij3} (\overline{U_{gj}} - \overline{U_j}) - \frac{\partial (\overline{u_i'w'})}{\partial z} \qquad (\text{for } i \neq 3) \tag{6.6a}$$

$$\frac{\partial \overline{\theta}}{\partial t} = - \frac{\partial (\overline{w'\theta'})}{\partial z}$$

$$\frac{\partial \overline{e}}{\partial t} = -\overline{u'w'} \frac{\partial \overline{U}}{\partial z} - \overline{v'w'} \frac{\partial \overline{V}}{\partial z} + \left(\frac{g}{\overline{\theta}}\right) \overline{w'\theta'} - \frac{\partial \left[\overline{w'((p'/\overline{\rho}) + e)}\right]}{\partial z} - \varepsilon$$

$$\frac{\partial \left(\overline{\theta'^2}\right)}{\partial t} = - 2 \overline{w'\theta'} \frac{\partial \overline{\theta}}{\partial z} - \frac{\partial (\overline{w'\theta'^2})}{\partial z} - 2\varepsilon_\theta - \varepsilon_R$$

$$\frac{\partial (\overline{u_i'u_j'})}{\partial t} = - \overline{u_i'w'} \frac{\partial \overline{U_i}}{\partial z} - \overline{u_j'w'} \frac{\partial \overline{U_i}}{\partial z} - \frac{\partial (\overline{u_i'u_j'w'})}{\partial z} + \left(\frac{g}{\overline{\theta}}\right)\left[\delta_{i3}\overline{u_j'\theta'} + \delta_{j3}\overline{u_i'\theta'}\right]$$

$$+ \left(\frac{\overline{p'}}{\rho}\right)\left[\frac{\partial u_i'}{\partial x_j} + \frac{\partial u_j'}{\partial x_i}\right] - 2\varepsilon_{u_iu_j}$$

$$\frac{\partial \overline{(u_i'\theta')}}{\partial t} = -\overline{w'\theta'}\frac{\partial \overline{U}_i}{\partial z} - \overline{u_i'w'}\frac{\partial \overline{\theta}}{\partial z} - \frac{\partial \overline{(u_i'w'\theta')}}{\partial z} + \delta_{i3}g\frac{\overline{\theta'^2}}{\overline{\theta}} + \overline{\left(\frac{1}{\overline{\rho}}\right)\left[p'\frac{\partial \theta'}{\partial x_i}\right]} - \varepsilon_{u\theta}$$

The unknowns in this set of equations include pressure-correlation terms: $(1/\overline{\rho})[\overline{p'\,\partial\theta'/\partial x_i}]$, $(\overline{p'/\rho})\,[\overline{\partial u_i'/\partial x_j + \partial u_j'\partial x_i}]$ and $\overline{w'p'/\rho}$; third moments: $\overline{w'e}$, $\overline{w'\theta'^2}$, $\overline{u_i'w'\theta'}$, and $\overline{u_i'u_j'w'}$; and dissipation terms: ε, ε_θ. Table 6-5 lists two different parameterizations (Deardorff, 1973; Donaldson, 1973) for these terms. Many other parameterizations have appeared in the literature (Launder, et al., 1975; Lumley and Khajeh-Nouri, 1974; Mellor and Yamada, 1974; Wyngaard, et al., 1974; Mellor and Herring, 1973; Hanjalic and Launder, 1972; Lumley and Mansfield, 1984; Rotta, 1951; Schumann, 1977; Wyngaard, 1982; Zeman, 1981; Wichmann and Schaller, 1986; Wai, 1987).

Table 6-5. Sample second-order closure parameterizations suggested by (A) Donaldson, and (B) Deardorff. (Reference: Workshop on Micrometeorology, 1973). The Λ_i are length scales, which are either held constant or based on mixing-length arguments.

1

$$\overline{u_i'u_j'u_k'} = -\Lambda_2\,e^{-1/2}\left[\frac{\partial \overline{u_i'u_j'}}{\partial x_k} + \frac{\partial \overline{u_i'u_k'}}{\partial x_j} + \frac{\partial \overline{u_k'u_j'}}{\partial x_i}\right] \quad (A)$$

$$= -\frac{3}{2}\left(\frac{\Lambda_2}{e^{-1/2}}\right)\left[\overline{u_k'u_m'}\frac{\partial \overline{u_i'u_j'}}{\partial x_m} + \overline{u_j'u_m'}\frac{\partial \overline{u_i'u_k'}}{\partial x_m} + \overline{u_i'u_m'}\frac{\partial \overline{u_k'u_j'}}{\partial x_m}\right] \quad (B)$$

2

$$\overline{u_i'u_j'\theta'} = -\Lambda_2\,e^{-1/2}\left[\frac{\partial \overline{u_i'\theta'}}{\partial x_j} + \frac{\partial \overline{u_j'\theta'}}{\partial x_i}\right] \quad (A)$$

$$= -\frac{3}{2}\left(\frac{\Lambda_2}{e^{-1/2}}\right)\left[\overline{\theta'u_m'}\frac{\partial \overline{u_i'u_j'}}{\partial x_m} + \overline{u_j'u_m'}\frac{\partial \overline{u_i'\theta'}}{\partial x_m} + \overline{u_i'u_m'}\frac{\partial \overline{\theta'u_j'}}{\partial x_m}\right] \quad (B)$$

3

$$\overline{u_i'\theta'^2} = -\Lambda_2\,e^{-1/2}\left[\frac{\partial \overline{\theta'^2}}{\partial x_i}\right] \quad (A)$$

$$= -\frac{3}{2}\left(\frac{\Lambda_2}{\overline{e}^{-1/2}}\right)\left[2\ \overline{\theta' u_m'}\ \frac{\partial \overline{u_i'\theta'}}{\partial x_m} + \overline{u_i'u_m'}\ \frac{\partial \overline{\theta'^2}}{\partial x_m}\right] \qquad \text{(B)}$$

4

$$\overline{\left(\frac{p'}{\rho}\right)\left[\frac{\partial u_i'}{\partial x_j} + \frac{\partial u_j'}{\partial x_i}\right]} = -\left(\frac{\overline{e}^{-1/2}}{\Lambda_1}\right)\left[\overline{u_i'u_j'} - \frac{2}{3}\delta_{ij}\ \overline{e}\right] \qquad \text{(Rotta, 1951)} \quad \text{(A)}$$

$$= -\left(\frac{\overline{e}^{-1/2}}{\Lambda_1}\right)\left[\overline{u_i'u_j'} - \frac{2}{3}\delta_{ij}\ \overline{e}\right] + \frac{2}{5}\overline{e}\left[\frac{\partial \overline{U_i}}{\partial x_j} + \frac{\partial \overline{U_j}}{\partial x_i}\right] \qquad \text{(B)}$$

5

$$\overline{\left(\frac{1}{\rho}\right)\left[p'\ \frac{\partial \theta'}{\partial x_i}\right]} = -\left(\frac{\overline{e}^{-1/2}}{\Lambda_1}\right)\overline{u_i'\theta'} \qquad \text{(A)}$$

$$= -\left(\frac{\overline{e}^{-1/2}}{\Lambda_1}\right)\overline{u_i'\theta'}\ -\frac{1}{3}\delta_{13}\ \frac{g}{\theta}\ \overline{\theta'^2} \qquad \text{(B)}$$

6

$$\frac{1}{\rho}\overline{\left(p'u_i'\right)} = -\left(\frac{\overline{e}^{-1/2}}{\Lambda_3}\right)\frac{\partial \overline{u_i'u_j'}}{\partial x_j} \qquad \text{(A,B)}$$

7

$$\varepsilon = \frac{\overline{e}^{-3/2}}{\Lambda_4} \qquad \text{(A,B)}$$

8

$$\varepsilon_\theta = \frac{\overline{e}^{-1/2}\ \overline{\theta'^2}}{\Lambda_4} \qquad \text{(A,B)}$$

Note: the $\overline{w'e}$ parameterization is the same as that for $\overline{u_i'u_j'u_k'}$.

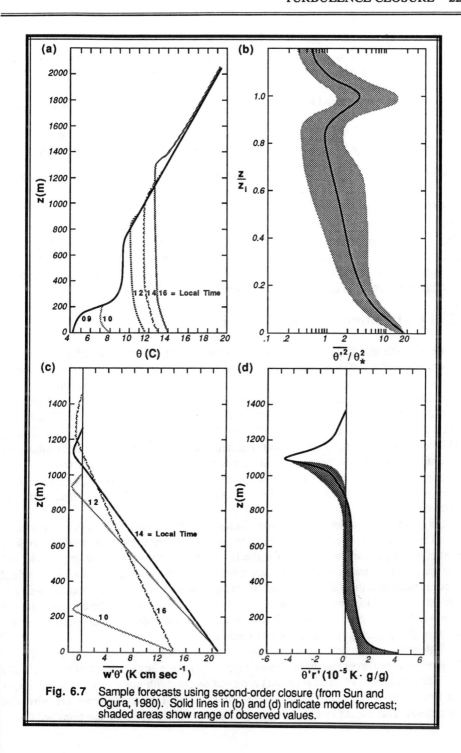

Fig. 6.7   Sample forecasts using second-order closure (from Sun and Ogura, 1980). Solid lines in (b) and (d) indicate model forecast; shaded areas show range of observed values.

There are three basic closure ideas contained in Table 6-5:
- *Down-gradient diffusion* (items 1-3 and 6 in the table), diffusion of the third-order statistics down the gradient of the second-order statistics;
- *Return to isotropy* (items 4 and 5), proportional to the amount of anisotropy;
- *Decay* (items 7 and 8), proportional to the magnitude of the turbulence.

A sample second-order closure model forecast is given in Fig 6.7, based on the moist convective boundary layer simulations of Sun and Ogura (1980). In addition to the equations listed above, they included prognostic equations for mixing ratio, $\bar{r}$, moisture variance $\overline{r'^2}$, moisture flux $\overline{w'r'}$, and temperature-moisture covariance, $\overline{r'\theta'}$. Using the full second-order set of equations, they could produce forecasts of mean variables (Fig 6.7a), as can be produced (with poorer accuracy) by first-order closure. They could forecast variances (Fig 6.7b), as can be produced (with poorer accuracy) by one-and-a-half-order closure. Most importantly, they can also produce forecasts of fluxes (Fig 6.7c) and other covariances (Fig 6.7d) that the lower-order schemes can not forecast.

6.7 Local Closure — Third Order

It is beyond the scope of this book to go into the details of third-order closure. In general, the prognostic equations for the triple-correlation terms are retained, while parameterizations are devised for the fourth-order correlations, for the pressure correlations, and for viscous dissipation. Some of the parameterizations presented in the literature (André, et al, 1978; Wyngaard, 1982; Moeng and Randall, 1984; Bougeault, 1981a, 1981b, 1985; Wichmann and Schaller, 1985; André and Lacarrere, 1985; Briere, 1987) assume that the fourth-order moments have a quasi-Gaussian probability distribution, and can be approximated as a function of second-moment terms.   Any unrealistic values for some of the third moments are truncated or *clipped* to remain within physically realistic ranges, and various eddy damping schemes are used to prevent negative variances.

It is generally assumed that equations for lower-order variables (such as mean wind or fluxes) become more accurate as the closure approximations are pushed to higher orders. In other words, parameterizations for the fourth-order terms might be very crude, but there are enough remaining physics (unparameterized terms) in the equations for the third moments that these third moments are less crude. The second moment equations bring in more physics, making them even more precise — and so on down to the equations for the mean wind and temperature, etc.   Based on the successful simulations published in the literature, this philosophy indeed seems to work.

Higher-order moments are extremely difficult to measure in the real atmosphere. Measurements of fluxes (second moments) typically have a large amount of scatter. Eddy correlation estimates of third moments are even worse, with noise or error levels larger than the signal level.   Accurate fourth-order moment measurements are virtually nonexistent. This means that we have very little knowledge of how these third and fourth moments behave; therefore, we have little guidance for suggesting good parameterizations for these moments.   Now we see why such crude approximations are made in third-order closure models.

Higher-order closure models have many parameters that can be adjusted advantageously to yield good forecasts.   These parameters are fine-tuned using special limiting case studies and laboratory flows where simplifications cause some of the terms to disappear, allowing better determination of the few remaining terms.

6.8 Nonlocal Closure — Transilient Turbulence Theory

Nonlocal closure recognizes that larger-size eddies can transport fluid across finite distances before the smaller eddies have a chance to cause mixing.   This advective-like concept is supported by observations of thermals rising with undiluted cores, finite size swirls of leaves or snow, and the organized circulation patterns sometimes visible from cloud photographs.

Two first-order nonlocal closure models will be presented here.   One, called *transilient turbulence theory*, approaches the subject from a physical space perspective.   The other, called *spectral diffusivity theory*, utilizes a spectral or phase-space approach.   Both allow a range of eddy sizes to contribute to the turbulent mixing process.

Two separate forms of transilient turbulence theory have evolved (Stull, 1984): one in discrete form for numerical modeling, and the other in analytical integral form for theoretical work.   We will start with the discrete form because it is easier to picture physically.

6.8.1 Discrete Form — Definition of Framework

Imagine a one-dimensional column of air that is split into separate equal-size grid boxes, as sketched in Fig 6.8a.   In a numerical model, grid point locations in the center of each box represent the average conditions within those boxes.   If we focus on just one particular (reference) grid box, we can identify those eddies that mix air into our reference box from other boxes above and below, and we can locate the destination boxes for air that leaves the reference box. This same procedure can be used to investigate mixing between all boxes in the column, as shown by the superposition of eddies in Fig 6.8b.

Turbulent mixing of fluid into our reference box (the box with index i) from the other boxes can change the state (temperature, humidity, tracer concentration, etc.) of that box.

For example, let $\overline{\xi}_i$ represent the average concentration of passive tracer in our reference box, i.   If c_{ij} represents the fraction of air in box i that came from box j during a time interval Δt, then we need only sum the mixing from over all N grid boxes in the column to find the new concentration at box i:

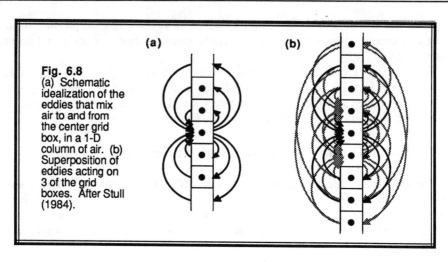

Fig. 6.8
(a) Schematic idealization of the eddies that mix air to and from the center grid box, in a 1-D column of air. (b) Superposition of eddies acting on 3 of the grid boxes. After Stull (1984).

$$\bar{\xi}_i(t + \Delta t) = \sum_{j=1}^{N} c_{ij}(t, \Delta t)\, \bar{\xi}_j(t) \qquad (6.8.1)$$

This equation says that when air is mixed into box i from box j, the air carries with it an amount c_{ij} of the tracer with concentration $\bar{\xi}_j$. The coefficient c_{ii} represents the fraction of air in box i that remains within box i.

Although (6.8.1) was developed for just one reference box, it is general enough to work for any and all equal-size boxes. If we recognize that c_{ij} is an NxN matrix of mixing coefficients (called a *transilient matrix*), and $\bar{\xi}_j$ is a Nx1 matrix (i.e., a vector) of concentrations, then it is obvious that (6.8.1) describes simple matrix multiplication.

When eddies move parcels of air from one grid box to another, the air will carry with it not only the tracer concentration, but the heat, moisture, momentum and other measures of the state of the fluid. Hence, (6.8.1) can be used for any of these variables, where the transilient matrix is the same for each variable (i.e., it changes only with time and timestep). Of course, the $\bar{\xi}_j$ vector is different for each variable.

A variety of physical processes can be modeled with the transilient scheme depending on the form of the transilient matrix. Examples include complete mixing, top-down / bottom-up mixing, small-eddy mixing (like K-theory), cloud top detrainment, a detraining updraft core, patchy turbulence, no turbulence, or eddies triggered by the surface layer. These are illustrated in Fig 6.9.

No Mixing

Grid Boxes   Transient Matrix

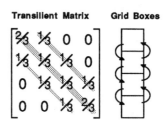

$$\begin{bmatrix} 1 & 0 & 0 & 0 \\ 0 & 1 & 0 & 0 \\ 0 & 0 & 1 & 0 \\ 0 & 0 & 0 & 1 \end{bmatrix}$$

Small Eddy Mixing (K-theory)

Transient Matrix   Grid Boxes

$$\begin{bmatrix} 2/3 & 1/3 & 0 & 0 \\ 1/3 & 1/3 & 1/3 & 0 \\ 0 & 1/3 & 1/3 & 1/3 \\ 0 & 0 & 1/3 & 2/3 \end{bmatrix}$$

Complete Mixing

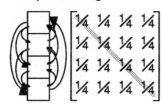

$$\begin{bmatrix} 1/4 & 1/4 & 1/4 & 1/4 \\ 1/4 & 1/4 & 1/4 & 1/4 \\ 1/4 & 1/4 & 1/4 & 1/4 \\ 1/4 & 1/4 & 1/4 & 1/4 \end{bmatrix}$$

Cloud-top Detrainment

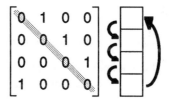

$$\begin{bmatrix} 0 & 1 & 0 & 0 \\ 0 & 0 & 1 & 0 \\ 0 & 0 & 0 & 1 \\ 1 & 0 & 0 & 0 \end{bmatrix}$$

Patchy Turbulence

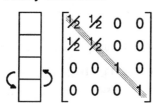

$$\begin{bmatrix} 1/2 & 1/2 & 0 & 0 \\ 1/2 & 1/2 & 0 & 0 \\ 0 & 0 & 1 & 0 \\ 0 & 0 & 0 & 1 \end{bmatrix}$$

Detraining Updraft Core

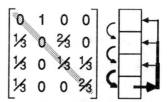

$$\begin{bmatrix} 0 & 1 & 0 & 0 \\ 1/3 & 0 & 2/3 & 0 \\ 1/3 & 0 & 1/3 & 1/3 \\ 1/3 & 0 & 0 & 2/3 \end{bmatrix}$$

Top - down, bottom up

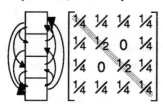

$$\begin{bmatrix} 1/4 & 1/4 & 1/4 & 1/4 \\ 1/4 & 1/2 & 0 & 1/4 \\ 1/4 & 0 & 1/2 & 1/4 \\ 1/4 & 1/4 & 1/4 & 1/4 \end{bmatrix}$$

Eddies Triggered by One Layer

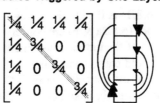

$$\begin{bmatrix} 1/4 & 1/4 & 1/4 & 1/4 \\ 1/4 & 3/4 & 0 & 0 \\ 1/4 & 0 & 3/4 & 0 \\ 1/4 & 0 & 0 & 3/4 \end{bmatrix}$$

Fig. 6.9   Examples of transilient matrices showing nonlocal mixing possibilities. The arrows are not physical eddies, but they represent the net mixing effect of many real eddies acting in 3-D space. Grid index conventions are indicated in the top left drawing.

6.8.2 Physical Constraints

If the total amount of air within the reference box does not change with time, then just as much air must leave the box during Δt as enters. Expressed in another way, if c_{ij} is the fraction of air entering box i from box j, then by definition the *conservation of air mass* requires that the sum over j of all mixing fractions be unity:

$$1 = \sum_{j=1}^{N} c_{ij} \tag{6.8.2a}$$

One way to visualize this is to picture a column of air with initially the same tracer concentration in each box. We known that after any amount of mixing, the final concentration in any of the boxes must equal the initial concentration: $\bar{\xi}_j = \bar{\xi}_i = \sum_{j=1}^{N} c_{ij} \bar{\xi}_j$. This is satisfied by (6.8.2a).

Also, if the amount of tracer initially in box j is conserved as it mixes out of j and into the other boxes, i, then *conservation of tracer amount* requires that:

$$1 = \sum_{i=1}^{N} c_{ij} \tag{6.8.2b}$$

One way to appreciate this constraint is to picture a column of air with no tracers in any box, except for a unit amount of tracer in one box ($\bar{\xi}_j = 1$). If all of the tracer is initially in box j, then after the mixing associated with one timestep, some of this tracer could be mixed into any or all of the other boxes. Nevertheless, the total amount of tracer in all boxes must still sum to unity ($\sum_{i=1}^{N} \bar{\xi}_i = 1$). The only way that this is possible given mixing defined by (6.8.1) is if (6.8.2b) is true.

We see that (6.8.2a & b) are conservation constraints for air mass and tracer mass. In addition, none of the individual elements c_{ij} should be allowed to be negative, otherwise turbulence would cause "unmixing" and would decrease randomness or entropy. Thus, each element of the matrix must be $0 \le c_{ij} \le 1$, and each row and each column must sum to one. Such a matrix is sometimes called a *doubly stochastic matrix*.

6.8.3 Numerical Constraints.

One ancillary feature of the physical constraints is that a numerical forecast based on transilient turbulence theory is *absolutely numerically stable* for any size timestep and any size grid spacing (Stull, 1986). No additional numerical constraints are required to achieve this characteristic.

To show this, we can employ Gershgorin's theorem of linear algebra (Pearson, 1974) that states that the largest modulus (magnitude) of any of the eigenvalues of the transilient matrix is no greater than the largest sum of any row or column of our non-negative matrix. Because each of our rows and columns sum to one, we know that the largest eigenvalue modulus is no greater than one. This latter condition means that numerical stability is insured. See Haltiner and Williams (1980) for details of this last statement.

There is one numerical constraint that is recommended: that no eigenvalue of the transilient matrix be negative. It can be shown that a transilient matrix with a negative eigenvalue causes the tracer concentration to oscillate from timestep to timestep. This characteristic is undesirable because the solution depends on the timestep rather than the physics. An example of such a bad matrix is one that causes convective overturning, as is shown below:

$$\begin{bmatrix} 0 & 0 & 1 \\ 0 & 1 & 0 \\ 1 & 0 & 0 \end{bmatrix}$$

As a general guideline, transilient matrices with large elements clustered around the cross-diagonal of the matrix rather than around the main diagonal are matrices likely to be timestep dependent.

6.8.4 Flux Determination

Turbulent kinematic fluxes are also easy to determine, because the transilient matrix tells us directly about the transport between grid boxes. The kinematic flux $\overline{w'\xi'}(k)$ across level k is given by:

$$\overline{w'\xi'}(k) = \left(\frac{\Delta z}{\Delta t}\right) \sum_{i=1}^{k} \sum_{j=1}^{N} c_{ij}\left(\bar{\xi}_i - \bar{\xi}_j\right) \tag{6.8.4a}$$

where Δz is the grid point spacing, and Δt is the timestep interval for the $c_{ij}(t,\Delta t)$ matrix. The level k is defined as the border between grid boxes k and $k+1$. Thus, although $\bar{\xi}_i$ is known at the center of a grid box, $\overline{w'\xi'}(k)$ is known at the edge of the grid box. This makes physical sense, because the flux represents the transport between grid boxes.

An outline is given here for the derivation of (6.8.4a). First, start with a simple conservation equation where the only forcing is vertical flux divergence: $\partial \bar{\xi} / \partial t$ = - $\overline{\partial w' \xi'} / \partial z$. Integrate this over one timestep, from t to t+Δt, and rewrite the result as

$\overline{\partial w' \xi'}^{\Delta t} / \partial z = [\bar{\xi}(t) - \bar{\xi}(t+\Delta t)] / \Delta t$. Next, integrate this from the surface (z=0) to the height of interest where the flux is desired (z=Z), where Z = k Δz. By splitting the integral of the right side of the equation into the sum of smaller integrals, where each small integral is over one grid box of depth Δz, our equation now becomes

$\overline{w' \xi'}^{\Delta t} (Z) - \overline{w' \xi'}^{\Delta t} (0) = (\Delta z / \Delta t) \sum_{i=1}^{k} [\bar{\xi}_i(t) - \bar{\xi}_i(t+\Delta t)]$. But the turbulent flux across a solid boundary, or into any nonturbulent part of the atmosphere, is zero by definition, making the second term on the left become zero. Also, the term in square brackets can be written using (6.8.1) and (6.8.2a) as $[\sum_{j=1}^{N} c_{ij} \bar{\xi}_i - \sum_{j=1}^{N} c_{ij} \bar{\xi}_j]$. A bit of algebraic manipulation puts the final result into the form of (6.8.4a).

As just stated, the turbulent flux across a solid boundary is zero by definition, even though the nonturbulent fluxes might not be zero (see Section 7.1 for a discussion of effective fluxes). Thus, $\overline{w' \xi'}(k=0) \equiv 0$. With this boundary condition, (6.8.4a) can be rewritten as a recursion relationship:

$$\overline{w' \xi'} (k) = \overline{w' \xi'} (k - 1) + \left(\frac{\Delta z}{\Delta t}\right) \sum_{j=1}^{N} c_{kj} \left(\bar{\xi}_k - \bar{\xi}_j\right) \tag{6.8.4b}$$

6.8.5 Example of Transilient Framework

Problem: Suppose that we start with a shallow mixed layer of depth 300 m within a 500 m column of air, with initial profiles of potential temperatures and winds as indicated below. Assume the column is divided into five equal-thickness (100 m) grid boxes.

Profile A:

Given:	Grid index =	1	2	3	4	5	
	z (m) =	50	150	250	350	450	(At the box centers)
	$\bar{\theta}$ (°C) =	15	15	15	16	18	
	$\bar{U}$ (m/s) =	5	5	5	7	6	

Next, assume that there are molecular (nonturbulent) fluxes of heat, $Q_H = 0.2$ K m/s, and momentum, $F = -0.15$ m^{2}s^{-2}, across the land surface into the air. During a 10 min timestep, we can assume that there is a surface molecular flux into the bottom of the bottom layer, but no molecular flux out of the top of the bottom layer, giving $\Delta\theta_1/\Delta t = Q_H/\Delta z$ and $\Delta U_1/\Delta t = F/\Delta z$. Neglecting other forcings such as radiation and Coriolis force leaves only the bottom layer altered:

Profile B:
Given:

Grid index =	1	2	3	4	5
z (m) =	50	150	250	350	450
$\bar{\theta}$ (°C) =	16.2	15	15	16	18
$\bar{U}$ (m/s) =	4.1	5	5	7	6

Notice that only the bottom one grid box has changed so far, because we have not yet applied the transilient turbulence to mix these heat and momentum changes higher in the mixed layer.

Next, assume that there is turbulent mixing during the 10 min period as specified by the transilient matrix below:

(j =	1	2	3	4	5)	
	0.590	0.236	0.118	0.056	0.000	(i=1)
	0.236	0.590	0.118	0.056	0.000	(2)
c (t, Δt = 10 min) =	0.118	0.118	0.708	0.056	0.000	(3)
	0.056	0.056	0.056	0.832	0.000	(4)
	0.000	0.000	0.000	0.000	1.000	(5)

Starting with Profile B, (a) determine and plot the final profile (let's call it Profile C) after turbulent mixing; and (b) determine and plot the fluxes of heat and momentum.

Solution to part (a):   First, it is easy to verify that the transilient matrix is a valid one, with each row and column summing to one, and no negative elements. Next, to illustrate the solution, look at the second row of the transilient matrix. The second row tells us about fluid that is mixing *into* the second grid box. For our case, it says that 23.6% comes *from* box 1, 59% stays in box 2, 11.8% comes from box 3, and so forth. Thus, the new state of box two after mixing is:

$$\bar{\theta}_2 = 0.236 \cdot (16.2) + 0.590 \cdot (15) + 0.118 \cdot (15) + 0.056 \cdot (16) + 0.000 \cdot (18) = 15.34 \; (°C)$$
and
$$\bar{U}_2 = 0.236 \cdot (4.1) + 0.590 \cdot (5) + 0.118 \cdot (5) + 0.056 \cdot (7) + 0.000 \cdot (6) = 4.90 \; (m/s)$$

In general we can set up the problem as a matrix multiplication, with the answer indicated below for potential temperature. The same procedure can be used for winds.

$$
\begin{bmatrix} 15.76 \\ 15.34 \\ 15.20 \\ 15.90 \\ 18.00 \end{bmatrix} = \begin{bmatrix} 0.590 & 0.236 & 0.118 & 0.056 & 0.000 \\ 0.236 & 0.590 & 0.118 & 0.056 & 0.000 \\ 0.118 & 0.118 & 0.708 & 0.056 & 0.000 \\ 0.056 & 0.056 & 0.056 & 0.832 & 0.000 \\ 0.000 & 0.000 & 0.000 & 0.000 & 1.000 \end{bmatrix} \begin{bmatrix} 16.2 \\ 15.0 \\ 15.0 \\ 16.0 \\ 18.0 \end{bmatrix}
$$

Profile C: The state of the air after 10 minutes.

Given:

Grid index =	1	2	3	4	5	
z (m) =	50	150	250	350	450	(At the box centers)

$\bar{\theta}$ (°C)   = 15.76 15.34 15.20 15.90 18.00

$\bar{U}$ (m/s)   = 4.58   4.90   5.01   6.61   6.00

Discussion of part (a): The new temperature in grid box 5 has not changed, because the fifth element along the main diagonal of the transilient matrix equalled one. Thus, 100% of the air in box 5 stayed in box 5, and no air was mixed in from other boxes. The initial and final potential temperature and wind profiles are plotted in Figs 6.10 a & b.

Some of the warm air from the surface has been mixed up into the mixed layer, resulting in warming of the mixed layer and a reduction of the static instability between the bottom two grid points. Also, there appears to be some entrainment into the top of the mixed layer, resulting in an increase in mixed layer depth.

Solution to part (b): For example, the flux at z=100 m (i.e., at k=1, between grid points 1 and 2) can be found using recursive relationship (6.8.4b):

$$
\overline{w'\theta'}(1) = \overline{w'\theta'}(0) + \frac{\Delta z}{\Delta t} \sum_{j=1}^{N} c_{1j} (\overline{\theta}_1 - \overline{\theta}_j)
$$

$\overline{w'\theta'}(1) = \quad 0 \quad + \quad (100 \text{ m} / 600 \text{ s}) \cdot [0.590 \cdot (16.2\text{-}16.2) + 0.236 \cdot (16.2\text{-}15.0) +$
$\qquad\qquad\qquad 0.118 \cdot (16.2\text{-}15.0) + 0.056 \cdot (16.2\text{-}16.0) + 0.000 \cdot (16.2\text{-}18.0)]$
$\qquad\quad = 0.0726 \text{ (K·m·s}^{-1})$

and for momentum

$\overline{u'w'}(1) = \quad 0 \quad + \quad (100 \text{ m} / 600 \text{ s}) \cdot [0.590 \cdot (4.1\text{-}4.1) + 0.236 \cdot (4.1\text{-}5.0) +$
$\qquad\qquad\qquad 0.118 \cdot (4.1\text{-}5.0) + 0.056 \cdot (4.1\text{-}7.0) + 0.000 \cdot (4.1\text{-}6.0)]$
$\qquad\quad = -0.080 \text{ (m}^2\text{s}^{-2})$

Similarly, the turbulent fluxes between the other grid points can be found:

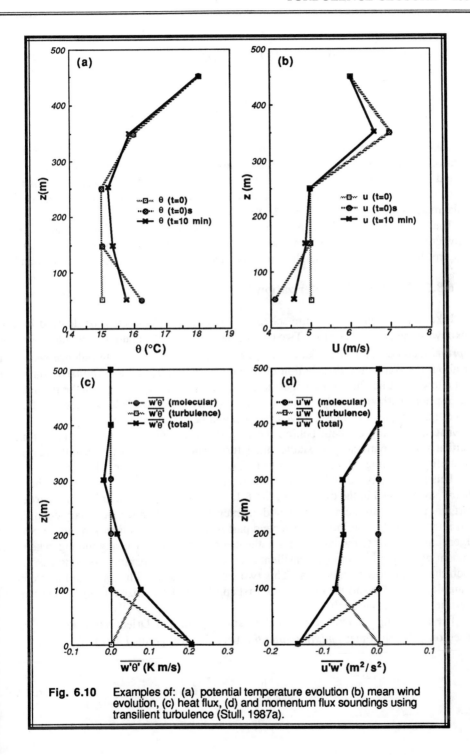

Fig. 6.10 Examples of: (a) potential temperature evolution (b) mean wind evolution, (c) heat flux, (d) and momentum flux soundings using transilient turbulence (Stull, 1987a).

Grid index =	0	1	2	3	4	5
z (m) =	0	100	200	300	400	500
$\overline{w'\theta'}$ (K m/s) =	0	0.0726	0.0160	-0.0169	0.0	0.0
$\overline{u'w'}$ (m²s⁻²) =	0	-0.0802	-0.0634	-0.0644	0.0	0.0

Discussion of part (b): These flux profiles are plotted in Figs 6.10 c & d. In addition to the turbulent fluxes in these figures, the molecular surface fluxes are also plotted. The total flux at any height is the sum of the turbulent and molecular fluxes, as indicated by the solid lines. It is this total flux that is usually plotted as the effective flux in the meteorological literature (see Chapter 7).

The heat flux profile shows the expected decrease of flux with height characteristic of convective mixed layers, with a negative value in the entrainment zone. The momentum flux profile is negative at all heights, implying a loss of momentum down to the ground.

6.8.6 Closure Parameterization

Like other turbulence closure schemes, a parameterization must be devised for the unknowns: the c_{ij} coefficients. Two closure methods have been used in the literature: an *a-priori* method that utilizes knowledge or assumptions about the turbulence spectrum or about the frequency distribution of turbulent velocities; and a **responsive** approach that allows the transilient coefficients to change in response to changes in the mean flow. The a-priori method is used with the spectral diffusivity nonlocal closure described at the end of this Chapter. The responsive approach will be described here.

The underlying concept behind the responsive closure is that if the fluid is made statically or dynamically unstable, then turbulence will form to partially undo the instability by mixing. This is analogous to LeChatelier's Principle of chemistry.

This principle is carried into the numerical implementation of transilient turbulence theory, as sketched in the time line of Fig 6.11. Each timestep is split into two parts, one part that includes all the nonturbulent dynamics, thermodynamics, boundary conditions, and internal (body) forcings. The second part is where the transilient turbulence closure algorithm responds to the instabilities in the mean flow field by causing mixing, where the amount of mixing is given by transilient coefficients that are parameterized to (partially) undo the mean flow instabilities. This two step process was illustrated in the previous example, where the surface fluxes were first applied to destabilize the flow, and then the transilient mixing was applied.

The following responsive parameterization (Stull and Driedonks, 1987) for c_{ij} is based on a dimensionless **mixing potential** between grid points i and j, Y_{ij}:

$$c_{ij} = \frac{Y_{ij}}{\|Y\|} \qquad \text{for } i \neq j \qquad (6.8.6a)$$

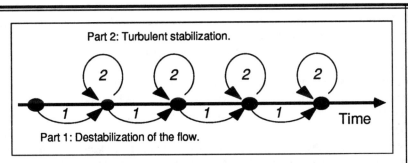

Fig. 6.11   Timeline showing the two parts of the responsive parameterization. Part 1 consists of the dynamics, thermodynamics, body forcing and boundary conditions. Part 2 consists of the transilient mixing to reduce any flow instabilities that might have developed during part 1. After Stull (1987).

where $\|Y\|$ is a scalar norm of matrix Y, and:

$$c_{ii} = 1 - \sum_{\substack{j=1 \\ j \neq i}}^{N} c_{ij} \qquad (6.8.6b)$$

to satisfy the conservation constraint discussed earlier.

The potential for mixing depends on the instability of the flow, so a natural starting point for the parameterization of Y is the TKE equation.   Start with (5.1b), neglect the turbulent transport and pressure correlation terms for now, allow for an arbitrary coordinate direction, integrate over time, and then normalize the resulting equation by dividing by the TKE. Next, and most importantly, we hypothesize that the result can be interpreted nonlocally, where gradients can be expressed as differences across finite distances. Let Δ_{ij} refer to a nonlocal difference between grid points i and j; for example,

$\Delta_{ij}\overline{U} = \overline{U}_i - \overline{U}_j$. We can finally write the equation for potential for mixing between grid points i and j, for i≠j:

$$Y_{ij} = \left[\frac{\Delta t\, T_o}{(\Delta_{ij}z)^2} \right] \left[(\Delta_{ij}\overline{U})^2 + (\Delta_{ij}\overline{V})^2 - g(\Delta_{ij}\overline{\theta}_v) \frac{(\Delta_{ij}z)}{(\theta_v R_c)} \right] - \frac{D_Y \Delta t}{T_o} \qquad (6.8.6c)$$

The first two terms on the right represent the mechanical production, the third term is the buoyant production/consumption, and the last term represents dissipation.

The Y_{ij} values found from (6.8.6c) are the preliminary off-diagonal elements of a **Y** matrix. We recommend that the mixing potential for eddies of any size be no less than the mixing potential for larger size eddies. In other words, if there is strong mixing between

100 m and 500 m, then there is at least as much mixing between 200 m and 300 m. It would be ridiculous to allow turbulence to mix fluid between two distant points across an interior region that was nonturbulent. Thus, we make the additional requirement that the elements of the Y matrix increase monotonically from the upper-right and lower-left corners toward the main diagonal. Any of the preliminary elements that do not satisfy this last requirement are increased to equal the largest elements further away from the diagonal. The elements along the main diagonal are set equal the the largest immediate element on the same row, plus a Y_{ref} value that represents the potential for subgrid scale mixing within one box.

If we assume that the mixing between i and j equals the mixing between j and i, then both the mixing potential matrix and the transilient matrix are symmetric. This reduces the degrees of freedom from $(N^2 - N)$ to $(N^2 - N)/2$. By using the mixing potential parameterization just described, we further reduce the degrees of freedom to just four, based on the four empirical parameters. Their values are recommended (based on atmospheric simulation tests) to be $T_o = 1000$ s, $D_Y = 1$ (dimensionless), the critical Richardson number is $R_c = 0.21$, and $Y_{ref} = 1000$ (dimensionless).

Finally, if each row of the final mixing potential matrix is summed, the scalar norm is set equal to the maximum sum. Namely,

$$\|Y\| = \max_i \left[\sum_{j=1}^{N} Y_{ij} \right] \qquad (6.8.6d)$$

Fig 6.12 shows an example of a boundary layer forecast made with the parameterization above. The growth of an entraining mixed layer is clearly evident, even though no explicit mixed layer or boundary layer specification was imposed other than the surface heating that acted on the bottom one grid point. The corresponding wind and flux forecasts for this case are shown by Stull and Driedonks (1987).

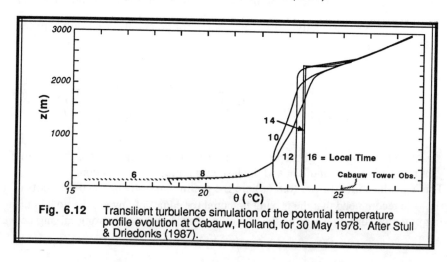

Fig. 6.12   Transilient turbulence simulation of the potential temperature profile evolution at Cabauw, Holland, for 30 May 1978. After Stull & Driedonks (1987).

6.8.7   Continuous Form

The analytical form of transilient turbulence theory can be written as

$$\frac{\partial \bar{\xi}(z,t)}{\partial t} = TM + Prod. - Loss \qquad (6.8.7a)$$

where Prod. and Loss are other production and loss terms, respectively, given by the conservation equations. The turbulent mixing (flux divergence) term, TM, is:

$$TM\,(z,t) = \int_{Z=z_b}^{z_t} \left[\,\bar{\xi}\,(Z,t) - \bar{\xi}\,(z,t)\right]\,\gamma_T(z,Z,t)\,dZ \qquad (6.8.7b)$$

where $\gamma_T(z,Z,t)$ is a transilient rate coefficient (units of $s^{-1}m^{-1}$) for mixing between level Z and level z. To help understand $\gamma_T(z,Z,t)$, we note that the kinematic mass flux of air moving from $Z+\Delta z$ to $z+\Delta z$ is $M = \iint \gamma(z,Z,t)dz\,dZ$. The top and bottom of the turbulent domain are at $z=z_b$ and z_t, respectively.

A typical shape for $\gamma_T\,(z,Z,t)$ is shown in Fig 6.13. This graph is similar to a graph of the magnitudes of the c_{ij} elements along any one row of the transilient matrix, where $i=j$ (main diagonal) corresponds to $z-Z=0$.

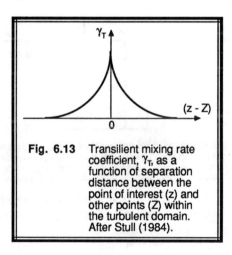

Fig. 6.13   Transilient mixing rate coefficient, γ_T, as a function of separation distance between the point of interest (z) and other points (Z) within the turbulent domain. After Stull (1984).

The turbulent kinematic flux, $\overline{w'\xi'} = F(z_1)$ at location z_1 is given by:

$$\overline{w'\xi'}(z_1,t) = \int\limits_{z=z_b}^{z_1} \int\limits_{Z=z_b}^{z_t} \left[\bar{\xi}(Z,t) - \bar{\xi}(z,t) \right] \underset{T}{\gamma}(z,Z,t) \ dZ \ dz \qquad (6.8.7c)$$

By definition, the turbulent flux at the top and bottom of the turbulent domain is zero.

6.8.8 Example of Continuous Form

Problem: Given the idealized transilient rate curve shown in Fig 6.14a, and the initial sounding of potential temperature shown in Fig 6.14b, calculate the initial tendency, $\partial\bar{\theta}/\partial t$ from (6.8.7a), neglecting other production and loss terms. The initial potential temperature sounding below the temperature jump (i.e., at $z<0$, assuming the origin is placed at the height of the jump) is constant and equal to $\bar{\theta}_o$, while above that height the temperature is $\bar{\theta}_o + \Delta\bar{\theta}$. The transilient rate curve is zero everywhere, except $\gamma_T = (A/L)(L - Z + z)$ for $0 < (Z-z) < L$, and $\gamma_T = (A/L)(L + Z - z)$ for $-L < (Z-z) < 0$.

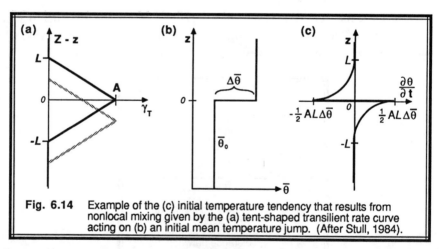

Fig. 6.14   Example of the (c) initial temperature tendency that results from nonlocal mixing given by the (a) tent-shaped transilient rate curve acting on (b) an initial mean temperature jump. (After Stull, 1984).

Solution: The transilient rate curve indicates that the largest eddies (i.e., large values of $|Z - z|$) present for this case are of size L. Thus, for $z < -L$ and $z > L$, mixing can not change the temperature because the smaller eddies (less than L) are mixing air of the same temperature. Air of different temperature from the other side of the temperature jump is out of reach of these small eddies. Thus, by inspection we can write the first part of our answer:

For $z < -L$ and $L < z$:

$$\frac{\partial \overline{\theta}}{\partial t} = 0 \qquad \text{for } \begin{cases} z > L \\ z < -L \end{cases}$$

For $-L < z < 0$:

We must integrate (6.8.7b), using the initial condition that $\overline{\theta}(Z) - \overline{\theta}(z)$ equals 0 for $Z < 0$, and equals $\Delta\overline{\theta}$ for $Z > 0$. Thus, the initial (t=0) tendency of potential temperature is:

$$\frac{\partial \overline{\theta}}{\partial t}(z) = TM = \int_{Z=z_b}^{z_T} \left[\overline{\theta}(Z) - \overline{\theta}(z) \right] Y_T(Z,z) \, dZ$$

$$= \frac{A}{L} \int_{Z=z_b}^{0} [0] \, Y_T \, dZ \; + \; \frac{A}{L} \int_{Z=0}^{z+L} [\Delta\overline{\theta}] \, (L - Z + z) \, dZ \; + \; \frac{A}{L} \int_{Z=z+L}^{z_T} [\Delta\overline{\theta}] \, (0) \, dZ$$

$$= \frac{A \, \Delta\overline{\theta}}{2 \, L} \, (L+z)^2$$

For $0 < z < L$:
Integrating in a similar manner we find:

$$\frac{\partial \overline{\theta}}{\partial t}(z) = - \frac{A \, \Delta\overline{\theta}}{2 \, L} \, (L-z)^2$$

Discussion: These initial tendencies are plotted in Fig 6.14c. Just above the temperature jump the tendency is negative, indicating cooling. Just below the jump, there is warming. The net result is that turbulence is tending to round the corners of the temperature sounding by mixing cooler air up and warmer air down. Also, these initial tendencies are nonzero a finite distance away from the jump, something that can not be modeled with a local closure theory such as K-theory.

6.8.9 Closure Parameterization of the Continuous Form

In order to close the continuous form of transilient turbulence theory, we need to find an analytical expression for Y_T as a function of z, Z, and t. One easy way to do this employs the a-priori method of closure: namely, assuming some turbulent state of the air rather than letting it respond to changes in the mean flow. For example, given the spectrum of turbulence with an inertial subrange, it can be shown (Stull, 1984) that

$$\gamma_T (z,Z,t) \propto \varepsilon^{1/3} |Z\text{-}z|^{-5/3} \tag{6.8.9}$$

where ε is the TKE dissipation rate.   Many other closure parameterizations can be formulated that lead to interesting theoretical results.

6.9 Nonlocal Closure — Spectral Diffusivity Theory

We saw in Section 6.4.1 that K-theory can be used to approximate a flux as $\overline{w'\xi'} = -K$

$\partial\overline{\xi}/\partial z$. When this is put back into the conservation equation for a passive tracer, $\partial\overline{\xi}/\partial t =$

$-\partial\overline{w'\xi'}/\partial z$, we arrive at the diffusion equation $\partial\overline{\xi}/\partial t = K \partial^2\overline{\xi}/\partial z^2$, where we have assumed that K is not a function of z for simplicity.   Suppose that K varies with eddy size (Berkowicz and Prahm, 1979; Prahm et al., 1979; Berkowicz, 1984).   If we let κ=wavenumber of the eddy, then we can spectrally decompose the diffusion equation to be

$$\frac{\partial\overline{\xi}(\kappa)}{\partial t} = K(\kappa)\frac{\partial^2\overline{\xi}(\kappa)}{\partial z^2} \tag{6.9a}$$

where $K(\kappa)$ is called the **spectral turbulent diffusivity**.
By integrating this equation over all wavenumbers, we again arrive at the forecast

equation for $\overline{\xi}$:

$$\frac{\partial\overline{\xi}(z,t)}{\partial t} = TM + Prod. - Loss$$

where the turbulent mixing term is now

$$TM(z,t) = \frac{\partial \int \left[\Xi(z,Z,t)\frac{\partial\overline{\xi}(Z)}{\partial Z} \right] dZ}{\partial z} \tag{6.9b}$$

and where the integration is over the domain of turbulence.   The similarity of this to (6.8.7b) is striking. $\Xi(z,Z,t)$, which has units of m/s, is called the **turbulent diffusivity transfer function**, and is defined by:

$$\Xi\,(z,Z,t) \;=\; \frac{1}{2\pi}\; \int K\,(\kappa,t)\ \exp\!\left[\,i\,\kappa\,(z-Z)\right] d\kappa \tag{6.9c}$$

where i is the square root of -1 (see Chapter 8 for more details about Fourier analysis), and where the integration is over all wavenumbers, from $\kappa = -\infty$ to $+\infty$.

To use this nonlocal approach, we must parameterize either $K(\kappa)$ or Ξ. Berkowicz & Prahm suggested that larger eddies are more effective at causing diffusion (i.e., have larger K) than smaller eddies (see Fig 6.15). Based on this concept, they suggested that

$$K(\kappa) \;=\; \frac{K_o}{\left[\,1 + B_o\,\kappa^{4/3}\,\right]} \tag{6.9d}$$

where the two parameters, K_o and B_o, are both assumed to be proportional to the 4/3 power of the wavelength at the peak in the TKE energy spectrum. In other words, this is an *a-priori* approach, where some knowledge of the spectrum of turbulence is assumed. K_o has units of $m^2\,s^{-1}$, while B_o has units of $m^{-4/3}$. This spectral dependence of K yields a plot of $\Xi(z,Z,t)$ vs. (z-Z) that looks very similar to the plot of $Y(z,Z,t)$ in Fig 6.13, with a peak of Ξ at z-Z = 0 and tails that approach zero at both sides.

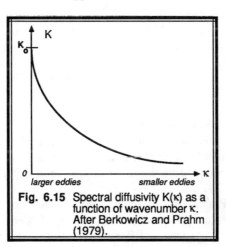

Fig. 6.15 Spectral diffusivity $K(\kappa)$ as a function of wavenumber κ. After Berkowicz and Prahm (1979).

When spectral diffusivity theory is used to predict the spread of smoke puffs within a homogeneous turbulent environment, the result shown in Fig 6.16 is a slow dispersion rate initially while the puff is small. Later, when the puff is larger, it grows at the same rate as would be expected for normal (nonspectral) K-theory. In the figure, $B_o = 0$ corresponds to the normal K-theory dispersion, while $B_o = 3\ m^{-4/3}$ is the dispersion with spectral diffusivity.

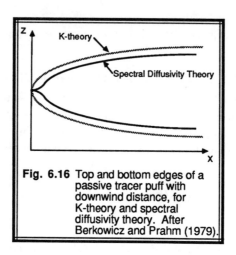

Fig. 6.16 Top and bottom edges of a passive tracer puff with downwind distance, for K-theory and spectral diffusivity theory. After Berkowicz and Prahm (1979).

The initial slower dispersion rate and non-Gaussian distribution of tracer concentration is common to both transilient theory and spectral diffusivity theory. Although some experimental evidence supports these results, they differ from those of statistical dispersion theory .

6.10 References

André, J.-C. & P. Lacarrere, 1985: Mean and turbulent structures of the oceanic surface layer as determined from one-dimensional third-order simulations. *J. Physical Ocean.*, **15**, 121-132.

André, J.-C., F. DeMoor, Pl. Lacerure, G. Therry, & R. DuVachat, 1978: Modeling the 24-hour evolution of the mean and turbulent structures of the planetary boundary layer. *J. Atmos. Sci.*, **35**, 1861-1883.

Beljaars, A.C.M., J.L. Walmsley, and P.A. Taylor, 1987: A mixed spectral finite-difference model for neutrally stratified boundary-layer flow over roughness changes and topography. *Bound.-Layer Meteor.*, **38**, 273-303.

Berkowicz, R., 1984: Spectral methods for atmospheric diffusion modeling. *Bound. Layer Meteor.*, **30**, 201-220.

Berkowicz, R. & L.P. Prahm, 1979: Generalization of K-theory for turbulent diffusion. Part 1: Spectral turbulent diffusivity concept. *J. Appl. Meteor.*, **18**, 266-272.

Bhumralkar, C.M., 1975: *A survey of Parameterization Techniques for the Planetary Boundary Layer in Atmospheric Circulation Models.* Defense Advanced Research Projects Agency R-1653-ARPA. ARPA order No. 189-1. 84pp.

Briere, S., 1987: Energetics of daytime sea-breeze circulation as determined from a two-dimensional, third-order turbulence closure model. *J. Atmos. Sci.*, **44**, 1455-1474.

Bougeault, P., 1981: Modeling the trade-wind cumulus boundary layer. Part 2: A high-order one-dimensional model. *J. Atmos. Sci.*, **38**, 2429-2439.

Bougeault, P., 1985: The diurnal cycle of the marine stratocumulus layer: A higher order model study. *J. Atmos. Sci.*, **42**, 2826-2843.

Businger, J.A., 1982: Equations and concepts. *Atmospheric Turbulence and Air Pollution Modelling*, Edited by F.T.M. Nieuwstadt and H. van Dop, D. Reidel Publishing Co., Dordrecht., 1 - 36.

Chou, P.Y., 1945: On velocity correlations and the solutions of the equations of turbulent fluctuation. *Quart. J. Appl. Math.*, **3**, 38-54.

Deardorff, J.W., 1973: Three-dimensional numerical modeling of the planetary boundary layer. *Workshop on Micrometeorology*, Amer. Meteor. Soc., Edited by D. Haugen. 271-311.

Delage, Y., 1974: A numerical study of the nocturnal atmospheric boundary layer. *Quart. J. Roy. Meteor. Soc.*, **100**, 351-364.

Detering, H.W. and D. Etling, 1985: Application of the E-ε turbulence closure model to the atmospheric boundary layer. *Bound.-Layer Meteor.*, **33**, 113-133.

Donaldson, C. duP., 1973: Construction of a dynamic model of the production of atmospheric turbulence and the dispersal of atmospheric pollutants. *Workshop on Micrometeorology*, Edited by D.A. Haugen. American Meteorological Society, Boston. 313-392.

Ekman, V.W., 1905: On the influence of the earth's rotation on ocean currents. *Arkiv. Math Astron. O. Fysik*, **2**, 11.

Estournel, C., and D. Guedalia, 1987: A new parameterization of eddy diffusivities for nocturnal boundary layer modeling. *Bound.-Layer Meteor.*, **39**, 191-203.

Haltiner, G.J. and R.T. Williams, 1980: *Numerical Prediction and Dynamic Meteorology*, 2nd ed. Wiley & Sons, 477pp.

Hanjalic, K. & B.E. Launder, 1972: A Reynolds' stress model of turbulence and its application to thin shear flows. *J. Fluid Mech.*, **52**, 609-638.

Keller, L.V. and A.A. Friedman, 1924: Differentialgleichung für die turbulente Bewegung einer Kompressiblen Flüssigkeit. *Proc. 1st Intern. Congr. Appl. Mech.*, Delft, 395-405.

Kitada, T., 1987: Turbulence transport of a sea breeze front and its implication in air pollution transport — application of k-ε turbulence model. *Bound. Layer Meteor.*, **41**, 217-239.

Lacser, A. and S.P.S. Arya, 1986: A comparitive assessment of mixing-length parameterizations in the stably stratified nocturnal boundary layer (NBL). *Bound.-Layer Meteor.*, **36**, 53-70.

Launder, B.E., G.J. Reece & W. Rodi, 1975: Progress in the development of a Reynolds-stress turbulence closure. *J. Fluid Mech.*, **68**, 537-566.

Lumley, J.L. & B. Khajeh-Nouri, 1974: Computational modeling of turbulent transport. *Adv. in Geophys.*, **18A**, Academic Press. 169-192.

Lumley, J.L. & P. Mansfield, 1984: Second-order modeling of turbulent transport in the boundary layer. ???

Mellor, G.L. & J.R. Herring, 1973: *A.I.A.A. J.*, **11**, 590-599.

Mellor, G.L. & T. Yamada, 1974: A hierarchy of turbulence closure models for the planetary boundary layer. *J. Atmos. Sci.*, **31**, 1791-1806.

Moeng, C.-H. & D.A. Randall, 1984: Problems in simulating the stratocumulus-topped boundary layer with a third-order closure model. *J. Atmos. Sci.*, **41**, 1588-1600.

O'Brien, J.J., 1970: A note on the vertical structure of the eddy exchange coefficient in the planetary boundary layer. *J. Atmos. Sci.*, **27**, 1213-1215.

Pearson, C.E. (Ed.), 1974: *Handbook of Applied Mathematics, Selected Results and Methods.* Van Nostrand Reinhold Co., 1265pp.

Prahm, L.P.,R. Berkowicz & O. Christensen, 1979: Generalization of K-theory for turbulent diffusion. Part 2: Spectral diffusivity model for plume dispersion. *J. Appl. Meteor.*, **18**, 273-282.

Prandtl, L., 1910: Bericht über Untersuchungen zur ausgebildeten Turbulenz. *Zs. angew. Math. Mech.*, **5**, 136-139.

Ray, D., 1986: Variable eddy diffusivities and atmospheric cellular convection. *Bound.-Layer Meteor.*, **36**, 117-131.

Rotta, J.C., 1951: Statistische theorie nichthomogener turbulenz. *Zeitschrift für Phys.*, **129**, 547-572.

Schumann, U. 1977: Realizability of Reynolds' stress turbulence models. *Phys. Fluids*, **20**, 721-725.

Stull, R.B., 1984: Transilient turbulence theory. Part 1: The concept of eddy mixing across finite distances. *J. Atmos. Sci.*, **41**, 3351-3367.

Stull, R.B., 1986: Transilient turbulence theory. Part 3: Bulk dispersion rate and numerical stability. *J. Atmos. Sci.*, **43**, 50-57.

Stull, R.B., 1987: Transilient turbulence algorithms to model mixing across finite distances. *Environmental Software*, **2**, 4-12.

Stull, R.B. and A.G.M. Driedonks, 1987: Applications of the transilient turbulence parameterization to atmospheric boundary-layer simulations. *Bound.-Layer Meteor.*, **40**, 209-239.

Sun, W-H. and Y.Ogura, 1980: Modeling the evolution of the convective planetary boundary layer. *J. Atmos. Sci.*, **37**, 1558-1572.

Wai, M.M.-K., 1987: A numerical study of the marine stratocumulus cloud layer. *Bound.-Layer Meteor.*, **40**, 241-267.

Wichmann, M. and E. Schaller, 1985: Comments on "Problems in simulating the stratocumulus-topped boundary layer with a third-order closure model". *J. Atmos. Sci.*, **42**, 1559-1561.

Wichmann, M. and E. Schaller, 1986: On the determination of the closure parameters in higher-order closure models. *Bound.-Layer Meteor.*, **37**, 323-341.

Wyngaard, J.C., O.R. Coté, K.S. Rao, 1974: Modeling of the atmospheric boundary layer. *Adv. in Geophys.*, **18A**, Academic Press, 193-212.

Wyngaard, J.C., 1982: Boundary layer modeling. *Atmospheric Turbulence and Air Pollution Modelling*, Edited by F.T.M. Nieuwstadt and H. van Dop, D. Reidel Publ. Co., Dordrecht, 69-106.

Yamada, T. & G. Mellor, 1975: A simulation of the Wangara atmospheric boundary layer data. *J. Atmos. Sci.*, **32**, 2309-2329.

Zeman, O., 1981: Progress in the modeling of planetary boundary layers. *Ann. Rev. Fluid Mech.*, **13**, 253-272.

6.11 Exercises

1) Which turbulence closure scheme would be most appropriate for the atmospheric boundary layers in each of the following applications? Why?
   a) Nested Grid Model (NGM) numerical forecast model (or other operational weather forecast model).
   b) Diffusion of smoke from a tall stack.
   c) Mesoscale numerical forecast model for the interior of continents.
   d) Sea-breeze numerical forecast model for your favorite coastline.
   e) Air flow through the turbine disks in a jet engine.
   f) 3-D numerical thunderstorm model.
   g) Global climate model
   h) Numerical model of individual thermals in the BL.
   i) Study of the air flow over complex terrain for wind energy siting.
   j) Study of heat, moisture, and mass transfer over a ginseng crop.
   k) Forecasting minimum daily temperatures at your town.
   l) Study the moisture budget of a hurricane.
   m) Study the turbulence structure of the Venusian atmosphere.
   n) Forecasting the nocturnal jet.
   o) Numerical simulation of the interaction between ocean currents and the atmosphere.

2) Very briefly define the following, and comment or give examples of their use in micrometeorology:
   a) Ekman spiral
   b) Turbulence closure problem
   c) K-theory

3) Write a correlation triangle for third-order closure similar to the lower-order triangles in Table 6-2.

4) What is the closure problem, and how does it affect the study of turbulent boundary layers?

5) Given:

z	12	8	2	$0.1=z_0$	(m)
$\overline{\theta}$	300	301	303	308	(K)
$\overline{U}$	5.4	5.0	3.4	0	(m/s)

Use first-order closure to find $\overline{w'\theta'}$ and $\overline{u'w'}$ at $z = z_0$ and at $z = 10$ m.

6) What advantages and disadvantages are there for using one-and-a-half-order closure, as compared to first-order closure?

7) Use K-theory to find the heat flux as a function of height, given a potential temperature profile of the shape $\theta = \theta_0 - a \ln(z/z_0)$, where $\theta_0 = 300$ K at $z = z_0 = 1$ mm, and a = 5 K. Assume that $K = k z u_*$ with $u_* = 0.1$ m/s and k=0.4 .

8) What order closure (zero, first, 1 1/2, second, third) is used in the derivation of the Ekman spiral?

9) Can transilient turbulence closure simulate K-theory diffusion? Can K-theory simulate looping smoke plumes?

10) For an eddy diffusivity of $K_m = 2$ m^{2}s^{-1} and a geostrophic vorticity of 0.0001 s^{-1} with an average tangential velocity of 8 m/s, calculate the vertical velocity at Madison, Wisconsin caused by "Ekman pumping" at the top of a 500 m thick boundary layer.

11) What is a common first-order closure method?

12) Given the following matrix of transilient coefficients:

$$c_{ij} \, (\Delta t = 10 \text{ min}) = \begin{bmatrix} 0.1 & 0.9 & 0. & 0. & 0. \\ 0.2 & 0.05 & 0.75 & 0. & 0. \\ 0.25 & 0.05 & 0.2 & 0.5 & 0. \\ 0.25 & 0. & 0.05 & 0.3 & 0.4 \\ 0.2 & 0. & 0. & 0.2 & 0.6 \end{bmatrix}$$

a) Verify that mass and state are conserved, and that entropy increases.

b) Given an initial state of pollution concentration in the boundary layer of

Grid index (i)	z (m)	S (g/m^3)
1	100	500
2	300	0
3	500	0
4	700	0
5	900	0

Forecast the pollution concentration at each of these heights every 10 min from t=0 to t=90 min. (Assume a 1km thick boundary layer, with 5 grid points centered at the heights indicated above, with $\Delta z = 200$ m.)

c) Plot a time height diagram of your results from (b), and draw isopleths for S = 90, 110, and 130 g/m^3. Comment on the behavior of the pollutants, and on what type of boundary layer (ML, NBL, RL) is probably there.

d) Find and plot the flux profile associated with the first and second timesteps.

e) If a timestep of Δt=30 min were desired instead, calculate the new matrix of transilient coefficients.

f) Using the answer from (b), make a new forecast of pollutant concentration from t=0 to t=90 min, using Δt=30 min and the same initial conditions as before. Compare your answer to that of part (b).

g) If the rows of the transilient coefficient matrix given at the top of this page were switched, such that the bottom row is at the top, the second to the bottom becomes second from the top, etc., then comment on the nature of smoke dispersion.

13) Given the following transilient coefficient matrix

$$c_{ij} \, (\Delta t = 10 \text{ min}) = \begin{bmatrix} 0.7 & 0.2 & 0.1 & 0.0 \\ 0.2 & 0.7 & 0.2 & 0.1 \\ 0.1 & 0.2 & 0.7 & 0.2 \\ 0.0 & 0.1 & 0.2 & 0.7 \end{bmatrix}$$

a) What is wrong with this matrix?

b) Make a forecast (only out to t=10 min) to demonstrate your conclusions from part
   (a). Start with an initial tracer concentration distribution given by

index	concentration
1	0
2	0
3	100
4	0

14) Let $K_m = 5$ m²s⁻¹ $=$ constant with height. Calculate and plot

   a) $\overline{u'w'}$ b) $\overline{w'\theta_v'}$

   from z=0 to 50 m using the data from problem 26 of Chapt. 5.

15) a) Using the answers from problem (14) above, find the initial tendency for virtual

   potential temperature $(\partial\overline{\theta_v}/\partial t)$ for air at a height of z=10 m.

   b) If this tendency does not change with time, what is the new $\overline{\theta_v}$ at z=10 m, one

   hour after the initial state (i.e., the state of problem 26, Chapt. 5)?

16) Verify the Ekman boundary layer solutions by plugging them back into the original
   differential equations, and by checking to see if they satisfy the boundary conditions.

17) Given $K_m = k z u_*$. Solve for $\overline{U}$ as a function of height in the SL.

18) For 1.5-order closure, indicate which prognostic equations are needed to forecast

   $\overline{\theta'^2}$? To save time, write only the storage term for each equation used. Hint, one of

   the equations is $\partial\,\overline{\theta'^2}\,/\partial t = \cdots$

19) Given the following matrix of transilient coefficients:

$$c_{ij}\,(\Delta t = 30\text{ min}) = \begin{bmatrix} 0.3 & 0.25 & 0.25 & 0.2 \\ 0.25 & 0.3 & 0.25 & 0.2 \\ 0.25 & 0.25 & 0.3 & 0.2 \\ 0.2 & 0.2 & 0.2 & 0.4 \end{bmatrix} \quad \begin{matrix} \text{index} \\ 1 \\ 2 \\ 3 \\ 4 \end{matrix}$$

a) Does this matrix satisfy mass and state conservation? Why?
b) Describe the nature of mixing (e.g., overturning, small-eddies, no turbulence, well
   mixed, sub-domain of turbulence, etc.)
c) Given the following initial concentrations, find the new concentrations after 30 min.

index	concentration
1	100
2	100
3	200
4	0

d) Find the flux profile for this one timestep, given $\Delta z = 100$ m.

20) Given the following transilient matrix:

$$c_{ij}\ (\Delta t = 10\ \text{min}) = \begin{bmatrix} 0.7 & 0.2 & 0.1 & 0. & 0. & 0 \\ 0.2 & 0.5 & 0.2 & 0.1 & 0. & 0 \\ 0.1 & 0.2 & 0.4 & 0.2 & 0.1 & 0 \\ 0. & 0.1 & 0.2 & ? & 0.2 & 0 \\ ? & ? & ? & ? & ? & ? \\ 0 & 0 & 0 & 0 & 0 & 1 \end{bmatrix}$$

a) Fill in the missing elements.
b) In the interior of the turbulent domain, is the turbulence homogeneous? Is it isotropic (in one dimension)?
c) Comment on the kind of turbulent boundary layer parameterized by the matrix above? (That is, what is the nature of the turbulence: diffusive, convective, other...)
d) Given the following initial distribution of pollutant concentration (S, in units of micrograms per cubic meter) at each of six grid points, forecast the concentration at t=10, 20, & 30 minutes using a Δt=10 min timestep.

Grid Index	z (m)	S
1	50	100
2	150	0
3	250	0
4	350	0
5	450	0
6	550	0

Check to be sure that pollutant mass is conserved at each time.
e) Find the flux as a function of height for the first timestep.
f) Comment on the centroid of pollutant concentration, and how it moves with time.
g) After 100 timesteps, qualitatively discuss the anticipated vertical distribution of pollutants. (i.e., you need not make an actual forecast).

21) Given the following transilient matrix:

$$c_{ij}\ (\Delta t = 15\ \text{min}) = \begin{bmatrix} 0.7 & 0 & 0.2 & 0.1 \\ 0.1 & 0.6 & ? & 0.2 \\ ? & ? & ? & ? \\ 0.1 & 0.2 & ? & 0.3 \end{bmatrix}$$

a) Fill in the missing elements to yield an allowable matrix.
b) Given an initial tracer distribution of 100 mg m^{-3} at the lowest grid box (i=1), 10 mg m^{-3} in the top grid box (i=4), and no tracer elsewhere, find the tracer distribution after 15 minutes.
c) Was tracer amount conserved? How do you know?
d) Does the transilient matrix represent homogeneous turbulence?
e) Does the transilient matrix represent isotropic turbulence?

22) If one considers only the momentum equations, how many total equations and unknowns (only of velocity correlations in this case) are there for fourth-order closure? (Hint, look at Table 6-1.)

23) Given an initial pollutant concentration distribution of

index	z(m)	concentration
1	10	0
2	30	0
3	50	0
4	70	100
5	90	0

and a transilient matrix of:

$$c_{ij} (\Delta t = 5 \text{ min}) = \begin{bmatrix} 0.6 & 0.4 & 0 & 0 & 0 \\ 0.4 & 0.3 & 0.3 & 0 & 0 \\ 0 & 0.3 & 0.6 & 0.1 & 0 \\ 0 & 0 & 0.1 & 0.9 & 0 \\ 0 & 0 & 0 & 0 & 1 \end{bmatrix}$$

a) Find the new concentrations S_i at $t = 10$ min.
b) When would pollutants first reach the lowest grid box?
c) Is there a way to find the concentrations at $t = 10$ min in just one timestep, starting from the concentration at $t = 0$?
d) Find and plot the flux profile.

24) Does the following closure method obey the rules of parameterization?
a) The first-order closure of equation (6.4.1b).
b) The one-and-a-half-order closure of equation (6.5b) for heat flux.

c) The second-order parameterizations by Deardorff for $\overline{\theta' u_i' u_j'}$ (see Table 6-5).

d) The second-order parameterization by Rotta for the pressure correlation (see Table 6-5).

25) Without using the assumption that the mixing length is proportional to height above the ground, rederive the mixing-length expression for moisture flux for the case where a solid boundary limits the maximum size of eddies.   Hint, use a statistical approach with a probability distribution that is zero beyond the solid boundary.

26) Given the heat flux profiles in Figs 3.1, 3.2, and 3.3, and the mean profiles of potential temperature in Figs. 3.4 and 3.5:
a) calculate the value for K_H as a function of height for :
   (1) Flight 2
   (2) Flight 3
   (3) Flight 13
b) Which parameterization in Table 6-4 best fits the calculated K values from part (a)?

27) For an eddy viscosity of 10 m^{2}s^{-1}, and a pressure gradient of 0.2 kPa $/ 100$ km in the atmosphere, plot the Ekman spiral winds as a function of height, and find the Ekman layer depth at:
a) Bergen, Norway
b) Madison, Wisconsin
c) Christmas Island (in the Line Islands)

28) Suggest a parameterization for $\overline{u_i'u_j'u_k'u_m'}$.

29) Given the γ_T and θ curves in the following figure, calculate the initial tendency (TM) due to turbulent mixing using the continuous form of transilient turbulence theory.

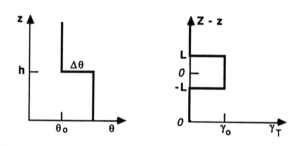

30) Use spectral diffusivity theory to solve problem (29), using $K_o = 2 \ m^2s^{-1}$ and $B_o = 3$. (i.e., do not use γ_T for this case). Also, plot K vs. κ.

31) Given the following heat flux and potential temperature gradients, find and plot K_H vs z/z_i . Comment on the meaning of K_H at each height.

z/z_i	$\overline{w'\theta'}$ (K m/s)	$\partial\overline{\theta}/\partial z$ (K/m)
0.0	0.20	-0.020
0.2	0.15	-0.005
0.4	0.10	-0.001
0.6	0.05	0.001
0.8	0.00	0.005
1.0	-0.05	0.020

7 Boundary Conditions and Surface Forcings

Without a bottom boundary on the atmosphere there would be no boundary layer. Friction at the surface slows the wind, and heat and moisture fluxes from the surface modify the state of the boundary layer. The heat and moisture fluxes are driven, in turn, by the external forcings such as radiation from the sun or transpiration from plants. Forcings across the top of the boundary layer also alter mean characteristics within it.

In this chapter we examine some of the external forcings, and show how those forcings and fluxes can be parameterized. Unfortunately, we cannot use any of the methods discussed in Chapter 6, because those methods apply only within a volume of air. For fluxes between a solid surface and air, we need to develop a different set of parameterizations, such as are discussed below.

7.1 Effective Surface Turbulent Flux

Turbulence by itself cannot transfer heat, momentum, or moisture across the interface from the ocean or from the earth. After all, it is rather infrequently that we see clods of soil jumping up and down in turbulent motion. Even ocean waves and turbulence have little direct coupling to atmospheric turbulence. Consequently, we must consider molecular effects, in addition to turbulent transport.

Molecular conduction of heat, molecular diffusion of tracers, and molecular viscous transfer of momentum cause transport between the surface and the lowest millimeters of air. Once in the air, turbulence takes over to transport momentum, heat and other constituents to greater depths in the atmosphere. The molecular and turbulent transport processes work together as sketched in Fig 7.1.

To simplify our equations for boundary layer evolution, we find it convenient to define an *effective turbulent flux* that is the sum of the molecular and turbulent fluxes. At the surface where there is no turbulent flux, the *effective surface*

251

turbulent flux has a magnitude equal to that of the molecular flux. Above the lowest few centimeters, however, the molecular contribution is so small that it can be neglected compared to the turbulent flux.

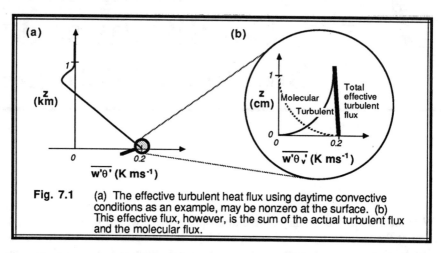

Fig. 7.1   (a) The effective turbulent heat flux using daytime convective conditions as an example, may be nonzero at the surface. (b) This effective flux, however, is the sum of the actual turbulent flux and the molecular flux.

The thin layer of air in which molecular processes dominate is called the *micro layer*. Within this layer, molecular transport, such as conduction of heat, can be described by:

$$Q_H = -\nu_\theta \frac{\partial T}{\partial z} \tag{7.1}$$

where ν_θ is the molecular thermal diffusivity (on the order of 2×10^{-5} m^{2}s^{-1} for air). For a typical kinematic heat flux of 0.2 K·m·s^{-1}, (7.1) tells us that a temperature gradient of 1×10^4 K·m^{-1} is required. This corresponds to a 10 °C temperature difference across a micro layer 1 mm thick.

Such large gradients are indeed observed. Those of you who have walked barefoot across a black asphalt road on a sunny summer day can testify that the surface "skin" temperature can become burning hot to the touch, even though the air temperature may be a pleasant 25 or 30 °C. The hot skin temperature can create a large temperature gradient in the lowest millimeters of air. These large gradients in the micro layer are also the cause of the mirage that we sometimes see over highways.

From this point on, whenever we refer to a surface turbulent flux in this text, we are really implying a surface *effective* flux. In this way, we can ignore the molecular processes, and just use turbulence equations such as (3.5.3) with the effective flux. As shown in Chapters 1, 3 and 4, the effective flux varies by only a small portion of its magnitude in the lowest tens of meters of the BL. Thus, the turbulent flux measured at the standard "surface" instrument shelter (screen) height of 2 m provides a good approximation to the effective surface turbulent flux.

7.2 Heat Budget at the Surface

Picture a layer of air with its top just above the highest trees (or ocean wave crests) in a region, and its base just below the earth's surface (or below the wave troughs), as sketched in Fig 7.2a.   This layer has turbulent energy transfer with the air above it, radiative transfer through the top of it, and molecular energy transfer into the soil (or sea) below.

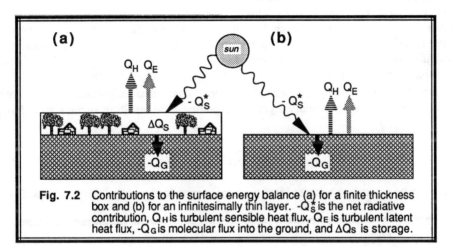

Fig. 7.2 Contributions to the surface energy balance (a) for a finite thickness box and (b) for an infinitesimally thin layer. $-Q_S^*$ is the net radiative contribution, Q_H is turbulent sensible heat flux, Q_E is turbulent latent heat flux, $-Q_G$ is molecular flux into the ground, and ΔQ_S is storage.

The energy budget for this layer, where upward fluxes are positive, is:

$$- \tilde{Q}_s^* \;=\; \tilde{Q}_H \;+\; \tilde{Q}_E \;-\; \tilde{Q}_G \;+\; \Delta \tilde{Q}_S \tag{7.2a}$$

or

$$- Q_s^* \;=\; Q_H \;+\; Q_E \;-\; Q_G \;+\; \Delta Q_S \tag{7.2b}$$

where
Q^*_s = net upward radiation at the surface
Q_H = represents the upward sensible heat flux out of the top
Q_E = represents the upward latent heat flux out of the top
Q_G = represents the upward molecular heat flux into the bottom
ΔQ_S = denotes the storage or intake of internal energy (positive for warming and for chemical storage by photosynthesis).

In simple terms, this is nothing more than energy *in* balancing energy *out* and *storage*. The external forcing is Q^*_s, and all the other terms are response terms.

Equation (7.2a) is in energy flux units such as $J \cdot m^{-2} s^{-1}$ or $W \cdot m^{-2}$, while (7.2b) is in kinematic units such as $K \cdot m \cdot s^{-1}$. Thus, the second equation is just the first equation divided by ρC_p. From section 2.6, we recall that $Q_H = (\overline{w'\theta'})_s$, and $Q_E = (L_v/C_p) \cdot (\overline{w'q'})_s$.

Very complex processes can occur within our imaginary layer (Geiger, 1965; Oke, 1978; Brutsaert, 1985), such as: radiation between leaves, plants, buildings, and animals; turbulent circulations different from those higher in the boundary layer; vertical variations of the sensible and latent heat flux associated with evaporation and condensation; and transpiration. Because of this complexity, we have employed the simplification of a layer into which the net effect of all of these processes can be lumped together as ΔQ_S.

Sometimes, we prefer to conceptually employ an infinitesimally thin layer, as sketched in Fig 7.2b. This is not a really layer, but a plane. The resulting **surface heat budget** is

$$- Q_s^* = Q_H + Q_E - Q_G \tag{7.2c}$$

There can be no storage because there is no mass contained within a zero thickness layer.

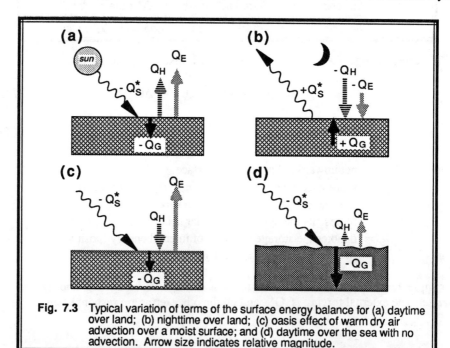

Fig. 7.3 Typical variation of terms of the surface energy balance for (a) daytime over land; (b) nighttime over land; (c) oasis effect of warm dry air advection over a moist surface; and (d) daytime over the sea with no advection. Arrow size indicates relative magnitude.

The neglect of the storage layer works well for quasi-steady-state situations where there is no appreciable change in the mean temperature of that layer. It also works well for flat barren land surfaces, and for waveless sea surfaces.

During a late morning sunny *day over land*, Q^*_s is negative because there is more downward radiation entering the layer than leaving upward. Q_H and Q_E are positive because of heat and moisture transport away upward from the surface. Q_G is negative when heat is conducted downward into the ground from the warm surface (see Fig 7.3a).

At *night over land*, Q^*_s is often positive because of the net upward longwave radiative cooling to space. Q_H is negative because of a downward heat flux from the air. Dew or frost formation makes Q_E negative. Conduction of heat from the warm ground up to the cooler surface makes Q_G positive. The release of stored heat from the layer makes ΔQ_S negative (see Fig 7.3b).

The daytime and nighttime examples above demonstrate a classical behavior where the terms were either all positive during the day, or all negative at night. Typical diurnal evolution of terms in (7.2a) is shown in Fig 7.4 for a vegetated land surface at mid latitudes.

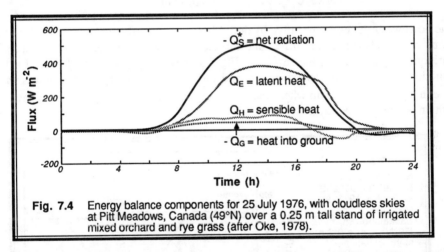

Fig. 7.4   Energy balance components for 25 July 1976, with cloudless skies at Pitt Meadows, Canada (49°N) over a 0.25 m tall stand of irrigated mixed orchard and rye grass (after Oke, 1978).

Other nonclassical situations occur in nature, such as the *oasis effect*. Picture warm dry air blowing over a cool moist oasis (Fig 7.3c). There is strong evaporation from the moist ground and plants into the air, resulting in latent cooling that keeps the oasis at a pleasant temperature. However, this upward latent heat flux is opposed by a downward sensible heat flux from the warm air to the cool ground. Thus, Q_E is positive, while Q_H and Q^*_s are negative. If we focus on just these three terms, we see that the latent heat flux can be greater in magnitude than the solar heating, because of the additional energy that is extracted from the warm air by evaporation.

The ocean budget behaves differently than the land budget because turbulence in the water can efficiently transport heat away from the surface and distribute it deeper in the water. Also, the heat capacity ($\rho \, C_v$) of water is about 4000 times larger than that of air, meaning that a lot of heat can be absorbed into water with little temperature change. Thus, the diurnal cycle of radiation is almost completely balanced by a corresponding diurnal variation of energy transport into the sea (Fig 7.3d). In addition, the nearly constant sea surface temperature with time results in a nearly constant heat and moisture flux, and associated slow temporal changes in air temperature and humidity.

7.3 Radiation Budget

It is often convenient to split the net radiation term into four components:

$$Q^* = K{\uparrow} + K{\downarrow} + I{\uparrow} + I{\downarrow} \qquad (7.3)$$

where $K{\uparrow}$ = upwelling *reflected* short wave (solar) radiation
    $K{\downarrow}$ = downwelling shortwave radiation *transmitted* through the air
    $I{\uparrow}$ = longwave (infrared, IR) radiation *emitted* up
    $I{\downarrow}$ = longwave *diffusive* IR radiation down

The downward fluxes are negative by definition, and upward are positive. Each of these terms represents the sum of direct and diffuse radiation components crossing a locally horizontal plane such as the surface.   Fig 7.5 shows a typical diurnal cycle for these radiation components in clear skies at a land surface. Although this equation was written here in kinematic flux form where each term has units of $K{\cdot}m{\cdot}s^{-1}$, it could also have been written in energy flux form.

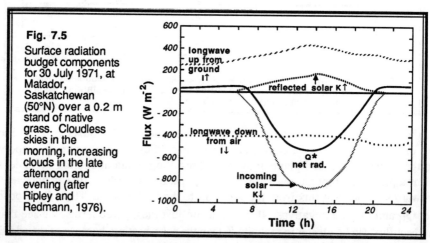

Fig. 7.5

Surface radiation budget components for 30 July 1971, at Matador, Saskatchewan (50°N) over a 0.2 m stand of native grass. Cloudless skies in the morning, increasing clouds in the late afternoon and evening (after Ripley and Redmann, 1976).

Splitting radiation into only two wavelength bands (short and longwave) is possible because the peak in the solar spectrum is at the normal visible light wavelengths, while the earth/atmosphere system is emitting infrared radiation characteristic of its absolute temperature (in the range of 280 K at the surface to about 245 K at the top of the atmosphere). Since there are no other bodies near the earth that contribute significantly to the radiation budget, we need only be concerned with those two bands.

7.3.1 Shortwave Radiation

The intensity of incoming solar radiation at the top of the atmosphere is called the *solar irradiance*, S. Although it was formerly known as the **solar constant**, this term is being used less frequently because of the realization that the solar irradiance is not constant — ranging from about -1360 to -1380 $W \cdot m^{-2}$. We will use a value of S = -1370 $W \cdot m^{-2}$ (Kyle, et al., 1985), or S = -1.127 $K \cdot m \cdot s^{-1}$ in kinematic units, where the density and specific heat of air in the boundary layer is used for the conversion to kinematic units.

Some of this radiation is attenuated by scattering, absorption, and reflection from clouds on the way down to the surface. When the sun is lower in the sky, the radiation will also be attenuated by its longer path through the atmosphere en route to the surface. Define T_K as the net sky *transmissivity*, or the fraction of solar radiation that makes it to the surface. Define Ψ as the solar elevation angle; namely, the angle of the sun above the local horizon. One simple parameterization (Burridge and Gadd, 1974) for the transmissivity is:

$$T_K = (0.6 + 0.2 \sin \Psi) \cdot (1 - 0.4 \, \sigma_{C_H}) \cdot (1 - 0.7 \, \sigma_{C_M}) \cdot (1 - 0.4 \, \sigma_{C_L}) \qquad (7.3.1a)$$

where σ_C represents the cloud-cover fraction, and where subscripts H, M, and L signify high, middle, and low clouds respectively. When the sun is directly overhead and there are no clouds, $T_K = 0.80$. If the sun is overhead but there are overcast clouds at all three levels, then $T_K = 0.086$.

The solar elevation angle is also important because when it is less than 90°, the radiation that does reach the surface is spread out over a larger area, reducing the radiation per unit surface area by a factor of $\sin \Psi$. The expression for downwelling radiation at the surface is approximately

$$K\!\downarrow_s = S \cdot T_K \cdot \sin \Psi \qquad \text{for daytime (i.e., } \sin \Psi \text{ positive)} \qquad (7.3.1b)$$
$$= 0 \qquad \text{for nighttime (i.e., } \sin \Psi \text{ negative)}$$

Determination of the local elevation angle is a straightforward exercise in geometry (Zhang and Anthes, 1982), resulting in:

$$\sin \Psi = \sin \phi \sin \delta_s - \cos \phi \cos \delta_s \, \cos\left[\left(\frac{\pi \, t_{UTC}}{12}\right) - \lambda_e\right] \qquad (7.3.1c)$$

where ϕ and λ_e are the latitude (positive north) and longitude (positive west) in radians,

δ_s is the *solar declination angle* (angle of the sun above the equator, in radians), and t_{UTC} is Coordinated Universal Time in hours. The solar declination angle is

$$\delta_s = \phi_r \cos \left[\frac{2\pi \, (d - d_r)}{d_y} \right]$$ (7.3.1d)

where ϕ_r is the latitude of the Tropic of Cancer ($23.45° = 0.409$ radians), d is the number of the day of the year (e.g., October 27 = day 300), d_r is the day of the summer solstice (173), and d_y is the average number of days per year (365.25).

Define the *albedo*, a, as the fraction of downwelling radiation at the surface that is reflected. The albedo varies from about 0.95 over fresh snow, 0.4 over light-colored dry soils, 0.2 over grass and many agriculture crops, 0.1 over coniferous forests, to 0.05 over dark wet soils. The upwelling (reflected) radiation is thus

$$K\!\uparrow_s = - a \, K\!\downarrow_s$$ (7.3.1e)

The albedo of water not only varies with wave state, but is a strong function of sun angle (Krauss, 1972). When the sun is directly overhead over a smooth water surface, the albedo is about 0.05, while it increases to nearly 1.0 at low elevation angles.

7.3.2 Longwave Radiation

As is obvious in Fig 7.5, the upward and downward longwave radiation terms are both large, but of opposite sign. Also, they do not vary much with time in clear sky conditions. As a result, the *net longwave radiation* ($I^* = I\!\uparrow + I\!\downarrow$) is approximately constant with time, and is often negative because of the net radiative loss from the earth/atmosphere system to space.

When clouds are present, much of the outgoing radiation can be balanced by downward radiation from the clouds. Low clouds are more effective at this than high clouds. For overcast clouds at all three levels, we might expect the net radiation to be approximately zero.

The net upward longwave radiation at the surface is sometimes approximated (Burridge and Gadd, 1974) by :

$$I^* = (0.08 \text{ Kms}^{-1}) \, (1 - 0.1\sigma_{C_H} - 0.3 \, \sigma_{C_M} - 0.6 \, \sigma_{C_L})$$ (7.3.2a)

This type of parameterization is known to be an oversimplification of the actual physics. Nevertheless, it is useful when detailed radiation parameterizations are not appropriate.

An alternative to parameterizing the net longwave radiation is to parameterize the $I\!\uparrow$ and $I\!\downarrow$ terms separately. The *Stefan-Boltzmann law* gives:

$$I\uparrow_s = \varepsilon_{IR}\, \sigma_{SB}\, T^4 \qquad (7.3.2b)$$

where $\sigma_{SB} = 5.67 \times 10^{-8}$ W·m^{-2}·K^{-4} is the **Stefan-Boltzmann constant**. The infrared *emissivity*, ε_{IR}, is in the range 0.9 to 0.99 for most surfaces.

The downward longwave radiation I↓ is much more difficult to calculate, because one must vertically integrate the radiative flux divergence equations. Simple radiation models are discussed in Chapters 12 and 13.

Putting together the various parameterizations for short and longwave radiation yields the following approximation for net radiative flux at the surface:

$$\begin{aligned} Q^* &= (1 - a)\, S\, T_K \sin\Psi + I^* & \text{during daytime} & \qquad (7.3.2c) \\ &= +\, I^* & \text{during nighttime.} & \end{aligned}$$

where S is negative. One must remember that this is just one possible parameterization. Other equally good approximations have appeared in the literature.

7.3.3   Radiation Budget Example

Problem: Calculate the radiation budget terms for every hour during a 24-hour cycle, given the following scenario:

location:	Madison, Wisconsin (latitude = 43.08°N, longitude = 89.42°W)
clouds:	none
date:	5 November (day 309)
start time:	midnight local time.
average albedo:	0.2

Solution: The following example was prepared, solved, and plotted using spreadsheet software, which is available for most microcomputer systems. Table 7-1 lists the spreadsheet results for this case.

First, the latitude and longitude was converted to radians, and the solar declination angle was found using (7.3.1d). This angle was -0.284 radians (-16.27°). The negative sign tells us that the northern hemisphere is in, or approaching, winter.

For each of the 24 hours, the local time was converted to UTC time. For this example, the conversion was done using longitudes rather than political time zones, knowing that it takes the earth 24 hours to rotate a full 360 degrees of longitude. Next, (7.3.1c) was used to find sin Ψ (sinphi) at each time, where sinphi was set to zero whenever the sun was below the horizon. Next, the transmissivity T_K was calculated using (7.3.1a). Then K_{down} was found using (7.3.1b), and K_{up} was found using (7.3.1e). The net longwave radiation, I^*, was found using (7.3.2a). Finally, the net radiation was calculated using $Q^* = K_{up} + K_{down} + I^*$, which is a form of (7.3).

Table 7-1. Example radiation calculation for Madison, Wisconsin.

Radiation Parameterization for: Madison, Wisconsin

Cloudcover:		Location	(deg)	(rad)
Low =	0	Lat.=	43.08	0.75189
Mid =	0	Long.=	89.42	1.56067
High =	0			

			Time Date Info:	
Solar constant =	-1.127	(Km/s)	Date:	5 Nov
Longwave max =	0.08	(Km/s)	Day #:	309
Solar declination =	-0.2843	(radians)	Local start h	0
Albedo =	0.2		Timestep h	1

Local (hr)	UTC (hr)	sinphi	Trans.	K down (Km/s)	Kup (Km/s)	I* (Km/s)	Q* (Km/s)
0	5.96	0	0.6	0	0	0.08	0.080
1	6.96	0	0.6	0	0	0.08	0.080
2	7.96	0	0.6	0	0	0.08	0.080
3	8.96	0	0.6	0	0	0.08	0.080
4	9.96	0	0.6	0	0	0.08	0.080
5	10.96	0	0.6	0	0	0.08	0.080
6	11.96	0	0.6	0	0	0.08	0.080
7	12.96	0	0.6	0	0	0.08	0.080
8	13.96	0.1589	0.6318	-0.1132	0.0226	0.08	-0.011
9	14.96	0.3041	0.6608	-0.2265	0.0453	0.08	-0.101
10	15.96	0.4155	0.6831	-0.3199	0.0640	0.08	-0.176
11	16.96	0.4856	0.6971	-0.3815	0.0763	0.08	-0.225
12	17.96	0.5095	0.7019	-0.4030	0.0806	0.08	-0.242
13	18.96	0.4856	0.6971	-0.3815	0.0763	0.08	-0.225
14	19.96	0.4155	0.6831	-0.3199	0.0640	0.08	-0.176
15	20.96	0.3041	0.6608	-0.2265	0.0453	0.08	-0.101
16	21.96	0.1589	0.6318	-0.1132	0.0226	0.08	-0.011
17	22.96	0	0.6	0	0	0.08	0.080
18	23.96	0	0.6	0	0	0.08	0.080
19	24.96	0	0.6	0	0	0.08	0.080
20	25.96	0	0.6	0	0	0.08	0.080
21	26.96	0	0.6	0	0	0.08	0.080
22	27.96	0	0.6	0	0	0.08	0.080
23	28.96	0	0.6	0	0	0.08	0.080
24	29.96	0	0.6	0	0	0.08	0.080

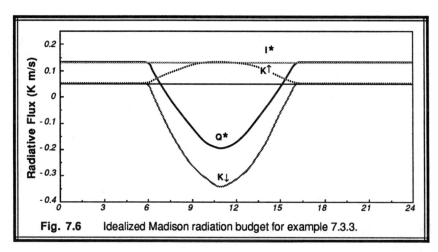

Fig. 7.6   Idealized Madison radiation budget for example 7.3.3.

Discussion: The radiation terms in this equation are plotted in Fig 7.6 for the 24-hour period. As expected, there is net heating during the day and cooling at night. Heating starts almost an hour after sunrise, and ends roughly an hour before sunset.

7.4 Fluxes at Interfaces

The vertical flux, F_ξ, of any variable ξ is assumed to be driven by the difference in $\overline{\xi}$ across the interface

$$F_\xi = - U_T (\overline{\xi}_{top} - \overline{\xi}_{bottom}) \tag{7.4a}$$

where U_T represents a *transport velocity* across that interface, and the $\overline{\xi}_{top}$ and $\overline{\xi}_{bottom}$ are the values just above and below the boundary. It can also be thought of as a *conductivity*, because a given $\overline{\xi}$-difference (voltage potential) yields a greater flux (current) if the conductivity is greater.

The transport velocity is usually parameterized as a function of some measure of turbulence appropriate to the type of interface:

$$U_T = C_D \cdot \overline{M} \qquad\qquad \text{at } z = 0 \tag{7.4b}$$

and

$$U_T = w_e \qquad\qquad \text{at } z = z_i \tag{7.4c}$$

where $\overline{M}$ is the mean horizontal wind magnitude at height z above the surface, C_D is the **bulk transfer coefficient** for the same height, and w_e is the **entrainment velocity** into the top of the mixed layer.

At the surface, the wind speed is zero, while at some height just above the surface there is a nonzero wind. The resulting shear is assumed to generate turbulence which supports the transport. This parameterization obviously fails in calm wind conditions.

The entrainment zone is an interface between the free atmosphere and the mixed layer. Entrainment brings air into the mixed layer from the free atmosphere, and creates a flux just within the top of the mixed layer. Just above the entrainment zone, however, the flux is often near zero, and is unrelated to the entrainment flux just below the entrainment zone interface (see Chapter 11 for more details).

7.4.1 Surface Fluxes — Drag and Bulk Transfer Methods

Definitions. In 1916, G.I. Taylor suggested that a velocity squared law might be used to describe the drag of the atmosphere against the earth's surface. Using u_*^2 as a measure of surface stress associated with drag, we find that

$$u_*^2 = C_D \overline{M}^2 \qquad (7.4.1a)$$

For momentum transfer, C_D is called the **drag coefficient.** Generically it is still a bulk transfer coefficient, and sometimes is written as C_M in the literature. The individual components of surface stress are correspondingly given by the **drag laws**:

$$\overline{(u'w')}_s = - C_D \overline{M}\,\overline{U} \qquad (7.4.1b)$$

$$\overline{(v'w')}_s = - C_D \overline{M}\,\overline{V} \qquad (7.4.1c)$$

At first glance (7.4.1b) does not appear to follow the form of (7.4a), but it turns out that the proper form is followed because the wind speed below the surface is zero. Thus, $\overline{U} = \overline{U} - 0 = \overline{U}_{air} - \overline{U}_{ground} = \overline{U}_{top} - \overline{U}_{bottom}$. The three factors C_D, $\overline{M}$ and $\overline{U}$ should all correspond to the same height above the surface. Often 10 m is assumed as the standard height, if not otherwise specified.

Similar expressions can be used to parameterize surface heat and moisture fluxes:

$$\overline{(w'\theta')}_s = - C_H \overline{M}\,(\overline{\theta} - \theta_G) \qquad (7.4.1d)$$

$$\overline{(w'q')}_s = - C_E \overline{M}\,(\overline{q} - q_G) \qquad (7.4.1e)$$

where the subscript $_G$ denotes "on the ground or sea surface".

The parameters C_H and C_E are the *bulk transfer coefficients* for heat and moisture, respectively, although sometimes C_H is called the **Stanton number**. For statically neutral conditions (subscript N), it is often assumed that

$$C_{HN} = C_{EN} = C_{DN} \qquad (7.4.1f)$$

Typical values range from 1×10^{-3} to 5×10^{-3} (dimensionless).

There is a subtle, but important, difference between subscripts $_G$ and $_s$. Subscript $_s$ represents values in the *air* near the surface, where "near" often means at 2 m or 10 m above the surface. The subscript $_G$ means the value in the top 1 mm of the soil or sea surface, or sometimes a *skin* value representative of just the top few molecules of the soil or sea.

One problem is how to specify the "surface" value, θ_G, or q_G. Over the sea, it is often assumed that the air in the micro layer is saturated. The saturation specific humidity is a well known function of temperature (Wallace and Hobbs, 1977). Over land, the soil surface temperature and moisture depend on many factors, and are not easily approximated. One approach is to include forecast equations for the temperature and moisture for a thin layer of soil, such as is described in more detail in section 7.6. Another approach, for temperature, is to radiometrically measure the surface skin temperature (Huband and Monteith, 1986).

Dependence on Surface Roughness. Rougher surfaces are likely to cause more intense turbulence, which increases the drag and transfer rates across the surface.

Over land, drag of the air can be caused by frictional skin drag, form drag, and wave drag. *Frictional skin drag* is related to the molecular diffusion of momentum across an interface, and applies equally well to transport of heat and scalars. Usually, drag associated with small size obstacles such as blades of grass or gravel or sometimes even trees are parameterized as skin drag in the atmosphere. *Form drag* is related to the dynamic pressure differential formed by the deceleration of air as it hits an obstacle such as a mountain or a building. There is no analogy to form drag for heat or scalars. *Wave drag* is related to the transport of momentum by buoyancy (gravity) waves in statically stable air. Mountain waves are a classic example.

Since waves and pressure fluctuations can transport momentum but not heat or scalars, we see that the momentum drag coefficient could be significantly different in magnitude from the bulk transfer coefficients for heat or moisture. Thus, one should be cautious when equating transfer coefficients.

When averaged over large horizontal regions such as continents, all drag processes can contribute. Table 7-2 indicates the magnitude such drag coefficients.

Table 7-2. Average values of drag coefficients (C_{DN}, for 10 m winds) over continents for neutral stability. Geostrophic drag coefficients (C_{GN}) for neutral stability over continents. After Garratt (1977).

Continent	C_{DN}	C_{GN}
North America	10.1×10^{-3}	1.89×10^{-3}
South America	26.6×10^{-3}	2.16×10^{-3}
Northern Africa	2.7×10^{-3}	1.03×10^{-3}
Southern Africa	12.9×10^{-3}	1.98×10^{-3}
Europe	6.8×10^{-3}	1.73×10^{-3}
U.S.S.R.	7.9×10^{-3}	1.83×10^{-3}
Asia (north of 20°N)	3.9×10^{-3}	1.31×10^{-3}
Asia (south of 20°N)	27.7×10^{-3}	2.18×10^{-3}
Australia	6.0×10^{-3}	1.50×10^{-3}

On a smaller scale, one measure of roughness is the spacing density of individual obstacles or *roughness elements*. For example, the leaves of many trees, plants, and crops can form a *canopy* elevated above the ground surface. If we imagine that a large box could be placed over one whole plant or tree that would just touch the top and sides of the plant, then the volume of this box represents the *space* taken by the plant. Of course, most of this space is filled by air between the leaves and branches. The total surface area of the plant, including the area of both sides of each leaf can theoretically be measured or estimated. The *area density of roughness elements*, S_r, is defined as the plant surface-area divided by the space volume. A *dimensionless canopy density*, C_*, can be defined by:

$$C_* = c_m S_r h^* \qquad (7.4.1g)$$

where c_m is the drag coefficient associated with an individual roughness element ($c_m = 0.05$ to 0.5 for typical plants and crops), and h^* is the average height of the canopy (Kondo and Kawanaka, 1986).

The variation of bulk transfer coefficients with canopy density is shown in Fig 7.7. As expected, the value of the transfer coefficients increase as the canopy density increases, corresponding to more roughness elements. For dimensionless canopy densities greater than about 0.4 to 1.0, however, the bulk transfer coefficients decrease. This happens when the roughness elements are so close together that they begin to appear to the wind as a solid smooth surface displaced above the true ground.

Another measure of the surface roughness is the *aerodynamic roughness length*, z_o, which will be discussed in more detail in Chapter 9. It is on the order of centimeters over grass and crops, and on the order of meters over sparse forests and towns. This roughness measure is based on the observed wind shear in the surface layer, and thus avoids the necessity of estimating the areas and spaces occupied by each roughness element.

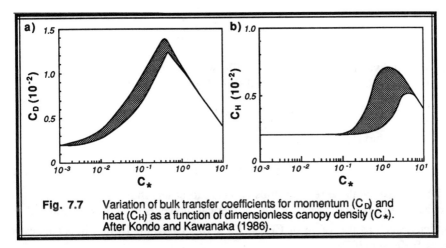

Fig. 7.7   Variation of bulk transfer coefficients for momentum (C_D) and heat (C_H) as a function of dimensionless canopy density (C_*). After Kondo and Kawanaka (1986).

Over oceans, the drag laws are a bit easier to parameterize, because the roughness length associated with ocean wave height is a known function of surface stress or wind speed:

$$z_o = 0.015 \frac{u_*^2}{g} \tag{7.4.1h}$$

which is known as **Charnock's relation** (1955).   Stronger wind stress make higher waves, which results in a greater roughness length.   The application of roughness length to bulk transfer is tied to the topic of measurement heights, which is discussed next.

Dependence on Measurement Height.   As introduced earlier, the factors C_H , $\overline{M}$ and $\overline{\theta}$ should all correspond to the same heights, z, above the surface. Unfortunately, no standard has been set on which height to use.   One obvious height is instrument shelter height.   A problem is that "surface" temperature and moisture are routinely measured at z = 2 m, while "surface" winds are measured at z = 10 m.   In numerical models, an obvious height would be the height of the lowest grid point, even though it may be hundreds of meters above the surface.   The height used has a dramatic affect on the value of the drag coefficient, because the wind speed and the difference between surface and elevated values of temperature or humidity increase as height increases for any given surface flux.   This requires that magnitude of the bulk transfer coefficient decreases with height z to yield the proper surface flux.

The relationship between drag coefficient, measurement height, and surface roughness under statically *neutral* conditions in the surface layer over either land or oceans is given by:

$$C_{DN} = k^2 \left[\ln\left(\frac{z}{z_o}\right) \right]^{-2} \tag{7.4.1i}$$

This can be derived from the log-wind profile, which is reviewed in Chapter 9.

For oceans, (7.4.1i) can be combined with Charnock's relationship to give:

$$C_{DN} = 4.4 \times 10^{-4} \, \overline{M}^{0.55} \tag{7.4.1j}$$

which yields a surface stress of:

$$u_*^2 = 4.4 \times 10^{-4} \, \overline{M}^{2.55} \tag{7.4.1k}$$

Fig 7.8 shows how these relationships compare to observations of stress and drag over the oceans and over ice. Occasionally, numerical modelers use constant drag values for simplicity, such as the value of $C_{DN} = 1.5 \times 10^{-3}$ based on 10 m winds suggested by Anthes & Keyser (1979).

Typical values for C_{DN} and C_D over land are listed in Table 7-3.

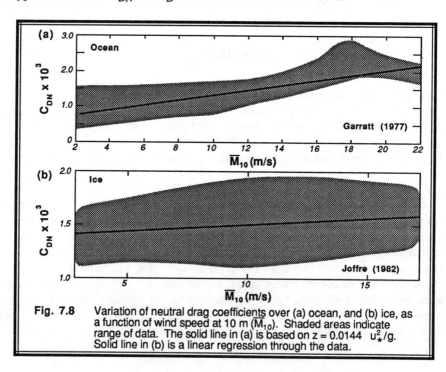

Fig. 7.8   Variation of neutral drag coefficients over (a) ocean, and (b) ice, as a function of wind speed at 10 m ($\overline{M}_{10}$). Shaded areas indicate range of data. The solid line in (a) is based on $z = 0.0144 \, u_*^2/g$. Solid line in (b) is a linear regression through the data.

Table 7-3. Sample drag and bulk-transfer coefficients. After Garratt (1977), Anthes and Keyser (1979), Gadd and Keers (1970), Deardorff (1968), Verma, et al. (1986), and Kondo and Yamazawa (1986a).

Coefficient	Ratio	Conditions
$C_{DN} = 2.6 \times 10^{-3}$		10 m winds over plains, nighttime
$C_D = 1.3$ to 1.5×10^{-3}	$C_D/C_{DN} = 0.5$	10 m winds over plains, nighttime
$C_{DN} = 1.4 \times 10^{-3}$		10 m winds over plains, daytime
$C_D = 1.8 \times 10^{-3}$	$C_D/C_{DN} = 1.3$	10 m winds over plains, daytime
$C_D = 16.0 \times 10^{-3}$		10 m winds over deciduous forest
$C_D = 40.0$ to 160.0×10^{-3}		10 m winds over coniferous forest
$C_H = 2.0 \times 10^{-3}$		10 m winds over snow surface
$C_D = [k^{-1} \cdot \ln(u_* z / v) + 5.5]^{-2}$		10 m winds over snow surface
$C_D = 5.0 \times 10^{-3} + (6.45 \times 10^{-3}) \cdot [z_T/(1+z_T)]$		for z_T = terrain height in km
$C_D = [1 + 2.5 \cdot z_T] \times 10^{-3}$		for z_T = terrain height in km
$C_{DN} = [1 + 0.07 \cdot M] \times 10^{-3}$		10 m winds over plains, nighttime
$C_{DN} = [0.75 + 0.067 \cdot M] \times 10^{-3}$		10 m winds over water
$C_{DN} = [1 + 0.07 \cdot M] \times 10^{-3}$		10 m winds over water
$C_D = 0.7 \times 10^{-3}$		10 m winds over water
$C_{DN} = 0.51 \times 10^{-3} \cdot M^{0.46}$		10 m winds over water

Dependence on Stability. Statically unstable flows generally cause a greater transport rate across an interface than statically neutral flows, which in turn transport more than stable flows. Sometimes Richardson numbers are used as a measure of stability, while at other times $\zeta = z/L$ is used. The dimensionless wind shear, $\phi_M(\zeta)$, and lapse rate, $\phi_H(\zeta)$, can be used with surface-layer similarity theory (to be described in Chapter 9) to give stability correction terms [$\psi_M(\zeta)$ and $\psi_H(\zeta)$]. This yields:

$$C_D = k^2 \left[\ln\left(\frac{z}{z_o}\right) - \psi_M(\zeta) \right]^{-2} \tag{7.4.11}$$

$$C_H = k^2 \left[\ln\left(\frac{z}{z_o}\right) - \psi_M(\zeta) \right]^{-1} \left[\ln\left(\frac{z}{z_o}\right) - \psi_H(\zeta) \right]^{-1} \tag{7.4.1m}$$

It is usually assumed that $C_E = C_H$. Figs 7.9 and 7.10, based on the work of Louis (1979), Garratt (1977), Joffre (1982) and Greenhut (1982), show the variation of bulk transfer coefficients with stability, sensor height, and roughness.

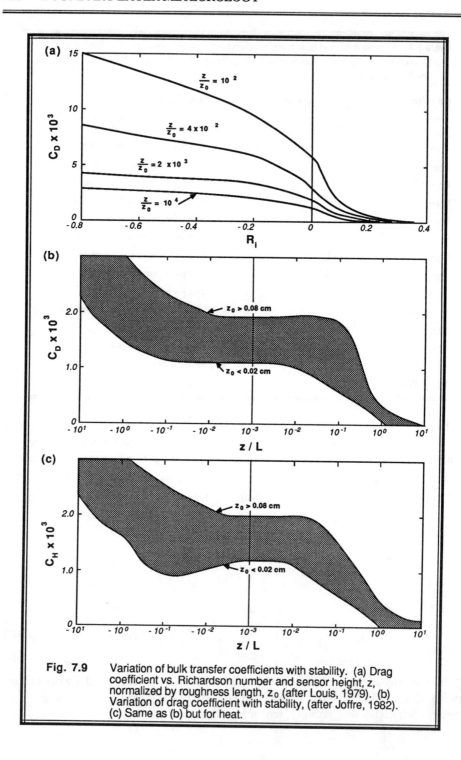

Fig. 7.9   Variation of bulk transfer coefficients with stability. (a) Drag coefficient vs. Richardson number and sensor height, z, normalized by roughness length, z_0 (after Louis, 1979). (b) Variation of drag coefficient with stability, (after Joffre, 1982). (c) Same as (b) but for heat.

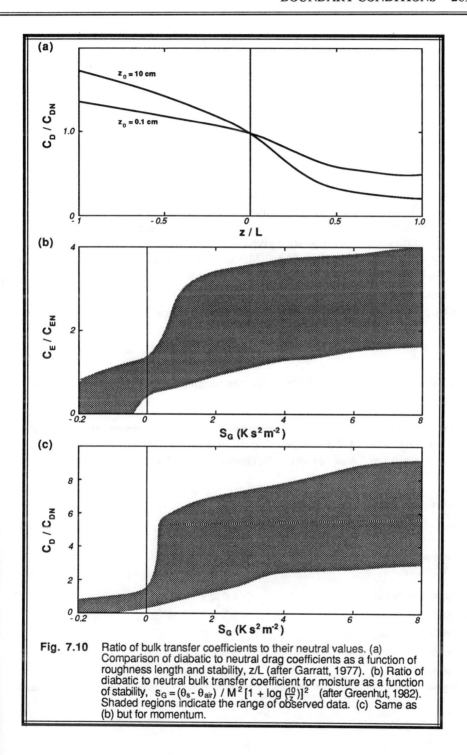

Fig. 7.10   Ratio of bulk transfer coefficients to their neutral values. (a) Comparison of diabatic to neutral drag coefficients as a function of roughness length and stability, z/L (after Garratt, 1977). (b) Ratio of diabatic to neutral bulk transfer coefficient for moisture as a function of stability, $s_G = (\theta_s - \theta_{air}) / M^2 [1 + \log (\frac{10}{z})]^2$ (after Greenhut, 1982). Shaded regions indicate the range of observed data. (c) Same as (b) but for momentum.

7.4.2 Surface Fluxes — Geostrophic Drag.   Sometimes in numerical or theoretical models, one does not know the winds anywhere near the surface, but one can calculate the geostrophic wind instead.   Deacon (1973), Clarke and Hesse (1974), Melgargejo and Deardorff (1974), Arya (1975), Nicholls (1982), and Grant and Whiteford (1987) parameterized surface fluxes in terms of geostrophic wind, using

$$F_\xi = -C_G \ \overline{G}\left(\overline{\xi}_{top} - \overline{\xi}_{bottom}\right) \tag{7.4.2a}$$

where $\overline{G}$ is the magnitude of the geostrophic wind, and C_G is called the *geostrophic drag coefficient*. This parameterization is usually not as accurate as the one using surface wind, and should be avoided unless you have no choice.

Typical magnitudes of C_G were given in Table 7-2 for continental areas. Table 7-4 lists suggested values over other areas. Although we are defining geostrophic drag based on $u_*^2 = C_G \ G^2$, some investigators use $u_* = C_G \ G$ instead. Be careful to check which definition is used for C_G when comparing values reported in the literature.

One parameterization for the neutral geostrophic drag coefficient is

$$C_{GN} = 0.0123 \ Ro^{-0.14} \tag{7.4.2b}$$

where $Ro = \overline{G}/(f_c \ z_o)$ is the *surface Rossby number*. Its variation with stability is given in Fig 7.11.

Table 7-4. Typical values of geostrophic drag coefficients (C_{GN}) for neutral stability over a variety of surfaces. After Garratt (1977), Grant and Whiteford (1987), and Kondo and Yamazawa (1986).

Surface	C_{GN}
sea	0.73×10^{-3}
rice paddies	1.1×10^{-3}
plains	2.2×10^{-3}
sparse houses or trees	2.0 to 2.5×10^{-3}
low mountains	2.5 to 2.7×10^{-3}
cities with tall buildings	2.7×10^{-3}
moderately high mountains	3 to 5×10^{-3}
very high mountains	8 to 10×10^{-3}

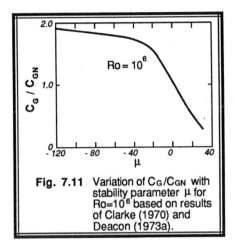

Fig. 7.11 Variation of C_G/C_{GN} with stability parameter μ for Ro=10^6 based on results of Clarke (1970) and Deacon (1973a).

Over oceans, Charnock's relationship gives (for geostrophic winds in m/s):

$$C_{GN} = 4.4 \times 10^{-4} \; \overline{G}^{\,0.16} \tag{7.4.2c}$$

$$u_*^2 = 4.8 \times 10^{-4} \; \overline{G}^{\,2.16} \qquad \text{(at 45°N)} \tag{7.4.2d}$$

7.4.3 Entrainment into the Top of the Mixed Layer

The top of the mixed layer sometimes behaves like a boundary, where the fluxes across it are controlled by the entrainment mechanism. When (7.4a) and (7.4c) are combined with the definition that $\Delta\xi = [\; \xi|_{\text{just above ML top}} - \xi|_{\text{within ML}}]$, then:

$$F_\xi = \overline{w'\xi'} = -w_e \, \Delta\overline{\xi} \tag{7.4.3a}$$

This expression can also be derived from Leibniz' theorem (see section 2.4.2). Using the mixing ratio example from that section (where z_i^+ denotes the value just: above z_i)

$$\frac{d\left[z_i \, \overline{r} \right]}{dt} = z_i \left[\frac{\overline{\partial r}}{\partial t} \right] + r\left(t, z_i^+ \right) \frac{dz_i}{dt}$$

The left hand side can be rewritten using the product rule of calculus as $\overline{r} \, dz_i/dt + z_i \, d\overline{r}/dt$,

where $\overline{r}$ represents an average within the mixed layer. Combining the first term from this expansion with the last term on the right hand side, and dividing by z_i gives:

$$\frac{d\bar{r}}{dt} = \left[\frac{\overline{\partial r}}{\partial t}\right] + \frac{w_e \Delta \bar{r}}{z_i} \qquad (7.4.3b)$$

using $w_e = d\, z_i/dt$ (in the absence of subsidence or clouds) and $\Delta \bar{r} = \bar{r}(t, z_i^+) - \bar{r}$.

But from (3.5.3) we know that the change in moisture in the mixed layer $d\bar{r}/dt$ is caused by various sources and sinks, $[\overline{\partial r/\partial t}]$, including flux divergence of surface flux and the flux divergence $\partial F/\partial z = F_{top}/z_i$ of flux from the top of the mixed layer. Hence:

$$\frac{d\bar{r}}{dt} = \left[\frac{\overline{\partial r}}{\partial t}\right] + \frac{\left[F_{top}\right]}{z_i} \qquad (7.4.3c)$$

Comparing (7.4.3b) and (7.4.3c), we conclude that $F_{top} = \overline{w'r'}|_{z_i} = -w_e\Delta\bar{r}$. One must remember that this interfacial flux occurs only when there is turbulent entrainment of air across the capping interface.

7.5 Partitioning of Flux into Sensible and Latent Portions

One alternative to the detailed modeling of surface temperature and soil moisture, as used in the previous section to find the heat, moisture, and momentum fluxes, is direct partitioning of the incoming solar radiative flux, described next.

Over land in clear-sky situations with weak or no advection, the fluxes of sensible and latent heat from the surface are governed by the diurnal cycle of solar radiation. Consider, for example, a situation where initially the sun is turned off, and the ground and air are at the same temperature. After turning on the sun with a constant net radiation of Q^*, the surface will warm. As it warms, a sensible heat flux will develop to remove some of the excess heat from the ground surface to the air. If the ground is moist, evaporation will also remove heat. Some heat will also be conducted into the ground.

If the removal of heat by sensible, latent, and ground fluxes is insufficient to balance the incoming Q^*, then the surface temperature will continue to warm. As it does, the sensible, latent, and ground fluxes will also increase until an equilibrium condition is finally reached where incoming radiative flux is balanced by outgoing turbulent and molecular fluxes. Although the surface temperature is important in creating this equilibrium, at equilibrium the fluxes must balance regardless of that temperature.

This argument suggests a parameterization where the net incoming radiation can be split or partitioned directly into the other fluxes of sensible, latent, and ground flux as governed by (7.2c), without requiring a forecast of ground surface temperature. In other words, the incoming radiation is an external forcing, while the sensible (Q_H), latent (Q_E) and ground fluxes (Q_G) are the response. The flux into the ground is often a small, but

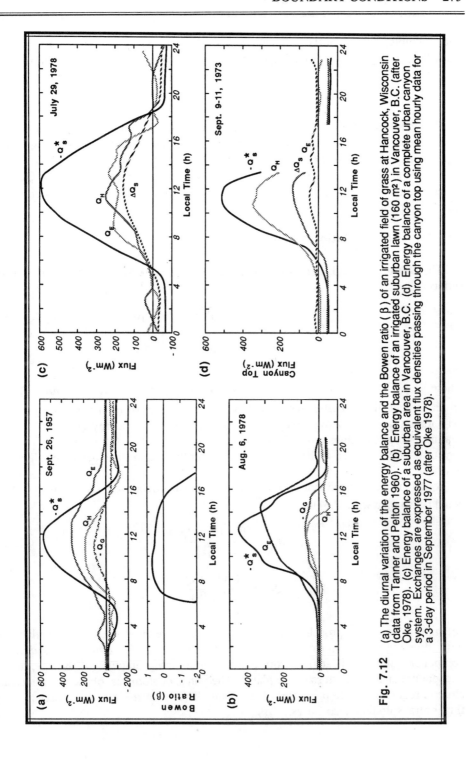

Fig. 7.12   (a) The diurnal variation of the energy balance and the Bowen ratio (β) of an irrigated field of grass at Hancock, Wisconsin (data from Tanner and Pelton 1960). (b) Energy balance of an irrigated suburban lawn (160 m²) in Vancouver, B.C. (after Oke, 1978). (c) Energy balance of a suburban area in Vancouver, B.C. (d) Energy balance of a complete urban canyon system. Exchanges are expressed as equivalent flux densities passing through the canyon top using mean hourly data for a 3-day period in September 1977 (after Oke 1978).

not negligible, fraction of the net radiation, as will be discussed in section 7.6. For now, we will treat $(-Q^*_s + Q_G)$ as the imposed forcing, which must be partitioned into Q_H and Q_E:

$$(-Q^*_s + Q_G) = Q_H + Q_E \tag{7.5}$$

As we saw in Fig. 7.4, both Q_H and Q_E reach a peak during midday at roughly the same time as the solar forcing peaks, and are small in the morning and evening.   This supports our idea of partitioning. However, a close examination of similar budget data from other sites (Fig. 7.12) suggests that the relative magnitudes of Q_H and Q_E vary depending on the wetness and vegetation of the surface. Any parameterization should take this variation into account.

7.5.1   Bowen Ratio Methods

The **Bowen ratio**, β, is defined as the ratio of sensible to latent heat fluxes at the surface:

$$\beta = \frac{Q_H}{Q_E} = \frac{\left(c_p\, \overline{w'\theta'}_s\right)}{\left(L_v\, \overline{w'q'}_s\right)} = \frac{\gamma\, \overline{w'\theta'}_s}{\overline{w'q'}_s} \tag{7.5.1a}$$

where $\gamma = C_p/L_v \cong 0.0004$ (g_{water}/g_{air})$\cdot K^{-1}$ is the **psychrometric constant**. As one might expect, β is smaller over moist surfaces where most of the energy goes into evaporation, and larger over dry surfaces where most of the energy goes into sensible heating.   Typical values range from 5 over semi-arid regions, 0.5 over grasslands and forests, 0.2 over irrigated orchards or grass, 0.1 over the sea, to some negative values over oases.

It was once suggested that if the Bowen ratio for a surface were known, then (7.5.1a) could be coupled with (7.5) to give:

$$Q_H = \frac{\beta\,(-Q^*_s + Q_G)}{(1 + \beta)} \tag{7.5.1b}$$

$$Q_E = \frac{(-Q^*_s + Q_G)}{(1 + \beta)} \tag{7.5.1c}$$

Attempts to use this approach have mostly failed, because the Bowen ratio usually varies with time and weather over each site.   Furthermore, the **evapotranspiration** component of latent heat flux from plants is a complex function of the age, health, temperature and water stress of the plant. The pores, or stomates, of the plant open and close to regulate the life processes of the plant. Thus, the **stomatel resistance** to water flux, or transpiration, also varies.

7.5.2 Priestley-Taylor Method

The next level of sophistication (Priestley and Taylor, 1972) comes by recognizing that to first order, we can use K-theory to approximate the fluxes by gradients of temperature and humidity in the Bowen ratio, assuming that $K_E = K_H$:

$$\beta = \gamma \frac{\partial \bar{\theta} / \partial z}{\partial \bar{q} / \partial z} = \frac{\left[\gamma (\partial \bar{T} / \partial z + \Gamma_d) \right]}{\partial \bar{q} / \partial z} \tag{7.5.2a}$$

where Γ_d is the dry adiabatic lapse rate of 9.8 K/km. If the air is saturated (i.e., if $q = q_{sat}$), then the change of specific humidity with temperature is given by the *Clausius-Clapeyron* equation:

$$\frac{dq_{sat}}{dT} = \frac{\varepsilon L_v \bar{q}_{sat}}{(\Re \bar{T}^2)} \equiv s_{cc}(\bar{T}) \tag{7.5.2b}$$

where $\varepsilon = 0.622$ (g_{water} / g_{air}) is the ratio of gas constants for air and water vapor.

We can use a variation of Teten's formula (Bolton, 1980) to find q_{sat} as a function of temperature (Fig. 7.13):

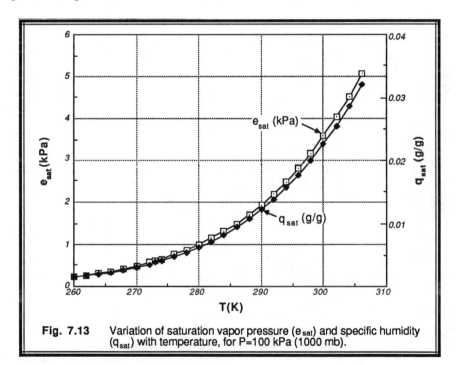

Fig. 7.13 Variation of saturation vapor pressure (e_{sat}) and specific humidity (q_{sat}) with temperature, for P=100 kPa (1000 mb).

$$q_{sat} = 0.622 \frac{e_{sat}}{P} \tag{7.5.2c}$$

where:

$$e_{sat} = (0.6112 \text{ kPa}) \cdot \exp\left[\frac{17.67 \cdot (T - 273.16)}{T - 29.66}\right] \tag{7.5.2d}$$

for T in degrees Kelvin.

Both $s_{cc}(\overline{T})$ and $\gamma(\overline{T}) = C_p/L_v$ are plotted in Fig. 7.14 for easy reference. If we replace $\partial\overline{q}/\partial z$ by $s_{cc}(\overline{T}) \cdot \partial\overline{T}/\partial z$, then:

$$\beta = \frac{\left[\gamma(\partial\overline{T}/\partial z + \Gamma_d)\right]}{(s_{cc} \partial\overline{T}/\partial z)} \tag{7.5.2e}$$

or

$$\beta = \frac{\gamma}{s_{cc}} + \frac{\gamma \cdot \Gamma_d}{s_{cc} \cdot (\partial\overline{T}/\partial z)} \tag{7.5.2f}$$

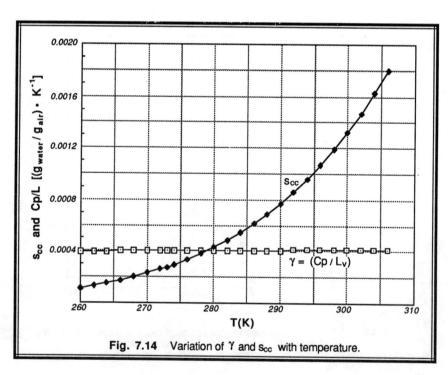

Fig. 7.14 Variation of γ and s_{cc} with temperature.

If one neglects the last term in (7.5.2f), then substitution of (7.5.2f) back into (7.5.1b & c) yields the *Priestly-Taylor* parameterization:

$$Q_H = \frac{\left[(1 - \alpha_{PT}) \, s_{cc} + \gamma\right]\left(- Q_s^* + Q_G\right)}{(s_{cc} + \gamma)} \qquad (7.5.2g)$$

$$Q_E = \alpha_{PT} \, s_{cc} \, \frac{(- Q_s^* + Q_G)}{(s_{cc} + \gamma)} \qquad (7.5.2h)$$

where the parameter α_{PT} is introduced to make up for neglecting the last term in (7.5.2f), and to allow the equations to be used for unsaturated conditions (DeBruin and Keijamn, 1979; DeBruin, 1983). For well-watered surfaces, $\alpha_{PT} \cong 1.25$.

The Priestley-Taylor method can yield incorrect fluxes when advection is happening. Although one possible fix is to increase the magnitude of α_{PT}, many investigators (McNaughton, 1976a & b; Singh and Taillefer, 1986) prefer to add an additional advection term, A, to the moisture flux equation to yield: $Q_E = [s_{cc}/(s_{cc}+\gamma)] \cdot (-Q^*_s + Q_G) + A$. This same advection term must be subtracted from the equation for heat flux.

7.5.3 Penman-Monteith Method

One way to include the evaporative cooling effects such as occurs during advection is via a correction term, F_w, added to the Priestly-Taylor parameterization for Q_E and subtracted from that for Q_H, to yield the *Penman-Monteith* (Penman, 1948; Monteith, 1965; deBruin and Holtslag, 1982) form:

$$Q_H = \frac{\left[\gamma(- Q_s^* + Q_G) - F_w\right]}{(X_G \, s_{cc} + \gamma)} \qquad (7.5.3a)$$

$$Q_E = \frac{\left[X_G \, s_{cc} (- Q_s^* + Q_G) + F_w\right]}{(X_G \, s_{cc} + \gamma)} \qquad (7.5.3b)$$

where X_G is like a relative humidity of the earth or plant surface. F_w is like a specific humidity flux, and is approximated by a bulk transfer law of the form: $F_w = C_E \, \overline{M} \, (X_G - X_s) \cdot \overline{q}_{sat}$, where X_s is the relative humidity of the air near the surface.

Remembering that $C_E \overline{M}$ is like a conductance for moisture, we see that surfaces with greater water conductivity have a greater latent heat flux and smaller sensible heat flux than less conductive surfaces.

Alternately, we can view the parameterization via the concept of a resistance to transfer. Physically, we would expect that surface/plant systems with less resistance to moisture transport would have greater evaporation, with the resulting reduction in sensible heat flux. For example, the *air resistance* , r_a, to transfer of water vapor away from a plant stomate into the air is just the inverse of the *conductance* ($C_E \overline{M}$) introduced earlier:

$$r_a = 1 / (C_E \overline{M})$$

The total resistance of the plant, r_p, is governed by the movement of water from the ground through the roots up into the plant to the cavity of the stomate where it is finally transpired into the air. We can combine the air and plant resistances to rewrite the correction term as: $F_w = (\overline{q}_{sat} - \overline{q}_{air}) / (r_a + r_p)$, where $\overline{q}_{air} = (X_s / X_G) \overline{q}_{sat}$, and $X_G = r_a/(r_a + r_p)$. This expression for F_w is equivalent to the one written earlier.

The topic of fluxes across vegetation systems is a complex one that we have barely parameterized above. A more detailed analysis should include factors such as canopy height, vegetation coverage, displacement heights, roughness lengths, plant reflectivity, plant architecture, root zone depth, ground water depth, heat conductivity, soil moisture, and stomatel resistance. These topics are beyond the scope of this book (see Verma, et al., 1986; Brutsaert, 1985; Mahrt and Ek, 1984; McNaughton and Jarvis, 1983; Deardorff, 1978; Geiger, 1965).

7.5.4   Flux Partitioning Example

Problem: Find Q_H and Q_E as a function of time for the same situation as presented in the radiation budget example of section 7.3.3. Compare the results of the Priestley-Taylor and the Penman-Monteith methods, given the typical diurnal cycle in temperature as listed in Table 7-5.

Let the surface pressure be constant at 100 kPa (1000 mb). For the first method, assume that $\alpha_{PT} = 1.25$, with no additional advection term. For the second method assume that $C_E = 0.002$, $X_s = 0.5$, $X_G = 0.9$, and $\overline{M} = 5$ m/s.

Solution: As in the radiation example, we will employ spreadsheet software to implement the solution. To find the imposed forcing ($-Q^*_s + Q_G$), we can use Q^* from the previous example, but we have not yet discussed parameterizations for Q_G. Assume for simplicity that $Q_G = 0.5 \ Q^*$ during nighttime, and $Q_G = 0.1 \ Q^*$ during the daytime (to be discussed in section 7.6.1). The latent heat of vaporization is approximated by a linear function of temperature: $L_v(J/kg) = [2.501 - 0.00237 \cdot T(°C)] \, 10^6$. The specific heat at constant pressure is assumed to be constant $C_p = 1005$ J/(kg K), and the gas constant for air is also assumed to be constant $\Re = 287.04$ J/(kg K).

Both the Priestley-Taylor and the Penman-Monteith methods are a function of s_{cc}, which in turn is a function of temperature. Table 7-5 lists the values of $L_v(T)$, $\gamma(T)$, $q_{sat}(T)$, $s_{cc}(T)$, and $(-Q^*_s + Q_G)$, and saturation vapor pressure $e_{sat}(T)$.

In the second page of Table 7-5, we finally compute Q_E and Q_H for both the Priestly-Taylor and the Penman-Monteith methods. Also computed in the table are the Bowen ratios for both methods, and the value of F_w [$(g_{water}/g_{air}) \cdot m/s$] for the Penman-Monteith method.

Discussion: Fig 7.15a-c shows the variations of the terms in the surface energy balance for both methods. Fig 7.15a indicates that all of the fluxes are in phase with the radiative forcing for the Priestely-Taylor method. In particular, all of the fluxes cross the zero line at the same times. The latent heat flux is larger than the sensible heat flux for this case. At night, all of the fluxes are negative, including the latent heat flux. This is related to dew or frost.

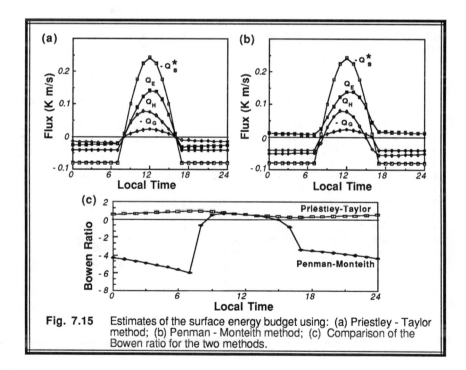

Fig. 7.15   Estimates of the surface energy budget using: (a) Priestley - Taylor method; (b) Penman - Monteith method; (c) Comparison of the Bowen ratio for the two methods.

Table 7-5. Flux partitioning example, displayed as output from a spreadsheet.

Flux Partition for: Madison, Wisconsin

P =   1000 mb   = 100 kPa
Rv =   287.04 J/(kg K)
Cp =   1005   J/(kg K)

Local	Q*	T	Lv	gamma (Cp/Lv)	e sat	q sat	Scc	Q*s-QG
(hr)	(Km/s)	(K)	(J/kg)	((g/g)/K)	(mb)	(g/g)	(q/K)	(K m/s)
0	0.080	280	2482500	0.00040	10.024	0.00626	0.00043	0.040
1	0.080	279	2485000	0.00040	9.356	0.00584	0.00040	0.040
2	0.080	278	2487500	0.00040	8.728	0.00545	0.00038	0.040
3	0.080	277	2490000	0.00040	8.137	0.00508	0.00036	0.040
4	0.080	276	2492500	0.00040	7.582	0.00473	0.00034	0.040
5	0.080	275	2495000	0.00040	7.060	0.00440	0.00031	0.040
6	0.080	274	2497500	0.00040	6.571	0.00410	0.00030	0.040
7	0.080	273	2500000	0.00040	6.112	0.00381	0.00028	0.040
8	-0.011	273	2500000	0.00040	6.112	0.00381	0.00028	-0.009
9	-0.101	274	2497500	0.00040	6.571	0.00410	0.00030	-0.091
10	-0.176	276	2492500	0.00040	7.582	0.00473	0.00034	-0.158
11	-0.225	278	2487500	0.00040	8.728	0.00545	0.00038	-0.203
12	-0.242	280	2482500	0.00040	10.024	0.00626	0.00043	-0.218
13	-0.225	282	2477500	0.00041	11.488	0.00718	0.00048	-0.203
14	-0.176	284	2472500	0.00041	13.138	0.00821	0.00055	-0.158
15	-0.101	286	2467500	0.00041	14.993	0.00938	0.00061	-0.091
16	-0.011	287	2465000	0.00041	16.004	0.01001	0.00065	-0.009
17	0.080	287	2465000	0.00041	16.004	0.01001	0.00065	0.040
18	0.080	286	2467500	0.00041	14.993	0.00938	0.00061	0.040
19	0.080	285	2470000	0.00041	14.038	0.00878	0.00058	0.040
20	0.080	284	2472500	0.00041	13.138	0.00821	0.00055	0.040
21	0.080	283	2475000	0.00041	12.289	0.00768	0.00051	0.040
22	0.080	282	2477500	0.00041	11.488	0.00718	0.00048	0.040
23	0.080	281	2480000	0.00041	10.734	0.00670	0.00046	0.040
24	0.080	280	2482500	0.00040	10.024	0.00626	0.00043	0.040

Table 7-5 (part 2). Flux partitioning example.

					Penman - Monteith Method			
					CE =	0.002		
					Xa =	0.5		
		Priestley - Taylor Method			Xo =	0.9		
alpha =		1.25			M(m/s)=	5		
time local	Q*s (K m/s)	QG (K m/s)	QE (K m/s)	QH (K m/s)	Bowen Ratio	Fw (g/g)(m/s)	QE (K m/s)	QH (K m/s)	Bowen Ratio
0	0.080	0.040	-0.0257	-0.0143	0.554	2.5E-05	0.0121	-0.0521	-4.31
1	0.080	0.040	-0.0250	-0.0150	0.601	2.3E-05	0.0115	-0.0515	-4.48
2	0.080	0.040	-0.0242	-0.0158	0.651	2.2E-05	0.0109	-0.0509	-4.68
3	0.080	0.040	-0.0235	-0.0165	0.704	2E-05	0.0103	-0.0503	-4.89
4	0.080	0.040	-0.0227	-0.0173	0.762	1.9E-05	0.0097	-0.0497	-5.12
5	0.080	0.040	-0.0219	-0.0181	0.824	1.8E-05	0.0092	-0.0492	-5.37
6	0.080	0.040	-0.0212	-0.0188	0.890	1.6E-05	0.0086	-0.0486	-5.64
7	0.080	0.040	-0.0204	-0.0196	0.961	1.5E-05	0.0081	-0.0481	-5.94
8	-0.011	-0.001	0.0048	0.0046	0.961	1.5E-05	0.0270	-0.0176	-0.65
9	-0.101	-0.010	0.0482	0.0429	0.890	1.6E-05	0.0608	0.0303	0.50
10	-0.176	-0.018	0.0899	0.0685	0.762	1.9E-05	0.0946	0.0637	0.67
11	-0.225	-0.023	0.1228	0.0799	0.651	2.2E-05	0.1221	0.0806	0.66
12	-0.242	-0.024	0.1404	0.0778	0.554	2.5E-05	0.1382	0.0800	0.58
13	-0.225	-0.023	0.1379	0.0648	0.470	2.9E-05	0.1391	0.0636	0.46
14	-0.176	-0.018	0.1134	0.0449	0.396	3.3E-05	0.1232	0.0351	0.28
15	-0.101	-0.010	0.0684	0.0227	0.331	3.8E-05	0.0915	-0.0004	0.00
16	-0.011	-0.001	0.0073	0.0022	0.302	4E-05	0.0460	-0.0365	-0.79
17	0.080	0.040	-0.0307	-0.0093	0.302	4E-05	0.0168	-0.0568	-3.38
18	0.080	0.040	-0.0300	-0.0100	0.331	3.8E-05	0.0161	-0.0561	-3.48
19	0.080	0.040	0.0294	-0.0106	0.363	3.5E-05	0.0154	-0.0554	-3.60
20	0.080	0.040	-0.0287	-0.0113	0.396	3.3E-05	0.0147	-0.0547	-3.72
21	0.080	0.040	-0.0279	-0.0121	0.432	3.1E-05	0.0140	-0.0540	-3.85
22	0.080	0.040	-0.0272	-0.0128	0.470	2.9E-05	0.0134	-0.0534	-3.99
23	0.080	0.040	-0.0265	-0.0135	0.511	2.7E-05	0.0127	-0.0527	-4.14
24	0.080	0.040	-0.0257	-0.0143	0.554	2.5E-05	0.0121	-0.0521	-4.31

Fig 7.15b indicates that the fluxes do not all cross the zero line at the same time for the Penman-Monteith method. This is indeed more realistic. Also note that at night there is still a positive latent heat flux. This is partially balanced by a greater negative sensible heat flux than in the previous method. Physically, this means that enough heat is lost from the air to the ground to balance both the radiative loss and to also support evaporation from the surface at night. During the day, the sensible heat flux becomes negative well before the radiative forcing becomes zero near sunset.

Finally, Fig 7.15c most graphically demonstrates the differences between the two methods. By looking at the Bowen ratio, we see that the two methods give close to the same answer only during mid-day. At other times, they differ significantly from each other, even having opposite signs at night. It appears that the Penman-Monteith method has the potential for giving better results, assuming that appropriate values can be estimated for the parameters such as relative humidity, ground moisture, etc.

7.6 Flux To and From the Ground

We will use the words *ground flux* to represent heat flux into the ground measured at the top of the soil. This flux is a small, but not insignificant, component of the surface energy budget. This flux is also related to the surface skin temperature, which is what the atmosphere "sees" when it radiatively looks down at the bottom boundary. If the flux is not known or measured directly, then we often need to parameterize it for PBL forecast models. Sample parameterizations are given below.

7.6.1 Simple Parameterizations

Averaged over a full 24-hour cycle, the net heat flux is often near zero. Namely, the heating of the ground during the day is nearly balanced by cooling at night, leading to little net change of heat in the soil. Thus, for some climate and general circulation models (*GCM*), it is assumed for simplicity that $Q_G = 0$.

If we choose to use a flux partitioning scheme such as discussed in Section 7.5, then we could partition some portion of flux to go into or from the ground. One simple scheme is to assume that the ground flux is a percentage, X, of the net radiation:

$$Q_G = X \ Q^* \tag{7.6.1a}$$

where $X = 0.1$ during the daytime, and $X = 0.5$ at night, for example. Fig 7.16 shows data from a number of daytime measurements over land, where most of the ground flux values are between 5% and 15% of the net radiation.

An alternative is to assume the ground flux is a percentage of the turbulent sensible heat flux into the air:

$$Q_G = 0.3 \ Q_H \tag{7.6.1b}$$

Both of these simple schemes assume that the sign of the ground flux is always the same as that of the net radiation or sensible heat flux. The latter scheme would fail in an oasis situation.

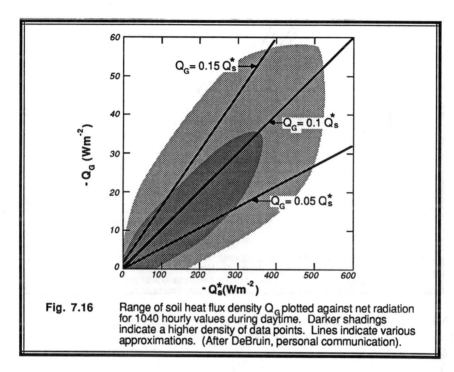

Fig. 7.16   Range of soil heat flux density Q_G plotted against net radiation for 1040 hourly values during daytime. Darker shadings indicate a higher density of data points. Lines indicate various approximations. (After DeBruin, personal communication).

7.6.2   Multilevel Soil Model

At the opposite end of sophistication is a multilevel or analytical model of the soil, where prognostic equations are solved for the temperature at a large number of depths in the soil. Fig 7.17 shows a typical temperature variation over a diurnal cycle in the soil and the atmosphere. Note the scale change for the height-depth axis. Most of the temperature variation in the soil happens within the top 20 centimeters. Below about one meter there is little temperature change associated with the daily cycle, although the annual cycle signal is much larger at that depth.

Since molecular conduction is the primary transport process, we can write the ground flux at any depth as:

$$Q_g = -k_g \frac{\partial T}{\partial z} \qquad (7.6.2a)$$

where k_g is the thermal molecular conductivity of the soil, and $Q_G = Q_g(z=0)$. Appendix C lists examples of the conductivy for a variety of soils.

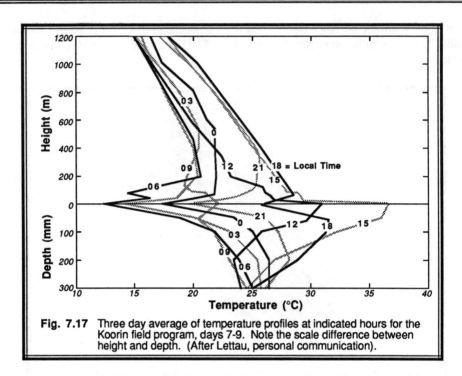

Fig. 7.17   Three day average of temperature profiles at indicated hours for the Koorin field program, days 7-9. Note the scale difference between height and depth. (After Lettau, personal communication).

Assuming that there are no other sources or sinks of heat in the soil, the second law of thermodynamics yields the simple prognostic equation:

$$\frac{\partial T}{\partial t} = -\left(\frac{1}{C_g}\right)\frac{\partial Q_g}{\partial z} \qquad (7.6.2b)$$

where C_g is the *soil heat capacity* (i.e., soil density times specific heat).

Combining this with the previous equation gives the classical **heat conduction equation**:

$$\frac{\partial T}{\partial t} = \nu_g \frac{\partial^2 T}{\partial z^2} \qquad (7.6.2c)$$

where $\nu_g = k_g/C_g$ is the *soil thermal diffusivity*. Typical values for ν_g are 2.7×10^{-7} (m²/s) for snow, 2.0×10^{-7} to 1×10^{-6} for farms, and 1.5×10^{-7} for water. A variety of solutions to this classic equation for various boundary conditions are given in applied mathematics, physics, and engineering texts.

For a soil with uniform thermal diffusivity with depth, the boundary conditions often used are: (1) periodic temperature variation at the surface, and (2) no temperature change at great depths.   The solution is a periodic temperature variation that decreases in amplitude with depth, and increases phase lag with depth, as is sketched in Fig 7.18.   If the period of the cycle is $\mathbb{P}$, then the amplitude ΔT of the wave changes with depth as:

$$\Delta T\,(z)\ =\ \Delta T_{surface}\ \cdot\ \exp\left[-z\left\{\frac{\pi}{v_g\,\mathbb{P}}\right\}^{1/2}\right] \tag{7.6.2d}$$

The time lag Δt associated with the phase shift across a depth range Δz is:

$$\Delta t\ =\ \left(\frac{\Delta z}{2}\right)\left[\frac{\mathbb{P}}{\pi\,v_g}\right]^{1/2} \tag{7.6.2e}$$

These equations can be applied to an annual ($\mathbb{P} = 1$ year) or daily cycle ($\mathbb{P} = 1$ day).

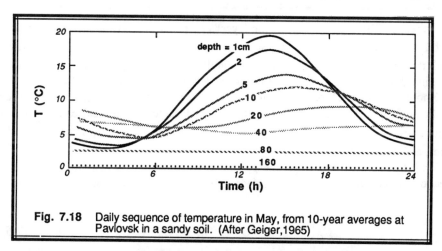

Fig. 7.18   Daily sequence of temperature in May, from 10-year averages at Pavlovsk in a sandy soil. (After Geiger,1965)

For more realistic nonperiodic forcings or nonuniform soil properties, (7.6.2c) can be solved with a variety of numerical schemes, analytical series expansions or Fourier decompositions.   In this way, variations in thermal diffusivity can be included to allow for changes in soil moisture. Since soil moisture depends on precipitation, evaporation, water table level, runoff, and the transpiration of plants, a full detailed parameterization of the soil process often becomes quite complicated.

7.6.3   Force-Restore Method

Because most of the soil temperature changes occur within a shallow layer near the surface, Blackadar (1976, 1979; Zhang and Anthes, 1982; Anthes, et al., 1987) suggested a two-layer approximation, where a shallow slab of soil is bounded below by a thick constant-temperature slab (see Fig 7.19).

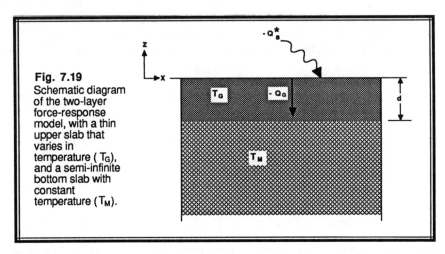

Fig. 7.19
Schematic diagram of the two-layer force-response model, with a thin upper slab that varies in temperature (T_G), and a semi-infinite bottom slab with constant temperature (T_M).

Concept. The depth, d_s, of the top layer is carefully chosen, based on knowledge of the full solutions from the previous section: $d_s = [(\nu_g \, \mathbb{P})/(4\pi)]^{1/2}$. This is the effective depth that "feels" the diurnal cycle, and is on the order of several centimeters for a daily cycle in a desert. We see that longer-period cycles are felt over a deeper depth.

Knowing this critical depth, a soil heat capacity per unit area, C_{GA}, can then be found for the surface soil slab: $C_{GA} = C_g \cdot d_s$. Typical values of the heat capacity per unit volume are $C_g = 1 \times 10^6$ (J m^{-3} K^{-1}) for snow, 1.3×10^6 for deserts, 3×10^6 to 5×10^6 for farms, and 1×10^6 for muddy water.

With these definitions, the energy balance (7.2b) can be rewritten as a forecast equation for the temperature of the surface slab (Anthes, et al., 1987):

$$C_{GA} \frac{\partial T_G}{\partial t} = -Q_s^* - Q_H - Q_E + Q_G \qquad (7.6.3a)$$

In Blackadar's original formulation, the heat fluxes into the ground and air were parameterized directly, and the latent heat flux was neglected:

$$\frac{\partial T_G}{\partial t} = \frac{-Q_s^*}{C_{GA}} + \left(\frac{2\pi}{\mathbb{P}}\right) \left[T_M - T_G \right] - a_{FR} \left[T_G - \overline{T}_{air} \right] \qquad (7.6.3b)$$

$$\quad\text{I}\qquad\quad\text{II}\qquad\qquad\text{III}\qquad\qquad\text{IV}$$

The temperature (term I) of the top slab is assumed to respond to the net radiation forcing (term II), to conduction from the deeper slab (term III) and to turbulent transport with the air (term IV). In other words, the flux from the deep soil tends to *restore* the surface slab temperature, opposing the radiative *forcing*. The temperature of the bottom slab, T_M, is assumed to be a quasi-constant boundary condition. The ground flux can be found from:

$$-Q_G = C_{GA}\left(\frac{\partial T_G}{\partial t}\right) + \left[2\pi\frac{C_{GA}}{\mathbb{P}}\right]\left[T_G - T_M\right] \qquad (7.6.3c)$$

Both of these equations can be solved numerically by stepping forward in time from some known initial condition.

We recognize "a_{FR}" as a form of "conductivity" between the ground and the air. Blackadar suggested a magnitude on the order of $a_{FR} = 2\pi/\mathbb{P}$, or 3×10^{-4} s^{-1} for $T_G > T_{air}$ (e.g., daytime), and 1×10^{-4} s^{-1} for $T_G < T_{air}$ (e.g., nighttime).

This two layer force-restore model has become a popular compromise between the more complex multilayer soil model of the previous section, and the overly simplistic approximations of Section 7.6.1. Deardorff (1978) extended this approach to model soil moisture in order to estimate the latent heat flux.

Example

Problem: Make a 24 hour forecast of Q_H, Q_G, T_G, and T_{air} for farmland near Wausau, Wisconsin, starting at 12 UTC on 25 April (day 116). Assume a sky covered with 10% low clouds, no middle clouds, and 60% high clouds. Assume that the albedo is 0.2, $T_G = 10°C$, and $z_i = 1000$ m and is constant. For initial conditions, start with an air temperature of 8.4°C and a soil surface-slab temperature of 4.5°C. Assume a dry soil (i.e., no latent heat flux).

Solution: To make the desired forecast, we must also specify the net radiative flux $-Q^*_s$. We can utilize the parameterization (7.3.2c) for $-Q^*_s$, because we know the latitude and longitude of Wausau (44.97N, 89.63W) and we know the cloud cover.

First, we estimate the soil diffusivity to be 5×10^{-7} m^2/s, and $C_g = 1.67\times10^6$ J m^{-3} K^{-1} for farmland, based on the tables in Appendix C. Next, we must calculate the values for d_s and C_{GA} based on the equations of the previous section. We find that the soil slab depth is $d_s = 5.9$ cm, and the soil heat capacity per unit area (converted to kinematic units) is $C_{GA} = 80.33$ m, assuming a 24-hour time period for the cyclic forcing.

Table 7-6 shows just the first portion of the resulting forecast of fluxes and temperatures, taking a 15 minute timestep. A longer timestep would have lead to errors for this case.

Table 7-6. Example of a ground flux forecast using the force-restore method.

```
Enter the Julian day of the year (Example: April 9, 1984 = 100):    100.000
Enter forecast start time (UTC) in hours:          12.0000
Enter forecast duration in hours:                  24.0000
Enter latitude in degrees (positive North):        44.9700
Enter longitude in degrees (positive West):        89.6300
Enter the low, mid, and high cloudcover fractions:    0.1000    0.0000    0.6000
Enter the albedo:                                   0.2000
Enter Cg, the soil heat capacity/area (m)          80.3300
Enter the deep ground temperature TM (C):          10.0000
Enter the initial surface soil temp TG (C):         4.5000
Enter the initial air temperature (C):              8.4000
Enter the boundary layer thickness (m):            1000.00
```

T (HR)	-QSTAR(KM/S)	-QG(KM/S)	QH(KM/S)	TG(C)	TAIR(C)
12.0000	-0.0322	0.0023	-0.0345	4.8505	8.3689
12.2500	-0.0127	0.0196	-0.0323	5.3625	8.3399
12.5000	0.0073	0.0359	-0.0286	6.0161	8.3142
12.7500	0.0277	0.0514	-0.0237	6.7937	8.2928
13.0000	0.0483	0.0660	-0.0177	7.6793	8.2769
13.2500	0.0690	0.0798	-0.0108	8.6581	8.2671
13.5000	0.0898	0.0862	0.0036	9.6463	8.2704
13.7500	0.1104	0.0824	0.0281	10.5353	8.2956
14.0000	0.1309	0.0814	0.0494	11.3619	8.3401
14.2500	0.1509	0.0822	0.0687	12.1480	8.4020
14.5000	0.1705	0.0840	0.0865	12.9064	8.4798
14.7500	0.1894	0.0863	0.1032	13.6435	8.5726
15.0000	0.2076	0.0887	0.1190	14.3619	8.6797
15.2500	0.2249	0.0909	0.1340	15.0617	8.8003
15.5000	0.2411	0.0929	0.1482	15.7411	8.9336
15.7500	0.2562	0.0946	0.1616	16.3976	9.0791
16.0000	0.2700	0.0958	0.1742	17.0279	9.2359
16.2500	0.2824	0.0965	0.1859	17.6285	9.4032
16.5000	0.2934	0.0967	0.1967	18.1956	9.5803
16.7500	0.3028	0.0964	0.2064	18.7256	9.7660
17.0000	0.3106	0.0955	0.2151	19.2150	9.9596
17.2500	0.3166	0.0941	0.2226	19.6604	10.1599
17.5000	0.3210	0.0921	0.2288	20.0588	10.3659
17.7500	0.3235	0.0897	0.2338	20.4076	10.5763
18.0000	0.3242	0.0867	0.2375	20.7043	10.7901
18.2500	0.3231	0.0832	0.2399	20.9471	11.0060
18.5000	0.3202	0.0793	0.2409	21.1344	11.2228
18.7500	0.3156	0.0750	0.2406	21.2652	11.4393
19.0000	0.3092	0.0703	0.2388	21.3387	11.6543
19.2500	0.3011	0.0653	0.2358	21.3550	11.8665
19.5000	0.2914	0.0600	0.2314	21.3142	12.0747
19.7500	0.2801	0.0544	0.2257	21.2172	12.2778
20.0000	0.2674	0.0486	0.2187	21.0651	12.4747
20.2500	0.2533	0.0427	0.2106	20.8595	12.6642
20.5000	0.2380	0.0367	0.2013	20.6025	12.8454
20.7500	0.2215	0.0306	0.1910	20.2963	13.0173
21.0000	0.2041	0.0244	0.1797	19.9437	13.1790
21.2500	0.1858	0.0183	0.1674	19.5477	13.3297
21.5000	0.1667	0.0123	0.1544	19.1116	13.4686
21.7500	0.1470	0.0064	0.1406	18.6388	13.5952
22.0000	0.1269	0.0006	0.1263	18.1331	13.7088
22.2500	0.1064	-0.0050	0.1114	17.5981	13.8091
22.5000	0.0857	-0.0104	0.0961	17.0378	13.8956

Discussion: Fig 7.20a &b show plots of the fluxes and temperature. We see that the ground flux changes sign well before sunset. The negative ground flux that follows implies an upward transport of heat from the ground to the surface. When this is added to the incoming radiation $-Q^*_s$, we see that the sensible heat flux Q_H becomes larger than $-Q^*_s$ in the late afternoon.

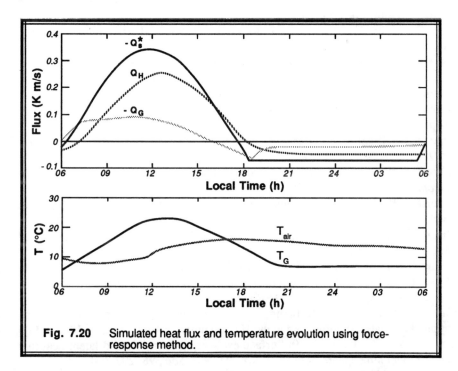

Fig. 7.20   Simulated heat flux and temperature evolution using force-response method.

The magnitude of the ground flux is larger near sunset than later at night. This is related to the fact that the ground has some thermal inertia, leaving the ground temperature warmer than the air temperature during a few hours near sunset when the ground flux has already become negative. During this short period, the ground is loosing heat not only to the cooler air, but also by radiation to space. This causes the temporary bulge in ground flux at night. Later at night, the ground finally becomes colder than the air, allowing some of the heat lost to space to be replaced by conduction from the air.

7.7 References

Anthes, R.A., E.-Y. Hsie, and Y.-H. Kuo, 1987: *Description of the Penn State/NCAR Mesoscale Model Version 4 (MM4)*. NCAR Tech. Note NCAR/TN-282+STR. Boulder, CO. 66pp.

Anthes, R.A. and D. Keyser, 1979: Tests of a fine-mesh model over Europe and the

United States. *Mon. Wea. Rev.*, **107**, 963-984.

Blackadar, A.K., 1976: Modeling the nocturnal boundary layer. *Third Symposium on Atmospheric Turbulence, Diffusion and Air Quality*, Raleigh, NC, Oct 19-22. Amer. Meteor. Soc., 46-49.

Blackadar, A.K., 1979: Modeling pollutant transfer during daytime convection. *Fourth Symposium on Turbulence, Diffusion and Air Pollution*. Reno, NV, Jan 15-18. Amer. Meteor. Soc., 443-447.

Bolton, D., 1980: The computation of equivalent potential temperature. *Mon. Wea. Rev.*, **108**, 1046-1053.

Brutsaert, W.H., 1985: *Evaporation and the Atmosphere*. Reidel. 299pp.

Burridge, D.M., and A.J. Gadd, 1974: The Meteorological Office Operational 10 Level Numerical Weather Prediction Model (December 1974). British Met. Office Tech. Notes Nos. 12 and 48. London Rd., Bracknell, Berkshire, RG12 2SZ, England. 57pp.

Charnock, H., 1955: Wind stress on a water surface. *Quart. J. Roy. Meter. Soc.*, **81**, 639.

Deardorff, J.W., 1968: Dependence of air-sea transfer coefficients on bulk stability. *J. Geophys. Res.*, **73**, 2549-2557.

Deardorff, J.W., 1978: Efficient prediction of ground surface temperature and moisture, with inclusion of a layer of vegetation. *J. Geophys. Res.*, **83**, 1889-1903.

DeBruin, H.A.R., 1983: A model for the Priestley-Taylor parameter a. *J. Clim. & Appl. Meteor.*, **22**, 572-578.

DeBruin, H.A.R. and A.A.M. Holtslag, 1982: A simple parameterization of the surface fluxes of sensible and latent heat during daytime compared with the Penman-Monteith concept. *J. Appl. Meteor.*, **21**, 1610-1621.

DeBruin, H.A.R. and J.Q. Keijamn, 1979: The Priestley-Taylor evaporation model applied to a large shallow lake in the Netherlands. *J. Appl. Meteor.*, **18**, 898-903.

Gadd, A.J., and F.J. Keers, 1970: Surface exchange of sensible and latent heat in a 10-level model atmosphere. *Quart. J. Roy. Meteor. Soc.*, **96**, 297-308.

Garratt, J. R., 1977: Review of drag coefficients over oceans and continents. *Mon. Wea. Rev.*, **105**, 915-929.

Geiger,R., 1965: *The Climate Near the Ground*. Harvard University Press, Cambridge. 611pp.

Grant, A.L.M. and R. Whiteford, 1987: Aircraft estimates of the geostrophic drag coefficient and the Rossby similarity functions A and B over the sea. *Bound.-Layer Meteor.*, **39**, 219-231.

Greenhut, G.K., 1982: Stability dependence of fluxes and bulk transfer coefficients in a tropical boundary layer. *Bound-Layer Meteor.*, **24**, 253-264.

Huband, N.D.S. and J.L. Monteith, 1986: Radiative surface temperature and energy balance of a wheat canopy. Parts 1 and 2. *Bound.-Layer Meteor.*, **36**, 1-17.

Joffre, S.M., 1982: Momentum and heat transfers in the surface layer over a frozen sea. *Bound.-Layer Meteor.*, **24**, 211-229.

Kondo, J. and A. Kawanaka, 1986: Numerical study on the bulk heat transfer coefficient for a variety of vegetation types and densities. *Bound.-Layer Meteor.*, **37**, 285-296.

Kondo, J. and H. Yamazawa, 1986a: Bulk transfer coefficient over a snow surface. *Bound.-Layer Meteor.*, **34**, 123-135.

Kondo, J. and H. Yamazawa, 1986b: Aerodynamic roughness over an inhomogeneous ground surface. *Bound.-Layer Meteor.*, **35**, 331-348.

Kraus, E.B., 1972: *Atmosphere-Ocean Interaction.* Oxford Monographs on Meteorology. Clarendon Press, Oxford. 275pp.

Kyle, H.L., P.E. Ardanuy, and E.J. Hurley, 1985: The status of the Nimbus-7 earth-radiation-budget data set. *Bull. Am. Meteor. Soc.*, **66**, 1378-1388.

Louis, J. F., 1979: A parametric model of vertical eddy fluxes in the atmosphere. *Bound.-Layer Meteor.*, **17**, 187-202.

Mahrt, L. and M. Ek, 1984: The influence of atmospheric stability on potential evaporation. *J. Clim. & Appl. Meteor.*, **23**, 222-234.

McNaughton, K.G., 1976: Evaporation and advection, Parts 1 and 2. *Quart. J. Royal Meteor. Soc.*, **102**, 181-191.

McNaughton, K.G. and P.G. Jarvis, 1983: Predicting effects of vegetation changes on transpiration and evaporation. In T.T. Kozlowski (Ed.), *Water Deficits and Plant Growth.* Academic Press. 1-47.

Monteith, J.L., 1965: Evaporation and environment. *Symp. Soc. Exp. Biol.*, **19**, 205-234.

Oke, T.R., 1978: *Boundary Layer Climates.* Halsted Press, NY. 372pp.

Penman, H.L., 1948: Natural evaporation from open water, bare soil, and grass. *Proc. Roy. Soc. London*, **A193**, 120-195.

Priestly, C.H.B. and R.J. Taylor, 1972: On the assessment of surface heat flux and evaporation using large-scale parameters. *Mon. Wea. Rev.*, **100**, 81-92.

Ripley, E.A. and R.E. Redmann, 1976: Grassland. *In Vegetation and the Atmosphere, Vol. 2, Case Studies.* (Ed. by J.L. Monteith). Academic Press, London. 349-398.

Singh, B. and R. Taillefer, 1986: The effect of synoptic-scale advection on the performance of the Priestley-Taylor evaporation formula. *Bound.-Layer Meteor.*, **36**, 267-282.

Taylor, G.I., 1916: Conditions at the surface of a hot body exposed to the wind, *Brit. Adv. Com. Aero. Rep. and Memor.*, 272.

Verma, S.B., D.D. Baldocchi, D.E. Anderson, D.R. Matt, R.J. Clement, 1986: Eddy fluxes of CO_2, water vapor, and sensible heat over a deciduous forest. *Bound.-Layer Meteor.*, **36**, 71-91.

Wallace, J.M. and P.V. Hobbs, 1977: *Atmospheric Science, An Introductory Survey.* Academic Press, NY. 467pp.

Zhang, D. and R.A. Anthes, 1982: A high-resolution model of the planetary boundary layer — sensitivity tests and comparisons with SESAME-79 data. *J. Appl. Meteor.*, **21**, 1594-1609.

7.8 Exercises

1) Derive equation (7.3.1c) using basic geometric principles.   (Note, there are other equivalent forms of this equation, based on various triginometric identities.)

2) What is the local sun angle above the horizon at 1900 UTC on 25 December at
   a) Madison, Wisconsin
   b) Christmas Island (Line Islands)
   c) Seattle Washington
   d) Tallahassee, Florida
   e) Munich, W.Germany
   f) Barrow, Alaska?

3) Estimate the net radiation, Q^*, at Madison, Wisconsin on 25 June at noon local daylight time, if there are 2/8 cirrus clouds and 3/8 cumulus clouds.

4) Based on the discussion of atmospheric conductance in section 7.4, if atmospheric resistance is defined as the inverse of conductance, then what are the units and typical magnitude of atmospheric resistance?

5) Estimate the 10 m drag (C_D) and heat transfer (C_H) coefficients over land, given $z_o = 1$ cm, and the Obukhov length = 20 m.   How would they differ if the static stability were neutral?

6) Given the following data, estimate the values of
   a. I^*
   b. Q^*
   c. Q_H
   d. Q_E
   e. Bowen ratio
   Given: location is Clam Lake, WI (a town with albedo = 20%)
   Latitude = 46 deg 10 min N
   Longitude = 90 deg 54 min W
   Local time of day is 10 AM CST (central standard time = local time = UTC - 6 h)
   Date is   April 10, 1981
   Cloud cover is 50% of sky covered by altocumulus
   Surface relative humidity is 80 %
   Anemometer-level humidity is 60 %
   wind is 5 m/s
   Surface temp is 15°C
   Use the methods discussed in the surface forcings section of boundary layer characteristics.

7) Using the desert Q^* data below, apply the force-restore method to calculate the heat flux into the ground and the ground temperature for at least the following times of day: 8, 12, 16, & 24 local time.   Justify all assumptions.   Assume that the deep soil temperature is 20°C.

Local time: 4   6   8   10   12   14   16   18   20   22   24   02
Q^* (J m^{-2}s^{-1}): -80 -20 100 300 500 600 400   50 -120 -110 -100 -90

8) Based on the following instantaneous measurements of potential temperature (θ), specific humidity (q), and vertical velocity (w), find the Bowen ratio.

θ (°C)	q (g/kg)	w (m/s)
22	10	0.1
21	12	0.5
18	12	-0.4
20	6	-0.2
19	10	0

9) State some reasons why one might need to use geostrophic drag relationships.

10) [Problems 10 and 11 are identical to the solved examples in the text. We recommend that the instructor modify the problem to any other location and time of interest to the students.]   What is the value of the net radiation (Q^*) absorbed at the earth's surface as a function of time (for a 24 hour period) under the following conditions:

albedo = 0.2
location = farmland near Wausau, Wisconsin
date = April 9, 1982
cloudcover: low= 10%   mid = 0   high = 60%
Plot your result (start the graph at 6 AM).

11) Given the following conditions:
same situation as problem (10).
deep soil temperature is 10°C.
soil is dry ($Q_E = 0$).
BL remains 1 km thick.
potential temperature is constant with height (but not time).
no advection or entrainment into the BL.
Initial conditions (at 6 AM): $T_{air} = 8.4$°C, and T_G = 4.5°C
Calculate and plot the following as a function of time for a 24 hour period starting at 6 AM.   Use the force-restore method for the ground temperature, and turbulence equations for mean flow for the air temperature.

a) T_{air}
b) T_G
c) Q_G
d) Q_H

12) SITUATION: Daytime boundary layer over land (N. America).
OBSERVATIONS:

Height , z (m)   12   8   2   0.1=z_o (m)

θ	300	301	303	308	(K)
U	5.4	5.0	3.4	0	(m/s)

What are the values of: a) C_{DN} b) C_D

13) For sandy clay with 15% moisture, at what depth below the surface will the diurnal temperature variations be 1 % of the surface temperature variations?

14) Given a 1km constant thickness boundary layer with initial $\theta = 10°C$ flowing at M=10 m/s over land, where the land has the same surface temperature as that of the air near the surface. At some point, the air leaves the land and flows over the ocean, where the ocean sea surface temperature is 20°C and the pressure is 100 kPa. Assume that the boundary layer is well mixed.   Calculate and plot the heat flux Q_H and the boundary layer temperature as a function of distance from the shoreline.

15) For a pressure gradient of 0.2 kPa / 100 km and a surface roughness length of $z_o = 2$ cm, find:
a) the value of the surface Rossby number
b) the value of the neutral geostrophic drag coefficient
c) u_* , assuming statically neutral conditions

16) Given a drag coefficient of 3×10^{-3} at $R_i = -0.4$ and $z_s = 10$ m, how would the drag coefficient change if $z = 100$ m?

17) What is the roughness length, z_o, over the ocean for a wind speed of 40 m/s ?

18) What is the time lag of the diurnal cycle of temperature at a depth of 15 cm in farmland? What will be the amplitude of the temperature wave at that depth?

19) Use equation (7.5.2b) to calculate and plot $s_{cc}(T)$ vs T, and compare your answer with Fig 7.13.

20) For a temperature of 20°C, dew point of 10°C, ground surface relative humidity of X = 80%, and wind speed of 5 m/s, find Q_H and Q_E using:
a) Priestley-Taylor method
b) Penman-Monteith method
(Hint: use the result from the previous question.)

21) Verify that the surface energy balance equation is satisfied using the Q_H and Q_E parameterizations from:
a) The Bowen ratio method (equations 7.5.1 b & c)
b) The Priestley-Taylor Method (equations 7.5.2e & f)
c) The Penman-Monteith method (equations 7.5.3a & b).

8 Some Mathematical & Conceptual Tools: Part 2. Time Series

Spectrum analysis is a statistical tool that we can employ to probe further into the workings of turbulence. By decomposing a series of measurements into frequency or wavenumber components, we can discover how eddies of different time and space scales contribute to the overall turbulence state.

In this chapter we review the computational techniques for the spectrum analysis of measured data. We also introduce related tools such as the autocorrelation function, structure function, and periodogram. Also discussed is the concept of a process spectrum, where mixing processes rather than turbulence states are decomposed into a spectrum of scales. Theoretical spectral decomposition of the TKE equation is briefly covered.

8.1 Time and Space Series

When measurements are taken at a fixed point over a period of time, the resulting series of data points is called a *time series*. Similarly, measurements at a fixed time over a series of locations in space is called a *space series*. Both series give measurements of a dependent variable such as temperature or humidity as a function of an independent variable, such as time, t, or location, x. Because of this similarity, we will discuss the two types of series interchangeably, and sometimes will use the generic name, *series*.

This review will be limited to *discrete* series; namely, measurements taken at regularly-spaced intervals that lead to a finite number, N, of data points. A discrete series represents a *sample* of the true, continuously-varying signal. Examples of discrete series include temperature or tracer concentration measurements made every second during the course of an hour at a fixed location such as an instrumented tower, or measurements of humidity taken every meter from an aircraft flying on a 25 km flight leg.

If A(t) represents the true signal as a continuous function of time, then we could sample that signal at evenly-spaced times: $t = t_o$, $t = t_o+\Delta t$, $t = t_o+2\Delta t$, $t = t_o+3\Delta t$, ..., $t = t_o+(N-1)\Delta t$, where the total number of data points is N. We will use an *index*, k, to denote the position within the time series. The kth data point corresponds to time $t_k = t_o + k\Delta t$, where $0 \le k \le (N-1)$. Sometimes the value of variable A at time t_k is represented by $A(t_k)$, but usually the shorthand notations A(k) or A_k is used. We will assume that the *sampling interval* is Δt, with no missing data and no changes of Δt within any one series. The *total period* of measurements is $\mathbb{P} = N\Delta t$, in the sense that each of the N data points represents a sample within an interval Δt.

8.2 Autocorrelation

In section 2.4.5 we discussed the covariance and the correlation coefficient, which quantify the amount of common variation between two different variables. Extending this idea, we could also ask about the degree of common variation between a variable (A) sampled at time t and that same variable sampled at a later time, t+L, where L is the time lag. Such a correlation of a variable with itself is called *autocorrelation*, $R_{AA}(L)$.

Consider a 12 hour time series that has a simple sinusoidal variation of unit amplitude with a 4 hour period. The wave equals 1.0 at regular intervals of 1, 5, and 9 h. Also, the wave equals -1.0 at 3, 7, and 11 h. In fact at ANY time, t, the series is perfectly correlated with itself (i.e., has the same value) at exactly times t + 4 h, t + 8 h, and t + 12 h. Similarly, we can show that the wave is negatively correlated with itself at t + 2 h, t + 6 h, and t + 10 h. We have, in essence, just determined the autocorrelation for this series at lags 2, 4, 6, 8, 10, and 12 h.

If our time series consists of a wave that varies in frequency during the duration of the series, then a wave at t_1 might be perfectly correlated with itself at $t_1 + 4$ h, but the value at t_2 might not be perfectly correlated with the value at $t_2 + 4$ h. When averaged over all possible pairs of data points with 4 h lag in this series, the result might NOT give a large correlation value at all.

In other words, the autocorrelation measures the persistence of a wave within the whole duration a time or space series. The capability to determine persistent waves or oscillations within a series is particularly valuable because the regular variation might be associated with a physical phenomenon such as an eddy. Alternately, when the autocorrelation becomes close to zero, it tells us that there is a random process (e.g. turbulence) occurring with no persistent or regularly-recurring structures.

8.2.1 Definition

The exact definition for the discrete autocorrelation is:

$$R_{AA}(L) = \frac{\displaystyle\sum_{k=0}^{N-j-1}\left[(A_k - \overline{A}_k)(A_{k+j} - \overline{A}_{k+j})\right]}{\left[\displaystyle\sum_{k=0}^{N-j-1}(A_k - \overline{A}_k)^2\right]^{1/2}\left[\displaystyle\sum_{k=0}^{N-j-1}(A_{k+j} - \overline{A}_{k+j})^2\right]^{1/2}} \qquad (8.2.1a)$$

where two different mean values are used depending on which portion of the whole series is being considered:

$$\overline{A}_k = \frac{1}{N-j}\sum_{k=0}^{N-j-1} A_k \qquad \text{and} \qquad \overline{A}_{k+j} = \frac{1}{N-j}\sum_{k=0}^{N-j-1} A_{k+j}$$

and where lag $= L = j\,\Delta t$. Notice that each of the square bracket terms in the denominator acts like a standard deviation over the portion of the data set being used.

We can approximate (8.2.1a) if it is assumed that the data is sufficiently stationary (or homogeneous for space series) that the mean values over each portion of the series is equal to the overall series mean, and that the standard deviations from each portion equal the overall series standard deviation. This results in:

$$R_{AA_{approx}}(L) = \frac{\overline{A_k' A_{k+j}'}}{\sigma_A^2} \qquad (8.2.1b)$$

This simple approximation works satisfactorily for small lags (i.e., small j) and large N, but is inadequate otherwise.

Autocorrelations are usually calculated for a range of lags, and the result plotted on a graph of R_{AA} vs L. For the special case of zero lag, the autocorrelation is identically equal to unity [$R_{AA}(0) = 1.0$] for all signals. The autocorrelation of an irregular signal such as turbulence approaches zero as L approaches infinity, although it may appear as damped oscillations about zero while L is small. Also, as the lag increases, the percentage of the time series used to calculate $R_{AA}(L)$ decreases. As a result, the statistical significance of R_{AA} decreases as lag increases, making R_{AA} unrepresentative when $j > (N/2)$.

Sample autocorrelation curves for convective turbulence measured at different heights in the ML are shown in Fig 8.1 (Deardorff and Willis, 1985).

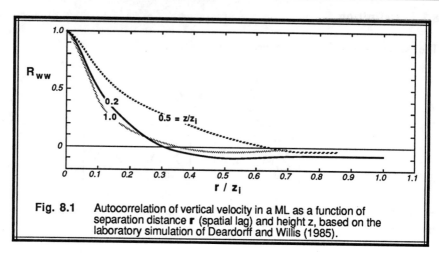

Fig. 8.1   Autocorrelation of vertical velocity in a ML as a function of separation distance **r** (spatial lag) and height z, based on the laboratory simulation of Deardorff and Willis (1985).

8.2.2 Example

Problem: Given the following series of measurements of relative humidity made every 3 h over a 96 h period (4 days) at a fixed point. Find the autocorrelation for relative humidity, $R_{rh,rh}(L)$, for time lags ranging from 0 to 48 hours, and plot the result.

Data:

		Relative humidity (percent)					
Day 1:	49	46	44	45	52	59	61	57
Day 2:	53	50	50	52	55	55	54	47
Day 3:	41	36	32	33	36	41	40	37
Day 4:	34	31	29	32	38	45	48	45

As can be seen in a plot of the time series (Fig 8.2a), there are regular diurnal cycle oscillations superimposed on longer period trends.

Solution: There are 32 data points, with $\Delta t = 3$ h. We must solve (8.2.1) 17 different times, for $j = 0$ through $j = 16$. The result is:

L(h)	$R_{rh,rh}$	L(h)	$R_{rh,rh}$	L(h)	$R_{rh,rh}$	L(h)	$R_{rh,rh}$	L(h)	$R_{rh,rh}$	L(h)	$R_{rh,rh}$
0	1.00	9	0.35	18	0.49	27	0.34	36	-0.13	45	0.40
3	0.67	12	0.34	21	0.53	30	0.09	39	-0.04	48	0.58
6	0.47	15	0.40	24	0.50	33	-0.12	42	0.17		

Discussion: Looking at Fig 8.2b, we see that the autocorrelation starts at 1.0 at zero lag, and quickly decreases. As is sometimes the case with weather data, the diurnal cycle shows up as an oscillation in the autocorrelation function with a 24 hour period. We could have anticipated this, because 12 h from any time, the humidity time series is in the opposite side of its oscillation. If the humidity is high in the early morning, then 12 h later it is drier. If the humidity is low in the afternoon, then 12 h later it is more humid. On the average, humidity is negatively correlated with itself 12 h later.   Over a 24 h period, however, like comparing a morning humidity with the next morning's humidity, or the afternoon humidity with the next afternoon's humidity, we anticipate a positive correlation.

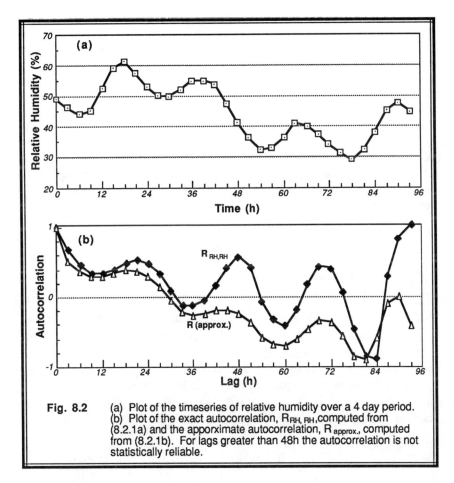

Fig. 8.2   (a) Plot of the timeseries of relative humidity over a 4 day period.
(b) Plot of the exact autocorrelation, $R_{RH, RH}$, computed from
(8.2.1a) and the apporximate autocorrelation, $R_{approx.,}$ computed
from (8.2.1b). For lags greater than 48h the autocorrelation is not
statistically reliable.

The initial drop off of the autocorrelation from 1.0 to smaller values is a measure of the accuracy of a persistence forecast. Namely, if we forecast the humidity 3 or less hours from now to be the same as the present humidity, we would probably be close to correct because the autocorrelation is 60% or higher. Longer forecasts would be less accurate.

8.3 Structure Function

8.3.1 Definition

An alternative statistic to view common variation is the *structure function*, $D_{AA}(L)$:

$$D_{AA}(L) = \frac{1}{N} \sum_{k=0}^{N-1} \left[A_k - A_{k+j} \right]^2 \qquad (8.3.1a)$$

For a time series, $L = j\, \Delta t$ is the time lag, while for a space series $L = r = j\, \Delta r$ represents the spatial separation, r, between the two measurements. The structure function uses the difference, rather than the product, of two different points in the series. The structure function has units of A^2 like variance, rather than being dimensionless like the autocorrelation.

For zero lag or zero separation distance, the structure function is identically zero. As L or r increases, so does the structure function for most turbulent flows. Within the inertial subrange of turbulence, similarity arguments (similarity theory is reviewed in Chapter 9) suggest that:

$$D_{AA}(r) = c_{A^2}\, r^{2/3} \qquad (8.3.1b)$$

where c_{A^2} is the *structure function parameter* for variable A. The four most common structure function parameters are c_{T^2} for temperature, c_{V^2} for velocity, c_{q^2} for moisture, and c_{nref^2} for the index of refraction, n_{ref}. These parameters are not dimensionless, but have units determined by the units of A and r to make (8.3.1b) dimensionally consistent (see example below).

Remote sensors such as radar (microwaves) or sodar (sound) can receive returns from clear air because some of the transmitted signal is scattered off of refractive index variations in the atmosphere. For example, the radar reflectivity η (radar cross section per unit volume) in clear air (no rain or other hydrometeors) is:

$$\eta = 0.38\, c_{nref^2}\, \lambda_R^{-1/3} \qquad (8.3.1c)$$

where λ_R is the wavelength of the radar. Since the index of refraction is related to temperature, moisture, and pressure to varying degrees depending on the type of remote sensor, equations can be derived relating c_{nref^2} to c_{T^2}, c_{q^2} and c_{V^2} (see review by Lenschow, 1986). Thus, the magnitude of the returned signal η gives c_{nref^2} from (8.3.1c), which yields estimates of c_{T^2}, c_{q^2} and c_{V^2}. Sample profiles of c_{T^2} and c_{V^2}

are shown in Fig 8.3.

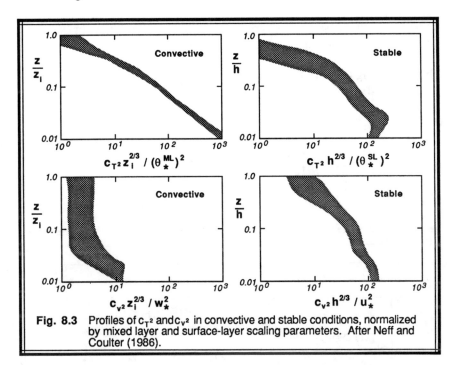

Fig. 8.3 Profiles of c_{T^2} and c_{V^2} in convective and stable conditions, normalized by mixed layer and surface-layer scaling parameters. After Neff and Coulter (1986).

Remote sensors can yield a variety of quantitative boundary layer information, utilizing the structure function statistic. We can estimate dissipation rates because similarity arguments suggest that:

$$c_{V^2} = 2\,\varepsilon^{2/3} \qquad \text{and} \qquad c_{T^2} = a^2\,\varepsilon_\theta\,\varepsilon^{-1/3} \qquad\qquad (8.3.1d)$$

for the inertial subrange, where a^2 is a constant of about 3.0 (estimates in the literature range from 2.8 to 3.2). Wyngaard, et. al. (1971) demonstrate that surface turbulent heat flux in the ML can be calculated from c_{T^2} :

$$c_{T^2} \; = \; 2.68 \left(\frac{g}{\overline{T}}\right)^{-2/3} \cdot \left(\frac{z}{\overline{w'\theta'}_s}\right)^{-4/3} \qquad\qquad (8.3.1e)$$

Wyngaard and LeMone (1980) further suggest that the magnitude of the temperature and moisture jumps across the capping inversion at the top of the ML can also be measured from c_{T^2} and c_{q^2} .

8.3.2 Example

Problem: Given the same time series of relative humidity (rh) as in example 8.2.2, calculate the structure function for lags from 0 - 48 hours.

Solution: Using (8.3.1a), the structure function values (in units of relative humidity percentage squared) are:

L (h)	$D_{rh,rh}$	L(h)	$D_{rh,rh}$	L(h)	$D_{rh,rh}$	L(h)	$D_{rh,rh}$
0	0	15	153	27	183	39	301
3	78	18	134	30	243	42	303
6	124	21	127	33	299	45	326
9	155	24	142	36	316	48	377
12	162						

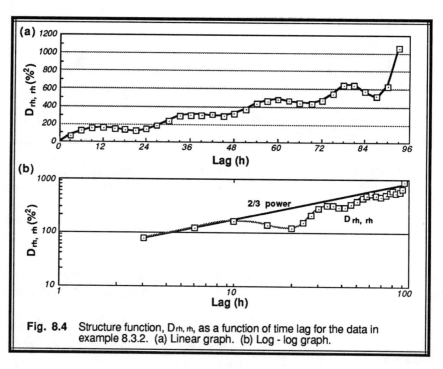

Fig. 8.4 Structure function, $D_{rh,rh}$, as a function of time lag for the data in example 8.3.2. (a) Linear graph. (b) Log - log graph.

Discussion: Fig 8.4a shows the resulting variation of the structure function on a linear scale. A similar plot on log-log graph is given in Fig 8.4b. The straight line on this latter graph is given by $D_{rh,rh} = c_{rh^2} L^{2/3}$, with the structure function for relative humidity $c_{rh^2} = 35\ (\%\ ^2 \cdot h^{-2/3})$.

8.4   Discrete Fourier Transform

From Fourier analysis in calculus we remember that any well-behaved continuous function can be described by an infinite Fourier series — namely, the sum of an infinite number of sine and cosine terms. In the case of a discrete time series with a finite number of points, we are required to have only a finite number of sine and cosine terms to fit our points exactly.

8.4.1 Definition

Using Euler's notation [$\exp(ix) = \cos(x) + i\,\sin(x)$, where i is the square root of -1] as a shorthand notation for the sines and cosines, we can write the discrete Fourier series representation of A(k) as:

$$\text{Inverse Transform:} \qquad A(k) = \sum_{n=0}^{N-1} F_A(n)\; e^{i\,2\pi nk/N} \qquad (8.4.1a)$$

where n is the frequency, and $F_A(n)$ is the *discrete Fourier transform*. We see that a time series with N data points (indexed from k=0 through N-1) needs no more than N different frequencies to describe it (actually, it needs less than N, as will be shown later).

There are a number of ways to describe frequency:

> n = number of cycles (per time period $\mathbb{P}$),
> ñ = cycles per second = $n/\mathbb{P}$,
> f = radians per second = $2\pi n/\mathbb{P} = 2\pi n/(N\Delta t)$.

A frequency of zero (n = 0) denotes a mean value. The *fundamental frequency*, where n = 1, means that exactly one wave fills the whole time period, $\mathbb{P}$. Higher frequencies correspond to *harmonics* of the fundamental frequency. For example, n = 5 means that exactly 5 waves fill the period $\mathbb{P}$.

$F_A(n)$ is a complex number, where the real part represents the amplitude of the cosine waves and the imaginary part is the sine wave amplitude. It is a function of frequency because the waves of different frequencies must be multiplied by different amplitudes to reconstruct the original time series. If the original time series A(k) is known, then these coefficients can be found from:

$$\text{Forward Transform:} \qquad F_A(n) = \sum_{k=0}^{N-1}\left[\frac{A(k)}{N}\right] e^{-i\,2\pi nk/N} \qquad (8.4.1b)$$

Notice the similarity between (8.4.1a) and (8.4.1b). These two equations are called *Fourier transform pairs*. The second equation performs the *forward transform*, creating a representation of the signal in *phase space* (another name for the frequency or spectral domain). This process is also known as *Fourier decomposition*. The first equation performs the *inverse transform*, converting from frequencies back into *physical space*.

8.4.2 Example

Problem: Given the following 8 data points of specific humidity, q, as a function of time:

Index (k):	0	1	2	3	4	5	6	7
Time (UTC):	1200	1215	1230	1245	1300	1315	1330	1345
q (g/kg):	8	9	9	6	10	3	5	6

Perform a forward Fourier transform to find the 8 coefficients, $F_q(n)$. To check your results, perform an inverse transform to confirm that the original time series is recreated. Remember that the $F_q(n)$ coefficients are complex, each having a real and an imaginary part: $F_q(n) = F_{real}(n) + i\, F_{imag}(n)$.

Solution: $N = 8$ and $\Delta t = 15$ min. Thus, the total period is $\mathbb{P} = N\Delta t = 2$ h. Equation (8.4.1b) must be used to find $F_q(n)$. For those computer languages that accept complex numbers, (8.4.1b) can be programmed directly, where each of the A(k) data points has a real part equal to the value listed in the table, and an imaginary part of zero.

For hand calculation, we can use Euler's formula to translate (8.4.1b) back into sines and cosines:

$$F_A(n) = \frac{1}{N} \sum_{k=0}^{N-1} A(k)\, \cos(2\pi nk/N) - \frac{i}{N} \sum_{k=0}^{N-1} A(k)\, \sin(2\pi nk/N)$$

As an example, for n = 0, all of the cosines of zero are unity and all of the sines are zero. This leaves:

$$F_A(0) = \frac{1}{N} \sum_{k=0}^{N-1} A(k)$$

which is just the mean of A. For our case: $F_q(0) = 7.0 - 0.0\,i$. For n = 1 we can't make such a simplification, so we are forced to sum over all k for both the real and imaginary parts. This gives us $F_q(1) = 0.28 - 1.03\,i$. Continuing this procedure for all other n yields:

n	$F_q(n)$	n	$F_q(n)$
0	7.0	4	1.0
1	0.28 - 1.03 i	5	-0.78 + 0.03 i
2	0.5	6	0.5
3	-0.78 - 0.03 i	7	0.28 + 1.03 i

This is the answer to the first part of the problem. Note that for frequencies greater than 4, the Fourier transform is just the complex conjugate of the frequencies less than 4.

As a check of our transform, we can perform the inverse transform using (8.4.1a) directly in a computer program. Otherwise, we can use Euler's formula to write it as:

$$A(k) = \sum_{n=0}^{N-1} F(n)_{(\text{real part})} \cdot \cos(2\pi nk/N) - \sum_{n=0}^{N-1} F(n)_{(\text{imag.part})} \cdot \sin(2\pi nk/N)$$

In actuality, there are four sums, not just the two listed above. The remaining sums consist of the real part of F times the imaginary factor $i \cdot \sin(...)$, and the imaginary part of F times the real factor $\cos(...)$. Because the last half of the Fourier transforms are the complex conjugates of the first half (not counting the mean), these two sums identically cancel, leaving the two listed above. Upon performing the calculations for $A(k)$, we do indeed reproduce the original time series.

Discussion: To graphically demonstrate that the sum of these sines and cosines does indeed equal our original series, Fig 8.5 shows each individual wave multiplied by its appropriate amplitude. As can be seen, the reconstructed time series fits perfectly the eight original data points. In between these points, however, the sum oscillates in a manner that is not necessarily realistic, but which is irrelevant because it occurs below the discretization resolution specified by the original data points.

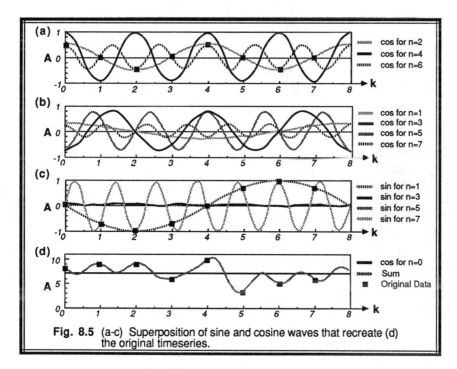

Fig. 8.5 (a-c) Superposition of sine and cosine waves that recreate (d) the original timeseries.

8.4.3   Aliasing and Other Hazards.

Measurements: A basic rule of discrete data analysis is that at least two data points are required per period or wavelength in order to resolve a wave. Since Fourier analysis involves splitting arbitrary signals into waves, the two data point requirement also holds for our arbitrary signals. For example, if we have a total of N data points, then the highest frequency that can be resolved in our Fourier transform is $n_f = N/2$, which is called the *Nyquist frequency*. These requirements apply to measurements; namely, if a wave period as small as 0.1 s must be measured while flying in an aircraft, then the physical signal must be digitized at least once every 0.05 s. Similarly, if a wavelength as small as 1 m must be measured, then the physical signal must be digitized at least once every 0.5 m.

What happens when there is a physical signal of high frequency that is not sampled or digitized frequently enough to resolve the signal? The answer is that the true high-frequency signal is *folded* or *aliased* into a lower frequency, creating an erroneous and deceiving Fourier transform. This is illustrated with aid of the example in the previous subsection. Look at the first graph in Fig 8.5, where the cosine waves for n = 2, 4 and 6 are plotted. Since we started with N = 8 data points, we can anticipate a Nyquist frequency of $n_f = 4$. Namely, the shortest period wave that can be resolved is one that has 4 cycles per period $\mathbb{P}$. Thus, the curve corresponding to n = 6 is greater than the Nyquist frequency, and is likely to cause problems.

Look closely at the curves for n = 2 and n = 6. They coincide exactly at the points k = 0, 1, 2, 3, 4, 5, 6 and 7. In other words, if there were a true signal of n = 6 that was sampled only at the integer k values listed above, then anyone connecting the resulting plotted points with a line or curve would find that they have drawn a wave with n = 2 cycles per period. In other words, the n = 6 signal was folded into the n = 2 frequency. Similarly, looking at the third graph in Fig 8.5, the n = 7 sine waves are folded into an n = 1 sine wave. In general, if n_h represents a frequency higher than the Nyquist frequency, then the signal or amplitude of that wave will be folded down to a frequency of $n = N - n_h$, where it will be added to any true amplitude that already exists at n.

Since this folding or reflection occurs around the Nyquist frequency, it is also known as the *folding frequency*. Such folding is readily apparent when wave amplitudes are plotted as spectral energies (to be discussed in Section 8.6). As illustrated in Fig 8.6, any nonzero wave amplitudes and spectral energies in the "true" signal at frequencies higher than the Nyquist frequency are folded back and added to the energies of the "true" signal at the lower frequencies, yielding an aliased (and erroneous) spectrum.

Aliasing is a problem whenever two conditions both occur: (1) the sensor can respond to frequencies higher than the rate that the sensor is sampled; and (2) the true signal has frequencies higher than the sampling rate. As we already know, there is a spectrum of wavenumbers and frequencies of turbulence in the atmosphere, some of which are very high. All measurement systems have limitations on the rate at which they can sample.

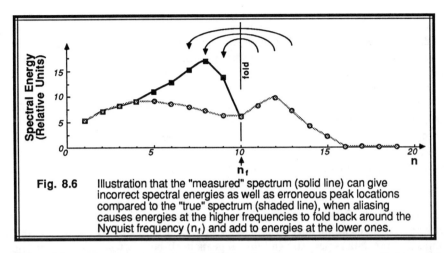

Fig. 8.6   Illustration that the "measured" spectrum (solid line) can give incorrect spectral energies as well as erroneous peak locations compared to the "true" spectrum (shaded line), when aliasing causes energies at the higher frequencies to fold back around the Nyquist frequency (n_f) and add to energies at the lower ones.

Sometimes this rate is given by limitations of the data logger or computer digitizer. If this is the case, then the raw electronic analog signal from the sensor (thermistor, thermocouple, gust probe accelerometer, etc) should be filtered by an <u>analog</u> electronic filter prior to the digitizing or sampling to remove frequencies higher than the Nyquist frequency. Sometimes the sensor itself has such a slow response that it performs the analog filtering automatically.

If the analog filtering is not performed, then there is no way to remove the erroneous aliased component from the resulting time series. Postprocessing of the discrete time series with digital filters will NOT work, because it is impossible to know what portion of the wave amplitude at the resolvable frequencies is real, and what is folded into it.

Digital averaging is sometimes successfully used for other reasons, however. Suppose that a sensor is designed with appropriate analog filters to yield unaliased data when sampled at a very high frequency. Next, suppose that the amount of this unbiased discrete data is too large to record in a convenient manner, or is coming in too fast to be processed. The stream of incoming discrete data values can be block averaged (e.g., average every 10 data points), or filtered with a variety of filters (e.g., Butterworth filters) before being recorded or processed further. This yields a lower-frequency sample without aliasing errors. If, however, one records only every fifth or tenth (or any interval) value from the sampled stream, then aliasing is again a problem.

Fourier analysis. Now that we are convinced that we can't resolve frequencies greater than the Nyquist frequency, why does the Fourier transform operation given by (8.4.1b) give amplitudes $F_A(n)$ up to the frequency $n = N - 1$? The answer is that it doesn't really. Looking at example 8.4.2 again, we again note that $F_q(n)$ for $n > n_f$ is just the complex conjugate of the $F_q(n)$ values for $n < n_f$. This is always the case and can be proved mathematically, assuming that the initial time series consists of only real numbers. Hence, the half of the $F_q(n)$ values for which $n > n_f$ give no new information.

Thus, the N different F_A values having both real and imaginary parts superficially gives 2N pieces of information, but since half of that is the complex conjugate of the other half, we are left with only n pieces of spectral information. It is reassuring that given only N data points in our original time series in physical space, we require only N pieces of information in phase space to precisely describe the data.

For an original time series consisting of complex numbers (2N pieces of data), the Fourier transform does not have the complex conjugate property described above, resulting in 2N pieces of information is phase space too. For meteorological data where the time series is usually real, we still need to utilize the whole Fourier transform with the complex conjugate information, because without it the inverse transform will produce complex numbers instead of our desired real number physical field.

Data Window. Fourier series apply to infinite-duration periodic data sets. Stated in other words, if we examine only a finite size record of data, the Fourier analysis implicitly assumes that the data is periodic and thus repeats itself both before and after our limited period of measurement.

In boundary layer meteorology, nothing is periodic for infinite time, or for infinite distance. Given a true signal (for example temperature) that varies as in Fig 8.7a, if we measure it over period $\mathbb{P}$ (our *data window*) as in Fig 8.7b, then we are left with the segment shown in Fig 8.7c. Given this segment, the Fourier analysis assumes that it is dealing with a periodic (repeating) signal as shown in Fig 8.7d.

In this example, a smoothly varying meteorological signal appears as a saw-tooth pattern to the Fourier analysis. From basic calculus, recall that a Fourier analysis can indeed describe series such as sawtooth or square wave patterns, but a wide range of frequencies are required to get the sums of all the sines and cosines to make the sharp bends at the points of the teeth. These spurious frequencies are called *red noise* by analogy to visible light because they appear at the low frequency end of the spectrum. To avoid red noise, we must at the very least *detrend* the data series by subtracting the straight line best-fit from the data segment (Fig 8.7c), leaving a modified time series as exemplified in Fig 8.7e.

In general, any very low frequency that has a period longer than our whole sampling period will also generate the noise. If we know a-priori the period of this frequency, such as diurnal or annual, then we can perform a least-squares fit of this frequency to the time series and subtract the result from the series. Otherwise, we might try to fit a simple polynomial curve to the data and subtract it to both detrend it and remove these low frequencies.

Even after detrending, the sharp edges of the data window cause what is known as *leakage*, where spectral estimates from any one frequency are contaminated with some spectral amplitude leaking in from neighboring frequencies. To reduce leakage, a modified data window with smoother edges is recommended, such as is shown in Fig 8.7f. Although a variety of smoothers can be used, a common one utilizes sine or cosine squared terms near the beginning and ending 10% of the period of record, and is known as a *bell taper*:

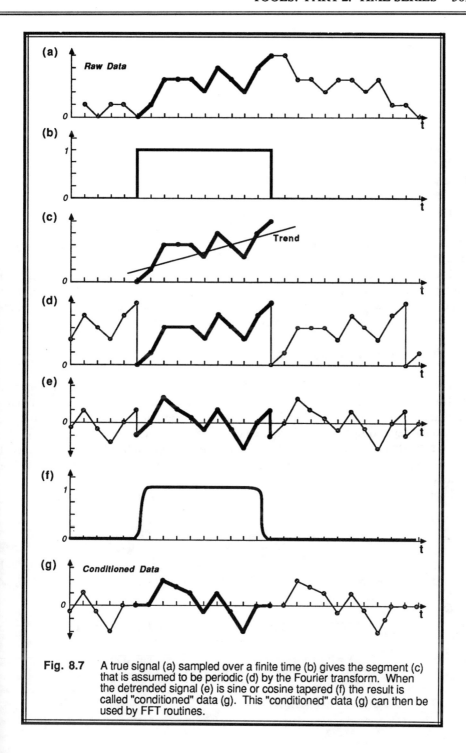

Fig. 8.7 A true signal (a) sampled over a finite time (b) gives the segment (c) that is assumed to be periodic (d) by the Fourier transform. When the detrended signal (e) is sine or cosine tapered (f) the result is called "conditioned" data (g). This "conditioned" data (g) can then be used by FFT routines.

$$W(k) = \begin{cases} \sin^2(5\pi k/N) & \text{for } 0 \le k \le 0.1N \\ 1 & \text{elsewhere} \\ \sin^2(5\pi k/N) & \text{for } 0.9\,N \le k \le N \end{cases} \quad (8.4.3)$$

When this window weight, $W(k)$, is multiplied by the time series, $A(k)$, the result yields a modified time series with fluctuations that decrease in amplitude at the beginning and end of the series (see Fig 8.7g). The Fourier transform can then be performed in this modified time series.

The bell taper data window is not without its problems. Although the tapered ends reduced the leakage, they also reduce our ability to resolve spectral amplitude differences between small changes in frequencies. Also, the tapered window reduces high-frequency noise at the expense of introducing low-frequency noise.

The process of detrending, despiking (removing erroneous data points), filtering, and bell tapering is known as *conditioning* the data. Conditioning should be used with caution, because anytime data is modified, errors or biases can be introduced. The best recommendation is to do as little conditioning as is necessary based on data quality.

8.5 Fast Fourier Transform

The fast Fourier transform, or *FFT*, is nothing more than a discrete Fourier transform that has been factored and restructured to take advantage of the binary computation processes of the digital computer. As a result, it produces the same output, and has the same limitations and requirements as the discrete transform. It can also be used for forward as well as inverse transforms. The description that follows is not meant to be a comprehensive review of FFT methods, but is designed to give an overview of the process.

In general, both the forward and the inverse discrete transform can be written as

$$X = \sum_{k=0}^{N-1} Y Z^{nk} \quad \begin{matrix} k=0 \ \ (\text{forward}) \\ n=0 \ \ (\text{inverse}) \end{matrix} \quad (8.5a)$$

where

Forward Transform	Inverse Transform
$X(n) = F_A(n)$	$X(k) = A(k)$
$Y(k) = A(k)/N$	$Y(n) = F_A(n)$
$Z_N = \exp(-i2\pi/N)$	$Z_N = \exp(i2\pi/N)$

The decimal numbers n and k can be represented by their binary equivalents:

$$n = \sum_{j=0}^{\infty} 2^j n_j \quad \text{and} \quad k = \sum_{j=0}^{\infty} 2^j k_j \tag{8.5b}$$

where n_j and k_j represent the individual bits of the number. For example, if $N = 8$, then we need only three bits ($j = 0$ to 2) to represent n and k, since they can take on values of only 0 to 7. Thus $n = 4 \cdot n_2 + 2 \cdot n_1 + 1 \cdot n_0$. For example, 101 is the binary representation of the decimal 5, giving $n_2 = 1$, $n_1 = 0$, and $n_0 = 1$.

Using this binary representation, any function of n is now a function of n_2, n_1, and n_0, with similar forms for functions of k. Thus, $X(n)$ becomes $X(n_2, n_1, n_0)$. Equation (8.5a) can now be rewritten, using the forward transform with $N = 8$ as the example, as:

$$X(n_2, n_1, n_0) = \sum_{k_2 = 0}^{1} \sum_{k_1 = 0}^{1} \sum_{k_0 = 0}^{1} Y(k_2, k_1, k_0) \, Z^{(4n_2 + 2n_1 + n_0)(4k_2 + 2k_1 + k_0)}$$

Performing the multiplications in the exponent of Z, rearranging terms, and remembering that Z to certain powers equals unity because of the nature of sines and cosines, we find:

$$X(n_2, n_1, n_0) = \sum_{k_0 = 0}^{1} \sum_{k_1 = 0}^{1} \sum_{k_2 = 0}^{1} Y(k_2, k_1, k_0) \, Z^{4n_0 k_2} Z^{4n_1 k_1} Z^{2n_0 k_1} Z^{4n_2 k_0} Z^{2n_1 k_0} Z^{n_0 k_0}$$

In this last equation the Z's are essentially weighting factors. To solve this equation, the inner sum is performed, using only the first weight because it is the only weight that is a function of k_2. When the next sum is performed, the additional two weights are included. Finally, the last sum uses the remaining three weights. This pattern of solving the sums, and gradually eliminating the k bits and replacing them with n bits can be programmed recursively, requires relatively little scratch storage, and is very efficient in computer time.

To a first approximation, the normal discrete Fourier transform requires N^2 operations, while the FFT requires only $(3N/2)\log_2 N$ operations. For small data sets (N < 100) the resulting computer time or cost difference is insignificant for all practical purposes, because of other overhead costs such as input and output. But for a data set of 1000 points, for example, the FFT computation takes 0.5% of the time that a traditional discrete transform computation would take. There is even some microprocessor hardware available that is specially configured to run FFTs. The bottom line is that the FFT is fast.

Most modern computer centers, and some statistical packages for microcomputers,

have "canned" FFT algorithms that users can access without having to write their own. Some of the early FFT packages were restricted to data sets with $N = 2^m$, where m was any integer. This meant that data sets slightly too long were truncated to the proper size, or data sets slightly too short we lengthened by adding bogus data (often zeros or the mean value). Both of these data mutilation tricks are not recommended. Modern FFTs factor the series into a variety of prime numbers in addition to the prime number 2, resulting in very little truncation of the time series.

One problem with all discrete Fourier transforms including FFTs, is that the input must consist of equally-spaced data points. No missing data is allowed. If the data set has gaps caused by instrument failures or by spurious data spikes that were removed, then artificial data points must be inserted to fill the gap. One is not allowed simply to close the gap by bringing the remaining parts of the data set together, because this alters the periods or wavelengths present in the original signal. The artificial data points must be chosen with care, otherwise this "fudge" can destroy an otherwise unbiased data set. Data with significant gaps can be analyzed with periodogram methods instead (see Section 8.9).

8.6 Energy Spectrum

8.6.1 Discrete Energy Spectrum

In meteorology we are frequently curious about how much of the variance of a time series is associated with a particular frequency, without regard to the precise phase of the waves. Indeed for turbulence, we anticipate that the original signal is not physically like waves at all, but we still find it useful to break the signal into components of different frequencies that we like to associate with different eddy sizes.

The square of the norm of the complex Fourier transform for any frequency n is:

$$|F_A(n)|^2 = [F_{\text{real part}}(n)]^2 + [F_{\text{imag. part}}(n)]^2 \qquad (8.6.1a)$$

When $|F_A(n)|^2$ is summed over frequencies $n = 1$ to N-1, the result equals the total biased variance of the original time series:

$$\sigma_A^2 = \frac{1}{N} \sum_{k=0}^{N-1} (A_k - \overline{A})^2 = \sum_{n=1}^{N-1} |F_A(n)|^2 \qquad (8.6.1b)$$

Thus, we can interpret $|F_A(n)|^2$ as the portion of variance explained by waves of frequency n. Notice that the sum over frequencies does not include n=0, because $|F_A(0)|$ is the mean value and does not contribute any information about the variation of the signal about the mean. To simplify the notation for later use, define: $G_A(n) = |F_A(n)|^2$. The ratio $G_A(n) / \sigma_A^2$ represents the fraction of variance explained by component n, and is very much like the correlation coefficient squared, r^2.

For frequencies greater than the Nyquist frequency the $|F_A(n)|^2$ values are identically equal to those at the corresponding folded lower frequencies, because the Fourier transforms of high frequencies are the same as those for the low frequencies, except for a sign change in front of the imaginary part. Also, since frequencies higher than the Nyquist cannot be resolved anyway, the $|F_A(n)|^2$ values at high frequencies should be folded back and added to those at the lower frequencies.

Thus, *discrete spectral intensity (or energy)*, $E_A(n)$, is defined as $E_A(n) = 2 \cdot |F_A(n)|^2$, for $n = 1$ to n_f, with $N = $ odd. For $N = $ even, $E_A(n) = 2 \cdot |F_A(n)|^2$ is used for frequencies from $n = 1$ to $(n_f - 1)$, along with $E_A(n) = |F_A(n)|^2$ (not times 2) at the Nyquist frequency. This presentation is called the *discrete variance (or energy) spectrum*. It can be used for any variable such as temperature, velocity, or humidity to separate the total variance into the components, $E_A(n)$, related to different frequencies. For variables such as temperature and humidity, however, we must not associate the resulting spectrum with concepts of eddy motions, because variations in these variables can persist in the atmosphere in nonturbulent flow as the "footprints" of formerly active turbulence.

The variance of velocity fluctuations, u', has the same units as turbulence kinetic energy per unit mass. Thus, the spectrum of velocity is called the *discrete energy spectrum*. As defined above, the name "energy spectrum" is sometimes used for all variance spectra.

8.6.2 Spectral Density

Although this chapter has dealt with discrete spectra, a number of theoretical concepts such as the spectral similarity discussed in the next chapter use continuous spectral representations. Namely, instead of summing the discrete spectral energy over all n to yield the total variance, these theories assume that there is a *spectral energy density*, $S_A(n)$ that can be integrated over n to yield the total variance.

$$\sigma_A^2 = \int_n S_A(n) \, dn \tag{8.6.2a}$$

The spectral energy density has units of A squared per unit frequency.
We can approximate the spectral energy density by

$$S_A(n) = \frac{E_A(n)}{\Delta n} \tag{8.6.2b}$$

where Δn is the difference between neighboring frequencies. When n is used to represent frequency, $\Delta n = 1$.   For other representations of frequency such as f, we will find that Δf is not necessarily equal to unity.

The $S_A(n)$ points estimated from (8.6.2b) can then be connected with a smooth curve to represent the shape of the spectrum. An example of this was shown in Chapter 2, Fig. 2.2. Thus, even with discrete meteorological data, we can estimate spectral densities that can be compared to theories.

8.6.3   Example

Problem: Use the results from the N = 8 data point example of section 8.4.2 to calculate the discrete spectral energies for all frequencies. Plot the result in the usual presentation format for discrete spectra. Show an additional graph of the estimate of spectral density.

Solution:

n	$F_q(n)$	$|F_q(n)|^2$	$E_q(n)$	$S_q(n)$
0	7.0 (= mean)			
1	0.28 - 1.03 i	1.14	2.28	2.28
2	0.5	0.25	0.5	0.5
3	-0.78 - 0.03 i	0.61	1.22	1.22
4 = n_f	1.0	1.0	1.0	1.0
5	-0.78 + 0.03 i	0.61		
6	0.5	0.25		
7	0.28 + 1.03 i	1.14		
	Sum =	5.0	= 5.0	

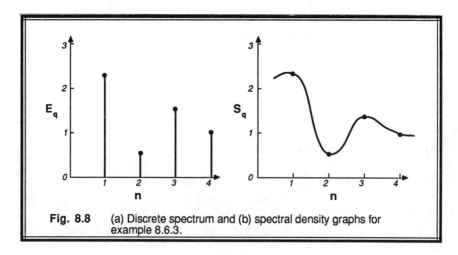

Fig. 8.8   (a) Discrete spectrum and (b) spectral density graphs for example 8.6.3.

where $E_q(n)$ has units of specific humidity squared, and $S_q(n)$ has units of specific humidity squared per unit frequency. Finally, the discrete spectrum is plotted in Fig 8.8a, and the spectral energy density is plotted in Fig 8.8b.

Discussion. The sum of the spectral energies equals the biased variance of the original signal, $\sigma_q^2 = 5.0$. This is always a good check of the FFT for you to perform.

8.6.4 Graphical Presentation of Atmospheric Spectra

A wide range of intensities are present in atmospheric turbulence spectra over an even larger range of frequencies. Atmospheric turbulence spectral energies characteristically peak at the lowest frequencies, namely at about 1 to 10 cycles per hour. At higher frequencies, the spectral energy decreases. For example, at frequencies of 10^4 cycles per hour the energy is one to two orders of magnitude smaller than at the peak.

We are often concerned about the full range of the spectrum: the peak is associated with the production of turbulence and usually the largest eddy sizes; the middle frequencies are associated with the inertial subrange, which is important for estimated dissipation rates; and the highest frequencies are associated with the dissipation of TKE into heat by viscous effects. Hence, we need a way to graphically present the spectral data in a form that not only highlights the important peaks and other characteristics, but which shows all portions of the wide range of data.

In the discussions that follow, a single idealized spectrum is presented in a variety of formats in Fig 8.9. The data for these plots is listed in Table 8-1.

Linear-linear presentation. When $S_A(f)$ is plotted vs. f on a linear-linear graph, the result has the desirable characteristic that the area under the curve between any pair of frequencies is proportional to the portion of variance explained by that range of frequencies. Unfortunately, the plot is useless to view because the wide range in values results in a compression of the data onto the coordinate axes (see Fig 8.9a). Alternatives include expanding the low frequency portion of the spectrum (Fig 8.9b) and plotting $f \cdot S(f)$ instead of just $S(f)$ on the ordinate (Fig 8.9c). Both techniques focus on the spectral peak at the expense of losing information at the higher frequencies.

Note that the $f \cdot S(f)$ plot causes the apparent peak to shift from the low frequency end of the spectrum towards the middle of the spectrum. Since $f \cdot S(f)$ is also used in a number of the other formats listed below, we should not be deceived into thinking that the middle frequencies are the ones with the most spectral energy.

Semi-log presentation. By plotting $f \cdot S_A(f)$ vs. log f, the low frequency portions of the spectra are expanded along the abscissa. Also, the ordinate for the high frequency portions are enhanced because the spectral density is multiplied by frequency (see Fig 8.9d). Another excellent quality is that the area under any portion of the curve continues to be proportional to the variance.

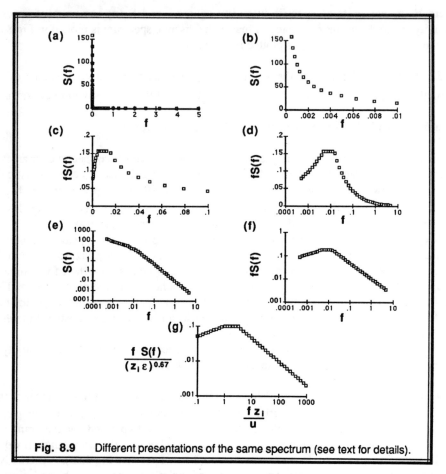

Fig. 8.9   Different presentations of the same spectrum (see text for details).

Log-log presentation. When $\log[S_A(f)]$ vs. log f is plotted, the result allows a wide range of frequencies and spectral densities to be displayed. Also, any power law relationships between $S_A(f)$ and f appear as straight lines on this graph. As will be discussed in more detail in the next chapter, $S_A(f)$ is proportional to $f^{-5/3}$ in the inertial subrange portion of the spectrum, which will appear as a straight line with -5/3 slope on a log-log graph (see Fig. 8.9e). Unfortunately, the area under the curve is no longer proportional to the variance.

Log f $S_A(f)$ vs. log f. A plot of $\log[f \cdot S_A(f)]$ vs. log f, has all of the desirable characteristics of the log-log presentation described above. In addition, the quantity $f \cdot S_A(f)$ has the same units as the variance of A, making scaling or normalization easier. Unfortunately, the area under the curve is also not proportional to variance (see Fig. 8.9f). Regardless of this problem, this presentation is the most used in the literature.

Table 8-1. Artificial data and spreadsheet calculations used to demonstrate various ways to present spectra.

	Variable	Value
This is assumed to be the spectrum of a	Zi (m)	1000
time series of velocity measurements.	U (m/s)	5
	Dissip.(m2 s-3)	0.002
	Size	21

___Logarithm of___						
Normalized	Normalized	Normalized	Normalized	f	S	f S
Frequency	Spectrum	Frequency	Spectrum	(1/s)	(m2/s3)	(m2/s2)
-1.0	-1.3010	0.1000	0.0500	0.0005	158.7401	0.0794
-0.8	-1.2412	0.1580	0.0574	0.0008	115.3005	0.0911
-0.6	-1.1807	0.2510	0.0660	0.0013	83.4309	0.1047
-0.4	-1.1204	0.3980	0.0758	0.0020	60.4486	0.1203
-0.2	-1.0602	0.6310	0.0871	0.0032	43.8016	0.1382
-0.0	-1.0000	1.0000	0.1000	0.0050	31.7480	0.1587
0.2	-1.0000	1.5850	0.1000	0.0079	20.0303	0.1587
0.4	-1.0000	2.5120	0.1000	0.0126	12.6385	0.1587
0.6	-1.0827	3.9810	0.0827	0.0199	6.5914	0.1312
0.8	-1.2175	6.3100	0.0606	0.0316	3.0495	0.0962
1.0	-1.3521	10.0000	0.0445	0.0500	1.4112	0.0706
1.2	-1.4868	15.8490	0.0326	0.0792	0.6530	0.0517
1.4	-1.6215	25.1190	0.0239	0.1256	0.3022	0.0379
1.6	-1.7562	39.8110	0.0175	0.1991	0.1398	0.0278
1.8	-1.8909	63.0960	0.0129	0.3155	0.0647	0.0204
2.0	-2.0255	100.0000	0.0094	0.5000	0.0299	0.0150
2.2	-2.1602	158.4890	0.0069	0.7924	0.0139	0.0110
2.4	-2.2949	251.1890	0.0051	1.2559	0.0064	0.0080
2.6	-2.4296	398.1070	0.0037	1.9905	0.0030	0.0059
2.8	-2.5643	630.9570	0.0027	3.1548	0.0014	0.0043
3.0	-2.6990	1000.0000	0.0020	5.0000	0.0006	0.0032

As will be discussed in the next chapter, both the abscissa and ordinate are often made dimensionless by normalizing with respect to scaling variables (see Fig 8.9g). The scaling variables used in this example are listed in Table 8-1.

8.7 Spectral Characteristics

Instead of discussing spectral behavior theoretically, this section demonstrates spectral behavior for a single variable through a series of examples with synthetic data. In each of the following cases, an artificial time series of 20 data points is plotted, along with the spectrum computed with an FFT program. The spectrum shows $E(n)$ normalized by the total biased variance, and shows the fraction of the total variance explained by each frequency. The Nyquist frequency is $n=10$ for all cases.

Case A (Fig 8.10a): Simple waves of one frequency. All of these examples show a wave having four cycles per time period. The first four examples in this case show that the spectrum is independent of the phase of the original time series. A single simple wave in physical space produces a single spike in the spectrum at $n=4$ that explains all the variance. The fifth example shows that if the spectrum is normalized by the total variance, we still have a single spike that explains 100% of the variance. If the spectrum had not been normalized, the spike for this fifth case would have been twice as large as the spikes for the other four cases, because the time series for the fifth case consisted of a wave with twice the amplitude.

Case B (Fig 8.10b): Simple waves of different frequencies. The first example shows a time series filled by one wave, resulting in a spectrum with a spike at $n = 1$. The next three examples show waves with 4, 8, and 10 cycles per period in the time series, resulting in spectra with frequency spikes at $n = 4$, 8, and 10 respectively. The fifth example shows a time series with a wave having 12 cycles per period, but the aliasing problem causes this signal to be folded back to $n = 8$, where it appears as a spike on the spectrum.

Case C (Fig 8.10c): Frequencies between resolvable frequencies. The FFT consists of waves of the fundamental frequency ($n = 1$) and only the exact harmonics ($n = 2, 3, 4,...$). But what happens if the real signal has a frequency of $n = 4.2$ or 4.5? These examples show that a wave of $n = 4.5$ appears as two large spikes at $n = 4$ and $n = 5$. The closer the signal is to an exact harmonic, the greater the spectral energy at that harmonic and the smaller the energy at the next nearest neighbor. Notice that for a signal with $n = 4.5$, the spectrum not only has the two large spikes described above, but there is also a leakage of some small amount of spectral energy to all the other frequencies. We might expect that a real turbulence signal consisting of a multitude of frequencies, many of which are not exact harmonics of the fundamental frequency, will result in a spectrum with a lot of leakage, making it difficult to separate the true signals from the underlying noise.

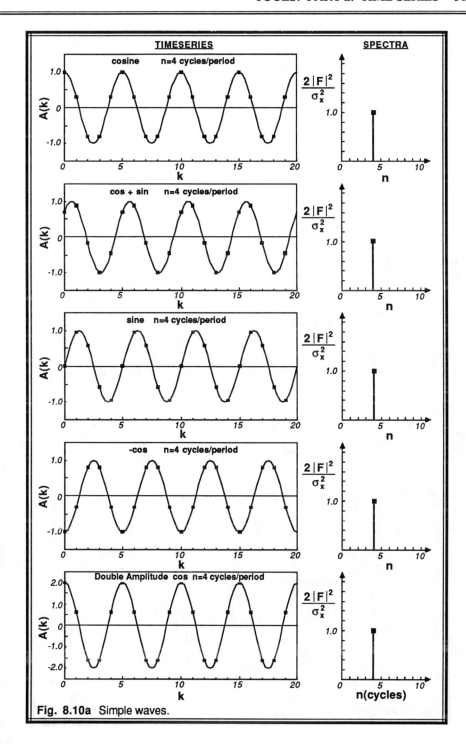

Fig. 8.10a Simple waves.

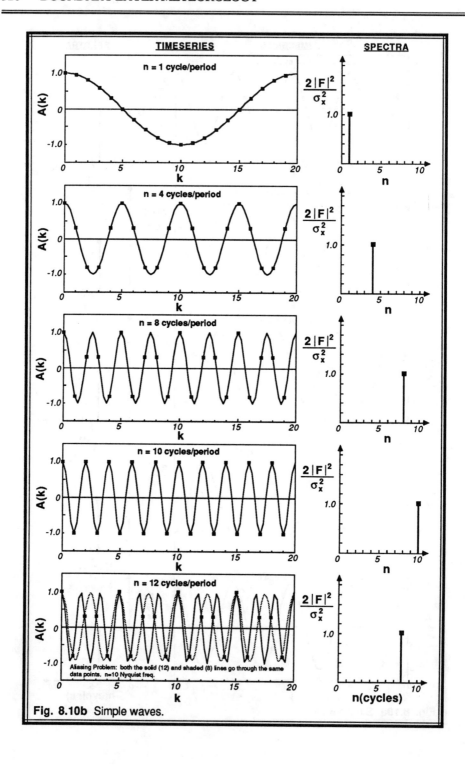

Fig. 8.10b Simple waves.

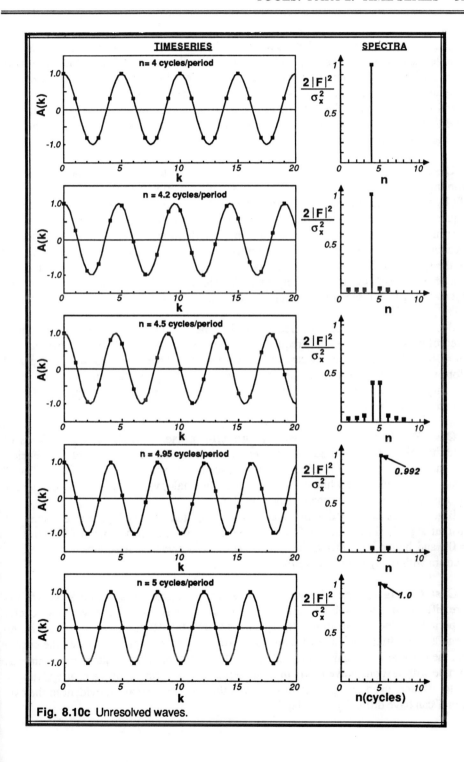

Fig. 8.10c Unresolved waves.

Case D (Fig 8.10d):   Unresolvable high and low frequencies. The first example shows that if the real signal has n = 0.5, then the computed spectrum has a spike at n = 1 with a significant amount of leakage to higher frequencies. This is called *red noise* and will be discussed in Case E. At n = 1.5, there is still significant red noise. It is as if the leakage from the left side of the spike is folded back around n = 0 to larger n values. At n = 8.5, the leakage off the right of the peak appears to fold back to the left, creating a *blue noise* signal.

Case E (Fig 8.10e):   Red noise. When signal with time period longer than the sampling period is truncated to fit within the sample window, the resulting periodic shape is fit by waves of the fundamental period and shorter. These waves are largest at the low-frequency end of the spectrum. As the wave period increases, this becomes more apparent. In the extreme case of a linear trend (which acts like an infinite period or wavelength wave), we find a purely red noise spectrum. Its name comes from the fact that the spectrum shows energy at the incorrect frequencies (i.e., error or noise), and that most of this noise is at the low frequencies (analogous to the red portion of the visible light spectrum). We see why it is important to detrend raw signals before computing the FFT.

Because of unresolvable low frequencies in general, and red noise in particular, most meteorologists do not consider frequencies of 3 or less as being reliable. Some use n=5 or n = 10 as the cut off. In any case, we look for at least three waves per sampling period before we are confident that the spectral results are telling us about the physics of the boundary layer. Often, these low frequencies are not even plotted on spectra that are presented in the literature.

Case F (Fig 8.10f): Red, white, and blue noise.   White noise consists of approximately equal amplitude spectral energies across the whole range of frequencies. This can be produced by a spike in the time series, or by completely random "hash" signal. If we could hear white noise (e.g., the audio analogy), it would sound like a hiss, like many leaves rustling or many waves breaking.

Blue noise is associated with larger spectral amplitudes at the higher frequencies. A constant signal, shown in the fourth example, consists of just a mean value (i.e., at n = 0), and hence has zero variance and no spectral energy. A square wave yields a spectrum with many peaks and zeros.

Case G (Fig 8.10g): Leakage. The shorter a signal lasts within a record, the more difficult it is to resolve it. Each of these examples shows a signal with five cycles per period. In the first example, the spectrum shows the desired spike at n = 5 with no energy at other frequencies. However, as the signal is cut shorter and shorter, the energy from the spike at n = 5 leaks more and more into the neighboring frequencies. In the last example with just one wave left in the time series, the spectrum shows a nearly Gaussian spread. Hence, even though certain signals in the time series may be evident to the eye, the FFT can have difficulty detecting it.

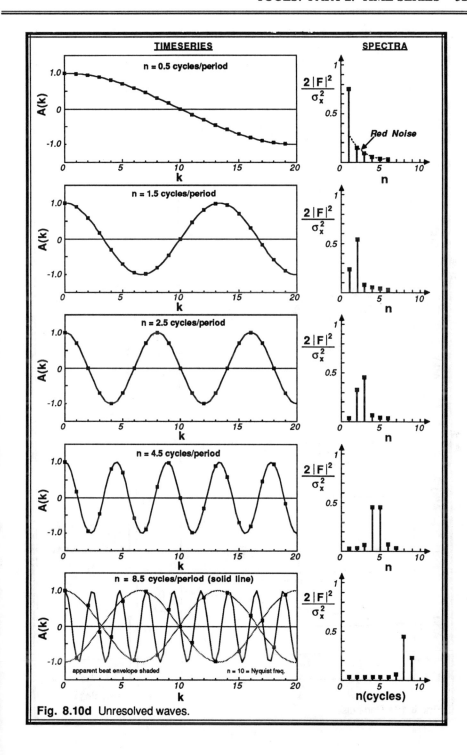

Fig. 8.10d Unresolved waves.

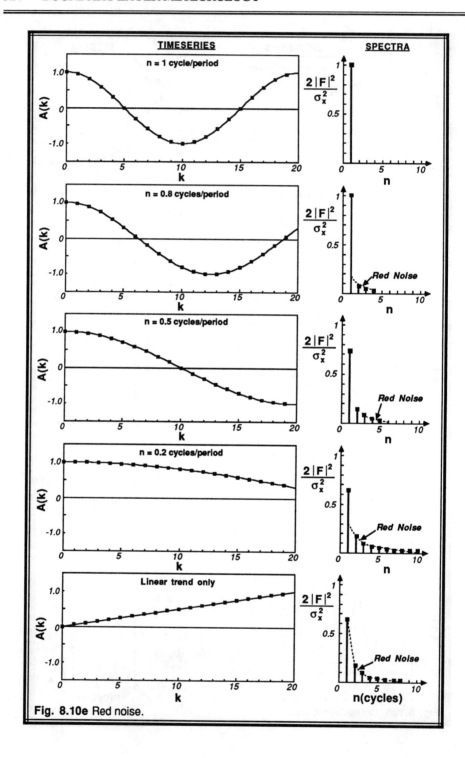

Fig. 8.10e Red noise.

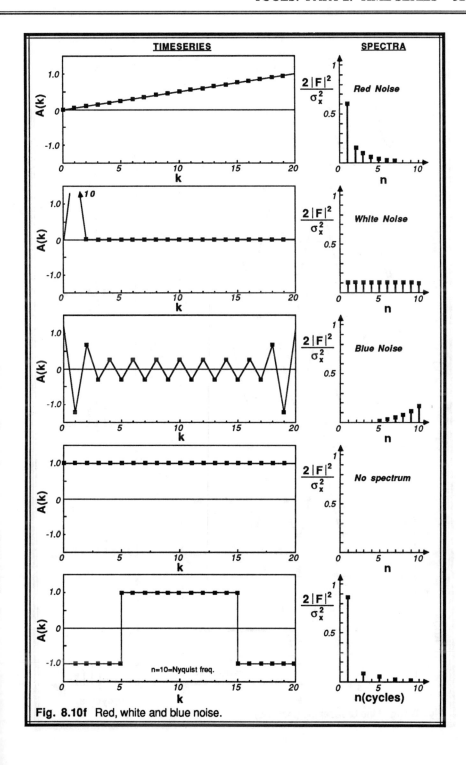

Fig. 8.10f Red, white and blue noise.

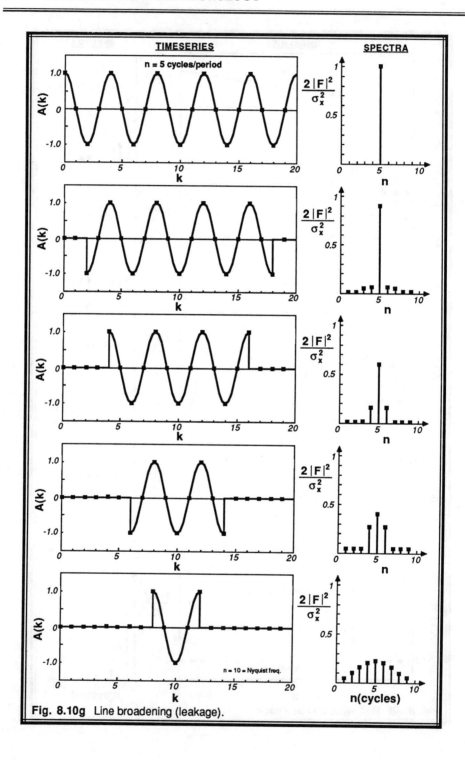

Fig. 8.10g Line broadening (leakage).

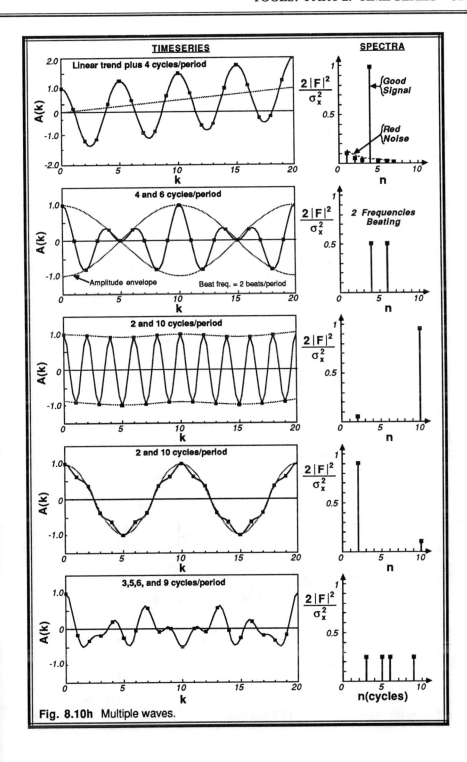

Fig. 8.10h Multiple waves.

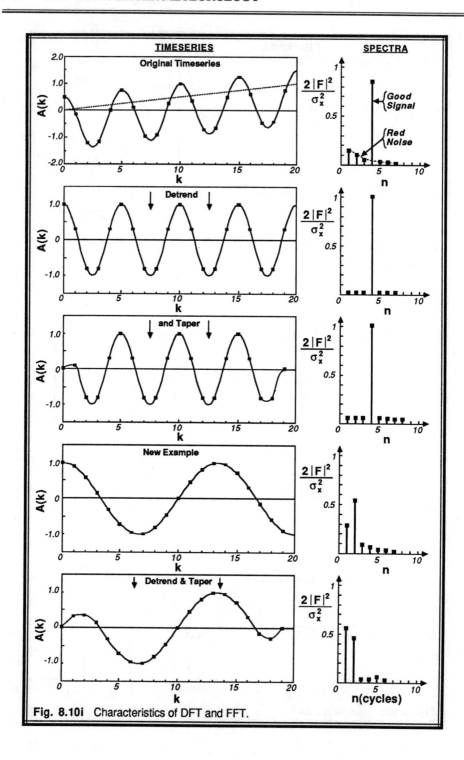

Fig. 8.10i Characteristics of DFT and FFT.

Case H (Fig 8.10h): Multiple Waves. These examples were constructed in reverse, where the spectrum was specified and the time series was generated with an inverse FFT. When the time series consists of the superposition of a number of different wave periods or wavelengths, the spectrum shows a number of spikes. If some of these waves are between harmonics or longer than the fundamental frequency, then the problems of spreading, leakage, and red noise are superimposed on the other resolvable signals. Also, for the second example, the two frequencies at n = 4 and 6 result in a beat frequency of 2, causing the amplitude envelope of the original time series to oscillate as shown.

Case I (Fig 8.10i):   Conditioning. The first three examples show one situation of an original time series that is superimposed on a trend.   Detrending the time series eliminates the red noise in the spectrum, and tapering the ends has little effect after that. The last two examples show a wave with n = 1.5, causing a significant amount of noise in the spectrum.   However, after detrending and tapering, the spectrum yields the desired spikes at n = 1 and n = 2.

8.8 Spectra of Two Variables

Just as we can find the spectrum for a single variable, we can also find a spectrum for a product of two variables. For example, given observations of w'(t) and θ'(t), we can create a new time series w'θ'(t) on which we can perform routine spectral analyses using an FFT. Occasionally it is useful to get more information about the spectrum of w'θ', such as how the phase of the w' fluctuations relate to the phase of the θ' fluctuations as a function of frequency. *Cross-spectrum analysis* relates the spectra of two variables.

8.8.1 Phase and Phase Shift

Phase refers to the position within one wave, such as at the crest or the trough (Fig 8.11a). It is often given as an angle. For example, the crest of a sine wave occurs at 90°, or at π/2 radians. *Phase shift* refers to the angle between one part of a wave like the crest and some reference point like a "start time" or the crest of another wave. For example, in Fig 8.11b the phase of the second wave is shifted 90° to the right of the first wave.

The equation for a single sine wave of amplitude C that is shifted by angle Φ to the right is:

$$A(k,n) = C(n) \cdot \sin\left(\frac{2\pi kn}{N} - \Phi(n)\right) \qquad (8.8.1a)$$

Through trigonometric identities, we can show that the same wave described above can also be written as the sum of one sine wave and one cosine wave:

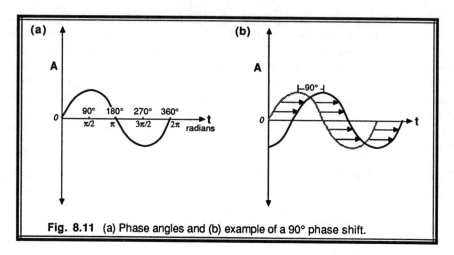

Fig. 8.11 (a) Phase angles and (b) example of a 90° phase shift.

$$A(k,n) = C_s(n) \cdot \sin\left(\frac{2\pi kn}{N}\right) + C_c(n) \cdot \cos\left(\frac{2\pi kn}{N}\right) \qquad (8.8.1b)$$

where $C_s = C \cdot \cos\Phi$ and $C_c = -C \cdot \sin\Phi$.

As shown in section 8.4.1 the Fourier transforms give the amplitudes of sine and cosine terms in the spectral decomposition of the original field. Thus, we can also interpret the spectra in terms of an amplitude and phase shift for waves of each frequency.

8.8.2 Cross Spectra

Define $G_A = |F_A(n)|^2$ as the unfolded spectral energy for variable A and frequency n. We can rewrite this definition as $G_A = F_A^* \cdot F_A$, where F_A^* is the complex conjugate of F_A, and where the dependence on n is still implied.

To demonstrate this last definition, let $F_A = F_{Ar} + i \cdot F_{Ai}$, where subscripts r and i denote real and imaginary parts respectively. Thus, the complex conjugate is simply $F_A^* = F_{Ar} - i \cdot F_{Ai}$. The expression for the spectral energy can now be written as:

$$
\begin{aligned}
G_A &= F_A^* \cdot F_A \\
&= (F_{Ar} - i F_{Ai}) \cdot (F_{Ar} + i F_{Ai}) \\
&= F_{Ar}^2 + i F_{Ai} F_{Ar} - i F_{Ai} F_{Ar} - i^2 F_{Ai}^2 \\
&= F_{Ar}^2 + F_{Ai}^2 \\
&= |F_A(n)|^2
\end{aligned}
$$

leaving the magnitude squared as a real number.

Similarly, define the spectral intensity $G_B = F_B^* \cdot F_B$, for a different variable B. We can now define the *cross spectrum* between A and B by

$$G_{AB} \quad = \quad F_A^* \cdot F_B \tag{8.8.2a}$$

$$= \quad F_{Ar} F_{Br} + i F_{Ar} F_{Bi} - i F_{Ai} F_{Br} - i^2 F_{Ai} F_{Bi}$$

Upon collecting the real parts and the imaginary parts, the real part is defined as the *cospectrum*, Co, and the imaginary part is called the *quadrature spectrum*, Q:

$$G_{AB} = Co - iQ \tag{8.8.2b}$$

where

$$Co = F_{Ar} F_{Br} + F_{Ai} F_{Bi} \tag{8.8.2c}$$

and

$$Q = F_{Ai} F_{Br} - F_{Ar} F_{Bi} \tag{8.8.2d}$$

Although not explicitly written in the equations above, F_A and F_B are functions of n, making both the cospectrum and quadrature spectrum functions of n too: Co(n) and Q(n).

The cospectrum is frequently used in meteorology, because the sum over frequency of all cospectral amplitudes, Co, equals the covariance between A and B, (i.e., $\sum_n Co(n) = \overline{a'b'}$). Note that the cospectrum computed as above is NOT equal to the spectrum of the time series of the product a'b'.

The quadrature spectrum is usually not used directly, but it too has a physical interpretation. The quadrature spectrum is equal to the spectrum of the product of b' times a phase shifted a', where a' is phase shifted a quarter period of n. In other words, the amount of time lag applied to a' depends on the frequency, n, such that the phase shift is always 90° for each n.

Three additional spectra can be constructed from the quad and co-spectra. An *amplitude spectrum*, Am, can be defined as

$$Am = G_{AB}^* \cdot G_{AB}$$

$$= Q^2 + Co^2 \tag{8.8.2e}$$

A large amplitude at any frequency n implies that A is very strongly correlated to B at that frequency, regardless of phase differences between A and B. In other words if both A and B have a strong amplitude component with frequency $n = 5$ even if A and B are out of phase, then Am will be large for $n = 5$. Also, if the amplitude is small for any frequency n, then coherence and phase spectra (described next) are not significant (i.e., unreliable) for that frequency.

The *coherence spectrum*, Coh, is defined by:

$$\text{Coh}^2 = \frac{G^*_{AB} G_{AB}}{G_A G_B} = \frac{Q^2 + Co^2}{G_A G_B} \tag{8.8.2f}$$

This is essentially a normalized amplitude, and is a real number in the range 0 to 1. It acts very much like a frequency dependent correlation coefficient.   Note that in some of the literature Coh^2 is defined as the coherence, rather than Coh.   Like the amplitude spectrum, it is not a function of phase shift.

Finally, a *phase spectrum*, Φ, can be defined as

$$\tan \Phi = Q / Co \tag{8.8.2g}$$

This can be interpreted as the phase difference between the two time series A and B that yielded the greatest correlation for any frequency, n.   The phase spectrum can be used to infer the nature of the physical flow.   For buoyancy waves, θ' is characteristically   90° out of phase with w'; while for turbulence, the two variables either in phase or 180° out of phase.

8.8.3 Example

Problem: Given the time series from section 8.4.2 for humidity, and the time series below for vertical velocity, w:

Index (k):	0	1	2	3	4	5	6	7
Time (UTC):	1200	1215	1230	1245	1300	1315	1330	1345
w (m/s):	0	-2	-1	1	-2	2	1	1

Find and plot:
   a) the discrete Fourier transform and the spectrum for w
   b) the cospectrum for w and q
   c) the quadrature spectrum
   d) the amplitude spectrum
   e) the coherence spectrum
   f) the phase spectrum.
Also find the discrete Fourier transform and the spectrum for the product w'q'.

Solution: The original time series are listed in Table 8-2 as a reference, along with the deviations squared and the series w'q'.   The Fourier transforms for both w and q are

Table 8-2. Spectra and cospectra data, computed with an FFT program, and then displayed here in spreadsheet form.

Timeseries:

k	w	q	w'2	q'2	w'q'
0	0	8	0	1	0
1	-2	9	4	4	-4
2	-1	9	1	4	-2
3	1	6	1	1	-1
4	-2	10	4	9	-6
5	2	3	4	16	-8
6	1	5	1	4	-2
7	1	6	1	1	-1
Sum:	0	56	Sum: 16	40	Sum: -24
Mean:	0	7	Variance: 2	5	Covar: -3

Simple Spectra:

n	Fw real	Fw imag	Gw	Ew/w'2	Fq real	Fq imag	Gq	Eq/q'2
0	0.000	0.000			7.000	0.000		
1	-0.104	0.604	0.375	0.375	0.280	-1.030	1.140	0.456
2	-0.250	0.250	0.125	0.125	0.500	0.000	0.250	0.100
3	0.604	0.104	0.375	0.375	-0.780	-0.030	0.610	0.244
4	-0.500	0.000	0.250	0.125	1.000	0.000	1.000	0.200
5	0.604	-0.104	0.375		-0.780	0.030	0.610	
6	-0.250	-0.250	0.125		0.500	0.000	0.250	
7	-0.104	-0.604	0.375		0.280	1.030	1.140	
Sum:			2.000	1.000			5.000	1.000

Cross-spectra (based on F & G values above):

n	Gwq Co	Q	Am	Coh2	Phase(°)	Fwq real	Fwq imag	Gwq	Ewq/(w'q')'2
0						-3.000	0.000		
1	-0.651	0.062	0.428	1.000	174.52	1.104	-0.354	1.343	0.398
2	-0.125	0.125	0.031	1.000	135.00	-0.250	1.250	1.625	0.481
3	-0.474	-0.063	0.229	1.000	187.52	0.396	-0.354	0.282	0.084
4	-0.500	0.000	0.250	1.000	180.00	0.500	0.000	0.250	0.037
5	-0.474	0.063	0.229	1.000	172.48	0.396	0.354	0.282	
6	-0.125	-0.125	0.031	1.000	225.00	-0.250	-1.250	1.625	
7	-0.651	-0.062	0.428	1.000	185.48	1.104	0.354	1.343	
Sum:	-3.000	0.000						6.750	1.000

Simple Spectrum of w'q' timeseries:

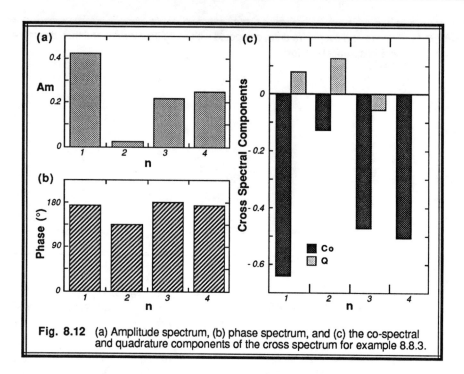

Fig. 8.12 (a) Amplitude spectrum, (b) phase spectrum, and (c) the co-spectral and quadrature components of the cross spectrum for example 8.8.3.

then found using an FFT program, and are listed in Table 8-2 along with their corresponding unfolded spectral intensities, G_w and G_q, and the fraction of variance explained, E_w / s_w^2 and E_q / s_q^2, where s^2 represents the variance.

Also listed is a subtable with co- and quad- spectral components of G_{wq}, the resulting values of Am, Coh2, and the phase angles in degrees. These are plotted in Fig 8.12. Finally, the simple spectrum of the w'q' time series is listed.

Discussion: The biased variances of the w and q time series are 2.0 and 5.0, respectively. From Table 8-2, we see that the sum of the G_w and G_q spectral components equals their respective variances. This is always a good check to do with the analysis. The associated normalized spectral components, E_w / s_w^2 and E_q / s_q^2, sum to unity as

desired. Also, the covariance $\overline{w'q'} = -3.0$, which agrees with the sum of the Co cospectral components.

Looking at the original time series, we see that w' is usually positive when q' is negative, as confirmed by the negative covariance. Thus, we anticipate that w' and q' are 180° out of phase. The phase spectrum supports this. In fact, the only phase values which are substantially different from 180° are those for which the amplitude (Am) values are small, suggesting that these phase values can't be trusted.

It is surprising to find that the coherence is 1.0 for all frequencies. This indicates that there is a very close relationship between w and q for all frequencies or wavelengths, for this contrived example. For real turbulence data the coherence would not equal 1.0 for all frequencies.

Next, look at the individual q series. There is an obvious oscillation with three cycles within the whole period of record. In addition there is a background low frequency change of the time series. Looking at the simple spectrum for q, the spectral intensity is indeed large for $n = 3$ and $n = 1$. A similar conclusion can be reached for w. For both of these series, there is a distinct spectral minimum at $n = 2$.

This minimum shows up in the cospectrum at $n = 2$. Thus, waves with two cycles per period contribute little to the total covariance $\overline{w'q'}$. This is in sharp contrast with the w'q' time series itself, which shows a very definite $n = 2$ wave. The simple spectrum analysis of w'q' also yields the largest spectral component at $n = 2$. This tells us that the variance (not covariance) of the w'q' time series has a large contribution at $n = 2$, even though the covariance itself, $\overline{w'q'}$, has a minimum at $n = 2$.

In the discussion presented above, it was easy to compare the spectra with features in the original time series, because the series were so short. For real turbulence data consisting of thousands of data points, it is not so easy to pick out features by eye. For these situations, spectral analysis is particularly valuable.

8.9 Periodogram

The periodogram is just a least squares best fit of sine and cosine waves to the original signal (i.e., to the time series). Because the original time series need not consist of evenly spaced data points for the periodogram to work, it has a very distinct advantage over the discrete Fourier transform. In fact, for some data sets with data gaps or missing data, it is the only method to calculate spectral information short of making up bogus data to fill the gaps. The prime disadvantage of the periodogram is that it takers longer to compute than an FFT.

First, the mean of the original time series of variable A is subtracted from each A(k) data point to yield a modified time series for A'(k). For each frequency (n) a wave of the following form is fitted to the data:

$$A' = a_1 \cos\left[\frac{2\pi kn}{N}\right] + a_2 \sin\left[\frac{2\pi kn}{N}\right] \tag{8.9a}$$

where A' is the deviation of A from the mean, and where a_1 and a_2 are the best-fit coefficients to be determined. Solving for a_1 and a_2 (both a function of n) in the least-squares sense gives:

$$a_1 = \frac{\overline{A's'}\,\overline{s'c'} - \overline{A'c'}\,\overline{s'^2}}{(\overline{s'c'})^2 - \overline{c'^2}\,\overline{s'^2}} \quad \text{and} \quad a_2 = \frac{\overline{A'c'}\,\overline{s'c'} - \overline{A's'}\,\overline{c'^2}}{(\overline{s'c'})^2 - \overline{c'^2}\,\overline{s'^2}} \qquad (8.9b)$$

where

$$s' = \sin[2\pi nk/N] - \overline{\sin[2\pi nk/N]} \quad \text{and} \quad c' = \cos[2\pi nk/N] - \overline{\cos[2\pi nk/N]}$$

and where the overbar denotes an average over all N data points of the original time series. In the definitions of s' and c' above, note that the overbar terms in each expression are identically equal to zero only for n equal to an integer value.

Given the best fit from above, we can compute the correlation coefficient squared, $r^2(n)$:

$$r^2(n) = \frac{\text{explained variance}}{\text{total variance}} = \frac{a_1\,\overline{A'c'} + a_2\,\overline{A's'}}{\overline{A'^2}} \qquad (8.9c)$$

If waves at frequency n explain a lot of variance in the original signal, then $r^2(n)$ is close to 1. Otherwise, it is closer to zero. For integer values of n, $r^2(n)$ is equal to the normalized spectral intensity, $E_A(n)/s_A^2$, of the FFT. This is where the spectral information comes from.

We must solve (8.9b & c) many times, for each different value of n that we are interested in. To calculate a complete spectra, we should solve the equation for at least the N different integer values of n. Thus, when $r^2(n)$ is plotted vs. n, the result is a spectrum. As an example, the spectra plotted in section 8.7 could have been labeled as r^2 vs n, where r^2 would have been computed using periodogram methods. Note that we can also solve the equations for noninteger values of n, and for n in the range 0 to 1, if desired, although sine waves of noninteger values are not mutually orthogonal.

8.10   Nonlocal Spectra

Contained within the nonlocal-closure transilient matrix is information about the amount of fluid that mixes between each pair of grid-point locations in a column of air during a finite time. This information can be extracted and grouped to yield spectral information about the contributions of different wavelengths to the overall mixing *process*. Such a spectrum differs from the spectrum of the fluid *state* obtained from harmonic analysis (FFT) of measurements of temperature, velocity or other state variables, as described in all the previous sections of this chapter. Tennekes (1976) pointed out that the FFT Fourier modes do not have a one-to-one correspondence with eddies.

8.10.1   Transport Spectra

The turbulent flux is the result of a turbulent process. Without the process of eddies moving and mixing various air parcels, there would be no turbulent flux regardless of the state of the air. Transport spectra are based on fluxes, while state spectra (FFTs) are based on variances.

To see how transport spectra (TS) are related to the transilient turbulence theory, start with the definition for kinematic flux given by (6.8.4a). Instead of summing over the contributions from ALL pairs of grid points within the domain of turbulence, as specified in (6.8.4a), we can selectively sum over only those pairs of points having a specified separation distance (i.e., wavelength). If we let $m \cdot \Delta z$ equal the wavelength of interest, then the portion of flux associated with wavelength-index m that contributes to the total flux at height-index k (i.e., at height $z = k \cdot \Delta z$) is:

$$TS(k,m) = \frac{\Delta z}{\Delta t} \sum_{i=1}^{k} \sum_{j=1}^{N} \delta_{m,|i-j|} \; c_{ij}(t,\Delta t) \cdot \left[\bar{\xi}_i - \bar{\xi}_j \right] \qquad (8.10.1a)$$

where δ_{ij} is the usual Kronecker delta, Δz is the vertical grid increment, Δt is the timestep increment, and m is an integer between 1 and N.

The total flux at location k is given by the sum of transport spectra over all wavelengths:

$$\overline{w' \xi'}(k) = \sum_{m=1}^{N} TS(k,m) \qquad (8.10.1b)$$

It is always a good check to use (8.10.1b) confirm that the sum of the process-spectral components does indeed equal the total flux.

A case-study example of a transport spectrum is shown in Fig 8.13 for the kinematic heat flux at two different heights within the turbulent boundary layer near the Cabauw tower in the Netherlands at 1500 UTC, 30 May 1978 (Stull and Driedonks, 1987). A 3 km column of air near the tower is modeled using 30 grid boxes, each 100 m thick. The mixed layer within this column at 1500 UTC was about 2100 m thick.

At a height of $z = 100$ m (i.e., at $k = 1$) we find that the smallest resolved wavelengths contribute most the the heat flux for this case, while the wavelengths of the range 1000 m to 2000 m make a smaller, but yet significant, contribution. At a height of 500 m above ground the smallest wavelengths contribute virtually nothing to the total flux, while wavelengths in the 500 m to 1700 m range dominate. At both of these heights, it is interesting to see that the largest wavelength within the turbulent domain has a negative contribution to the total heat flux. This is associated with the entrainment of warm air downward by the thermal-scale eddies.

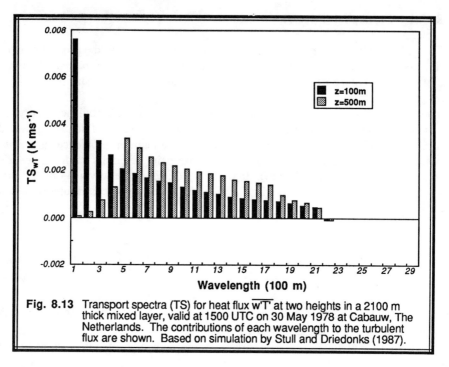

Fig. 8.13 Transport spectra (TS) for heat flux $\overline{w'T'}$ at two heights in a 2100 m thick mixed layer, valid at 1500 UTC on 30 May 1978 at Cabauw, The Netherlands. The contributions of each wavelength to the turbulent flux are shown. Based on simulation by Stull and Driedonks (1987).

8.10.2   Process Spectra

One limitation of transport spectra is that the flux contribution can be small even if the mixing process is vigorous. This can happen when the difference of the variable values between the heights being mixed is small or zero. To help focus on the mixing process alone, we can define a *process spectrum* that does not use the values of the variable being mixed:

$$PS(k,m) = \sum_{i=1}^{k} \sum_{j=1}^{N} \delta_{m,|i-j|} \; c_{ij} \, (t,\Delta t) \qquad (8.10.2a)$$

One expects this spectrum to be the same for heat, moisture, tracers, and maybe for momentum (neglecting pressure effects and waves), because it describes the mixing process rather than the effect of the mixing on the fluid state.

Fig 8.14 shows a process spectrum for the same Cabauw case as Fig 8.13. As before, the smallest wavelengths are the most important at the lowest height. As height increases, the peak in the spectrum becomes broader, less peaked, and shifts to longer wavelengths.

This behavior is related to mechanisms that generate turbulence. Near the surface, strong superadiabatic lapse rates and wind shears create strong dynamic instabilities across short distances, to which the transilient parameterization described in Chapter 6 responds with vigorous mixing at small wavelengths. Near the entrainment zone, however, the lapse rate is locally stable, suppressing generation of small-scale mixing except by wind shear. The largest scales continue to be slightly unstable near the entrainment zone, associated with warm thermals rising from the surface layer.

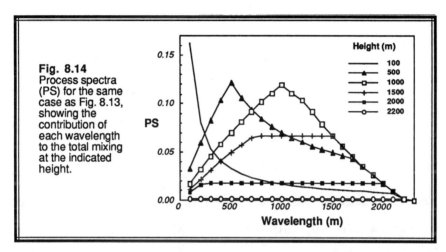

Fig. 8.14 Process spectra (PS) for the same case as Fig. 8.13, showing the contribution of each wavelength to the total mixing at the indicated height.

Prandtl (1925) recognized that eddies of a variety of sizes can simultaneously operate at any point in the turbulent domain. Due to lack of computer power in the early 1900s, he suggested averaging over the spectrum of eddy sizes to yield one mixing length for each point in space. At the time he had no real measure of the relative importance to the overall mixing length of the various eddy sizes. Instead, he parameterized the mixing length directly as a function of boundary layer and turbulence scales, as discussed in Chapter 6.

We can use the process spectrum to examine the relative importance of various scales of mixing to the overall mixing length by applying the process-spectral amplitudes as weights for their respective wavelengths. The resulting weighted wavelength is like a mixing length, l, which we can find at each height ($z = k \cdot \Delta z$):

$$l(k) = \Delta z \cdot \frac{\sum_{m=1}^{N} m \cdot PS(k,m)}{\sum_{m=1}^{N} PS(k,m)} \qquad (8.10.2b)$$

Fig 8.15 shows mixing lengths as a function of height for the same Cabauw case study. The mixing length increases with height in the bottom of the mixed layer, but becomes nearly constant with height in the top half of the mixed layer. The mixing length is undefined in the nonturbulent air above the mixed layer, because the process spectral amplitudes are zero.

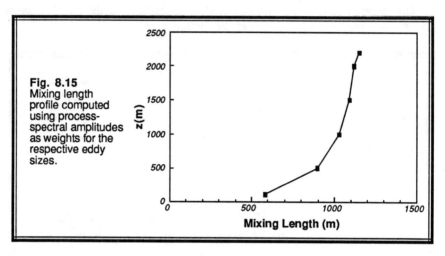

Fig. 8.15
Mixing length profile computed using process-spectral amplitudes as weights for the respective eddy sizes.

8.11 Spectral Decomposition of the TKE Equation

8.11.1 Spectral Decomposition Methods

Although we have concentrated on discrete spectral methods in this chapter, an obvious extension is to use integrals to describe the Fourier transform pair for continuous functions. In the following example, we will decompose the original function $A(t,x)$ into a Fourier integral in a single spatial direction, x, but will not perform a similar decomposition in time. For this case, the Fourier transform pair is:

$$\text{Inverse Transform} \qquad A(t,x) = \int_{\kappa = -\infty}^{\infty} F(t,\kappa)\, e^{i\kappa x}\, d\kappa \qquad (8.11.1a)$$

$$\text{Forward Transform} \qquad F(t,\kappa) = \frac{1}{2\pi} \int_{x=-\infty}^{\infty} A(t,x)\, e^{-i\kappa x}\, dx \qquad (8.11.1b)$$

where κ is wavenumber.

Substitution of (8.11.1a) into a term like $\partial A(t,x)/\partial x$ yields:

$$\frac{\partial A(t,x)}{\partial x} = \frac{\partial}{\partial x} \int_{\kappa} F(t,\kappa)\, e^{i\kappa x}\, d\kappa = \int_{\kappa} F(t,\kappa)\, \frac{\partial e^{i\kappa x}}{\partial x}\, d\kappa = \int_{\kappa} F(t,\kappa)\, i\, \kappa\, e^{i\kappa x}\, d\kappa$$

On the other hand, a term like $\partial A(t,x)/\partial t$ becomes:

$$\frac{\partial A(t,x)}{\partial t} = \int_{\kappa} \frac{\partial F(t,\kappa)}{\partial t} e^{i\kappa x} d\kappa$$

As an example of how this is used in complete equations, start with the simple advection equation with constant mean wind,

$$\frac{\partial A(t,x)}{\partial t} = -\overline{U} \frac{\partial A(t,x)}{\partial x}$$

and spectrally decompose it to yield:

$$\int_{\kappa} \frac{\partial F(t,\kappa)}{\partial t} e^{i\kappa x} d\kappa = -\overline{U} \int_{\kappa} F(t,\kappa) \, i \, \kappa \, e^{i\kappa x} d\kappa$$

An integral over wavenumbers appears in every term of the above equation. Thus, we could focus on the contribution of any ONE wavenumber to the whole equation by looking at the respective integrands (moving $\overline{U}$ under the integral because it is not a function of κ):

$$\frac{\partial F(t,\kappa)}{\partial t} e^{i\kappa x} = -\overline{U} \quad F(t,\kappa) \, i \, \kappa \, e^{i\kappa x}$$

The factor $\exp(i\kappa x)$ appears in every term of the above equation, and can be cancelled out. This leaves use with a spectral representation of the advection equation:

$$\partial F(t,\kappa) / \partial t = -\overline{U} \, i \, \kappa \, F(t,\kappa)$$

As you have probably anticipated, we could let $A(t,x)$ represent a variable like perturbation velocity, $u'(t,x)$, or perturbation potential temperature, $\theta'(t,x)$. Thus we can use the same general approach as show above to spectrally decompose the TKE equation.

8.11.2 Spectral Representation of the TKE Equation

Since the spectral decomposition of the TKE equation is somewhat complex, the reader is referred to Batchelor (1953) and Borkowski (1969) for the details. The end result for homogeneous turbulence, where S is the spectral energy, is:

$$\frac{\partial S(t,\kappa)}{\partial t} = \frac{g}{\theta_v} \gamma(t,\kappa) - \phi(t,\kappa)\frac{\partial \overline{U}}{\partial z} + \frac{\partial Tr(t,\kappa)}{\partial \kappa} - 2\nu \kappa^2 S(t,\kappa) \quad (8.11.2)$$

$$\text{I} \qquad\qquad \text{III} \qquad\qquad \text{IV} \qquad\qquad \text{V}(\kappa) \qquad\qquad \text{VII}$$

Term I   The local time tendency of the κth spectral component of the TKE

Term III   Buoyant production or loss associated with the κth component of $\overline{w'\theta_v'}$

Term IV   Mechanical (shear) production associated with the κth component of $\overline{u'w'}$

Term V(κ) Convergence of TKE transport across the spectrum

Term VII   Viscous dissipation of the κth component of TKE

This equation was the one referenced in section 5.3, where the relative contributions of the four terms on the right of (8.11.2) were plotted. When integrated over wavenumbers from 0 to ∞, the result is the TKE equation (5.1b).

The individual factors are defined as follows: TKE $= \int S(t,\kappa)\, d\kappa$, $\overline{w'\theta_v'} = \int \gamma(t,\kappa)\, d\kappa$, $\overline{u'w'} = \int \phi(t,\kappa)\, d\kappa$, and $\varepsilon = 2\nu \int \kappa^2 S(t,\kappa)\, d\kappa$, where the integrals are from 0 to ∞. The transport term $\partial Tr/\partial \kappa$ becomes zero when integrated across the spectrum, because it represents the transport of existing TKE from the low wavenumber portions of the spectrum (where energy is produced) to the high wavenumber regions (where it is dissipated).

In the inertial subrange portion of the spectrum where there is neither production nor dissipation, we would expect that the transport across the spectrum, $Tr(t,\kappa)$ would be equal in magnitude to the total dissipation rate: $Tr(t,\kappa) = \varepsilon$. This transport is nothing more than the ***energy cascade*** that was introduced early in the text.

8.12 References

Batchelor, G.K., 1953: *The Theory of Homogeneous Turbulence*. Cambridge Univ. Press, London.

Bergland, G.D., 1969: A guided tour of the fast Fourier transform. *IEEE Spectrum*, **6**, 41-51.

Borkowski, J., 1969: Spectra of anisotropic turbulence in the atmosphere. Proceedings of Colloquium on Spectra of Meteorological Variables, Stockholm, June 9-19, 1969. *Radio Science* (Dec 1969), **4**, 1351-1355.

Deardorff, J.W. and G.E. Willis, 1985: Further results from a laboratory model of the convective planetary boundary layer. *Bound.-Layer Meteor.*, **32**, 205-236.

Jenkins, G.M., and D.G. Watts, 1968: *Spectral Analysis and Its Applications*, Holden-Day, Oakland, CA. 523pp.

Lenschow, D.H., 1986: *Probing the Atmospheric Boundary Layer*. American Meteorological Society, Boston, MA. 269pp.

Lumley, J.L. and H.A. Panofsky, 1964: *The Structure of Atmospheric Turbulence*. Interscience Publishers (Wiley), New York. 239pp.

Neff, W.D. and R.L. Coulter, 1986: Acoustic remote sensing. *Probing the Atmospheric Boundary Layer*, D.H. Lenschow (Ed.), American Meteor. Society, Boston. 201-239.

Otnes, R.K. and L. Enochson, 1972: *Digital Time Series Analysis*. Wiley-Interscience, New York. 467pp.

Panofsky, H.A., and G.W. Brier, 1968: *Some Applications of Statistics to Meteorology*, Penn State Univ., University Park, PA. 224pp.

Panofsky, H.A. and J.A. Dutton, 1984: *Atmospheric Turbulence, Models and Methods for Engineering Applications*. John Wiley (Wiley-Interscience Pub.), New York. 397pp.

Prandtl, L., 1925: Bericht über Untersuchungen sur ausgebildeten Turbulenz. *Zs. angew. Math. Mech.*, **5**, 136-139.

Priestley, M.B., 1981: *Spectral Analysis and Time Series, Vols 1 & 2*. Academic Press, New York. 490 pp.

Stull, R.B. and A.G.M. Driedonks, 1987: Applications of the transilient turbulence parameterization to atmospheric boundary-layer simulations. *Bound.-Layer Meteor.*, **40**, 209-239.

Tennekes, H., 1976: Fourier transform ambiguity in turbulence dynamics. *J. Atmos. Sci.*, **33**, 1660-1663.

Wyngaard, J.C., Y. Izumi, and S.A. Collins, Jr., 1971: Behavior of the refractive index structure parameter near the ground. *J. Opt. Soc. Am.*, **61**, 1646-1650.

Wyngaard, J.C. and M.A. LeMone, 1980: Behavior of the refractive index structure parameter in the entraining convective boundary layer. *J. Atmos. Sci.*, **37**, 1573-1585.

8.13 Exercises

1) Given the following set of turbulence velocity measurements, find the autocorrelation function using both the exact and approximate formulae. These measurements were taken every second:

5.5, 6.3, 7.4, 3.3, 3.8, 5.9, 6.1, 5.7, 6.3, 7.1, 4.8, 3.1, 2.1, 2.4, 3.0

Plot this time series and your autocorrelation functions on separate graphs. What is the difference between the exact and approximate autocorrelation curves?

2) Compute the structure function vs lag, for the data of problem 1. Plot and discuss your results. Find the structure function parameter. Discuss the quality of the fit of (8.3.1b) to this data set.

3) Theoretically prove that if $A(t) = \sin(t)$, then $R(L) = \cos(L)$.

4) Prove analytically the following relationship between structure function and autocorrelation: $D(L) = \sigma_x^2 [1-R(L)]$. Assume stationary turbulence (i.e., σ_x^2 is constant).

5) Compute the exact autocorrelation and structure function for the following data set, and plot your results for lags from 0 through 180 s. Discuss how the shape of the original time series compares with the peaks in the autocorrelation curve.

t(s)	T(C)	w(m/s)	t(s)	T(C)	w(m/s)
0	25	2			
10	23	2	110	20	-4
20	21	-1	120	19	-1
30	21	1	130	20	1
40	30	4	140	25	3
50	20	-3	150	21	0
60	24	3	160	25	1
70	23	1	170	23	0
80	23	2	180	21	-2
90	24	3	190	20	-1
100	23	-1	200	19	-2

6) Compute the discrete Fourier transform for the time series from problem 1. Instead of using a canned FFT package, write your own program or develop a spreadsheet to solve it. Check your results by doing an inverse FFT to reconstruct the data set. Also check to see that the sum of the spectral energies equals the biased variance of the original signal. Compute and plot the spectrum.

7) Compute the discrete Fourier transform for the time series from problem 1, using a canned FFT program. If you solved problem 6, how do your answers compare.

8) Prove mathematically that sine waves of different integer frequencies, n, are orthogonal to each other. By orthogonal, we mean that a wave with one n value cannot be described by a sum of waves with other n values. As an alternative, test the orthogonality with some specific cases. For example, define $A(k)$ as a sine wave with n=2. Then see if the discrete Fourier transform of this wave has any energy at n=3.

9) Compute the individual FFTs for the two data sets from problem 5, using canned statistical packages. Compare the simple spectrum for T with that for w.

10) Using the FFT temperature output from problem 9, set the 5 frequencies nearest the Nyquist frequency to zero, on both sides of the Nyquist frequency. Do this to the Fourier transform itself, not to the spectrum. Then use this modified phase space data to compute an inverse transform. Plot the resulting physical space time series, along with the original temperature time series from problem 5. What you have just done was an ideal low-pass filter.

11) Use the FFT output from problem 9 to manually compute (or use a spreadsheet) the cross spectral information: cospectrum, quadrature spectrum, coherence, amplitude, and phase spectra. If you have a canned statistical package that also computes these values, compare the results.

12) Compute a new time series of $w'T'$ values from the data in problem 5. Calculate the covariance for this series, and then do an FFT on the series. Compare the resulting simple spectrum from this FFT output with the cross spectral results from problem 11.

13) Write a program to compute the periodogram. Use it to find the spectrum (r^2 values vs. n) for the data in problem 1, for n values ranging from 0.1 to 10, with increments of $\Delta n = 0.2$. Compare the r^2 values for integer n with the spectral output from problem 6 normalized by the total variance.

14) If you have an odd number of data points in a time series vs. an even number, how does that affect the Nyquist frequency?

15) When the spectral form of the TKE equation (8.6.2) is integrated over wavenumber from 0 to ∞, the result is close to the TKE equation (5.1b).
(a) Discuss the differences.

(b) Also, if (8.6.2) is integrated over wavenumbers from 0 to κ_{IS}, where κ_{IS} is in the middle of the inertial subrange, then describe the physical relationships and magnitudes of the resulting integrated terms.

16) Write a computer program to generate a time series consisting of 10 oscillations of a perfect sine wave within a 1-minute period. Sample this series 18 evenly-spaced times during that same period. Plot this sampled time series, and discuss the shape in relation to the Nyquist frequency and aliasing.

17) Prove mathematically that the Fourier transform results are also the best-fit sine and cosine waves in the least-squares sense.

18) Use the Profile B data and the transilient matrix from example 6.8.5 to calculate the the transport and process spectra for all wavelengths (m = 1 to 5) for heights z = 100, 200, and 300 m, and plot your results on separate graphs.

19) Spectrally decompose equation (3.5.3c). Comment on the wavenumber dependence of advection terms (both turbulent and mean).

9 Similarity Theory

For a number of boundary layer situations, our knowledge of the governing physics is insufficient to derive laws based on first principles. Nevertheless, boundary layer observations frequently show consistent and repeatable characteristics, suggesting that we could develop empirical relationships for the variables of interest. Similarity theory provides a way to organize and group the variables to our maximum advantage, and in turn provides guidelines on how to design experiments to gain the most information.

9.1 An Overview

9.1.1 Definitions and Methodology

Similarity theory is based on the organization of variables into *dimensionless groups*. Fortunately, there is a *dimensional-analysis* procedure called *Buckingham Pi theory* that aids us in forming dimensionless groups from selected variables. It is hoped that the proper choice of groups will allow empirical relationships between these groups that are *"universal"* — namely, that work everywhere all the time for the situation studied.

The four steps in developing a similarity theory are:

    (1) select (guess) which variables are relevant to the situation,

    (2) organize the variables into dimensionless groups,

    (3) perform an experiment, or gather the relevant data from previous experiments, to determine the values of the dimensionless groups,

    (4) fit an empirical curve or regress an equation to the data in order to describe the relationship between groups.

347

The result of this four-step process is an empirical equation or a set of curves which show the same shape — in other words the curves look self similar. Hence, the name *similarity theory*. If this empirical result is indeed universal, then we can use it on days and locations other than those of the experiment itself. Such expectations should be tested with an independent data set, before the results are disseminated to the rest of the scientific community.

If we selected in step (1) more variables than were necessary, the data will "tell us" of our mistake by indicating no change of the other dimensionless groups with respect to the group that is irrelevant. If we selected too few variables, or excluded an important variable, the data will also indicate our error by showing a large scatter or no repeatable patterns between the dimensionless groups.

Step (2) can often be performed by inspection of the relevant variables. In fact, based on the classes of similarity theory frequently used in meteorology (see Section 9.3.3), we can anticipate the dimensionless groups, although we can not always anticipate the relationship between the groups. For the very complex problems, we can employ Buckingham Pi Theory to identify the appropriate dimensionless groups.

Similarity theory does not tell us the form of the equation or the relationship between the dimensionless groups. Instead, we must use trial and error, physical insight, or automated techniques to select the form that qualitatively "looks the best". For example, we might express one group as a power law function of another group, as a logarithmic relationship, or as a constant that is not a function of other groups. The chosen equation usually contains unknown coefficients, which can then be solved by regression against the observed data.

The resulting equations are called *similarity relationships (or relations)*, or sometimes, improperly, *similarity laws*. Frequently, the dimensionless data graphs or curves are presented without a corresponding regression equation, because no simple equation could be found. For these cases, one can use the graph directly to determine the value of one dimensionless group as a function of the values of the other groups. We *normalize* an important variable when we divide it by other variables to make it dimensionless; hence, these graphs represent normalized data.

Similarity relationships are usually designed to apply to *equilibrium* (steady-state) situations. They are frequently used to yield equilibrium profiles of mean variables and turbulence statistics as a function of height or position. Rarely is time included as one of the relevant variables. Some variables, such as depth of the boundary layer, are so strongly dependent on time that no successful similarity expressions have been found to diagnose them. Instead, boundary layer depth must be calculated or measured using other techniques. This depth is used as input into dimensionless groups to diagnose other variables that do reach a quasi-steady state.

Finally, similarity theory is a type of zero-order closure. Once the similarity relationships have been identified, they can be used to diagnose equilibrium values of mean wind, temperature, moisture, and other variables as a function of height without any turbulence closure assumptions being made.

9.1.2 Example

Problem. Find a similarity relationship for the buoyancy flux, $\overline{w'\theta_v'}$, as a function of height in the convective mixed layer.

Solution. First (step 1), guess the relevant variables. Based on the problem statement, we already know that two of the variables of interest are $\overline{w'\theta_v'}$ and z.   The depth of the mixed layer, z_i, and the strength of the heat flux near the surface, $\overline{w'\theta_{v\,s}'}$, might also influence the flux within the interior of the mixed layer.   Thus, we will use four variables for this analysis.

Step (2), group these four variables into dimensionless groups.   By inspection, we can easily produce two dimensionless groups: $[z/z_i]$ and $[\overline{w'\theta_v'}/\overline{w'\theta_{v\,s}'}]$.   We have thus reduced our degrees of freedom from four to two.

In performing our experiment for step (3), dimensional analysis tells us that we need not measure all combinations of z, z_i , $\overline{w'\theta_v'}$, and $\overline{w'\theta_{v\,s}'}$.   Instead, we need only measure various combinations of the two groups: $[z/z_i]$ and $[\overline{w'\theta_v'} / \overline{w'\theta_{v\,s}'}]$.   This greatly simplifies the design and conduct of our experiment.

Suppose the heat flux data from Fig 3.7 (reproduced as Fig 9.1a) represents the results of our experiment.   The curves in this data set exhibit a common shape: there is a nearly-linear decrease of heat flux from the surface value to a small negative value near the top of the mixed layer.   Above that, the flux reduces toward zero.   As we shall soon learn in Chapter 11, the average depth of the mixed layer is frequently taken as the height where the heat flux is most negative.   When each of the data curves is replotted in terms of the two dimensionless groups, as shown in Fig 9.1b, we happily find that all of the data is closely clustered around a single curve.

For step (4), an obvious choice of curve is a straight line between the surface and the top of the mixed layer.   By definition we want the intercept of this line to equal 1, and by inspection it looks like the slope is roughly 1.2.   This results in:

$$\frac{\overline{w'\theta_v'}}{\overline{w'\theta_{v\,s}'}} = 1 - 1.2 \left[\frac{z}{z_i} \right] \qquad \text{for } 0 \le (z/z_i) \le 1$$

which is also plotted in Fig 9.1b.   As an independent test, the buoyancy flux data from Figs 3.1b, 3.2b, and 3.3b confirm the validity of our curve.

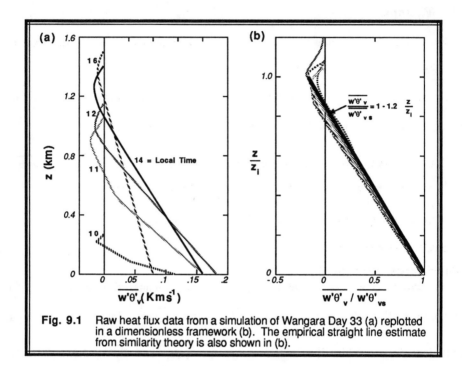

Fig. 9.1   Raw heat flux data from a simulation of Wangara Day 33 (a) replotted in a dimensionless framework (b). The empirical straight line estimate from similarity theory is also shown in (b).

Discussion. We hope to be able to use this equation to diagnose the value of the buoyancy flux at any height within the interior of a convective mixed layer on any other day at any other location, assuming we know the surface flux and the mixed layer depth. Even without this equation, we could use Fig 9.1b to determine the flux at any height.

For example, suppose the aircraft-measured buoyancy flux of $\overline{w'\theta_v}' = 0.1$ K·ms^{-1} at a height of $z = 200$ m when the mixed layer depth was $z_i = 1300$ m. We can used this data along with the similarity relation above to estimate the surface buoyancy flux: $\overline{w'\theta_v}'_s$ $= 0.123$ K·ms^{-1}.

9.2 Buckingham Pi Dimensional Analysis Methods

In 1914 Buckingham proposed a systematic approach for performing dimensional analysis. He called the resulting dimensionless groups *Pi groups*, which later caused the theory to be known as Buckingham Pi Theory (Perry, et al., 1963).

For each step of his approach we will first define the general procedure (in *boldface*), and then give an example (in normal type face). The example is that of fluid flow through a pipe, and involves the question: "How does the shear stress, τ, vary?"

Step 1. Hypothesize which variables could be important to the flow.
Example: stress, density, viscosity, velocity, pipe diameter, pipe roughness

Step 2. Find the dimensions of each of the variables in terms of the fundamental dimensions. The fundamental dimensions are:
$$L = length$$
$$M = mass$$
$$T = time$$
$$K = temperature$$
$$A = electric\ current$$
$$I = luminous\ intensity$$
The dimensions of any variable can be broken into these fundamental dimensions.

Example:	variable	name	fundamental dimensions
	ρ	fluid density	$M\,L^{-3}$
	μ	dynamic viscosity	$M\,L^{-1}\,T^{-1}$
	U	velocity	$L\,T^{-1}$
	τ	shear stress	$M\,L^{-1}\,T^{-2}$
	D	pipe diameter	L
	z_0	pipe roughness length	L

The first two variables describe fluid characteristics, the next two describe flow characteristics, and the last two describe pipe characteristics.

Step 3: Count the number of fundamental dimensions in our problem.
Example: There are 3 dimensions: L, M, T.

Step 4: Pick a subset of your original variables to become "key variables", subject to the following restrictions:
(a) The number of key variables must equal the number of fundamental dimensions.
(b) All fundamental dimensions must be represented in the key variables.
(c) No dimensionless group must be possible from any combination of these key variables.

Example: Pick 3 variables: ρ, D, and U to be the key variables.
Note that there are many other equally valid choices for key variables, such as:
ρ, z_0, U; or τ, μ, D, etc. It does not matter which three are picked, assuming that all of the above restrictions are satisfied. An invalid set would be U, D, z_0, because D/z_0 is dimensionless, and also because the fundamental dimension M is not represented. Another invalid set is τ, ρ, U, because $\tau/(\rho U^2)$ is dimensionless.

Step 5.   Form dimensionless equations of the remaining variables in terms of the key variables.

Example: $\tau = (\rho)^a \; (D)^b \; (U)^c$

$\mu = (\rho)^d \; (D)^e \; (U)^f$

$z_0 = (\rho)^g \; (D)^h \; (U)^i$

where a-i are unknown powers.

Step 6.   Solve for the powers a, b, c, . . . to yield dimensionally consistent equations.

Example: Solve each equation independently. For the first equation:

$$\tau = (\rho)^a \; (D)^b \; (U)^c$$

or $M \, L^{-1} \, T^{-2} = (M \, L^{-3})^a \; (L)^b \; (L \, T^{-1})^c$

or $M \, L^{-1} \, T^{-2} = M^a \quad L^{-3a+b+c} \quad T^{-c}$

The dimensions on the left hand side must equal the dimensions on the right. Thus:

M: $1 = a$

L: $-1 = -3a + b + c$

T: $-2 = -c$

These three equations can be solved for the three unknowns, yielding:

$a = 1 \qquad b = 0 \qquad c = 2$.

Thus, a dimensionally consistent equation is: $\tau = (\rho)^1 \; (D)^0 \; (U)^2$, or $\tau = \rho \, U^2$.

Similarly, we find that $d = 1$, $e = 1$, $f = 1$: yielding $\mu = \rho \, U \, D$.

Also: $g = 0$, $h = 1$, $i = 0$: yielding $z_0 = D$.

Step 7.   For each equation, divide the left hand side by the right hand side to give dimensionless (Pi) groups.   The number of Pi groups will always equal the number of variables minus the number of dimensions.

Example:

$$\pi_1 = \frac{\tau}{\rho \, U^2} \qquad\qquad \pi_2 = \frac{\mu}{\rho \, U \, D} \qquad\qquad \pi_3 = \frac{z_0}{D}$$

We started with 6 variables in our example, and reduced our degrees of freedom down to 3 dimensionless groups.

Step 8. (Optional)   If desired, alternative Pi groups can be formed from the ones derived in the previous step, as long as: the total number of Pi groups does not change, all variables are represented, and no one Pi group can be formed from any combination of the remaining groups.

Example: One alternative set of Pi groups might be: π_1 , π_4 $(=\pi_2/\pi_3)$, π_5 $(=1/\pi_3)$.

This new set is:

$$\pi_1 = \frac{\tau}{\rho U^2} \qquad \pi_4 = \frac{\mu}{\rho U z_o} \qquad \pi_5 = \frac{D}{z_o}$$

In fact, regardless of which set of primary variables were chosen, we can always arrive at the same set of Pi groups via this Pi-group manipulation process.

You might ask which set of Pi groups is the "correct" set. They are all equally valid, although some Pi groups have become more popular than others in the literature. Our pipe example is a case in point. We can recognize π_1 as identical to the definition of *drag coefficient*, C_D, while π_2 is just the inverse of the *Reynolds number*, Re = $\rho U D / \mu$. The π_3 group is called the *relative roughness*.

This is the end of the formal cookbook procedure for Buckingham Pi Theory. Of course, it is really only the second step of the overall similarity procedure. The next step would be to perform the necessary experiments to discover the relationships between the Pi groups. An example of laboratory pipe flow data is shown in Fig 9.2.

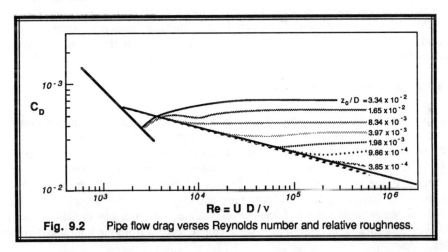

Fig. 9.2   Pipe flow drag verses Reynolds number and relative roughness.

Discussion: Several very important facts can be learned from this data. First, the stress decreases as the Reynolds number increases, until a critical Reynolds number of about 2100 is reached. This critical Reynolds number marks the transition from laminar to turbulent flow. At lower Reynolds number (laminar flow), the stress is NOT dependent on the relative roughness. As suggested in Section 9.1, the data is telling us that pipe roughness is not relevant for laminar flow.

Second, at Reynolds numbers just larger than critical, the stress increases again. Third, as Reynold's number increases further, the stress again decreases with Reynolds number, independent of the pipe roughness.   Fourth, this roughness independence fails when some **roughness Reynolds number** (given by π_4) is reached.   Fifth, at even larger Reynolds number, the stress is a constant depending only on relative roughness and not on the Reynolds number itself.

This last observation is of important consequence for the atmosphere. As previously discussed, the Reynolds number for the atmosphere is very large, on the order of 10^6 to 10^8, even within the boundary layer. Fig 9.2 shows us that **large Reynolds number flow** is independent of the Reynolds number!   Hence, we can usually ignore molecular viscosity and the associated Reynolds number in descriptions of the boundary layer. However, for the very smallest size eddies and in the very thin microlayer near the surface, molecular viscosity continues to be important for TKE dissipation and transport across the surface, respectively.

9.3   Scaling Variables

9.3.1 Choice of Key Variables

Within the constraints of Buckingham Pi theory, there is a wide variety of variables that could be chosen as the key variables. Usually, it is better to pick variables that represent forcings on the boundary layer, or variables that reflect aspects of the non-steady condition of the boundary layer. For example, most surface fluxes represent forcings that are (partially) controlled by external factors. The depth of the boundary layer, as mentioned before, is one important non-steady condition of the boundary layer.

9.3.2 Lists of Scaling Variables

Experience has shown that some key variables frequently appear in common classes of similarity problems, and hence are known as **scaling variables** for that class. As you might guess, a large variety of scaling variables have been suggested over the years (see Table 9-1). In any dimensional analysis problem, you must select only those scales appropriate to the situation. Recommendations for relevant scales were given in Figs. 5.26 and 5.27. Usually, you should pick only one length scale, one velocity scale, and if needed one temperature scale and one humidity scale to be your key variables. No time scale is usually picked, because a time scale can be formed from the length and velocity scales.

9.3.3 Combining Variables to Make New Scales

Some variables always appear grouped in the same arrangements, allowing us to define new scaling variables based on the combination of variables. For example, we have already encountered the friction velocity and other scales in Section 2.10 for the class of problems relating to the surface layer. In Section 4.2 we discussed the convective velocity and other scales related to the class of mixed-layer problems.

Table 9-1.   Summary of boundary layer scales.

Length:

z	= height above the surface
h or z_i	= depth of the boundary layer (or mixed layer)
H	= SBL integral length scale = heat-flux-history scale
L	$= -[\overline{u'w'}_s^2 + \overline{v'w'}_s^2]^{3/4}/[k\cdot(g/\overline{\theta}_v)\cdot(\overline{w'\theta_v'}_s)] =$ Obukhov length
L_L	$= -[\overline{u'w'}^2 + \overline{v'w'}^2]^{3/4}/[k\cdot(g/\overline{\theta}_v)\cdot(\overline{w'\theta_v'})] =$ local Obukhov length
h_e	$= u_*/f_c =$ Ekman layer depth
λ_{max}	= Wavelength corresponding to peak in turbulence spectrum
H	= height of obstacle
W	= width of obstacle
z_o	= aerodynamic roughness length
Z_s	= scale of surface features or roughness

Velocity:

u_*	$= [\overline{u'w'}_s^2 + \overline{v'w'}_s^2]^{1/4} =$ friction velocity
w_*	$= [(g/\overline{\theta}_v)\cdot\overline{w'\theta_v'}_s\cdot z_i]^{1/3} =$ convective velocity scale
w_{Lf}	$= [(g/\overline{\theta}_v)\cdot\overline{w'\theta_v'}\cdot z]^{1/3} =$ local free convection velocity scale
u_L	$= [\overline{u'w'}^2 + \overline{v'w'}^2]^{1/4} =$ local (friction) velocity scale
V_B	$= [(g/\Delta\overline{\theta}_{v\,s})\cdot\overline{w'\theta_v'}\cdot H]^{1/3} =$ SBL buoyancy velocity scale
V_M	$= (Z_s/\rho)^{1/2}[(\partial P/\partial x)^2 + (\partial P/\partial y)^2]_s^{1/4} =$ mechanical forcing scale
u_*^{ML}	$= u_*^2/w_* =$ convective stress scale velocity
$\overline{G}$ or $\overline{U}_g$	= geostrophic wind speed
$\overline{G}_s$	= geostrophic wind speed at the surface
$\overline{G}_{z_i}$	= geostrophic wind at the top of the boundary layer
$<\overline{G}>$	= geostrophic wind speed averaged over the boundary layer
$\overline{U}$ or $\overline{M}$	= wind speed
$\overline{M}_s$	= wind speed at the surface

$\overline{M}_{z_i}$ = wind speed at the top of the boundary layer

$\langle\overline{M}\rangle$ = wind speed averaged over the boundary layer

s_u or σ_u = standard deviation of U-wind

$(TKE)^{1/2}$ or $\overline{e}^{1/2}$ = square root of turbulence kinetic energy

$(k\,z\,\varepsilon)^{2/3}$ = dissipation velocity scale in the surface layer

Time: $1/f_c$ = inertial period, where f_c is the Coriolis parameter

$1/N_{BV}$ = buoyant period, where N_{BV} is the Brunt-Väisälä frequency

$1/f_{max}$ = eddy period, where f_{max} is the frequency at the peak in the turbulence spectrum

t_*^{ML} $= z_i / w_*$ = convective (ML) time scale

t_*^{SL} $= z / u_*$ = surface-layer time scale

$x/\overline{U}$ = time required for wind to move distance x

Temperature: θ_*^{ML} $= \overline{w'\theta_v'}_s / w_*$ = convective (ML) temperature scale

θ_*^{SL} $= -\overline{w'\theta_v'}_s / u_*$ = surface-layer temperature scale

θ_{Lf} $= \overline{w'\theta_v'} / w_{Lf}$ = local free-convection temperature scale

θ_L $= -\overline{w'\theta_v'} / u_L$ = local temperature scale

θ_* $= \overline{w'\theta_v'}_s$ /(any other velocity scale)

$\langle\overline{\theta_v}\rangle$ = mixed-layer average of $\overline{\theta_v}$

$\Delta\theta_s$ $= \langle\overline{\theta_v}\rangle - \overline{\theta}_{vs}$ = SBL surface cooling (inversion strength)

Moisture: q_*^{ML} $= \overline{w'q'}_s / w_*$ = convective (ML) humidity scale

q_*^{SL} $= -\overline{w'q'}_s / u_*$ = surface-layer humidity scale

q_{Lf} $= \overline{w'q'} / w_{Lf}$ = local free-convection humidity scale

$$q_L \qquad = -\overline{w'q'}/u_L = \text{local humidity scale}$$

$$q_* \qquad = \overline{w'q'}_s /(\text{any other velocity scale})$$

As an example of the power of combining variables into new scales, we might expect z_i, $\overline{w'\theta_v'}_s$, and maybe $g/\overline{\theta_v}$ to always be important in convection, because heating, convection, and buoyancy cause thermals to rise to the top of the mixed layer. In fact, these key variables are often grouped as $[(g/\overline{\theta_v})\cdot\overline{w'\theta_v'}_s\cdot z_i]$ during a dimensional analysis. We recognize this group to be the basis for the definition of the convective velocity scale w_* ; namely, $w_*^3 = [(g/\overline{\theta_v})\cdot\overline{w'\theta_v'}_s\cdot z_i]$ as defined in Section 4.2.

By using such a scaling variable in dimensional analysis, we can often reduce the total number of variables in our problem and greatly simplify the analysis. For example, if we wish to find a relationship for $\overline{w'^2}$ as a function of height in the ML, we might choose $\overline{w'^2}$, z, $\overline{w'\theta_v'}_s$, z_i, and $g/\overline{\theta_v}$ as the relevant variables for the first step in the analysis.

Alternately, by using the combined scaling variables, we would choose $\overline{w'^2}$, w_*, z, and z_i instead. With this last set of variables, we can easily identify the dimensionless groups by inspection: $[\overline{w'^2}/w_*^2]$ and $[z/z_i]$.

9.3.4 Classes of Similarity Scales

The most common classes of similarity scaling are *Monin-Obukhov similarity, mixed-layer similarity, local similarity, local free convection,* and *Rossby-number similarity*. When dealing with one of these well-defined classes of problems, it is appropriate to use the associated scaling variables as the key variables in a dimensional analysis.

Monin-Obukhov Similarity. This class is usually applied to the surface layer (Monin and Obukhov, 1954; Wyngaard, 1973; Sorbjan, 1986), and hence is sometimes called *surface-layer similarity*. Earlier we defined the surface layer as that part of the boundary layer where the fluxes vary by less than 10% of their magnitude with height. To a first order approximation, this layer is a *constant flux layer*. We can thus simplify our description of the surface layer by utilizing the flux at just one height — usually the surface.

Monin-Obukhov similarity works only when the winds are not calm, and u_* not zero. Relevant scales based on these surface fluxes and their typical orders of magnitude are listed here:

L	Order (1 m to 200 m)
z_0	Order (1 mm to 1 m)
u_*	Order (0.05 to 0.3 m/s)
θ_*^{SL}	Order (0.1 to 2.0 °C)
q_*^{SL}	Order (0.1 to 5 g_{water}/kg_{air})

Scales for pollutant concentration can be patterned after the humidity scale. Lists of Monin-Obukhov similarity relationships are tabulated in Sections 9.4 to 9.6, and a more detailed analysis of the log-wind profile in the surface layer is given in Section 9.7.

Mixed-Layer Similarity. This class is applied to mixed layers that are in a state of free convection (Deardorff, 1972; Deardorff, et al., 1980; Sorbjan, 1986), assuming calm or light winds. Free convection conditions can occur during cold air advection over a warmer surface, or with solar heating of the land during the daytime in light wind conditions. The relevant scales and typical orders of magnitude for the mixed layer are:

z_i	Order (0.2 to 2 km)
w_*	Order (2 m/s)
θ_*^{ML}	Order (0.1 K)
q_*^{ML}	Order (0.1 g/m^3)
u_*^{ML}	Order (0.02 m/s)

Other scales, such as for pollutant concentration, can be defined in analogy to the moisture scale. More details are discussed in Section 9.6.

Local Similarity. For statically stable boundary layers, this class recognizes that turbulence in the mid and upper SBL may not be in equilibrium with the surface fluxes (Wyngaard, 1973; Nieuwstadt, 1984; Sorbjan, 1987). Hence, local fluxes, shears and stability are more important than surface fluxes. The relevant scales are:

L_L	Order (0 to 50 m)
u_L	Order (0 to 0.3 m/s)
θ_L	Order (0 to 2.0 °C)
q_L	Order (0 to 5 g_{water}/kg_{air})

Dimensionless groups formed with the above scales are not a function of height above ground; hence, this scaling is also called *z-less* scaling. Although the dimensionless groups are independent of height, the individual variables that make up these groups (including the scaling variables above) vary significantly with height.

Local Free Convection Similarity. In statically unstable surface layers, buoyancy is the driving force behind the turbulence. However, turbulence in the surface layer might "feel" the influence of the ground more than the influence of the capping inversion (Wyngaard, et al., 1971; Tennekes, 1973; Wyngaard, 1973; Caughey and Palmer, 1979; Sorbjan, 1986). As a result, z_i is not a relevant parameter, but z is. The list of relevant scales becomes:

z	Order (0 to 50 m)
w_{Lf}	Order (0 to 0.5 m/s)
θ_{Lf}	Order (0 to 2.0 °C)
q_{Lf}	Order (0 to 5 g_{water}/kg_{air})

This similarity approach is useful for surface layers in conditions of calm mean winds. In that case the Obukhov length is zero, and is not an appropriate measure of the amount of turbulence being generated. Thus, Monin-Obukhov similarity will not work.

Rossby-number Similarity. In some situations such as large-scale modeling, it is desirable to relate surface fluxes to external forcings. In this regard, it is necessary to *match* the wind and temperature profiles higher in the boundary layer with those in the surface layer (Tennekes, 1973; Yamada, 1976). As a result, relevant scales include both surface scales (z_o, L, u_*, θ_*^{SL}, q_*^{SL}) and scales appropriate to the upper boundary layer (h_2, G_2, θ_2, q_2).

In early work (see review by Tennekes, 1982), it was suggested that the h_2 scale be described by G/f_c. When this "outer" scale is combined with the "inner" scale, z_o, the result is the *surface Rossby number*, $G/(f_c z_o)$. Unfortunately, this approach was not completely successful, resulting in a search for better outer scales. Although still not perfect, the following outer scales are now more widely accepted (z_i, <G>, $\Delta\theta_s$, Δq_s). Details of this approach are reviewed in Section 9.8.

9.3.5 Similarity Relationships

Many of the figures in Chapters 4 and 5 are already presented in similarity form using the above scales, with one dimensionless group plotted as a function of other dimensionless groups. These curves represent graphical representations of similarity relationships, and can be used directly to estimate the values of variables at any height, within the limitations of the data (e.g., free convection within the mixed layer). Similarity graphs are presented for unstable, neutral and stable boundary layers in those chapters.

Analytical equations have been suggested in the literature for some of the graphical similarity results. In some of these cases, curves are fit to only a subdomain of the turbulence, or apply under only certain conditions. Many of the analytical similarity relationships that have proved successful are summarized in Sections 9.4 through 9.6.

We will start in Section 9.4 by listing similarity relationships for stable (nocturnal) boundary layers, and then proceed to the neutral and unstable (convective) boundary layers in Sections 9.5 and 9.6, respectively. Subsections for mean variables, fluxes, variances, and other miscellaneous variables will be presented. Each of these subsections will include relationships for both the boundary layer, and the surface layer.

Occasionally, different investigators have suggested different values for the regression coefficients (i.e., the universal constants). For these situations a variety of values are listed, separated by commas.

Finally, an example will be presented in each Section to demonstrate the application of a similarity relationship.

9.4 Stable Boundary Layer Similarity Relationship Lists

In the stable surface layer, Monin-Obukhov similarity has allowed us describe the vertical profiles of some variables as a function of the dimensionless group z/L (Businger, et al., 1971; Wyngaard, et al., 1971; Caughey, et al., 1979). Higher in the SBL, z-less scaling is more appropriate (Wyngaard, 1973; Nieuwstadt, 1984; Lascer and Arya, 1986; Sorbjan, 1986, 1987). Sorbjan (1987) has shown how it is possible to develop SBL similarity expressions from the corresponding surface layer expressions.

All of these relationships assume that the SBL is continuously turbulent in time and space, with no gaps or patches of nonturbulent air. Since real SBLs can have sporadic, patchy turbulence, we must recognize the limitations of the expressions below.

9.4.1 Mean Variables and Their Gradients

In many cases it is difficult to describe the profiles of mean variables because of the influence of initial and boundary conditions. However, the profiles often exhibit a common shape, allowing similarity expressions to be derived for the gradients of mean variables.

Boundary-Layer Relationships:

$$\frac{k L_L}{u_L}\left[\left(\frac{\partial \overline{U}}{\partial z}\right)^2 + \left(\frac{\partial \overline{V}}{\partial z}\right)^2\right]^{1/2} = 2.5,\quad 4.7,\quad 5.22 \qquad (9.4.1a)$$

$$\frac{k L_L}{\theta_L}\frac{\partial \overline{\theta}}{\partial z} = 4.23,\quad 4.7,\quad 5.0 \qquad (9.4.1b)$$

Surface-Layer Relationships:

$$\phi_M = \frac{k z}{u_*}\left[\left(\frac{\partial \overline{U}}{\partial z}\right)^2 + \left(\frac{\partial \overline{V}}{\partial z}\right)^2\right]^{1/2} = 1 + 4.7\frac{z}{L} \qquad (9.4.1c)$$

$$\phi_H = \frac{kz}{\theta_*^{SL}} \frac{\overline{\partial \theta}}{\partial z} = 0.74 + 4.7 \frac{z}{L} \qquad (9.4.1d)$$

These last two expressions are known as **Businger-Dyer** **flux-profile** **relationships**, because they relate mean profile gradients to the fluxes in u_* and θ_*^{SL}. These relationships are plotted in Fig 9.9, and are discussed in more detail in Section 9.7.

9.4.2 Fluxes

Boundary-Layer Relationships:

$$\frac{\overline{u'v'}}{u_L^2} = -1.0 \qquad (9.4.2a)$$

$$\frac{\left[\overline{u'\theta'}^2 + \overline{v'\theta'}^2 \right]^{1/2}}{u_L \, \theta_L} = 3.5, \quad 4.0, \quad 5.0 \qquad (9.4.2b)$$

The above expressions are z-less. In order to make these more useful, some parameterizations have been proposed for the fluxes upon which u_L and q_L are based:

$$\frac{-\overline{u'w'}}{u_*^2} = 1 - \left(\frac{z}{h} \right)^{0.7} \qquad (9.4.2c)$$

$$\frac{-\overline{u'w'}}{u_*^2} = \left[1 - \frac{z}{h} \right]^{1 \text{ or } 2} \qquad (9.4.2d)$$

$$\frac{\overline{w'\theta'}}{\overline{w'\theta'}_s} = \left[1 - \left(\frac{z}{h} \right)^{0.5} \right]^{1.5} \qquad (9.4.2e)$$

$$\frac{\overline{w'\theta'}}{\overline{w'\theta'}_s} = \left[1 - \frac{z}{h} \right]^{1 \text{ to } 3} \qquad (9.4.2f)$$

$$\frac{-\overline{u'\theta'}}{\overline{w'\theta'}_s} = 3.2 \left[1 - \left(\frac{z}{h}\right)^{0.7} \right]^{1.7} \tag{9.4.2g}$$

These last five expressions are obviously not z-less, and are likely to be valid for only a small subset of real SBLs.

Surface-Layer Relationships:

$$\frac{\overline{u'v'}}{u_*^2} = \text{constant} \tag{9.4.2h}$$

$$\frac{\left(\overline{u'\theta'}^2 + \overline{v'\theta'}^2\right)^{1/2}}{u_* \, \theta_*^{SL}} = 4 \tag{9.4.2i}$$

9.4.3 Variances

Boundary-Layer Relationships:

$$\frac{\left[\overline{u'^2} + \overline{v'^2}\right]^{1/4}}{u_L} = 2.6, \qquad 3.1, \qquad 4 \tag{9.4.3a}$$

$$\frac{\left[\overline{w'^2}\right]^{1/2}}{u_L} = 1.5, \qquad 1.6, \qquad 2.0 \tag{9.4.3b}$$

$$\frac{\left[\overline{\theta'^2}\right]^{1/2}}{\theta_L} = 2.4, \qquad 3.5, \qquad 4.0 \tag{9.4.3c}$$

$$\frac{\overline{e}}{u_L^2} = \text{constant} \tag{9.4.3d}$$

As before, the above z-less expressions are difficult to use unless the local scaling variables are known as a function of height. As an alternative, the following height-dependent relationships have been suggested:

$$\frac{\overline{u'^2}}{u_*^2} = 6 \left[1 - \left(\frac{z}{h} \right)^{1/2} \right] \tag{9.4.3e}$$

$$\frac{\overline{w'^2}}{u_*^2} = 2.5 \left[1 - \left(\frac{z}{h} \right)^{0.6} \right] \tag{9.4.3f}$$

$$\frac{\overline{\theta'^2}}{\theta_*^{SL\,2}} = 6.0 \left[1 - \left(\frac{z}{h} \right)^{0.4} \right]^{1.5} \tag{9.4.3g}$$

Surface-Layer Relationships:

$$\frac{\overline{u'^2} + \overline{v'^2}}{u_*^2} = 8.5 \tag{9.4.3h}$$

$$\frac{\overline{w'^2}}{u_*^2} = 2.5 \tag{9.4.3i}$$

$$\frac{\overline{\theta'^2}}{\theta_*^{SL\,2}} = 4.0 \tag{9.4.3j}$$

$$\frac{\overline{e}}{u_*^2} = \text{constant} \tag{9.4.3k}$$

9.4.4 Miscellaneous

Boundary-Layer Relationships:

$$\frac{k\,L_L\,\varepsilon}{u_L^3} = 3.7 \tag{9.4.4a}$$

Surface-Layer Relationships:

$$\frac{k\, L\, \varepsilon}{u_*^3} = 3.7 \qquad\qquad (9.4.4b)$$

9.4.5   Example

Problem: Find the local value of the heat flux at a height of $z = 50$ m in a SBL, where the local lapse rate is 0.025 K/m and the local scaling velocity is $u_L = 0.1$ m/s.

Solution: Equation (9.4.1b) can be manipulated, using the definition for L_L, to read

$$\overline{w'\theta'}^2 = \frac{u_L^4}{(g/\theta) \cdot 4.7} \, \frac{\partial \overline{\theta}}{\partial z}$$

where the value of the universal constant was taken as 4.7. Assuming that $(g/\theta) = 0.0333$ m·s^{-2}·K^{-1}, we find that $\overline{w'\theta'} = 3.4 \times 10^{-3}$ K m/s.

Discussion: Such a small value of heat flux is typical of the upper SBL.

9.5 Neutral Boundary Layer Similarity Relationship Lists

Although boundary layers are rarely exactly neutral, there are situations such as strong winds and overcast skies where the boundary layer is approximately neutral. In a neutral boundary layer the only (or dominant) TKE generation mechanism is mechanical, associated with wind shear and surface stress. Thus, we expect u_* to be important. Rarely is the Obukhov length used, because it is infinite in statically neutral conditions.

Some investigators (Sorbjan, 1986) have suggested similarity parameterizations based on pseudo-local scaling (using scales u_L, θ_L, but including z instead of L_L),while others (Nicholls and Readings, 1979; Grant, 1986) have applied surface-layer similarity relationships higher in the boundary layer (using scales u_*, θ_*, and z,). Occasionally, it is assumed that a well defined top of the turbulent boundary layer can be identified, allowing z_i to be used.

Surface layer parameterizations are often based on the limiting cases of diabatic similarity relationships, for the case where z/L goes to zero (Wyngaard and Coté, 1971; Merry and Panofsky, 1976; Panofsky et al, 1977; Nicholls and Readings, 1979; Smith, 1980; Grant, 1986; and Sorbjan, 1986).

9.5.1 Mean Variables and Their Gradients

Boundary-Layer Parameterizations:

$$\frac{k z}{u_L} \frac{\partial \overline{U}}{\partial z} = 1 \tag{9.5.1a}$$

$$\frac{k z}{\theta_L} \frac{\partial \overline{\theta}}{\partial z} = 0.74 \tag{9.5.1b}$$

Surface-Layer Parameterizations:

$$\frac{k z}{u_*} \frac{\partial \overline{U}}{\partial z} = 1 \tag{9.5.1c}$$

$$\frac{k z}{\theta_*^{SL}} \frac{\partial \overline{\theta}}{\partial z} = 0.74 \tag{9.5.1d}$$

9.5.2 Fluxes (boundary layer)

$$\frac{\left[\overline{u'\theta'}^2 + \overline{v'\theta'}^2 \right]^{1/2}}{- \overline{w'\theta'}} = 4 \tag{9.5.2a}$$

$$\frac{\overline{u'w'}}{\overline{u'w'}_s} = 1 - \frac{z}{z_i} \tag{9.5.2b}$$

9.5.3 Variances

Boundary-Layer Relationships:

$$\frac{\overline{u'^2} + \overline{v'^2}}{u_L^2} = 8.5 \tag{9.5.3a}$$

$$\frac{\overline{w'^2}}{u_L^2} = 2.5 \tag{9.5.3b}$$

$$\frac{\overline{\theta'^2}}{\theta_L^2} = 4 \tag{9.5.3c}$$

$$\frac{\overline{u'^2}}{u_*^2} = 6\left(1 - \frac{z}{z_i}\right)^2 + \frac{z}{z_i}\frac{\overline{u'^2}_{top}}{u_*^2} \tag{9.5.3d}$$

$$\frac{\overline{v'^2}}{u_*^2} = 3\left(1 - \frac{z}{z_i}\right)^2 + \frac{z}{z_i}\frac{\overline{v'^2}_{top}}{u_*^2} \tag{9.5.3e}$$

$$\frac{\overline{w'^2}}{u_*^2} = \left(1 - \frac{z}{z_i}\right)^{1/2} \tag{9.5.3f}$$

Equations (9.5.3 d and e) include an additional ratio for the variance at the top of the boundary layer (assuming a well-defined top) normalized by the surface stress. Although this ratio is expected to vary from situation to situation, during the KONTUR experiment (Grant, 1986) it was found to equal 2.0 for both equations.

Surface-Layer Relationships:

$$\frac{\overline{u'^2}}{u_*^2} = 6.1, \quad 6.2, \quad 6.5 \tag{9.5.3g}$$

$$\frac{\overline{v'^2}}{u_*^2} = 2.9, \quad 3.0, \quad 4.3, \quad 6.1 \tag{9.5.3h}$$

$$\frac{\overline{u'^2} + \overline{v'^2}}{u_*^2} = 8.5 \tag{9.5.3i}$$

$$\frac{\overline{w'^2}}{u_*^2} = 1.0, \quad 1.7, \quad 2.5 \tag{9.5.3j}$$

$$\frac{\overline{\theta'^2}}{\theta_*^{SL\,2}} = 4 \tag{9.5.3k}$$

9.5.4 Miscellaneous

Boundary-Layer Relationships:

$$\frac{k z \varepsilon}{u_L^3} = 1 \tag{9.5.4a}$$

Surface-Layer Relationships:

$$\frac{k z \varepsilon}{u_*^3} = 1 \tag{9.5.4b}$$

$$\frac{k z \varepsilon}{u_*^3} = \left(1 + 0.5 \left| \frac{z}{L} \right|^{2/3} \right)^{3/2} \tag{9.5.4c}$$

$$\frac{\overline{w'e}}{u_*^3} = -2.3 \frac{z}{L} \tag{9.5.4d}$$

These last two relationships are designed for near neutral situations.

9.5.5   Example

Problem: In a neutral surface layer with $u_* = 0.2$ m/s, find the TKE dissipation rate as a function of height.

Solution: Rearranging equation (9.5.4b), we can see that the dissipation rate decreases inversely with height:

$$\varepsilon = \frac{u_*^3}{k z} = \frac{0.04}{z} \ \text{m}^2\text{s}^{-3} \quad \text{for } z \text{ in meters.}$$

Discussion: As verified by the TKE budget figures in Chapter 5, the dissipation rate is indeed very large near the ground. However, the equation above suggests that the dissipation rate is infinite at the surface, which is clearly unrealistic.

9.6 Convective Boundary Layer Similarity Relationship Lists

When turbulence in a mixed layer is driven by buoyancy and capped at a well defined height, it is obvious that w_* and z_i are important scales for all variables. In addition, when considering heat, moisture, and momentum fluxes we should include the θ_*^{ML}, q_*^{ML}, and u_*^{ML} scales, respectively. Many investigators have examined the convective mixed layer and unstable surface layer (Businger, et.al., 1971; Lenschow, 1974; Lenschow, et al., 1980; Caughey and Readings, 1974, 1975; Caughey and Palmer, 1979; Kaimal, et al., 1976; Smedman and Högström, 1983; Wyngaard, et.al., 1971; LeMone and Pennell, 1976; Brost, et.al., 1982; Berkowicz and Prahm, 1984; Webb, 1982; Zhou, et.al., 1985; and Sorbjan, 1986).

In the surface layer, local free-convective or Monin-Obukhov similarity can be applied depending on the relative importance of surface heating and stress (Wyngaard, et.al., 1971; Berkowicz and Prahm, 1984; Sorbjan, 1986).

9.6.1 Mean Variables and Their Gradients

Mixed-Layer Relationships:

$$\frac{z_i}{w_*} \frac{\partial \overline{U}}{\partial z} = 0 \qquad \text{for } 0.1\, z_i \leq z \leq 0.9\, z_i \qquad (9.6.1a)$$

$$\frac{z_i}{\theta_*^{ML}} \frac{\partial \overline{\theta}}{\partial z} = 0, \quad 1.4 \qquad \text{for } 0.1\, z_i \leq z \leq 0.9\, z_i \qquad (9.6.1b)$$

$$\frac{z_i}{q_*^{ML}} \frac{\partial \overline{q}}{\partial z} = -5 \qquad \text{for } 0.1\, z_i \leq z \leq 0.9\, z_i \qquad (9.6.1c)$$

Investigators (Mahrt and André, 1983) have become more aware that forcings can occur at the top of the mixed layer that are quasi-independent of the forcings at the ground. Entrainment can introduce fluxes at the top of the mixed layer, and shear across the entrainment zone can generate additional turbulence that does not scale with w_*.

Wyngaard and Brost (1984) and Moeng and Wyngaard (1984) have suggested a conceptual model called *top down - bottom up* diffusion, which models the contributions of mixing down from the top of the ML separately from mixing up from the ground. One result of this approach is a similarity relationship for gradients of mean variables within the middle 80% of the ML. The relationship for pollutant concentration is given below, but it can be applied just as easily to temperature, moisture, or wind gradients. More details of this method are discussed in Chapter 11.

$$\frac{\partial \overline{C}}{\partial z} = -0.4 \frac{\overline{w'c'}_s}{w_* z_i} \left(\frac{z}{z_i}\right)^{-3/2} + \frac{\overline{w'c'}_{top}}{w_* z_i} \left(1 - \frac{z}{z_i}\right)^{-3/2} \qquad (9.6.1d)$$

The gradient on the left of the equal sign is not normalized into a dimensionless group, because there is no single pollutant scale. Pollutant fluxes at both the top and bottom of the ML are relevant, and would yield two different pollutant scales.

Surface-Layer Relationships:

$$\phi_M = \frac{k z}{u_*} \frac{\partial \overline{U}}{\partial z} = \left(1 - 15 \frac{z}{L}\right)^{-1/4} \qquad (9.6.1e)$$

$$\phi_H = \frac{k z}{\theta_*^{SL}} \frac{\partial \overline{\theta}}{\partial z} = 0.74 \left(1 - 9 \frac{z}{L}\right)^{-1/2} \qquad (9.6.1f)$$

The above two flux-profile relationships apply when surface stress is nonzero, and have been plotted in Fig 5.24. Alternative relationships that have been proposed for zero stress situations are:

$$\frac{z_i}{w_*} \frac{\partial \overline{U}}{\partial z} = 1.2 \frac{\left[1 - \alpha (z/z_i)\right]^{2/3}}{\left[(z/z_i)^{4/3} (-\mu_i)^{2/3}\right]} \qquad (9.6.1g)$$

$$\frac{z_i}{\theta_*^{ML}} \frac{\partial \overline{\theta}}{\partial z} = \frac{\left[1 - \alpha (z/z_i)\right]^{2/3}}{(z/z_i)^{4/3}} \qquad (9.6.1h)$$

where α is a universal constant in the range 1.2-1.5, and $\mu_i = z_i / L$.

9.6.2 Fluxes for Both the Mixed Layer and Surface Layer

Because the flux profiles are linear with height in the ML, we propose here the following similarity relationships:

$$\frac{\overline{u'w'}}{\overline{u'w'}_s} = 1 - \frac{z}{z_i} + \frac{\overline{u'w'}_{top}}{\overline{u'w'}_s} \frac{z}{z_i} \qquad (9.6.2a)$$

$$\frac{\overline{w'\theta'}}{\overline{w'\theta'}_s} = 1 - \frac{z}{z_i} + \frac{\overline{w'\theta'}_{top}}{\overline{w'\theta'}_s} \frac{z}{z_i} \qquad (9.6.2b)$$

$$\frac{\overline{w'q'}}{\overline{w'q'}_s} = 1 - \frac{z}{z_i} + \frac{\overline{w'q'}_{top}}{\overline{w'q'}_s} \frac{z}{z_i} \tag{9.6.2c}$$

In the absence of shear or other independent forcings in the entrainment zone, the flux at the top of the mixed layer can be related to the flux at the bottom, resulting in the following simplified similarity relationships:

$$\frac{\overline{w'\theta'}}{\overline{w'\theta'}_s} = 1 - \alpha \frac{z}{z_i} \qquad \text{where } \alpha = 1.2 \text{ to } 1.5 \tag{9.6.2d}$$

$$\frac{\overline{u'\theta'}}{\overline{w'\theta'}_s} = 0.5 \left(1 - 2.2 \frac{z}{z_i} \right) \tag{9.6.2e}$$

9.6.3 Variances

Mixed-Layer Relationships: All of the following expressions fail in and near the entrainment zone at the top of the mixed layer, where locally generated turbulence and buoyancy waves can contribute to the variance.

$$\frac{\overline{u'^2}}{u_*^{ML\,2}} = \text{constant} \tag{9.6.3a}$$

$$\frac{\overline{v'^2}}{u_*^{ML\,2}} = \text{constant} \tag{9.6.3b}$$

$$\frac{\overline{w'^2}}{w_*^2} = 1.8 \left(\frac{z}{z_i} \right)^{2/3} \cdot \left(1 - 0.8 \frac{z}{z_i} \right)^2 \tag{9.6.3c}$$

$$\frac{\overline{\theta'^2}}{\theta_*^{ML\,2}} = 1.8 \left(\frac{z}{z_i} \right)^{-2/3} \tag{9.6.3d}$$

$$\frac{\overline{q'^2}}{q_*^{ML^2}} = 1.8 \left(\frac{z}{z_i}\right)^{-2/3} \tag{9.6.3e}$$

One might expect that relationships similar to (9.5.3 d & e) would work well for the mixed layer, assuming that the variance at the top of the mixed layer could be parameterized. This approach has not been explored in the literature.

Surface-Layer Relationships: When stress is nonzero, the following expressions are appropriate:

$$\frac{\left(\overline{w'^2}\right)^{1/2}}{u_*} = 1.9 \cdot \left(-\frac{z}{L}\right)^{1/3} \tag{9.6.3f}$$

$$\frac{\left(\overline{\theta'^2}\right)^{1/2}}{\theta_*^{SL}} = -0.95 \cdot \left(-\frac{z}{L}\right)^{-1/3} \tag{9.6.3g}$$

In cases of local free convection (calm winds) the following expressions are useful:

$$\frac{\overline{\theta'^2}}{\theta_{Lf}^2} = 1.85, \quad 2 \tag{9.6.3h}$$

$$\frac{\overline{w'^2}}{w_{Lf}^2} = 1.21, \quad 1.6 \tag{9.6.3i}$$

The following expressions combine mixed layer scaling with local free convection:

$$\frac{\overline{w'^2}}{w_*^{ML^2}} = 1.6 \left(\frac{z}{z_i}\right)^{2/3} \left(1 - 1.2\frac{z}{z_i}\right)^{2/3} \tag{9.6.3j}$$

$$\frac{\overline{\theta'^2}}{\theta_*^{ML^2}} = 2 \frac{(1 - 1.2 z/z_i)^{4/3}}{(z/z_i)^{2/3}} \tag{9.6.3k}$$

$$\frac{\overline{\theta'^2}}{\theta_*^{ML^2}} = \frac{\overline{q'^2}}{q_*^{ML^2}} = 1.8 \left(\frac{z}{z_i}\right)^{-2/3} \qquad (9.6.31)$$

9.6.4 Miscellaneous

Mixed-Layer Relationships:

$$\frac{\overline{w'^3}}{w_*^3} = 0.8 \frac{z}{z_i} \left(1 - \frac{1.1 z}{z_i}\right) \qquad (9.6.4a)$$

$$\frac{\varepsilon z_i}{w_*^3} = \text{constant with height above the surface layer} \qquad (9.6.4b)$$

$$\frac{\overline{w'e}}{w_*^3} = 0.8 \frac{z}{z_i} \left(1 - 0.9 \frac{z}{z_i}\right)^2 \qquad (9.6.4c)$$

$$\frac{z_i \, \overline{w'\theta'}}{w_* \, \theta_*^{ML^2}} \left(\frac{\partial \overline{\theta}}{\partial z}\right) = 1.4 - 2 \frac{z}{z_i} \qquad (9.6.4d)$$

$$\frac{\overline{w'\theta'^2}}{w_* \, \theta_*^{ML^2}} = 3.1 \left(1 - \frac{z}{z_i}\right)^3 \qquad (9.6.4e)$$

$$\frac{\overline{w'^2 \theta'}}{w_*^2 \, \theta_*^{ML}} = 0.5 \left(1 - 1.2 \frac{z}{z_i}\right) \qquad (9.6.4f)$$

$$\frac{z_i \, \overline{w'^2}}{w_*^2 \, \theta_*^{ML}} \left(\frac{\partial \overline{\theta}}{\partial z}\right) = 2.5 \left(\frac{z}{z_i}\right)^{2/3} \left(1 - 0.8 \frac{z}{z_i}\right)^2 \qquad (9.6.4g)$$

$$\frac{z_i \, \overline{w'q'}}{w_* \, (q_*^{ML})^2} \left(\frac{\partial \overline{q}}{\partial z}\right) = -5 + 2.5 \frac{z}{z_i} \qquad (9.6.4h)$$

$$\frac{\overline{w'^2 q'}}{w_*^2 q_*}^{ML} = 0.6 \left(1 - 0.4 \frac{z}{z_i} \right) \tag{9.6.4i}$$

$$\frac{z_i \overline{w'^2}}{w_*^2 q_*}^{ML} \left(\frac{\partial \bar{q}}{\partial z} \right) = -9 \left(\frac{z}{z_i} \right)^{2/3} \left(1 - 2 \frac{z}{z_i} \right)^2 \tag{9.6.4j}$$

$$\frac{\overline{\theta' q'}}{\theta_*^{ML} q_*^{ML}} = 1.8 \left(\frac{z}{z_i} \right)^{-2/3} \left(1 - 2 \frac{z}{z_i} \right) \tag{9.6.4k}$$

This last equation has also been applied to the surface layer.

Surface-Layer Relationships: The first equations apply to surface layers with nonzero stress.

$$\frac{\overline{w'e}}{u_*^3} = -2.3 \frac{z}{L} \tag{9.6.4l}$$

$$\frac{\overline{w'^2 \theta'}}{u_*^2 \theta_*^{SL}} = -1.3 \left(-\frac{z}{L} \right)^{1/3} \tag{9.6.4m}$$

$$\frac{k z \overline{w'^2}}{u_*^3} \left(\frac{\partial \bar{U}}{\partial z} \right) = 1.2 - 0.5 \frac{z}{L} \tag{9.6.4n}$$

$$\frac{k z \overline{w'^2}}{u_*^2 \theta_*^{SL}} \left(\frac{\partial \bar{\theta}}{\partial z} \right) = 0.9 \tag{9.6.4o}$$

The next equations work best when the stress is very small and buoyancy dominates:

$$\frac{\overline{w'^2 \theta'}}{w_*^2 \theta_*^{ML}} = \frac{\overline{w'^2 q'}}{w_*^2 q_*^{ML}} = 0.9 \left(\frac{z}{z_i} \right)^{1/3} \tag{9.6.4p}$$

$$\frac{\overline{w' \theta'^2}}{w_* \theta_*^{ML2}} = \left(-\mu_i \frac{z}{z_i} \right)^{-1/3} \left(1 - 1.2 \frac{z}{z_i} \right)^{5/3} \tag{9.6.4q}$$

$$\frac{\varepsilon_q \, z_i}{w_* \, q_*^{ML^2}} = 0.43 \left(\frac{z}{z_i}\right)^{-4/3} \qquad (9.6.4r)$$

9.6.5   Example

Problem: Use similarity theory to develop an expression for vertical velocity variance, $\overline{w'^2}$, as a function of height given the (synthetic) measurements in Table 9-2.

Table 9-2. Synthetic vertical velocity variance data.

z (m)	$\overline{w'^2}$ (m^2s^{-2})			
	Day 1	Day 2	Day 3	Day 4
1500				0.4
1400				0.6
1300				0.8
1200				1.0
1100				1.1
1000		0.10		1.3
900		0.16		1.4
800		0.22		1.4
700	0.6	0.30		1.5
600	0.9	0.37		1.6
500	1.2	0.39		1.6
400	1.5	0.40		1.5
300	1.6	0.40	0.20	1.4
200	1.5	0.36	0.36	1.2
100	1.2	0.28	0.40	1.0
0	0.8	0.20	0.20	0.8
z_i (m)	750	1000	350	1500
$\overline{w'\theta_v'}_s$ (K m/s)	0.33	0.03	0.09	0.16

Solution: Although this data set exhibits a variety of magnitudes over a range of heights, each of the individual data curves has the same shape (see Fig 9.3a) — a clue that they are created by a common physical process that could possibly be described empirically.

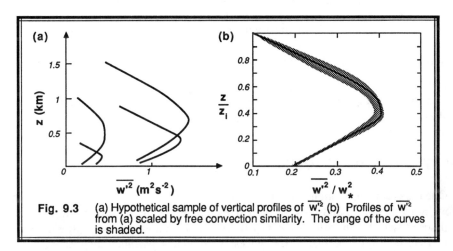

Fig. 9.3   (a) Hypothetical sample of vertical profiles of $\overline{w'^2}$ (b) Profiles of $\overline{w'^2}$ from (a) scaled by free convection similarity. The range of the curves is shaded.

The tabulated data includes mixed layer depth and surface heat flux. From these, we can calculate the length and velocity scales, z_i and w_*, assuming $g/\overline{\theta}_v = 0.0333$ ms^{-2}K^{-1}. Using Buckingham Pi analysis, or by inspection for this simple case, we can create the following dimensionless groups: z/z_i, and $\overline{w'^2}/w_*^2$. When the original data is replotted in this dimensionless framework, lo and behold most of the data points collapse into a single curve, as plotted in Fig 9.3b. Thus, all of the data are similar, allowing us to use similarity theory.

By trial and error using simply power laws, we find that the following equation approximates the shape of the data, and is plotted as the curve in Fig 9.3b:

$$\frac{\overline{w'^2}}{w_*^2} = 1.7 \left(\frac{z}{z_i}\right)^{2/3} \cdot \left(1 - 0.8\frac{z}{z_i}\right)^2$$

This equation is almost identical to (9.6.3c), except for the value of the regression coefficient (i.e., 1.7 vs. 1.8).

Discussion: The hope is that this equation, and the corresponding curve in Fig 9.3b, are "universal"; that is, they should work just as well for other free convection situations. For example, on a different day with $z_i = 1200$ m and $\overline{w'\theta_v'}_s = 0.2$ K m/s, we might wish to know the vertical velocity variance at $z = 500$ m without performing an experiment to measure it ourselves. At that height, $z/z_i = 0.42$, which can be used in the above equation or with Fig 9.3b directly to give us $\overline{w'^2}/w_*^2 = 0.42$. Since $w_* = 2$ m/s based on flux and z_i data given at the start of this example, we can easily solve for $\overline{w'^2} = 1.68$ m^2 s^{-2}.

The above example was more than just a contrived didactic case. If you look back at Fig 4.2a, you will see that vertical velocity variance measurements do indeed vary with height as described here.

9.7 The Log Wind Profile

One important application of similarity theory is to the mean wind profile in the surface layer. Since people spend most of their lives within the surface layer, the variation of wind speed with height affects their daily lives. The nature of this profile dictates the structure of buildings, bridges, snow fences, wind breaks, pollutant dispersion, and wind turbines, for example. Also, the surface layer wind profile has been studied extensively because of its accessibility to surface-based measurements.

As shown in Fig 9.4, the wind speed usually varies approximately logarithmically with height in the surface layer. Frictional drag causes the wind speed to become zero close to the ground, while the pressure gradient forces cause the wind to increase with height.

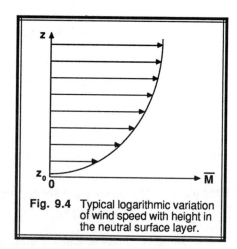

Fig. 9.4 Typical logarithmic variation of wind speed with height in the neutral surface layer.

When plotted on semi-log graph paper (Fig 9.5), a logarithmic relationship such as the wind profile in statically neutral situations appears as a straight line.   For non-neutral situations, the wind profile deviates slightly from logarithmic.   In stable boundary layers, the wind profile is concave downward on a semi-log plot, while unstable boundary layers are concave upward (see Fig 9.5).

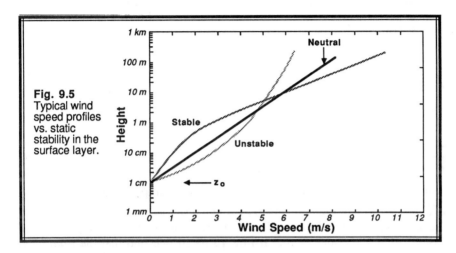

Fig. 9.5
Typical wind speed profiles vs. static stability in the surface layer.

9.7.1   Wind Profile in Statically Neutral Conditions

To estimate the mean wind speed, $\overline{M}$, as a function of height, z, above the ground, we speculate that the following variables are relevant: surface stress (represented by the friction velocity, u_*),   and surface roughness (represented by the *aerodynamic roughness length*, z_0).   Upon applying Buckingham Pi Theory, we find the following two dimensionless groups: $\overline{M}/u_*$, and z/z_0.   Based on the data already plotted in Figs 9.4 and 9.5, we might expect a logarithmic relationship between these two groups:

$$\frac{\overline{M}}{u_*} = \left(\frac{1}{k}\right) \ln\left(\frac{z}{z_0}\right) \tag{9.7.1a}$$

where (1/k) is a constant of proportionality.   As discussed before, the von Karman constant, k, is supposedly a universal constant that is not a function of the flow nor of the surface.   The precise value of this constant has yet to be agreed on, but most investigators feel that it is either near k = 0.35 or k = 0.4.

For simplicity, meteorologists often pick a coordinate system aligned with the mean wind direction near the surface, leaving $\overline{V} = 0$ and $\overline{U} = \overline{M}$. This gives the form of the log wind profile most often seen in the literature:

$$\overline{U} = \left(\frac{u_*}{k}\right) \ln\left(\frac{z}{z_0}\right) \tag{9.7.1b}$$

An alternative derivation of the log wind profile is possible using mixing length theory. Recall from Chapter 6 that the momentum flux in the surface layer is: $\overline{u'w'} =$ $-k^2 z^2 |\partial \overline{U}/\partial z| \, \partial \overline{U}/\partial z$. But since the momentum flux is approximately constant with height in the surface layer, $\overline{u'w'}(z) = \overline{u'w'}(z=0) = u_*^2$. Substituting this into the mixing length expression and taking the square root of the whole equation gives

$$\frac{\partial \overline{M}}{\partial z} = \frac{u_*}{k z} \tag{9.7.1c}$$

When this is integrated over height from $z = z_0$ (where $\overline{M}=0$) to any height z, we again arrive at (9.7.1b). This derivation is more sound than that of Buckingham Pi, because it predicts a log wind profile theoretically, without resorting to empirical arguments.

If we divide both sides of (9.7.1c) by $[u_*/(kz)]$, we find that the **dimensionless wind shear** (ϕ_M , see Chapter 5 or Appendix A) is equal to unity in the neutral surface layer:

$$\phi_M = \left(\frac{k z}{u_*}\right) \frac{\partial \overline{M}}{\partial z} = 1 \tag{9.7.1d}$$

This result was previously listed as (9.5.1c). Equations (9.4.1c) and (9.6.1e) also approach the above expression in the neutral limit of (z/L) approaching zero. In essence, each of these equations describes a log wind profile.

9.7.2   Aerodynamic Roughness Length

The aerodynamic roughness length, z_0, is defined as the height where the wind speed becomes zero. The word **aerodynamic** comes about because the only true determination of this parameter is from measurements of the wind speed at various heights. Given observations of wind speed at two or more heights, it is easy to solve for z_0 and u_* . Graphically, we can easily find z_0 by extrapolating the straight line drawn through the wind speed measurements on a semi-log graph (see Fig 9.5) to the height where $\overline{M} = 0$ (i.e., extrapolate the line towards the ordinate axis).

Although this roughness length is NOT equal to the height of the individual *roughness elements* on the ground, there IS a one-to-one correspondence between those roughness elements and the aerodynamic roughness length. In other words, once the aerodynamic roughness length is determined for a particular surface, it does not change with wind speed, stability, or stress. It can change if the roughness elements on the surface change, such as caused by changes in the height and coverage of vegetation, erection of fences, construction of houses, deforestation or lumbering, etc.

Typical values of the roughness length are indicated in Fig 9.6 (Smedman-Högström & Högström, 1978; Hicks, et al, 1975; Garratt, 1977; Nappo, 1977; Thompson, 1978; and Kondo and Yamazawa, 1986). As expected, higher roughness elements are associated with larger aerodynamic roughness lengths. In all cases, however, the aerodynamic roughness length is smaller than the physical height of the roughness element.

Lettau (1969) suggested a method for estimating the aerodynamic roughness length based on the average vertical extent of the roughness elements (h^*), the average silhouette or vertical cross-section area presented to the wind by one element (s_s), and the lot size per element [S_L = (total ground surface area / number of elements)]

$$z_o = 0.5 \, h^* \left(\frac{s_s}{S_L} \right) \qquad (9.7.2a)$$

This relationship is acceptable when the roughness elements are evenly spaced, not too close together, and of similar height and shape.

Kondo and Yamazawa (1986) proposed a similar relationship, where variations in individual roughness elements were accounted for. Let s_i represent the actual horizontal surface area occupied by element i, and h_i be the height of that element. If N elements occupy a total area of S_T, then the roughness length can be approximated by:

$$z_o = \frac{0.25}{S_T} \sum_{i=1}^{N} h_i \, s_i = \frac{0.25}{L_T} \sum_{i=1}^{N} h_i \, w_i \qquad (9.7.2b)$$

An approximation of the aerodynamic roughness can also be made by summing over the individual roughness elements encountered while traveling along a straight line of total length L_T. For this case, one must consider the longitudinal width, w_i, of each element in the direction of travel. These expressions have been applied successfully to buildings in cities.

We have already discussed Charnock's relationship for the roughness length of the sea surface, which can also be applied to blowing sand and blowing snow (Chamberlain, 1983) with appropriate change in parameter, α_c:

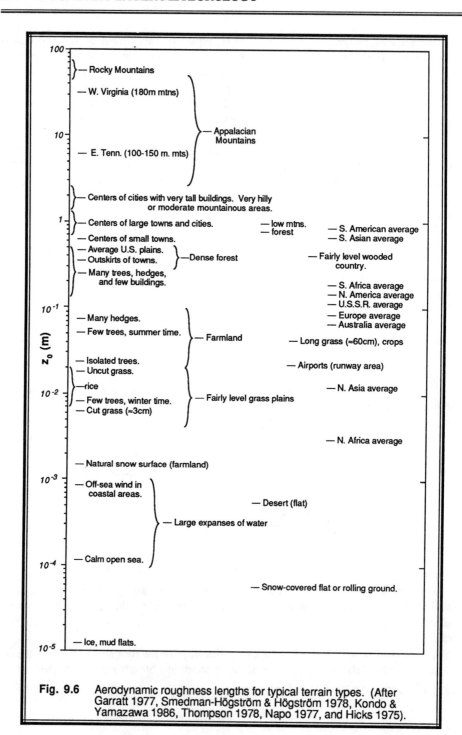

Fig. 9.6   Aerodynamic roughness lengths for typical terrain types. (After Garratt 1977, Smedman-Högström & Högström 1978, Kondo & Yamazawa 1986, Thompson 1978, Napo 1977, and Hicks 1975).

$$z_o = \frac{\alpha_c u_*^2}{g} \qquad (9.7.2c)$$

For the sea, $\alpha_c = 0.016$.

For many large-scale numerical weather-forecast models the lowest grid-points (at height z_1 above the surface) are so high that the surface layer is not resolved. Nevertheless, it is important to account for varying roughness in the model forecast. André and Blondin (1986) suggested that the *effective roughness length* (z_{0eff}) to be used in the model decreases as the altitude of the lowest grid point increases.   In particular, the ratio (z_{0eff})/h^* decreases from about 0.1 to 0.01 as z_1 increases from 0.1 km to 1 km. Taylor (1987), however, suggests that z_{0eff} is independent of z_1.

9.7.3   Displacement Distance

Over land, if the individual roughness elements are packed very closely together, then the top of those elements begins to act like a displaced surface. For example, in some forest canopies the trees are close enough together to make a solid-looking mass of leaves, when viewed from the air. In some cities the houses are packed close enough together to give a similar effect; namely, the average roof-top level begins to act on the flow like a displaced surface.

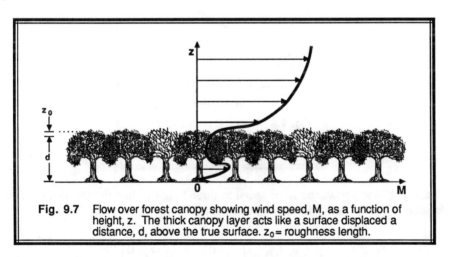

Fig. 9.7   Flow over forest canopy showing wind speed, M, as a function of height, z. The thick canopy layer acts like a surface displaced a distance, d, above the true surface. z_0 = roughness length.

Above the canopy top, the wind profile increases logarithmically with height, as shown in Fig 9.7. Thus, we can define both a displacement distance, d, and a roughness length, z_0, such that:

$$\overline{M} = \left(\frac{u_*}{k}\right) \ln\left[\frac{(z-d)}{z_o}\right]$$ (9.7.3a)

for statically neutral conditions, where we now define $\overline{M} = 0$ at $z = d + z_o$. Given wind speed observations in statically neutral conditions at three or more heights, it is easy to use computerized non-linear regression algorithms such as the Marquardt Method or the Gauss-Newton Method to solve for the three parameters, u_*, z_o, and d.

If you are unsure whether a nonzero displacement distance is appropriate to your situation, one approach is to plot $\overline{M}$ vs. $(z-d)$ for neutral conditions on a semi-log graph as shown in Fig 9.8. As a first guess, try d = 0. If your selected d is too small, then the plotted profile will curve concave upward. Use the intercept of this curve on the ordinate to provide the next guess for d. If d is too large, then the curve will be concave downward. Iterate until the plotted data shows no curvature. This trick will not work for non-neutral cases, nor for profiles that cross through internal boundary layers.

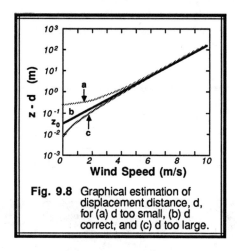

Fig. 9.8 Graphical estimation of displacement distance, d, for (a) d too small, (b) d correct, and (c) d too large.

Finally, if one knows the wind speed at three heights, then the following algebraic expression is easy to derive for the displacement distance.

$$\frac{(\overline{M}_2 - \overline{M}_1)}{(\overline{M}_3 - \overline{M}_1)} \ln\left(\frac{z_3 - d}{z_1 - d}\right) = \ln\left(\frac{z_2 - d}{z_1 - d}\right)$$ (9.7.3b)

The disadvantage of this expression is that it is not explicit in d. However, this equation can be iteratively solved for d.

9.7.4 Surface Stress

In Fig 9.5, u_* is proportional to the slope of the line, in statically neutral conditions. In fact, once the roughness length and displacement distance have been determined, it is easy to find the u_* from:

$$u_* = \frac{k\,\overline{M}}{\ln{(z/z_0)}} \qquad (9.7.4)$$

where $\overline{M}$ is the wind speed at height z. The magnitude of the surface stress in kinematic form is then u_*^2.

9.7.5 Wind Profile in Non-neutral Conditions

Expressions such as (9.7.1b) or (9.7.1d) for statically neutral flow relate the momentum flux, as described by u_*^2, to the vertical profile of $\overline{U}$-velocity. Hence, those expressions can be called *flux-profile relationships*. These relationships can be extended to include non-neutral (*diabatic*) surface layers.

Businger-Dyer Relationships. In non-neutral conditions, we might expect that the buoyancy parameter and the surface heat flux are additional relevant variables. When these are used with the variables from Section 9.7.1, Buckingham Pi analysis gives us three dimensionless groups (neglecting the displacement distance for now): $\overline{M}/u_*$, z/z_0, and z/L, where L is the Obukhov length. Alternatively, if we consider the shear instead of the speed, we get two dimensionless groups: ϕ_M and z/L. Based on field experiment data, Businger, et al., (1971) and Dyer (1974) independently estimated the functional form to be:

$$= 1 + \left(\frac{4.7\,z}{L} \right) \qquad \text{for } \frac{z}{L} > 0 \ \text{(stable)} \qquad (9.7.5a)$$

$$\phi_M \quad = 1 \qquad\qquad\qquad \text{for } \frac{z}{L} = 0 \ \text{(neutral)} \qquad (9.7.5b)$$

$$= \left[1 - \left(\frac{15z}{L} \right) \right]^{-1/4} \qquad \text{for } \frac{z}{L} < 0 \ \text{(unstable)} \qquad (9.7.5c)$$

These are plotted in Fig 9.9a, where Businger, et al., have suggested that k = 0.35 for their data set.

Similar expressions have been estimated for the heat flux vs. the virtual potential temperature profile:

$$= \frac{K_m}{K_H} + \frac{4.7\,z}{L} \qquad\qquad \text{for } \frac{z}{L} > 0 \text{ (stable)} \qquad (9.7.5d)$$

$$\phi_H = \frac{K_m}{K_H} \qquad\qquad \text{for } \frac{z}{L} = 0 \text{ (neutral)} \qquad (9.7.5e)$$

$$= \frac{K_m}{K_H} \left[1 - \frac{9\,z}{L} \right]^{-1/4} \qquad\qquad \text{for } \frac{z}{L} < 0 \text{ (unstable)} \qquad (9.7.5f)$$

where (K_m / K_H) is the ratio of eddy diffusivities of heat and momentum. This ratio equals 0.74 in neutral conditions. The curves corresponding to the above equations are plotted in Fig 9.9b. It is often assumed that the flux profile relationships for moisture or pollutants are equal to those for heat.

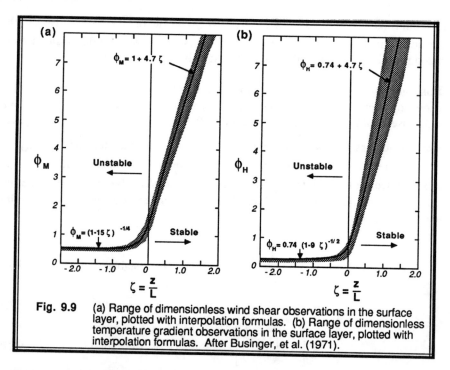

Fig. 9.9   (a) Range of dimensionless wind shear observations in the surface layer, plotted with interpolation formulas. (b) Range of dimensionless temperature gradient observations in the surface layer, plotted with interpolation formulas. After Businger, et al. (1971).

Diabatic Wind Profile.   The Businger-Dyer relationships can be integrated with height to yield the wind speed profiles:

$$\frac{\overline{M}}{u_*} = \left(\frac{1}{k}\right) \left[\ln\left(\frac{z}{z_o}\right) + \Psi_M\left(\frac{z}{L}\right) \right] \qquad (9.7.5g)$$

where the function $\Psi(z/L)$ is given for stable conditions $(z/L > 0)$ by:

$$\Psi_M\left(\frac{z}{L}\right) = \frac{4.7\,z}{L} \qquad (9.7.5h)$$

and for unstable $(z/L < 0)$ by:

$$\Psi_M\left(\frac{z}{L}\right) = -2\ln\left[\frac{(1+x)}{2}\right] - \ln\left[\frac{(1+x^2)}{2}\right] + 2\tan^{-1}(x) - \frac{\pi}{2} \qquad (9.7.5i)$$

where $x = [1 - (15z/L)]^{1/4}$. This last equation was presented by Paulson (1970), although alternative expressions that are more easily solved on computer were presented by Nickerson and Smiley (1975) and Benoit (1977). In the limit of statically neutral flow $(z/L = 0)$, both of these relationships reduce to the log wind profile.

When (9.7.5g and h) are combined, the resulting equation describes a *log-linear profile,* because $\overline{M}$ depends on both $\ln(z)$ and linearly on z/L. As plotted in Fig 9.5, the linear term causes the winds in the surface layer to increase with height faster than those of a neutral profile. This feature is expected on the underside of the nocturnal jet. Clearly the equation fails near the top of the nocturnal boundary layer, where the wind speed reaches a maximum and then frequently decreases with height. Thus, we must be content with applying the log-linear profile only within the stable surface layer.

Fluxes and Scaling Parameters. If the stability and the flux or stress is known in advance, then the integrated Businger-Dyer relationships can be solved directly for the wind speed or the potential temperature at any height.

Often, these equations are used in reverse, to estimate the flux knowing the mean wind or temperature profile. This is much more difficult. For example, u_*, appears in a number of places in the right hand side of (9.7.5g-i): once explicitly, and additional times hidden in L. Furthermore, L is a function of the heat flux, which must simultaneously be estimated from the temperature profile. The resulting coupled set of equations is very difficult to solve, and often involves an iterative approach. One way around this problem is to simplify the flux profile relationships.

For statically unstable conditions, Businger, et.al. (1971) found:

$$\frac{z}{L} = Ri, \tag{9.7.5j}$$

where Ri is the gradient Richardson number (see Fig 5.23 of Chapt. 5). Since Ri is based on gradients of mean potential temperature and wind, it is easy to calculate directly from measurements of those mean variables. Thus, the calculation of the u_* or $\theta_*{}^{SL}$ in (9.7.5i) is much easier.

For statically stable conditions, Arya (1981) suggested that the shape similarity of the temperature and wind profiles be utilized, to yield $u_*/\theta_*{}^{SL} = \Delta\overline{M}/\Delta\overline{\theta}$, where the differences Δ are taken vertically within the surface layer. These simplifications lead to

$$L \cong \frac{u_* \overline{\theta} \, \Delta\overline{M}}{(k \, g \, \Delta\overline{\theta})} \tag{9.7.5k}$$

where

$$u_* \cong \frac{\left[k \overline{M} - \left\{ \dfrac{4.7 \, k \, g \, \Delta\overline{\theta} \, z}{(\overline{\theta} \, \Delta\overline{M})} \right\} \right]}{\ln\left(\dfrac{z}{z_o} \right)} \tag{9.7.5l}$$

9.8 Rossby-number Similarity and Profile Matching

As introduced earlier, it is often necessary to be able to approximate the surface stress and fluxes in terms of mean variables at the grid points in numerical models. Unfortunately, in some models the lowest grid point is well above the surface layer, making it impossible to use the flux-profile relationships described earlier. By matching surface layer profiles to flows in the mid-boundary layer, surface fluxes can be related to conditions higher in the boundary layer.

The *profile matching* technique uses two separate similarity approximations: one that describes the departure of the actual wind from geostrophic (i.e., the *geostrophic departure*) in the (outer) mid-boundary layer:

$$\frac{(\overline{M} - M_2)}{u_*} = \left(\frac{1}{k}\right) A\left(\frac{z}{h_2}, \frac{h_2}{L}\right)$$

and the other for the log profile lower in the (inner) surface layer:

$$\frac{\overline{M}}{u_*} = \left(\frac{1}{k}\right)\left[\ln\left(\frac{h_2}{z_o}\right) + \Psi_M\left(\frac{z}{L}\right)\right]$$

where $A(z/h_2, h_2/L)$ is some universal function to be determined empirically, and $\overline{M}_2$ and h_2 are velocity and height scales in the outer layer. The flow at the bottom of the mid-boundary layer is matched or fitted to agree with that at the top of the surface layer. The resulting relationship relates the surface stress to the mean flow higher in the interior of the boundary layer.

When this procedure is also performed for heat and moisture, we arrive at the following set of equations:

$$\frac{U_2}{u_*} = \left(\frac{1}{k}\right)\left[\ln\left(\frac{h_2}{z_o}\right) - A\left(\frac{h_2}{L}\right)\right]$$

$$\frac{V_2}{u_*} = -\left(\frac{1}{k}\right)B\left(\frac{h_2}{L}\right) \text{ sign } (f_c)$$

$$\frac{\Delta\theta_s}{\theta_*} = \left[\frac{K_m}{(k\,K_H)}\right]\left[\ln\left(\frac{h_2}{z_o}\right) - C\left(\frac{h_2}{L}\right)\right]$$

$$\frac{\Delta q_s}{q_*} = \left[\frac{K_m}{(k\,K_H)}\right]\left[\ln\left(\frac{h_2}{z_o}\right) - D\left(\frac{h_2}{L}\right)\right] \tag{9.8a}$$

where A, B, C and D are "universal" functions, and $[K_m/(k\,K_H)] \cong 2$. Yamada (1976) tested a number of velocity scales, and found that $U_2 = \langle\overline{U}_g\rangle$, $V_2 = \langle\overline{V}_g\rangle$ worked best. The length scale, h_2, was best modeled using z_i during the day, and the depth of the turbulent layer at night. With these scales, his analysis of field experiment data yielded the functional forms for A, B and C plotted in Fig 9.10.

These equations can be combined and solved to give the bulk transfer (or drag) coefficients, which allow us to find the surface flux given knowledge of mean variables aloft and at the surface:

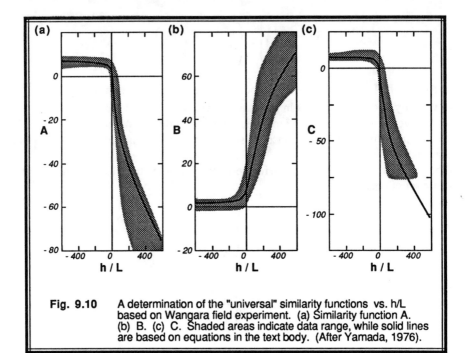

Fig. 9.10   A determination of the "universal" similarity functions vs. h/L based on Wangara field experiment. (a) Similarity function A. (b) B. (c) C. Shaded areas indicate data range, while solid lines are based on equations in the text body. (After Yamada, 1976).

$$C_D = k^2 \left\{ \left[\ln\left(\frac{h_2}{z_o}\right) - A\left(R_B\right) \right]^2 + B^2 \right\}^{-1}$$

$$\alpha_{ws} = -\frac{B \cdot (R_B) \, \text{sign} \, (f_c)}{\left[\ln\left(\frac{h_2}{z_o}\right) - A(R_B) \right]}$$

$$C_H = \left[\frac{k \, K_H}{K_m} \right] \left[\ln\left(\frac{h_2}{z_o}\right) - C\left(R_B\right) \right]^{-1}$$

$$C_E = \left[\frac{k \, K_H}{K_m} \right] \left[\ln\left(\frac{h_2}{z_o}\right) - D\left(R_B\right) \right]^{-1} \tag{9.8b}$$

where α_{ws} is the angle between the scale-wind direction and the surface stress, $R_B = g\, h_2$ $\Delta\theta_s / (\theta\, \overline{M}^2)$ is the bulk Richardson number, and $\Delta\theta_s$ is the temperature difference between the air and the ground (see Table 9-1). Figs 9.11 shows the resulting bulk transfer coefficients for momentum and heat. Because geostrophic wind has proved to be the best scale velocity in these expressions, the equations above are also known as *geostrophic drag laws*.

Although it is a worthwhile goal to estimate surface flux based on measurements of mean variables aloft, estimates of the universal functions and the bulk transfer coefficients have yielded much scatter. The "universality" of the functions are thus still in question, making this matching scheme of dubious reliability.

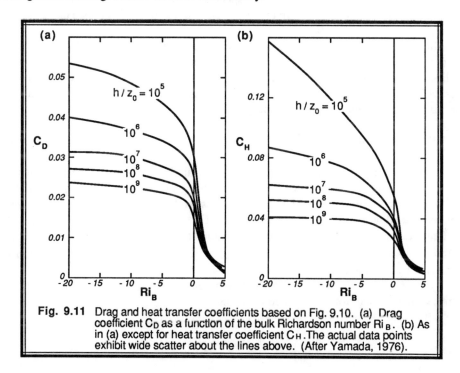

Fig. 9.11   Drag and heat transfer coefficients based on Fig. 9.10. (a) Drag coefficient C_D as a function of the bulk Richardson number Ri$_B$. (b) As in (a) except for heat transfer coefficient C_H. The actual data points exhibit wide scatter about the lines above. (After Yamada, 1976).

9.9 Spectral Similarity

Spectral analysis of atmospheric turbulence data is a powerful tool to help probe deeper into the workings of turbulent flow. Interest in this method has always been high, as indicated by an issue of Radio Science (1969) dedicated to the spectra of meteorological variables. Nevertheless, there are some fundamental questions regarding the correspondence of Fourier modes to physical eddies (Tennekes, 1976). Similarity theory has been applied to spectra to help organize the spectral results, and to help focus our understanding about turbulence.

As discussed in the previous chapter, the discrete power spectral intensity measures how much of the variance of a signal is associated with a particular frequency, f. If ξ represents any variable, then the discrete power spectral intensity $E_\xi(f)$ has units of ξ^2. An obvious way to make the spectral intensity dimensionless is to divide it by the total variance $\overline{\xi'^2}$. A continuous spectrum with power spectral density of $S_\xi(f)$ has the same units as ξ^2/f , and can be made dimensionless by dividing by $\overline{\xi'^2}/f$. Analogous expressions can be made for wavenumber spectra instead of frequency spectra. In both of these cases, the result is a spectrum that gives the fraction of total variance explained by a wavelength or wavelength band.

Alternately, if the turbulence is driven or governed by specific mechanisms, such as wind shear, buoyancy, or dissipation, then the spectral intensities can be normalized by scaling variables appropriate to the flow. The next three Sections show normalized spectra for the inertial subrange, for surface layer turbulence generated mechanically, and for mixed layer turbulence generated buoyantly.

9.9.1 Inertial Subrange

As discussed in Chapter 5, there are many situations where middle size turbulent eddies "feel" neither the effects of viscosity, nor the generation of TKE. These eddies get their energy inertially from the larger-size eddies, and lose their energy the same way to smaller-size eddies. For a steady-state turbulent flow, the cascade rate of energy down the spectrum must balance the dissipation rate at the smallest eddy sizes. Hence, there are only three variables relevant to the flow: S, κ, and ϵ. This similarity approach was pioneered by Kolmogorov (1941) and Obukhov (1941).

By performing a Buckingham Pi dimensional analysis, we can make only one dimensionless group from these three variables:

$$\pi_1 = \frac{S^3 \, \kappa^5}{\epsilon^2}$$

We know that this Pi group must be equal to a constant, because there are no other Pi groups for it to be a function of.

Solving the above equation for S yields:

$$S(\kappa) = \alpha_k \, \epsilon^{2/3} \, \kappa^{-5/3} \tag{9.9.1}$$

where the α_k is known as the **Kolmogorov constant.** The value of this constant has yet to be pinned down (Gossard, et.al., 1982), but it is in the range of $\alpha_k = 1.53$ to 1.68.

One of the easiest ways to determine whether any measured spectrum has an inertial subrange is to plot the spectrum (S vs. κ) on a log-log graph. The inertial subrange portion should appear as a straight line with a -5/3 slope (see Fig 9.12). The demonstration spectra plotted in Fig 8.9 all have an inertial subrange at normalized frequencies greater than 2.5, assuming that Taylor's hypothesis can be used to relate frequencies to wavenumbers via $f = \overline{M} \cdot \kappa$.

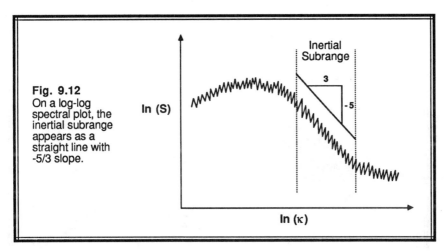

Fig. 9.12 On a log-log spectral plot, the inertial subrange appears as a straight line with -5/3 slope.

9.9.2   Surface Layer Spectra

Suppose that the velocity spectra $f S_u(f)$ for a surface layer in a state of forced convection were likely to be affected by the following variables: u_*, $\overline{w'\theta_v'}_s$, z, $\overline{U}$ (or $\overline{M}$), f, and ε. Buckingham Pi analysis of the above variables gives three dimensionless groups: $\pi_1 = f S_u(f) / (k z \varepsilon)^{2/3}$, $\pi_2 = f z / \overline{M}$, and $\pi_3 = z / L$.

Fig 9.13a shows the result when these π groups are plotted (Kaimal, et al, 1972). We see some important characteristics: (1) The peak spectral intensity is reduced as the static stability is increased, because stability is opposing turbulent motions. (2) The peak is shifted to higher frequencies as stability is increased, possibly because the lower frequencies are more strongly damped by the buoyancy forces. (3) At high frequencies, the spectral intensity is no longer dependent on the static stability (at least for the weak stabilities plotted), suggesting that the smaller size eddies in the inertial subrange receive all of their energy via the cascade process from larger eddies, with no direct interaction with the mean flow or the mean stratification. (4) Finally, there is a curious occurrence of an *excluded region* in the spectral plot near neutral stratification (lightly shaded in the figure).

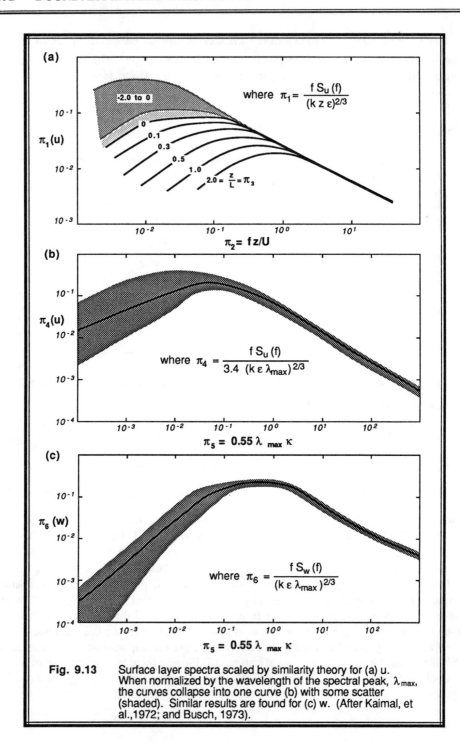

Fig. 9.13   Surface layer spectra scaled by similarity theory for (a) u. When normalized by the wavelength of the spectral peak, λ_{max}, the curves collapse into one curve (b) with some scatter (shaded). Similar results are found for (c) w. (After Kaimal, et al.,1972; and Busch, 1973).

Remember that the correct scaling variables are not specified from first principles, but must be determined from empirical data. For some cases u_*^2 might work, while for other cases $(kz\varepsilon)^{2/3}$ might be better. In a surface layer with turbulence generated mechanically, velocity spectral intensities can be normalized with respect to u_*^2, temperature spectra with respect to θ_*^{SL2}, and moisture spectra with respect to q_*^{SL2}. Frequency can be normalized by $\overline{M}/z$, u_*/z, or by frequency, f_{max}, or wavelength, λ_{max}, corresponding to the peak in the spectrum.

When these various spectra are normalized with respect to λ_{max}, all of the curves collapse onto one curve, as shown in Fig 9.13b, where $\pi_4 = f \cdot S_u(f)/[3.4 \, (k \, \varepsilon \, \lambda_{max})^{2/3}]$, $f = \kappa \overline{M}$, and $\pi_5 = 0.55 \, \lambda_{max} \, k$. A similar result is found for vertical velocity spectra (Fig 9.13c), where $\pi_6 = f \cdot S_w(f)/[(k \, \varepsilon \, \lambda_{max})^{2/3}]$.

9.9.3 Mixed Layer Spectra

We might speculate that the following variables affect the velocity spectrum, $fS_u(f)$, during convective conditions: $(g/\overline{\theta}_v)\overline{w'\theta_v'}_s$, f, z, z_i, $\overline{U}$ (or $\overline{M}$), and ε. Buckingham Pi

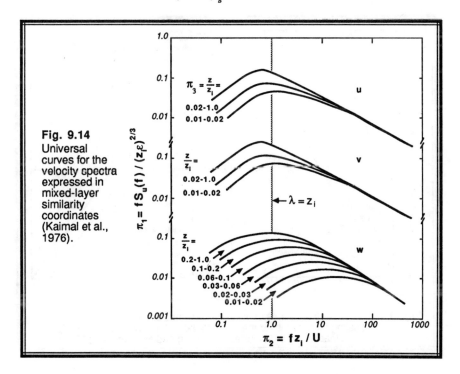

Fig. 9.14 Universal curves for the velocity spectra expressed in mixed-layer similarity coordinates (Kaimal et al., 1976).

analysis gives: $\pi_1 = f\, S_u(f)\, /\, (z_i\, \epsilon\,)^{2/3}$, $\pi_2 = f\, z_i\, /\, \overline{U}$, and $\pi_3 = z\, /\, z_i$. These are plotted in Fig 9.14 for the u, v, and w components of the one-dimensional spectra (Kaimal, et al., 1976).

Notice that the peak in these curves corresponds to a wavelength approximately equal to the ML depth. This confirms our earlier statement that the most energetic eddies are the large ones that are produced on the scale of the boundary layer. In situations such as flow over complex terrain, new length scales might be introduced into the flow based on the scales of the terrain irregularities (Panofsky, et al., 1982).

9.10 Similarity Scaling Domains

In different parts of the boundary layer, we typically find that the nature of the flow is dependent on some scaling parameters, and not dependent on others. For example, in the surface layer we expect turbulence and mean profiles to be related to z/L, but in the ML we find that z/z_i is more appropriate than z/L. To help organize or knowledge about the relevant scaling parameters, Fig 9.15 relates parameters to identifiable parts of the *unstable* BL, while Fig 9.16 does the same for the *stable* BL.

In Fig 9.15a, the regions that are independent of L or z_i are indicated. In Fig 9.15b, the regions are listed along with the variables that are relevant for each region. We find

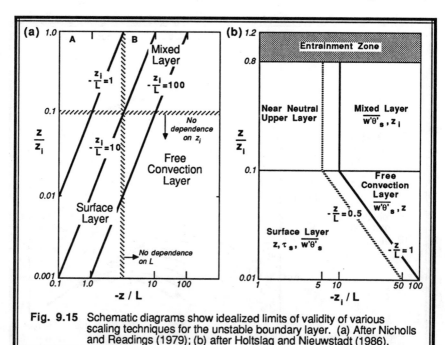

Fig. 9.15 Schematic diagrams show idealized limits of validity of various scaling techniques for the unstable boundary layer. (a) After Nicholls and Readings (1979); (b) after Holtslag and Nieuwstadt (1986).

that in addition to the ML and SL, there are two new regions that can be identified. One layer is the *free convection layer* that forms in strongly convective situations near the ground. It can be thought of as the region between the top of the SL and the bottom of the ML, where neither L nor z_i length scales are relevant. The other region is a *near neutral upper layer* that is similar to the residual layer except that it is still turbulent and still feeling the effects of the surface. These conditions might happen on a windy day with clear skies over land, where both buoyant and mechanical generation of turbulence are present. They might also exist in stratocumulus-topped mixed layers.

For stable conditions, Fig 9.16 shows a region in the upper right portion of the graph that corresponds to strongly stable air that is in the top of the SBL. Turbulence in this region is likely to be *intermittent*, because of the strong stability suppressing turbulence. In the middle of the SBL is a region that might be continuously turbulent, but which is independent of height above ground and of surface fluxes. In this *z-less* region, only the magnitudes of the local fluxes are important. Below this region *local scaling* continues to be important for more neutral stability, but now the turbulence senses the bottom boundary and is dependent on z. Finally, adjacent to the ground is the usual surface layer, where surface fluxes and z are important. Note that the *near neutral upper layer* defined in this graph is within the SBL, and is not the residual layer that lies above the SBL.

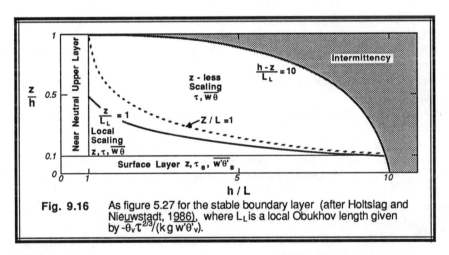

Fig. 9.16   As figure 5.27 for the stable boundary layer (after Holtslag and Nieuwstadt, 1986), where L_L is a local Obukhov length given by $-\theta_v \tau^{2/3}/(k\,g\,\overline{w'\theta'_v})$.

9.11   References

American Geophysical Union, 1969: Spectra of Meteorological Variables. *Radio Science*, **4**, No. 12, 1099-1397.

André, J.-C., and C. Blondin, 1986: On the effective roughness length for use in numerical 3-D models. *Bound.-Layer Meteor.*, **35**, 231-245.

Arya, S.P.S., 1981: Parameterizing the height of the stable atmospheric boundary layer. *J. Appl. Meteor.*, **20**, 1192-1202.

Benoit, R., 1977: On the integral of the surface layer profile-gradient functions. *J. Appl. Meteor.*, **16**, 859-860.

Berkowicz, R. and L.P. Prahm, 1984: Spectral representation of the vertical structure of turbulence in the convective boundary layer. *Quart. J. Roy. Meteor. Soc.*, **110**, 35-52.

Brost, R.A., J.C. Wyngaard and D.H. Lenschow, 1982: Marine stratocumulus layers. Part II: Turbulence budgets. *J. Atmos. Sci.*, **39**, 818-836.

Buckingham, 1914: On physically similar systems: illustrations of the use of dimensional analysis, *Phys. Rev.*, **4**, 345.

Busch, N.E., 1973: On the mechanics of atmospheric turbulence. *Workshop on Micrometeorology*. (Ed. by D.A. Haugen). Amer. Meteor. Soc., Boston. 1-65.

Businger, J.A., J.C. Wyngaard, Y. Izumi and E.F. Bradley, 1971: Flux profile relationships in the atmospheric surface layer. *J. Atmos. Sci.*, **28**, 181-189.

Caughey, S.J. and S.G. Palmer, 1979: Some aspects of turbulent structures through the depth of the convective boundary layer. *Quart. J. Roy. Meteor. Soc.*, **105**, 811-827.

Caughey, S.J. and C.J. Readings, 1974: The vertical component of turbulence in convective conditions. *Adv. in Geophys.*, **18A**, 125-130.

Caughey, S.J. and C.J. Readings, 1975: Turbulent fluctuations in convective conditions. *Quart. J. Roy. Meteor. Soc.*, **101**, 537-542.

Caughey, S.J., J.C. Wyngaard and J.C. Kaimal, 1979: Turbulence in the evolving stable boundary layer. *J. Atmos. Sci.*, **36**, 1041-1052.

Chamberlain, A.C., 1983: Roughness length of sea, sand and snow. *Bound. Layer Meteor.*, **25**, 405-409.

Charnock, H., 1955: Wind stress on a water surface. *Quart. J. Roy. Meteor. Soc.*, **81**, 639-640.

Deardorff, J.W., 1972: Numerical investigation of neutral and unstable planetary boundary layers. *J. Atmos. Sci.*, **29**, 91-115.

Deardorff, J.W., G.E. Willis and B.H. Stockton, 1980: Laboratory studies of the entrainment zone of a convectively mixed layer. *J. Fluid Mech.*, **100**, 41-64.

Dyer, A.J., 1974: A review of flux-profile relations. *Bound. Layer Meteor.*, **1**, 363-372.

Garratt,J.R., 1977: Review of drag coefficients over oceans and continents. *Mon. Wea. Rev.*, **105**, 915-929.

Gossard, E.E., R.B. Chadwick, W.D. Neff, and K.P. Moran, 1982: The use of ground-based Doppler radars to measure gradients, fluxes and structure parameters in elevated layers. *J. Appl. Meteor.*, **21**, 211-226.

Grant, A.L.M., 1986: Observations of boundary layer structure made during the 1981 KONTUR experiment. *Quart. J. Roy. Meteor. Soc.*, **112**, 825-841.

Hicks, B.B., P. Hyson and C.J. Moore, 1975: A study of eddy fluxes over a forest. *J. Appl. Meteor.*, **14**, 58-66.

Holtslag, A.A.M. and F.T.M. Nieuwstadt, 1986: Scaling the atmospheric boundary layer. *Bound.-Layer Meteor.*, **36**, 201-209.

Kaimal, J.C., J.C. Wyngaard, D.A. Haugen, O.R. Coté, Y. Izumi, S.J. Caughey, and

C.J. Readings, 1976: Turbulence structure in the convective boundary layer. *J. Atmos. Sci.*, **33**, 2152-2169.

Kaimal, J.C., J.C. Wyngaard, Y. Izumi and O.R. Coté, 1972: Spectral characteristics of surface layer turbulence. *Quart. J. Roy. Meteor. Soc.*, **98**, 563-589.

Kolmogorov, A.N., 1941: Energy dissipation in locally isotropic turbulence. *Doklady AN SSSR*, **32**, No. 1, 19-21.

Kondo, J. and H. Yamazawa, 1986: Aerodynamic roughness over an inhomogeneous ground surface. *Bound.-Layer Meteor.*, **35**, 331-348.

Lascer, A. and S.P.S. Arya, 1986: A numerical model study of the structure and similarity scaling of the nocturnal boundary layer. *Bound.-Layer Meteor.*, **35**, 369-386.

LeMone, M.A. and W.T. Pennell, 1976: The relationship of the trade wind cumulus distribution to subcloud layer fluxes and structure. *Mon. Wea. Rev.*, **104**, 524-539.

Lenschow, D.H., 1974: Model of the height variation of the turbulence kinetic energy in the unstable planetary boundary layer. *J. Atmos. Sci.*, **31**, 465-474.

Lenschow, D.H., J.C. Wyngaard, and W.T. Pennell, 1980: Mean field and second moment budgets in a baroclinic convective boundary layer. *J. Atmos. Sci.*, **37**, 1313-1326.

Lettau, H., 1969: Note on aerodynamic roughness-parameter estimation on the basis of roughness-element description. *J. Appl. Meteor.*, **8**, 828-832.

Mahrt, L. and J.-C. André, 1983: On the stratification of turbulent mixed layer. *J. Geophys. Res.*, **88**, 2662-2666.

Merry, M. and H.A. Panofsky, 1976: Statistics of vertical motion over land and water. *Quart. J. Roy. Meteor. Soc.*, **102**, 225-260.

Moeng, C.-H. and J.C. Wyngaard, 1984: Statistics of conservative scalars in the convective boundary layer. *J. Atmos. Sci.*, **51**, 3161-3169.

Monin, A.S. and A.M. Obukhov, 1954: Basic laws of turbulent mixing in the atmosphere near the ground. *Tr. Akad. Nauk., SSSR Geophiz. Inst.*, No. 24 (151), 1963-1987.

Nappo,C.J., Jr., 1977: Mesoscale flow over complex terrain during the Eastern Tennessee Trajectory Experiment (ETTEX). *J. Appl. Meteor.*, **16**, 1186-1196.

Nicholls, S., and C.J. Readings, 1979: Aircraft observations of the structure of the lower boundary layer over the sea. *Quart. J. Roy. Meteor. Soc.*, **105**, 785-802.

Nickerson, E.C. and V.E. Smiley, 1975: Surface layer and energy budget parameterizations for mesoscale models. *J. Appl. Meteor.*, **14**, 297-300.

Nieuwstadt, F.T.M., 1984: The turbulent structure of the stable, nocturnal boundary layer. *J. Atmos. Sci.*, **41**, 2202-2216.

Obukhov, A.M., 1941: Energy distribution in the spectrum of a turbulent flow. *Izvestiya AN SSSR, Ser. Geogr. Geofiz.*, No. 4-5, 453-466.

Panofsky, H.A., D. Larko, R. Lipschutz, G. Stone, E.F. Bradley, A.J. Bowen and J. Højstrup, 1982: Spectra of velocity components over complex terrain. *Quart. J. Roy. Meteor. Soc.*, 108, 215-230.

Panofsky, H.A., H. Tennekes, D.H. Lenschow, and J.C. Wyngaard, 1977: the characteristics of turbulent velocity components in the surface layer under convective conditions. *Bound.-Layer Meteor.*, **11**, 355-361.

Paulson, C.A., 1970: The mathematical representation of wind speed and temperature in the unstable atmospheric surface layer. *J. Appl. Meteor.*, **9**, 857-861.

Perry, R.H., C.H. Chilton and S.D. Kirkpatrick, (Eds.), 1963: *Perry's Chemical Engineer's Handbook, 4th Ed.*. McGraw Hill, NY. 2-87 to 2-90.

Smedman-Högström, A.-S., and U. Högström, 1978: A practical method for determining wind frequency distributions for the lowest 200 m from routine meteorological data. *J. Appl. Meteor.*, **17**, 942-954.

Smith, S.D., 1980: Wind stress and heat flux over the ocean in gale force winds. *J. Phys. Ocean.*, **10**, 709-726.

Sorbjan, Z., 1986: On similarity in the atmospheric boundary layer. *Bound. Layer Meteor.*, **34**, 377-397.

Sorbjan, Z., 1987: An examination of local similarity theory in the stably stratified boundary layer. *Bound.-Layer Meteor.*, **38**, 63-71.

Stull, R.B., 1983: Integral scales for the nocturnal boundary layer. Part I: Empirical depth relationships. *J. Clim. Appl. Meteor.*, **22**, 673-686.

Taylor, P.A., 1987: Comments and further analysis on effective roughness lengths for use in numerical 3-D models. *Bound.-Layer Meteor.*, **39**, 403-418.

Tennekes, H., 1973: Similarity laws and scale relations in planetary boundary layers. *Workshop on Micrometeorology* (Ed., D.A. Haugen), Am. Meteor. Soc., 177-216.

Tennekes, H., 1976: Fourier-transform ambiguity in turbulence dynamics. *J. Atmos. Sci.*, **33**, 1660-1663.

Tennekes, H., 1982: Similarity relations, scaling laws and spectral dynamics. *Atmospheric Turbulence and Air Pollution Modelling.* (Ed. by F.T.M. Nieuwstadt and H.van Dop). Reidel. 37-68.

Thompson, R. S., 1978: Note on the aerodynamic roughness length for complex terrain, *J. Appl. Meteor.*, **17**, 1402-1403.

Webb, E.K., 1982: Profile relationships in the superadiabatic surface layer. *Quart. J. Roy. Meteor. Soc.*, **108**, 661-688.

Whitaker, S., 1968: *Introduction to Fluid Mechanics.* Prentice-Hall, Englewood Cliffs. 457pp.

Wyngaard, J.C., 1973: On surface layer turbulence. *Workshop on Micrometeorology.* (Ed. D.A. Haugen). Am. Meteor. Soc. 101-148.

Wyngaard, J.C. and R.A. Brost, 1984: Top-down and bottom-up diffusion of a scalar in the convective boundary layer. *J. Atmos. Sci.*, **41**, 102-112.

Wyngaard, J.C. and O.R. Coté, 1971: Budgets of turbulent kinetic energy and temperature variance in the atmospheric surface layer. *J. Atmos. Sci.*, **28**, 190-201.

Wyngaard, J.C., O.R. Coté, and Y. Izumi, 1971: Local free convection, similarity, and the budgets of shear stress and heat flux. *J. Atmos. Sci.*, **28**, 1171-1182.

Yamada, T., 1976: On the similarity functions A, B, and C of the planetary boundary layer. *J. Atmos. Sci.*, **33**, 781-793.

Zhou, M.Y., D.H. Lenschow, B.B. Stankov, J.C. Kaimal, and J.E. Gaynor, 1985: Wave and turbulence structure in a shallow baroclinic convective boundary layer and overlying inversion. *J. Atmos. Sci.*, **42**, 47-57.

9.12 Exercises

1) Suppose that the wind speed, $\overline{M}$, near the surface at night is a function of $g/\overline{\theta}_v$, $\overline{w'\theta_v'}$, $\partial z/\partial x$, $\partial\overline{\theta}_v/\partial z$, z_0, and $\overline{U}_g$. Use Buckingham Pi dimensional analysis to determine the relevant Pi groups.

2) You were recently hired to make background environmental measurements at the site of a proposed power plant. The site happens to be of uniform roughness in all directions, and the land use in this area does not change with the seasons (i.e., no harvesting, logging, or other changes).

a) On one particular overcast day, you measured the following wind speed profile as a function of height using your instrumented 100 m tower and a rawinsonde that you launched: Find z_0 and u_*.

z (m)	1	10	100	200	500	1000
$\overline{M}$ (m/s)	1	2	3	3	3	3

b) Later in the year, you sold the tall tower and donated the funds to your favorite meteorological charity. But your contract with the power company still required you to make background measurements. Therefore, you erected a shorter, 10 m tower at the same site. Then came another overcast day with the wind from the same direction as before. You measured a speed of 4 m/s at the top of your short tower. What is value of u_* on this day, and what is the wind speed at z = 50 m (the height of the proposed smoke stack from the power plant)?

3) If an orchard is planted with 1000 trees per square kilometer, where each tree is 4 m tall and has a vertical cross-section area (effective silhouette to the wind) of 5 m^2, what is the aerodynamic roughness length? Assume d=0.

4) Given the following wind speed data for a neutral surface layer, find the roughness length (z_0), displacement distance (d), and friction velocity (u_*):

z(m):	5	8	10	20	30	50
$\overline{M}$(m/s):	3.48	4.43	4.66	5.50	5.93	6.45

5) Suppose that the following was observed on a clear night (no clouds) over farmland having $z_0 = 0.067$ m (assume k = 0.4), L (Obukhov length) = 30 m, $u_* = 0.2$ m/s.

Find and plot $\overline{M}$ as a function of height up to 50 m.

6) a) If the displacement distance is zero, find z_0 and u_* , given the following data in statically neutral conditions at sunset:

z(m)	M(m/s)
1	4.6
3	6.0
10	7.6
30	9.0

b) Later in the evening at the same site, $\overline{w'\theta_v'} = -0.01$ K m/s, and $g/\overline{\theta_v} = 0.03333$ m s^{-2} K^{-1}. If $u_* = 0.3$ m/s, calculate and plot the wind speed profile ($\overline{U}$ vs z) up to z = 50 m.

7) Assume that the following variables are relevant for flow over an isolated hill:

$(g/\overline{\theta_v})\,\partial\overline{\theta_v}/\partial z$ = stability parameter

$\overline{M}$ = wind speed
D   = diameter of hill
H   = height of hill

Use Buckingham Pi methods to find the dimensionless groups for this problem.

8) Given: A SBL with $z_o = 1$ cm, $\overline{\theta_v}$ (at z = 10 m) = 294 K, $\overline{w'\theta_v'}_s = -0.02$ K·ms^{-1}, $u_* = 0.2$ m/s, k = 0.4. Plot the mean wind speed as a function of height on a semi-log graph for $1 \le z \le 100$ m.

9) Given the following wind speeds measured at various heights in a neutral boundary layer, find the aerodynamic roughness length (z_o), the friction velocity (u_*), and the shear stress at the ground (τ). What would you estimate the wind speeds to be at 2 m and at 10 cm above the ground? Use semi-log paper. Assume that the von Karman constant is 0.35 .

z (m)	U (m/s)
2000	10.0
1000	10.0
500	9.5
300	9.0
100	8.0
50	7.4
20	6.5
10	5.8
4	5.0
1	3.7

10) Consider the flow of air over a housing development with no trees and almost identical houses. In each city block (0.1 km by 0.2 km), there are 20 houses, where each house has nearly a square foundation (10 m on a side) and has an average height of 5 m. Calculate the value of the surface stress acting on this neighborhood when a wind speed of 10 m/s is measured at a height of 20 m above ground in statically neutral conditions. Express the stress in Pascals.

11) Derive an expression for the kinematic heat flux ($\overline{w'\theta'}$) in terms of the dimensionless wind shear (ϕ_M) and the dimensionless lapse rate (ϕ_H). Then, given a wind shear of

0.02 1/s and a lapse rate of -0.012 K/m at a height of 20 m above ground, calculate the value of the heat flux at that height using Businger's flux-profile relationships. Assume a mean potential temperature of 21 °C. To simplify your calculations, recall that z/L is approximately equal to the Richardson number for unstable conditions.

12) State some reasons why one might need to use geostrophic drag relationships.

13) Use the definition of the drag coefficient along with the neutral log wind profile equation to prove that $C_{DN} = k^2 \ln^{-2}(z/z_o)$.

14) Given the definition for eddy diffusivity $\overline{u'w'} = - K \, \partial \overline{U}/\partial z$, solve for K as a function of height in the neutral surface layer, assuming a log wind profile.

15) Given the answer from the previous question, and the definition for Ekman layer depth $h_E = (2 \pi^2 K/f_c)^{1/2}$, show that the Ekman layer depth is proportional to u_*/f_c.

16) Given the following variables and their dimensions:

z	height	L
g/θ	buoyancy parameter	L T^{-2} K^{-1}
TKE	turbulence KE per unit mass	L^2 T^{-2}
Zi	depth of the mixed layer	L
w'θ'	surface kinematic heat flux	L T^{-1} K

Perform a dimensional analysis to find Pi groups for z and TKE, using the remaining variables as the primary variables.

17) Using the Businger-Dyer flux-profile relationship for statically stable conditions:
   a) Derive an equation for the drag coefficient, C_D, as a function of the following 4 parameters:
      z : height above ground
      z_o : roughness length
      L : Obukhov length
      k : von Karman constant (= 0.35)
   b) Find the resulting ratio of C_D/C_{DN}, where C_{DN} is the neutral drag coefficient.
   c) Given z = 10 m and z_o = 10 cm, calculate and plot C_D/C_{DN} for a few different values of stability: 0 < z/L < 1. How does this compare with Fig 7.10a?

18) Given the surface layer profile plotted at right: List 2 different reasons why the profile might look like this.

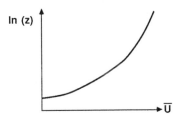

19) Given the following wind profile:

z (m)	U (m/s)
0.3	5.0
0.7	6.0
1.0	6.4
2.0	7.2
10.0	9.0
50.0	10.0
100.0	10.2
1000.0	10.4

and density $\bar{\rho} = 1.25$ kg/m^3. Assume the displacement distance d = 0.
a) Find z_0.
b) Find u_*
c) Find the surface stress (in units of N/m^2)
d) Given a different day over the same land surface, but with z/L = -1.0 and $u_* = $ 1.0 m/s. Is the wind fast enough to 'knock your socks off' ? (Assume that sock height = 25 cm, and that it takes a wind speed of 10 m/s at that height to 'knock your socks off'.)

20) Given: $\partial \bar{U}/\partial z = u_*/kz$, and $K_m = kzu_*$ for simplicity (although this is not necessarily a neutral surface layer). Given also: $u_* = 0.3$ m/s, $\bar{\theta} = 280 + 1.0$ ln(z/z$_0$) in degrees Kelvin, and z$_0$ = 1 cm.

a) Find $\partial \bar{\theta}/\partial t$.
b) Calculate the numerical value of the flux Richardson number.
c) Is the flow turbulent? Why?

21) During the night, assume that the TKE is a function of the following parameters:

(z$_0$, z, g/$\bar{\theta}$, $\Delta \bar{\theta}_s$, $\bar{U}_g$, v) where v has units of length times velocity.

Use dimensional analysis to find the dimensionless PI groups.

22) Given the following wind profile in statically neutral conditions:

z (m)	$\bar{U}$ (m/s)
0.95	3
3.0	4
9.5	5
30.0	6

Find the numerical value of:
a) z$_0$
b) u_*
c) K_m at 3 m
d) C_D.

23) Given $z_0 = 1$ cm = constant and $u_* = 0.25$ m/s = constant. Find and plot the ratio of 10 m mean wind speed for diabatic conditions to that for neutral conditions, as a function of stability for $0 \le (z/L) \le 2.0$. Briefly describe the significance of your result.

24) It has been suggested that in regions of strong static stability the lower (long wavelength, small wavenumber) end of the inertial subrange occurs at a wavenumber, κ_b, given by $\kappa_b = N_{BV}^{2/3}\, \varepsilon^{-1/2}$, where N_{BV} is the Brunt-Vaisala frequency, and ε is the turbulence dissipation rate. Use dimensional analysis to arrive at the above expression.

25) Given the following mean wind speed profile, find the roughness length (z_0) and the friction velocity (u_*). Assume that the surface layer is statically neutral, and that the displacement distance $d = 0$. Use $k = 0.35$.

z (m)	$\overline{M}$ (m/s)
1	3.0
3	4.0
10	5.0
20	5.6
50	6.4
100	6.8
500	7.0
1000	7.0

26) Knowing the shear ($d\overline{M}\,/\,dz$) at any height z is sufficient to determine the friction velocity (u_*) for a neutral surface layer:

$$u_* = k\ z\ d\overline{M}/dz$$

If, however, you do not know the local shear, but instead know the value of the wind speed $\overline{M}_2$ and $\overline{M}_1$ at the heights z_2 and z_1 respectively, then you could use the following alternative expression to find u_*: $u_* = k\,Z^*\,\Delta\overline{M}/\Delta z$

Derive the exact expression for Z*.

27) Given the following wind speed data for a neutral surface layer, find the roughness length (z_0), the displacement distance (d), and the friction velocity (u_*):

z (m)	$\overline{M}$ (m/s)
5	3.48
8	4.34
10	4.66
20	5.50
30	5.93
50	6.45

28) Suppose that you have made micrometeorological measurements over a wheat field, where $z_0 = 1$ cm. Assume $(g/\overline{\theta}) = 0.0333$ m/(s^2 K).

a) One afternoon, $u_* = 0.36$ m/s and $\overline{w'\theta'} = 0.20$ K m/s at the surface. What is the value of the Obukhov length? Plot the mean wind speed as a function of height from the surface to 50 m.

b) Later, during the night when $\overline{w'\theta'} = -0.05$ K m/s at the surface, you measured the same wind speed at a height of 20 m as you observed during the afternoon (from your answer to part a). Find the friction velocity u_* and the Obukhov length (L) for this nighttime situation, and plot the resulting wind speed profile between the surface and 50 m .

29) a) Given the following was observed over farmland on an overcast day:

$u_* = 0.4$ m/s , $d = 0$, $\overline{M} = 5$ m/s at $z = 10$ m . Find z_0 .

b) Suppose that the following was observed on a clear night (no clouds) over the same farmland: $L = 30$ m, $u_* = 0.2$ m/s . Find $\overline{M}$ at $z = 1$, 10, and 20 m.

c) Plot the wind speed profiles from (a) and (b) on semi-log graph paper.

30) Given the following data: $\overline{w'\theta'} = 0.2$ K m/s, $z_i = 500$ m, $g/\overline{\theta} = 0.0333$ m s^{-2} K^{-1}

$u_* = 0.2$ m/s, $k = 0.4$, $z_0 = 0.01$ m, $z = 6$ m

Find:

a) L (the Obukhov length)

b) z/L

c) $\partial \overline{U}/\partial z$ at z=6 m. (Hint: use diabatic surface layer similarity)

31) The wind speed $= 3$ m/s at a height of 4 m. The ground surface has a roughness length of $z_0 = 0.01$ m. Find the value of u_* for

a) A convective daytime boundary layer where Ri $= -0.5$.

b) A nocturnal boundary layer where Ri $= 0.5$.

10 Measurement and Simulation Techniques

Our fundamental understanding of the boundary layer comes from measurements. Most measurements are made in the *field*, some are made in *laboratory tank* or *wind tunnel* simulations, and some are samples from *numerical simulations*. Theories and parameterizations, such as presented in earlier chapters, are valuable only if they describe the observed boundary layer behavior.

In this chapter we will outline the measurement categories, systems, field experiments, and techniques that are available to acquire a variety boundary layer data. Details of instrument design and operation will not be covered here, but they can be found in other references (see review by Lenschow, 1986).

10.1 Sensor and Measurement Categories

To make boundary layer measurements, we need three components: (1) a *detector* or *sensor*, (2) an *encoder* or *digitizer*, and (3) a *data logger*. Most detectors are devices for which the physical characteristics (size, resistance, etc) change as a function of the variable being measured. Thus, virtually all detectors are *analog* in nature, providing a continuously varying output as a function of continuously varying meteorological conditions. This output signal must then be sampled to produce a discrete digital record, using some sort of encoder or *analog-to-digital converter*. The resulting discrete series of data must be recorded, often on magnetic tape, magnetic disks, or optical disks. *Instrument system* or *instrument package* is the name given to the set of all three components listed above.

Additional components are also required: (4) an *instrument platform*, (5) a means of *calibration*, and (6) *display devices*. Platforms such as a tower or aircraft are complex and expensive in their own right, and can often hold or carry many instrument systems. Power supplies, cooling, sheltering or shielding, and accommodations for humans are sometimes part of the platform. Calibration against known standards either should be performed periodically during the measuring program, or should be accomplished continuously as an intrinsic function of the sensor or instrument package. Uncalibrated data is worthless data. Finally, the measured values should be displayed on printers, plotters, or video displays in order to confirm the operation of the instrument. Display can be real time showing the raw or digitized values, pseudo-real-time (within 5 to 30 minutes) showing summary tables and average statistics, or daily post-processing. It would be embarrassing to perform a long field experiment only to discover in the post experiment analysis phase that all the data is bad, missing, improperly recorded or uncalibrated.

Boundary layer data are split into two categories: those obtained from *mean value* sensors, and those from *fast-response* sensors. Fast-response sensors are used for measuring the turbulent fluctuations, from which we directly calculate turbulence kinetic energies, fluxes, and higher moments. These instruments are often complex, delicate and expensive. In the surface layer, turbulent eddies are relatively small and short lived, requiring sensors that have faster responses than ones used to measure eddies in the middle or top of the convective boundary layer. At nighttime, turbulence is sometimes weaker, requiring sensitive sensors that have a good *signal-to-noise ratio*.

Fast-response instruments are also generally more costly to maintain, and require more expensive data logging equipment. The sensors are often small and delicate in order to achieve the fast response, and are thus vulnerable to damage by insects, precipitation, and mishandling. Either protective shrouds are needed, or replacements must be available. Since virtually all modern data logging is digital, the fast response signals are digitized at rates of once per second up to 50 or 100 times per second, resulting in data sets of thousands to hundreds of thousands of values for just a 20-minute record. The data logger must be able to ingest data at these high rates, and be able to store the large volume of data that accumulates.

If only mean values are required, then less-expensive, slower-response, more-durable instruments can be used. Based on profiles of mean variables, we can often indirectly calculate turbulence energies and fluxes. Most field experiments involve a mixture of mean and fast-response sensors, depending on the budget and the goals. Many of the fast-response sensors can also be used to find mean values, but not all.

Direct sensors are ones that are placed on some instrument platform to make *in-situ* measurements of the air at the location of the sensor. *Remote sensors* measure waves that are generated by, or modified by, the atmosphere at locations distant from the sensor. These waves propagate from the generation or modification point back to the sensor. *Active remote sensors* generate their own waves (sound, light, microwave), and have *transmitter* and *receiver* components. *Passive remote sensors* have only receiver components, and measure waves generated by the earth (infrared, microwave), the atmosphere (infrared), or the sun (visible).

Disadvantages of direct sensors include modification of the flow by the sensor or its platform, the fact that it gives a potentially unrepresentative point value, and the requirement to physically position the sensor in the part of the boundary layer where the measurement is to be made. Disadvantages of remote sensors include their size (large antennas or receiving dishes), cost and complexity, inability to measure certain boundary layer characteristics, their small signal to noise ratio which necessitates averaging over relatively large volumes (i.e., point values are not reliable), and the fact that the waves may be modified as they propagate back to the sensor. At present satellite-based remote sensors do not have adequate vertical resolution to provide much data in the boundary layer.

Advantages of direct sensors include sensitivity, accuracy, and simplicity. Advantages of remote sensors include the fact that they can quickly scan a large volume, area, or line of the atmosphere while remaining stationary on the ground, or located at convenient aircraft altitudes.

An additional trade-off between direct and remote sensors is the space/time required to acquire *statistically stable turbulent moments* (fluxes, variances). Lenschow and Stankov (1986) have shown that measurements by airborne direct sensors in convective boundary layers theoretically must be made over distances of about 10 to 100 km for turbulent variances, about 100-10,000 km for scalar fluxes, and 1000 to 100,000 km for stress, in order to achieve 10% accuracy. In many cases, it is impossible for flight legs of the required distance to be flown. Measurements over shorter distances would yield statistically nonrobust data that has a lot of scatter, such as demonstrated by the fluxes plotted in Figs 3.1 - 3.3 that were computed from data over 20 km flight legs. Remote sensors, which can scan large volumes in short times, can potentially achieve a much better statistical average, and are becoming more popular.

10.2   Sensor Lists

The large variety of sensors that have been developed is staggering. In the lists below, sensors that have been frequently used for fast-response measurements are flagged with (*). After each sensor is a brief explanation of the physical properties involved.

10.2.1 Temperature - thermometers

<u>Direct Sensors:</u>
liquid (mercury or alcohol) in glass - liquid expands higher into glass tube
wax thermostat - wax expands; e.g., automobile thermostats
bimetallic strip - two metals expand at different rates, causing bimetallic strip to bend
thermocouple * - junctions between different wires (copper & constantan) generate voltage
thermistor * - semiconductor electrical characteristics change
resistance wires/hot wires * - electrical resistance changes
sonic anemometer/thermometer * - speed of sound varies
melting/freezing points for different chemicals - e.g., wax melts on paper, changing color
quartz crystals - electronic resonance frequency changes

liquid crystals - optical characteristics, like color or darkness, change
crickets - chirp rates change
chemicals - reaction rates change

Remote Sensors:
microwave sounders - microwave propagation/ refractive index
DIAL lidar - dual wavelength laser radar
mirages - refraction of visible images
scintillation * - refraction of laser light beam in regions of temperature variation
radiometers * - emitted (black body) radiation
sodar * - scattering of sound off of regions of temperature variation

10.2.2 Humidity - hygrometers

Direct Sensors:
psychrometers - water evaporates from wick over thermometer, causing cooling
hair hygrometer - absorption of water causes organic fiber to expand
chilled mirror (dew pointer) - surface cooled to cause condensation
hygristor - electrical resistance of a chemical on a glass plate changes with humidity
Lyman-alpha and IR hygrometers * - absorption of radiation by water vapor
microwave refractometer * - index of refractivity changes
nephelometer - visibility changes
chemical reactions and alterations in biological life forms

Remote Sensors:
lidar - aerosol swelling in humid air
radar - variations in refractive index

10.2.3 Wind: Velocity - anemometers
Direction-vanes

Direct Sensors:
cup - drag against cups causes rotation on axis perpendicular to wind
propeller - turns blades on axis parallel with wind
Gill * - this is a lightweight variety of propeller anemometer
hot wire *- electrical current needed to maintain temp. of wire against cooling of wind
sonic * - speed of sound
pitot/static * - dynamic pressure increase associated with deceleration of wind into orifice
aspirator - pressure decrease associated with air flow across an orifice
drag sphere - drag force experienced by sphere, measured by strain gauges
pivot arm - hanging rod or plate that is blown against gravity or spring
gust vane * - measures v' or w' via lateral forces on vane (strain gauges or movement)
pibal - pilot balloon tracked by theodolite or radar
venturi - pressure drop related to Bernoulli effect

rotometer - fluid moving vertically in conical tube lifts a ball in tube
Beaufort scale - wave height or sea state
Fujita scale - damage assessment
wind vane - points in horizontal compass direction wind comes from
bivane - pivots up/down as well as left-right to give elevation angle and compass direction

Remote Sensors:
Doppler radar - Doppler frequency shift parallel to beam of microwaves
Doppler sodar - Doppler frequency shift parallel to path of sound
Doppler lidar - Doppler frequency shift parallel to laser beam
lidar - manual/statistical (autocorrelation) tracking of aerosol structures, any angle to
beams

10.2.4 Pressure - barometers & microbarographs

Direct Sensors:
aneroid elements * - compression of evacuated container against restoring force of spring
capacitive elements * - capacitance of sealed volume changes as container walls move
mercury in glass - weight of mercury balances pressure

Remote Sensors:
None that use wave propagation directly, but some that measure temperature and velocity
    fluctuations as mentioned above, and infer pressure perturbations as residual from
    governing equations.

10.2.5   Radiation - radiometers

Most radiometers measure the radiative warming of a tinted surface, where the surface
is protected from conductive and convective heating inside a glass or plastic hemispherical
or spherical chamber. Depending on the type of tinted surface and the exposure, the
radiometer can be designed to measure radiation in specific frequency bands coming from
specific directions:

radiometer - total radiation (all wavelengths) from all directions within one hemisphere
                         (upward or downward) hitting a plane surface that is parallel to
                         local horizontal
net radiometer - difference between top and bottom hemispheres
pyranometer - short wave (solar) radiation, one hemisphere, plane horizontal surface
net pyranometer - difference between top and bottom hemispheres, short wave only
pyrheliometer - short wave direct beam radiation normal to surface (shielded from diffuse)
diffusometer - pyranometer with a device to shade it from direct sunlight
prygeometer - long wave (IR), one hemisphere, plane horizontal surface.

10.3 Active Remote Sensor Observations of Morphology

Radars (radio detection and ranging) transmit microwaves, *sodars* sound, and *lidars* light. Some use pulsed transmissions, while some are continuous. By measuring the time from when a pulse was transmitted until a signal is returned, the distance to the feature that returned the signal can be calculated. If pulses or *shots* are aimed in different directions, then the resulting signals can be constructed into *scans*. By measuring the frequency (Doppler) shift in the signal, radial velocities can be found.

Active remote sensors can scan planes or volumes in the atmosphere, allowing us to compose "snapshot" like pictures of the boundary layer. With this capability we can examine the shape and form of turbulent eddies and convective elements (i.e., the *morphology* of turbulence structures). By *looping* through a series of successive snapshots in the cinematographic sense, we can watch the evolution of these elements. Noonkester (1979) has suggested a way for cataloging the different types of structures visible using remote sensors.

For stationary sensors on the ground, vertical scans at a fixed azimuth angle give what is called an *RHI (range-height indicator)* display. This is essentially a vertical cross section of the boundary layer. Scans through a range of azimuth angles, but with the elevation angle fixed, gives the *PPI (plan-position indicator)* display. For very low elevation angles, the PPI display is almost the same as a horizontal cross section. Modern data analysis algorithms allow volume scans to be remapped into horizontal or vertical cross sections.

Pseudo-RHI scans are made by fixed sensors looking only vertically. As the wind blows boundary layer structures over the sensor, the resulting sequence of shots with time can be interpreted as a vertical cross section by employing Taylor's hypothesis. True vertical cross sections are obtained from downward or upward-looking sensors mounted on aircraft.

In the absence of precipitation or chaff particles, radar and sodar detect variations in the refractive index of the air that are on the same order as the wavelength of the sensor. For radar, these fluctuations are most strongly related to moisture fluctuations of centimeter scales, while for sodar they are related to temperature fluctuations of meter scales. If there are regions in the boundary layer where such moisture or temperature variations exist, whether as active turbulence or as footprints of former turbulence, then some of the wave energy transmitted from the remote sensor will be scattered off of this region back to the sensor. So little energy is scattered back to the detector that very powerful pulses of energy must be transmitted, and very sensitive receivers must be designed to measure the returns. In the simplest displays, the regions of higher-intensity returned signals are plotted as a bright spots, relative to the dark areas of little returned signal.

10.3.1 Radar

Sketches of returned signal for radar RHI returns are shown in Fig 10.1a. In fair weather conditions, the boundary layer is often more humid than the free atmosphere. Thus, centimeter-scale eddies on the interface between the mixed layer and the free atmosphere create strong returns. Within the mixed layer there is little returned energy in spite of the strong turbulence, because the humidity is high everywhere. Similarly, there is usually little returned energy in the free atmosphere, because the humidity is low everywhere. Sometimes, however, layers of moist air aloft allow clear air turbulence regions to be visible on radar, exhibiting a characteristic "cats eye" pattern.

Normal weather radar is neither powerful nor sensitive enough to detect most clear-air boundary layer phenomena; so only a smaller number of research radars have been used for this work. These research radars can detect turbulence from near the surface up to the top of the troposphere. A variety of ground-based automated microwave *atmospheric sounders* have also been developed for deployment and operational gathering of wind and other meteorological data.

Fig 5.8 shows the difference between mechanical turbulence (forced convection) and convective turbulence (free convection) as viewed by radar. Free convection exhibits the characteristic upside-down "U" shaped returns associated with the tops of thermals. Mechanical turbulence creates a more random and torn return, with tilts in the displayed structures giving evidence of wind shear. Radar returns for stable boundary layers are plotted in Fig 10.1d. Fig 10.1g shows PPI scans of free convection, where the horizontal slices through thermals show circular or "donut"-like patterns around the perimeter of the thermals.

Even in cloud-free boundary layers, there may be enough insects to enhance the scattering back to the radar. In some field experiments, chaff is dispersed to create even more returns. Chaff are fine hair-like strands of plastic, coated with a thin layer of aluminum or other conductor. These strands are cut to the wavelength (or fraction or multiple of the wavelength) of the microwaves, and are dispensed in the air either from aircraft or from the ground. Their light weight causes their terminal velocity to be almost zero, and allows them to be carried aloft and dispersed by the turbulence.

10.3.2 Sodar

Sodars (sometimes called *acoustic sounders*) send an audible "beep" of sound, generated by powerful loudspeaker horns. The returned sound is usually focused by a parabolic dish into a sensitive microphone, which serves as the detector. To reduce the amount of ambient background noise entering the system, shelters and acoustic screens are usually erected around the dish/microphone receiver, or else the receiver is placed in a hole below ground. Otherwise, noise from rustling leaves, wind noise, traffic, and conversation can contaminate the desired signal.

Sodar detects the interface at top of the mixed layer, using variations between the cooler mixed layer air and the warmer temperature inversion air capping the mixed layer.

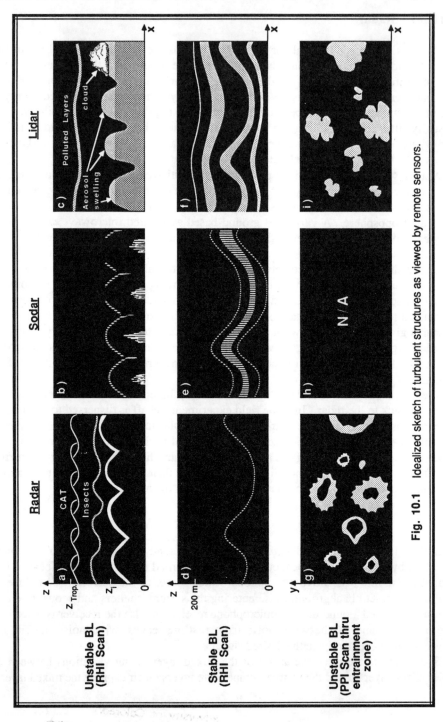

Fig. 10.1   Idealized sketch of turbulent structures as viewed by remote sensors.

Unfortunately, sound is attenuated so rapidly in the atmosphere that it is difficult to detect structures beyond a range of about 1 km. Since many mixed layers grow to heights above 1 km in the afternoons, the sodar is useless for determining z_i except in the early morning. Sodars can see the roots of thermals in the surface layer (see Fig 10.1b). Sodar is particularly valuable at night, where it detects buoyancy waves and turbulence in the stable boundary layer (Fig 10.1e).

10.3.3   Lidar

Lidars transmit laser light, which is scattered off of air molecules, cloud droplets and aerosols in the boundary layer. The returned light is collected in a telescope and focused on a photomultiplier detector, after which it is amplified, digitized and recorded. Because the source of many aerosols is the earth's surface, the boundary layer often has a higher concentration than the free atmosphere. As a result, the whole boundary layer often provides strong lidar returns, which stands out in the display as a deep white area, compared to the darker free atmosphere. Thus, the lidar sees the boundary layer air, rather than just the interface at the top of the boundary layer.

Fig 10.1c shows sketches of RHI lidar returns for a convective mixed layer, where individual thermals are often visible within the polluted mixed layer. Sometimes elevated haze layers are also visible. Aerosol swelling in the high relative humidities at the top of the mixed layer cause enhanced lidar returns, which show up brighter on the display. Clouds have such strong reflection and absorption that a strong return is generated at the base or the side of the cloud illuminated by the light. After a few tens of meters into the cloud, the light is so attenuated that there are nor more returns. Thus, the cloud appears opaque, and leaves a shadow behind it. Pollutants are also easily tracked by lidar.

In statically stable conditions, there is frequently a change of wind direction with height within the lowest few hundred meters of the ground. As a result, air from different source regions with different aerosol contents are advected into the region. This frequently causes a layer-cake appearance, with thin horizontal strata of different brightness on the RHI display (Fig 10.1f). When buoyancy waves exist, the strata also look wavy.

Lidar PPI displays of convective mixed layers (Fig 10.1i) show irregular perimeters to the thermals. Time lapse loops indicate lateral entrainment, or *intromission*, into the thermals analogous to the vertical entrainment at the top of the whole mixed layer. Near the center of the thermals there are sometimes undiluted cores.

10.4 Instrument Platforms

Fig 10.2 shows a sketch of the instrument platforms to be described below. These platforms provide the physical/structural support for the sensor, shielding for the sensor, electrical power, and sometimes chemicals necessary for the measurements.

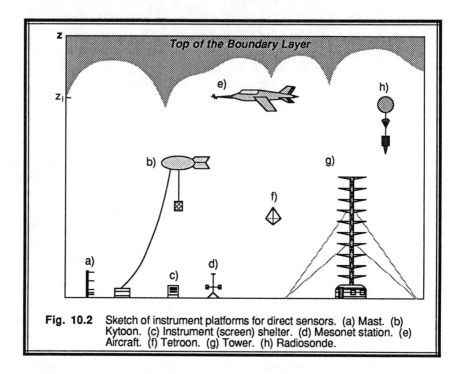

Fig. 10.2   Sketch of instrument platforms for direct sensors. (a) Mast. (b) Kytoon. (c) Instrument (screen) shelter. (d) Mesonet station. (e) Aircraft. (f) Tetroon. (g) Tower. (h) Radiosonde.

10.4.1   Instrument Shelters

The classical white louvered instrument shelter, called the *Stevenson screen*, is mounted on a stand to place the instruments inside at a height of about 2 m above the local ground level. The shelter protects the instruments from the sun, but it also filters out much of the smaller scale turbulence. As a result, we can expect to measure mean quantities such as temperature, humidity, and pressure within the shelter. Otherwise, this platform is not used very often for micrometeorological measurements.

10.4.2   Masts

A mast is a pole upon which instruments can be placed at a variety of heights. Sometimes scaffolding is used to erect a short tower. A typical mast height is 10 m to 50 m. It is relatively inexpensive, and can be erected with simple equipment. Because of its limited height, it is primarily useful for surface layer measurements. Sensors are often placed closer together at the bottom of the mast than at the top, because profiles of temperature, wind, and humidity vary approximately logarithmically with height.   For example, at the Minnesota experiment (see Table 10-1), temperature sensors were placed at $z = 0.5, 1, 2, 4, 8, 16,$ and 32 m. Such masts can be easily transported and erected for a field program, and then dismantled when the experiment ends. Wires carry the signals down the mast to a data logger or data trailer close by.

10.4.3   Portable Mesonet Stations

A variety of quasi-portable surface stations exist that can be deployed for field experiments. A typical implementation will consist of temperature, humidity, rain, pressure, and radiation measurements at $z = 2$ m, and a 10 m mast to measure wind speed and direction. Some stations have additional sensors to measure turbulent fluxes directly, while others make measurements of mean variables at two or more heights to infer fluxes (see Section 10.7). Data is transmitted to a central recording site via satellite, radio, or telephone lines.

10.4.4   Towers

Tall expensive towers have been erected at a few sites for permanent use. Examples include the 213 m tower near Cabauw, about 50 km southeast of the North Sea coastline in The Netherlands. Another is the 300 m Boulder Atmospheric Observatory (BAO) tower in Colorado, about 25 km east of the Rocky Mountains. Occasionally, existing television transmitting towers are instrumented, such as the 444 m KYTV tower in Oklahoma City. Towers are useful for studying the nighttime and early morning boundary layers that are shallow.

These are large structures with built-in elevators and many support guy wires. Because the tower is so large, it disturbs the flow close to it and downwind of it. For this reason, these towers have large horizontal booms that project horizontally away from the tower at different heights, upon which sensors are mounted. At each height, there are often booms projecting in 2 or 3 different compass directions, with the expectation that at least one of the booms will be in the upwind direction. Permanent buildings housing the communications, data logging, maintenance, and computer facilities are sometimes built near the tower.

10.4.5   Kytoons

A kytoon is an aerodynamically shaped helium-filled plastic balloon that is tethered to a winch on the ground. Instead of being blown down by the wind, the shape allows it to soar upward like a kite — hence the name kytoon. In typical applications, a sensor package is suspended a short distance below the balloon on lines different from the tether line. To make measurements at a variety of heights, the winch is used to draw in or feed out more line until the desired height is reached. The balloon is kept at each height of interest for 5 to 30 minutes to get a statistically stable sample, before changing its altitude. Also, the balloon can make measurements while it is rising or descending, allowing soundings to be recorded. In some cases, the instrument package sends its signals down electrical wires attached to the tether cable, while for other kytoons a transmitter radios the information to the ground. Although tethers on the order of 1.5 to 2 km are available, flight regulations sometimes restrict deployment to altitudes below about 800 m. Kytoons are much more portable than tall towers, and can easily be used at temporary field experiments, but they are limited to light winds.

Occasionally, tethered balloons are flown with instruments deployed at a variety of heights along the tether cable. Although these offer the advantage of simultaneous measurements at a number of heights, they are more difficult to deploy and pull back in.

10.4.6   Free balloons

A *radiosonde* is an expendable instrument and transmitter package attached below a free-flying helium balloon that measures temperature, humidity, and pressure. *Rawinsondes* also provide wind information, either by a tracking antenna on the ground, or via LORAN or OMEGA receiving systems in the instrument package itself. Observations from radiosondes or rawinsondes are called *raobs*. Special balloons and instrument packages are designed for boundary layer experiments, where the balloon rises more slowly, and the instrument package makes measurements at more heights within the boundary layer. Although each balloon package is not too expensive, the cost during a field program can be surprisingly expensive if balloons are launched frequently. For boundary layer work, the raob is difficult to utilize because it is often an unrepresentative point measurement that provides neither a time nor a space average. These balloons drift away from the launch site and rise until they burst, allowing the instrument package to parachute to the ground.

10.4.7   Tetroons

A special class of free balloon is the constant pressure balloon. Instead of being made of latex like radiosonde balloons, these are made of an unstretchable material such as mylar plastic. These balloons are often constructed in the shape of a tetrahedron — hence the name tetroon. When inflated and properly balanced, the balloon will rise to an altitude where the air density matches the balloon system overall density, allowing the balloon to theoretically stay at a constant (pressure) altitude. In reality, the balloon will often make large amplitude vertical oscillations about its mean altitude. Sometimes instrument systems are suspended from the balloon, while for other experiments a radar reflector or transmitter is suspended, to allow the balloon to be tracked for tracer dispersion studies.

10.4.8   Aircraft

A variety of aircraft ranging from model drones, ultralights, gliders, motorgliders, light single and multiengine propeller aircraft, multiengine turboprop, multiengine jet, and cabin-class civilian and military transports have been used as platforms for boundary layer study. Typically, instruments are mounted on special booms projecting forward from the nose or the wings of the aircraft, in order to get the sensors out of the flow disturbed by the aircraft itself. Some sensors are built into the radome, while others use remote sensing techniques to detect the flow in front of the aircraft. The larger aircraft allow on-board computer equipment to process the data, and airborne scientists can view real-time displays to help direct the experiment. Turbulence sensors must have a very fast response, because typical aircraft speeds are 50 to 100 m/s.

Typical boundary layer flight patterns consist of level horizontal flight legs of 10 to 20 minute duration in order to get a large enough sample to generate stable turbulence statistics. These legs are flown either over fixed tracks on the ground, or are flown in "L" patterns with one leg parallel to the current wind direction and the other leg in the cross-wind direction. In addition to legs flown at a variety of heights in the boundary layer, some ascent or descent soundings are made that yield much more representative vertical profiles than raob soundings.

As described in section 10.1, it might be impossible to fly a pattern of the proper length to gather data with robust statistics. Flight legs that are too short have potentially large sampling errors, yielding large scatter in the data. Flight legs that are long enough for stable statistics can consume so much flight time that stationarity within the BL might be violated. The way around this problem is to fly a faster aircraft, but the airborne sensors must have a correspondingly faster response, which might be difficult or impossible.

10.4.9 Platforms for remote sensors

As of the writing of this text, most remote sensing systems were so large that they were limited to either the larger aircraft, or to fixed or mobile surface sites.

10.5 Field Experiments

Because fast-response turbulence data is not routinely collected by most operational weather services, it is necessary to conduct special field programs. Some of the early experiments focused on surface layer data over flat uniform terrain or oceans. Later experiments probed higher in the boundary layer and/or picked a nonuniform site or shorelines. Table 10-1 summarizes the major experiments that were either completely devoted to boundary layer measurements, or had a major boundary layer subprogram.

Many of these experiments have resulted in the publication of data books, which are available from the lead principal investigator, the sponsor, or government printing offices. These field experiment names appear frequently in the literature, because they provide the data set against which theories and models are tested.

Table 10-1. Chronological summary of field programs that gathered boundary layer data. Abbreviations used for data collected: A=aircraft, B=raobs, C=chemistry, D=dispersion, E=surface energy budget, F=turbulent fluxes, G=(below) ground & soil data, g=geostrophic, H=hydrologic & precipitation, K=radiation, K2=kytoons, L=lidar, O=ocean, P=pressure, p=vertical profile, R=radar, S=sodar, s=surface, T=temperature, TKE=turbulence, U=uniform land, V=wind, WX=weather & clouds.   (* = incomplete information)

Short Name	Date	Location	Data Book or Reference, Long Name, Data Gathered
Great Plains	Aug-Sep 53	O'Neill, Nebraska, USA	Lettau & Davidson (1957). All except R,S,L,O.
Wangara	Jul-Aug 67	Hay, Australia	Clarke et al (1971). All except TKE,H,F,D,C,L,R,S,A,O
Kansas	Jul-Aug 68	SW Kansas, USA	Izumi (1971). 32m mast - surface layer data: F,TKE,U, Tp,Vp.
BOMEX	May-Jul 69	Barbados, Caribbean	Kuettner & Holland (1969). BOMEX=Barbados Oceanog. & Meteor. Exp. All except U,S,L,C,D
Marsta	69-71	Marsta, Sweden	Smedman-Högström ... (1973), Högström 1974), (DeHeer-Amissah (1981). All but H,D,C,A,K2,L,R,S,O
METROMEX	Summers 71-76	St.Louis, Missouri, USA	Changnon (1981). METROMEX = Metropolitan Meteor. Exp. All except O,U,L
VIHMEX	May-Sep 72	Carrizal, Venezuela	Betts & Miller (1975). VIMHEX=Venezuelan Internatl. Meteor. & Hydrolog. Exp. R,B,Ts,WX,Vs
Minnesota	Sep 73	NW Minnesota, USA	Izumi & Caughey (1976). 32m mast-sfc layer:F,TKE,U, Tp,Vp,K2.
Cabauw/C	1973-1984 (Continuous)	Cabauw tower, The Netherlands	Monna and Van der Vliet (1987), Wessels (1984). All except TKE, F, D, A, K2, L, R, O
Koorin	Jul-Aug 74	Daly Waters, Australia	Clarke & Brook (1979). All except TKE,H,D,C,L,R,S,O
GATE	Summer 74	Tropical Atlantic & W.Africa	Multinational, organized by WMO. Kuettner & Parker (1976); GATE=GARP Atlantic Tropical Exp., GARP = Global Atmos. Res. Prog. All but U, L, D, G.
AMTEX	Feb 74, 75	East China Sea, off Japan	Lenschow & Agee (1976). AMTEX=Air Mass Transformation Exp. All except G,D,C,K2,L,S,U
BLS77	May-Jun 77	Chickasha, Olkahoma, USA	Hildebrand (1980). BLS77=Bound. Layer Structure 1977. T,V,P,p,s,TKE,K,WX,F,R
VOVES	Jul 77	Voves-Villeau, Central France	André & Lacarrere (1980). All except H,D,C,K,L,O
Cabauw/E	1977-1979	Cabauw tower, The Netherlands	Monna and Van der Vliet (1987), Nieuwstadt (1984). All except D, A, K2, L, R, O
PHOENIX	Sep 78	At BAO tower Boulder, Colorado, USA	Hooke (1979). All except H,D,C,U,O
SESAME	Apr-Jun 79	Oklahoma, USA	AMS SESAME News (1979-82): SESAME=Severe Environ. Storms & Mesoscale Exp. All except O,S,L,D

CCOPE	May-Aug 81	Miles City, Montana, USA	Knight (1982). CCOPE=Cooperative Convective Precip. Exp. All except O,S,L,K2,C,D,G
PUKK/ KONTUR	Sep-Oct 81	German Bight, North Sea coast	Hoeber(1982), Kraus(1982), Kontur(1985).KONTUR =convect. &turb. over sea. All except F,C,K,L,R,D,G,H
COAST	May 83	Dutch coast	Weill, et.al. (1985), Desbraux & Weill (1986). Coastal meteor. All except D,C,K,L,R
BLX83	May-Jun 83	Chickasha, Oklahoma, USA	Stull & Eloranta (1984), Stull et.al. (1988). BLX83=Bound. Layer Exp. 1983. All except H,D,O
Øresund	Summer 84	Denmark & Sweden	Gryning (1985, 1986). Mesoscale dispersion land-water-land. All except C,K,L
PHOENIX II	May-Jul 84	At BAO, Colorado, USA	Lilly(1984). BL evolution. All except G,g,H,D,C,K,L,S,U,O
MESO-GERS	Sep-Oct 84	SW France	Weill, et.al. (1987). Mesoscale. All except G,D,C,K,L,O
PRE-STORM	May-Jun 85	Oklahoma & Kansas, USA	Cunning, J.B. (1986). PRE=Preliminary Regional Exp. of STORM-Central. All except G,D,C,K,L,S,O
ABLE	Jul-Aug 85 Apr-May 87	Manaus, Brazil	Harris (1987), also J. Geophy. Res. ABLE=Amazon . Boundary Layer Exp. All except D,S,O
DYCOMS	Jul-Aug 85	East Pacific off California, USA	DYCOMS=Dynamics and Chemistry of Marine Stratocumulus. All except s,G,g,H,K,D,L,R,S,U,E*
GALE	Jan-Mar 86	New England USA & Atlantic	AMS (1985). GALE=Genesis of Atlantic Low Exp. All except U,D,G
FASINEX	Feb-Mar 86	Atlantic near Bermuda	Stage & Weller (1985,86). FASINEX=Frontal Air Sea Interaction Exp. All except U,C,D,G
HAPEX	May-Jul 86	SW France	André, et.al. (1986, 1988). HAPEX = Hydrologic-Atmos. Pilot Exp. All except O,D,L,C
HEXOS	Autumn 86	North Sea, The Netherlands	Katsaros, et.al. (1987). HEXOS=Humidity EXchange Over the Sea. All except G,K,U,C,D.
TAMEX	May-Jun 87	Taiwan	TAMEX=Taiwan Area Mesoscale Exp.*
FIFE	May-Nov 87	Manhattan, Kansas, USA	FIFE=First ISLSCP Field Exp., ISLSCP=Internatl. Satellite Land Surface Climatology Proj.*

10.6 Simulation Methods

Although field measurements represent "truth" by definition, it is a truth that is composed of the superposition of many simultaneous effects and processes. For a few situations, we can attempt to isolate or focus on one specific process by the careful selection of a field site (e.g., uniform terrain) or weather pattern (e.g., fair weather). However, we can never fully isolate any one process, and the weather is rarely reproducible. Also, it is difficult to do sensitivity studies by systematically altering certain physical parameters, and then measuring the effects.

We can partially circumvent these difficulties by creating an artificial turbulent domain where only a limited number of processes or boundary conditions acts on the flow. Such simulations can be created physically in laboratory tanks and wind tunnels, or numerically via computer models. In both cases, we must (1) create a model of the real atmospheric situation, (2) run the model to generate the turbulence; (3) sample the turbulent field much the same way the the real atmosphere is sampled; and (4) compute turbulence statistics (fluxes and means) from the sampled field. If sufficient care is taken to create a realistic model, then we can often learn more, faster, and with less expense than by conducting a field program.

10.6.1 Numerical

In Chapters 2-5 the equations of motion were Reynolds averaged to statistically separate the mean flow from the turbulent perturbations.   Instead of taking that route, suppose we use the raw equations of motion to forecast velocities and temperatures at every point in a (hypothetical) modeled turbulent domain; then we should be able to forecast turbulent motions *deterministically*. For example, in certain regions we might see warm updrafts simulating convective thermals, other points simulating wave motions, and yet others exhibiting apparently random variations characteristic of isotropic turbulence.

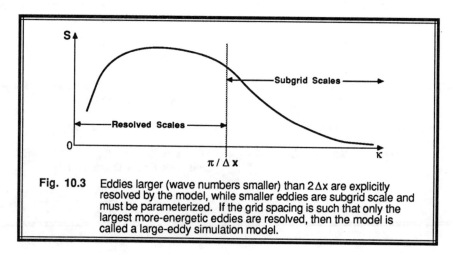

Fig. 10.3   Eddies larger (wave numbers smaller) than $2\Delta x$ are explicitly resolved by the model, while smaller eddies are subgrid scale and must be parameterized. If the grid spacing is such that only the largest more-energetic eddies are resolved, then the model is called a large-eddy simulation model.

Predictability, however, must be split into two categories: (1) *pattern predictability* and (2) *statistical results from pattern forecasts*. Studies have suggested that individual patterns or structures, whether it be a synoptic-scale cyclone or a turbulent-scale eddy, can be accurately predicted in an Eulerian model out to times no greater than the Lagrangian lifetime of the pattern or structure (Stull, 1985). For example, if we were to take our hypothetical numerical model and initialize it with the exact observed wind field including the precise variations in wind field associated with each of the turbulent eddies present, then we could anticipate forecasting the evolution of those initialized eddies out to about 5 to 15 min for the thermals, but only out to a few seconds for the smaller-sized eddies. This is pattern or structure predictability.

If we continue the forecast beyond the limit of accurately forecast deterministic patterns, then we will continue to generate eddies and thermals in the model, except these eddies will be happening at different times and places with different intensities than the true observed eddies. If the model was designed correctly, then the <u>statistics</u> (TKE,

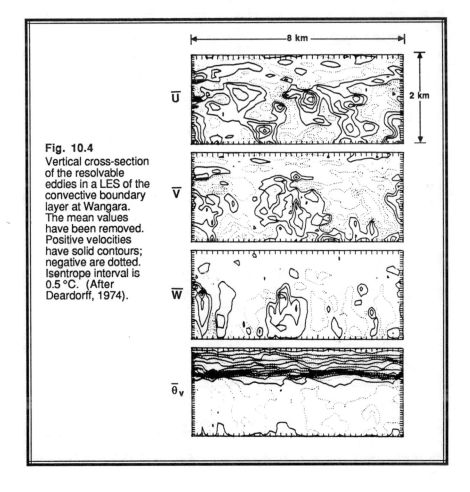

Fig. 10.4
Vertical cross-section of the resolvable eddies in a LES of the convective boundary layer at Wangara. The mean values have been removed. Positive velocities have solid contours; negative are dotted. Isentrope interval is 0.5 °C. (After Deardorff, 1974).

turbulent fluxes, etc.) calculated from this forecast will be close to those observed. In other words, we will have accurate statistical results from deterministic pattern forecasts, even though the simulated patterns themselves do not have a one-to-one correspondence with the true patterns. Such statistical accuracy can be maintained for days of simulated turbulence forecasts.

For the remainder of this section we will focus on such statistically-accurate pattern forecasts. In fact, most modelers do not have observed wind and temperature fields with a resolution that includes the effects of all the observed eddies. Instead, they (1) initialize the model with the observed mean wind and temperature and mean boundary forcings, (2) "kick" start turbulent motions by imposing a pseudo-random temperature and/or velocity perturbations on the mean field for just the first one or two timesteps, and (3) hope that statistically realistic structures develop. Thus, there is no intention of making precise pattern forecasts, even over the first few minutes.

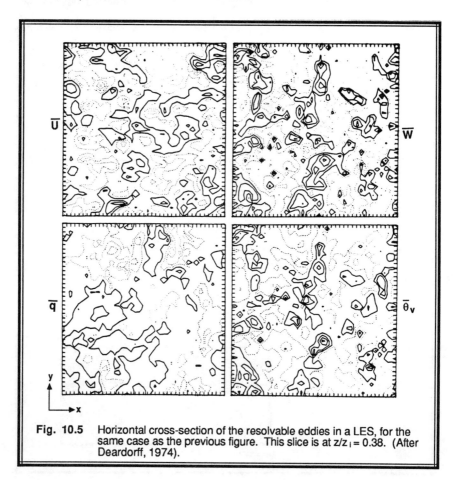

Fig. 10.5   Horizontal cross-section of the resolvable eddies in a LES, for the same case as the previous figure. This slice is at $z/z_i = 0.38$. (After Deardorff, 1974).

Because of computer limitations, numerical models are limited to a finite number of grid points or wavenumbers for all practical purposes, instead of the infinite number of points in our hypothetical model.   Thus, each modeled grid point must represent the average of a small volume of fluid in its immediate neighborhood. Given a grid point spacing of Δx, we know from the Nyquist frequency arguments in the previous chapter that the smallest feature that can be represented has a size of $2\Delta x$. Features of this size and larger are said to be *resolved* by the model, and represent *grid-scale* phenomena. Features smaller than $2\Delta x$ are *sub-grid scale*; their affect on the flow must be parameterized because they can not be explicitly resolved.

Deardorff (1972) pioneered the use of such numerical models for boundary layer simulations, while Wyngaard and Brost (1984) and Moeng (1984a, 1986, 1987) have updated versions. Computer limitations in those models restricted grid-point spacings to about 150 m in the horizontal and 50 m in the vertical, within a domain of $(x,y,z) =$ (5,5,3) km. These models were called *large-eddy simulation* (LES) models (see Fig 10.3), because only the largest-size turbulent eddies could be resolved.   At each of the over 60,000 grid points in a typical model, forecasts were made for θ, q, u, v and w, with additional diagnostic equations for p. At the time this text was written, such a model would take about an hour of computer CPU time for each hour simulated. The subgrid-scale turbulence was parameterized in these models using schemes ranging from K-theory to second-order closure.

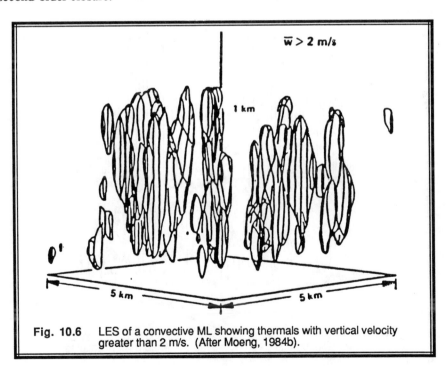

Fig. 10.6   LES of a convective ML showing thermals with vertical velocity greater than 2 m/s. (After Moeng, 1984b).

By averaging over a series of horizontal planes, we can compute mean vertical profiles of various variables. For ***area-averaged*** turbulence statistics such as TKE or fluxes, the subgrid parameterized component must be added to the resolvable-scale values to yield the total value at any height. Such area averages are analogous to area averages produced from remote sensor output, assuming the sensor can scan a volume. In a similar way, statistics can be found by volume-averaging over limited subvolumes of the domain. These ***volume averages*** are more closely related to the volume averages of real data that to the ensemble averages discussed in Chapter 2.

Convective Mixed Layer. Figs 10.4 and 5 show examples of vertical and horizontal cross sections through Deardorff's LES model, where resolvable-scale turbulent structures that look like thermals are revealed. Fig 10.6 shows Moeng's (1984b) simulation of thermal updrafts in a convective boundary layer, while Fig 10.7 shows an analysis of the temperature structure of thermals. It is obvious that these models forecast pseudo-random turbulent structures in both space and time. In fact, the variation of temperature or velocity with time at any one grid point within the turbulent domain would exhibit a trace similar to Fig 2.1. An example of the area-averaged mean potential temperature, humidity, and TKE profiles are shown in Fig 10.8.

Many of the figures in Chapters 3-5 are composites of real data with numerical simulations, because the simulations filled in the gaps that were not (or could not be) measured in the field.

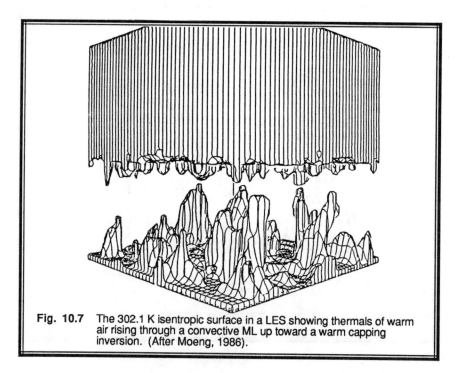

Fig. 10.7   The 302.1 K isentropic surface in a LES showing thermals of warm air rising through a convective ML up toward a warm capping inversion. (After Moeng, 1986).

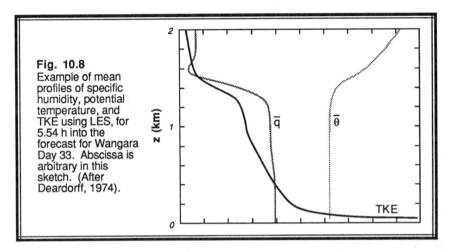

Fig. 10.8
Example of mean profiles of specific humidity, potential temperature, and TKE using LES, for 5.54 h into the forecast for Wangara Day 33. Abscissa is arbitrary in this sketch. (After Deardorff, 1974).

Neutral Boundary Layer. Mason and Thomson (1987) have used LES to simulate structures in neutral boundary layers. Their model shows that smaller eddies dominate near the surface, while larger eddies are more important aloft. These large eddies are often elongated in the direction of the shear vector.

Stable Boundary Layer. Large eddies are suppressed by static stabilities, leaving only the smaller eddies that cannot be resolved by a LES model (as of the date this book was written). Thus, we will have to wait for better computers before we can expect successful LES of the SBL.

10.6.2 Laboratory Tank

When a fluid such as water is used to simulate atmospheric turbulence, one must take care to insure that the simulation has the same dimensionless scales as the atmosphere (Willis and Deardorff, 1974; Deardorff and Yoon, 1984). The atmosphere is a high Reynolds number flow — so high that the Reynolds number is not a governing parameter. However, the dimensions of some laboratory tanks are small enough that the Reynolds number is not very large — meaning that viscosity causes the tank flow to differ from the atmospheric flow. Other numbers, such as the Rayleigh, Nussult, Richardson and Prandtl numbers should be considered. The surface heating rate or mechanical stirring rate must be chosen to make the proper convective or friction velocity scales, respectively, given the higher density of the water.

Convective Turbulence: A large number of simulations of convective mixed layers have been performed by Willis and Deardorff (1974; Deardorff and Willis, 1985, 1987) using water as the working medium. They used a tank that was approximately cubical, with dimensions of about 1 m on each side. The tank would initially be filled with cooler water on the bottom, smoothly varying to warmer water at the top, thereby simulating a stably stratified temperature inversion as is typically observed in the real atmosphere in the early mornings. The bottom of the tank was a metal plate that could be heated to generate convection in the tank. Convection has been induced by other investigators using other techniques such as cooling water below 4°C, or introducing denser brines into the top of the tank.

Using a variety of dyes and neutrally buoyant oil droplets, they could photogrametrically analyze the evolution, turbulent structure, and dispersive characteristics of the mixed layer. Illuminating the tank with thin planes of light allowed cross sections to be photographed, and illuminating the top part of the tank simulated features similar to small cumulus clouds. In situ measurements of temperature and velocity were also made with probes that could be slowly moved through the fluid. This technique was extended to study buoyancy waves and forced convection. Some of their discoveries concerning the entrainment zone and dispersion were later verified in the real atmosphere.

Mechanical Turbulence: The simulation of realistic shear-driven turbulent flows in the laboratory has proved to be very difficult. Circular or racetrack-shaped annulus tanks are often employed to avoid the acceleration/deceleration problems of a long-straight tank. Regardless of whether the fluid is put into motion by pumps or whether the boundaries are moved relative to the fluid, there appear to be major boundary problems associated with fluid circulating around the annulus (Scranton and Lindberg, 1983; Deardorff and Yoon, 1984). Some of the published results from such shear-driven flows are of questionable quality.

For some ocean mixed layer simulations, mechanical turbulence has been generated in tanks using oscillating grids. These mechanically stir the fluid. Although such stirring might be an approximate representation of the breaking of waves and wind-induced mixing at the top of the ocean mixed layer, it is difficult to find an analogy in the atmospheric mixed layer.

Stable Stratification: The non-turbulent flow of stably stratified fluid over terrain and around obstacles has been simulated using *towing tanks*. These are long narrow tanks filled with stably stratified (using salt brine or temperature to vary the density) water, where the mean water flow is zero. A scale model of the terrain or obstacle is then placed upside-down in the top of the tank and towed along the tank's length, while tracer dies are released from tiny holes in the model. Cameras translating at the same speed as the model record the flow, which then appears as fluid moving past the obstacle when the films are played back. These tanks have been particularly valuable for pollution dispersion studies.

10.6.3 Wind Tunnel

Wind tunnels have the advantage of using air as the working medium. A disadvantage is that it is difficult to stratify the flow. For neutral stratification, wind tunnels have been used successfully to study dispersion and flow over hills and around obstacles. In particular, complex physical models of buildings, smoke stacks, and terrain can be constructed and placed in the wind tunnel. Such a wind tunnel simulation is often superior to the corresponding numerical simulation, which is often too difficult to design.

One variation on the wind tunnel is the wind-wave tunnel, where the bottom half of the channel is filled with water and the top is air. This is useful in studying air-sea interaction. There are also many other good uses for wind tunnels in meteorology, such as calibrating anemometers.

10.7 Analysis Methods

Values of mean wind, temperature and humidity can be measured directly. Fluxes, TKE, and dissipation rates can be calculated in a variety of ways, as we discuss in detail here.

10.7.1 Eddy Correlation Method — for various statistical moments such as fluxes, variances, TKE

Fast-response measurements of state variables generate time series of data that we can statistically analyze; however, problems with the series must be fixed before performing the eddy-correlation calculations. Sometimes there are spikes in the data (unrealistically large or small values) associated with nonmeteorological events (bug strikes on the sensor, voltage surges in the power supply). These spikes must be removed and replaced with good or bogus data. The data should be detrended (or at least demeaned) and high-pass filtered to remove wavelengths longer than about 1/3 to 1/5 of the length of the series. This latter procedure insures that enough complete cycles of the retained wavelengths are averaged to yield significant statistics. If spectra are to be found, then additional conditioning is sometimes necessary, as described in Chapter 8. Finally, we are left with a "clean" series that can be used in the eddy-correlation analysis.

The first step in the eddy-correlation process is to calculate the perturbation values of the data points. For example, given a time series of measurements of potential temperature, we can subtract the mean potential temperature from each data point to yield the time series of perturbations [c.g., $\theta'(t)$, $\theta'(t+\Delta t)$, $\theta'(t+2\Delta t)$, $\theta'(t+3\Delta t)$, ...]. We can similarly find a time series of vertical velocity perturbations [e.g., $w'(t)$, $w'(t+\Delta t)$, $w'(t+2\Delta t)$, $w'(t+3\Delta t)$, ...]. Multiplying the respective values together yields a time series of $w'\theta'$: [e.g., $w'\theta'(t)$, $w'\theta'(t+\Delta t)$, $w'\theta'(t+2\Delta t)$, $w'\theta'(t+3\Delta t)$, ...]. The average of this series, $\overline{w'\theta'}$, is the definition of kinematic turbulent heat flux in the vertical, as was shown in Chapter 2.

Once time series of θ', q', u', v', and w' are calculated from a data set, we can find via simple multiplication and averaging quantities such as fluxes [e.g., $\overline{u'w'}$, $\overline{w'q'}$,

$\overline{w'\theta'}$], variances [e.g., $\overline{w'^2}$, $\overline{q'^2}$, $\overline{u'^2}$, $\overline{\theta'^2}$], turbulence kinetic energy [TKE = $0.5 \cdot (\overline{u'^2}$ + $\overline{v'^2}$ + $\overline{w'^2}$)], fluxes of variances [e.g., $\overline{w'q'^2}$, $\overline{w'\theta'^2}$, $\overline{u'^2 w'}$, $\overline{w'^3}$], fluxes of fluxes [e.g., $\overline{w'^2\theta'}$, $\overline{w'^2 q'}$, $\overline{u'w'^2}$], fluxes of energies [e.g., $\overline{w'e}$ = $0.5 \cdot \overline{w'(u'^2 + v'^2 + w'^2)}$], and higher moments [e.g., $\overline{w'^2\theta'^2}$, $\overline{u'^3 q'}$, $\overline{w'^4}$].

An advantage of this method is that it is direct and simple, and fluxes can be calculated at whatever height or location that the original time series was measured. A disadvantage is that expensive fast-response sensors must be used. If slower response sensors were substituted, the perturbation values will be filtered by the instrument response to smaller magnitudes, resulting in incorrect fluxes and other eddy correlations. Another disadvantage is that errors in the original series compound themselves as higher and higher moments are calculated, so that by the third or fourth moments the noise (error) is as large or larger than the signal.

In Sections 2.6 and 3.2.5 we gave only the briefest overview of the relationship between actual sensible heat flux, $\tilde{Q}_H$, and the kinematic flux, $\overline{w'\theta'}$. The precise relationship is (Brook, 1978; Riehl, et al., 1978):

$$\tilde{Q}_H = \overline{\rho} \, \overline{w'(C_p T)'} \tag{10.7.1a}$$

But the value of specific heat for air, C_p, varies with humidity approximately as

$$C_p = C_{pd} (1 + 0.84 \, q) \tag{10.7.1b}$$

where {[$C_{p \text{ water vapor}}$ - C_{pd}] / C_{pd} } = 0.84. Combining these two equations and neglecting the higher-order terms yields:

$$\tilde{Q}_H \cong \overline{\rho} \, C_{pd} \left(\overline{w'T'} + 0.84 \, \overline{T} \, \overline{w'q'} \right) \tag{10.7.1c}$$

The last term can cause about a 10% change in the estimate of sensible heat flux compared to that using only the dry specific heat, and thus should not be neglected. The corresponding expression for latent heat flux is:

$$\tilde{Q}_E = \overline{\rho} \, L_v \, \overline{w'q'} \tag{10.7.1d}$$

10.7.2   Flux-Profile Relationship Method — for surface fluxes

In Section 9.7.5 the Businger-Dyer relationships between the magnitude of the surface flux and the shape of the vertical profile of a mean variable in the surface layer were presented. To use this technique, either iterative or nonlinear regression techniques must be used to estimate the flux that yields the best fit of the theoretical mean profile to the data. For statically non-neutral situations, both the wind and temperature profiles must be simultaneously fit, because the Obukhov length that appears in the theoretical equations for each of those profiles is a function of both $\overline{w'\theta'}_s$ and u_*. Although this iterative approach is somewhat tricky, it nevertheless allows calculation of fluxes from measurements of mean-variable profiles.

Advantages of flux-profile methods are that less-expensive slow-response sensors can be used to measure the mean profiles, from which fluxes are then inferred. Disadvantages are that the relationship between flux and mean profile is an empirical parameterization. Also, sometimes the shape of the profile is altered by other factors, such as change in roughness or displacement distance. Another disadvantage is that only fluxes at the surface can be found, because surface-layer similarity is used.

Although the Businger-Dyer flux profile relationships are preferred for this calculation, one could also try to use K-theory. Thus, measurement of the temperature difference between two heights could be used to estimate the heat flux at the average height, for example,

$$\overline{w'\theta'} = -K_H \, \Delta\overline{\theta} / \Delta z \qquad (10.7.2)$$

assuming that the value for K is known. This offers the advantage that fluxes can be determined at any height where the local gradient of the corresponding mean variable can be measured. Disadvantages include the fact that the K-theory approach assumes local down-gradient diffusion, and neglects the contribution to flux from larger eddies. We do not recommend use of the K-theory method because of the uncertainty about the value of K.

10.7.3 Profile Similarity Method — for fluxes

Using a K-theory approximation like (10.7.2) for two different fluxes, we can write the ratio of fluxes as:

$$\frac{\overline{w'q'}}{\overline{u'w'}} = \frac{-K_E \ (\Delta\overline{q}/\Delta z)}{-K_m \ (\Delta\overline{U}/\Delta z)} = \left(\frac{K_E}{K_m}\right) \frac{\Delta\overline{q}}{\Delta\overline{U}} \qquad (10.7.3)$$

Thus, if only one of the fluxes is known from some other measurement technique, then the second flux can be estimated by simply comparing the ratio of the differences of the two mean variables across some height. Furthermore, even though the exact value for K is not well known, the ratio of K values is more well known. For example, it is often assumed that K for all scalars and heat are equal to each other, while $K_m = 0.74\ K_H$ for statically neutral conditions (see Section 9.7.5).

Advantages of this approach are that simple mean measurements at two different heights and knowledge of any one flux is sufficient to determine the other fluxes. This is particularly valuable for moisture flux, which is notoriously difficult to measure directly. The disadvantage is that one of the fluxes must still be measured by other means.

10.7.4 Bowen Ratio Method — for surface fluxes

An equation like (10.7.3) can be written for the Bowen ratio:

$$\beta = \frac{Q_H}{Q_E} = \frac{C_p\ K\ \partial\overline{\theta}/\partial z}{L_v\ K\ \partial\overline{q}/\partial z} = \gamma\frac{\Delta\overline{\theta}}{\Delta\overline{q}} \qquad (10.7.4a)$$

We are neglecting for simplicity the variation of C_p with humidity in the development that follows; a more accurate approach should employ (10.7.1c). At the surface, we can relate Q_H and Q_E via the energy balance (7.5): $(-Q^*_s + Q_G) = Q_H + Q_E$. These two equations can be solved for Q_H and Q_E :

$$Q_H = \frac{-Q^*_s + Q_G}{\dfrac{\Delta\overline{q}}{\gamma\,\Delta\overline{\theta}} + 1} \qquad \text{and} \qquad Q_E = \frac{-Q^*_s + Q_G}{\dfrac{\gamma\,\Delta\overline{\theta}}{\Delta\overline{q}} + 1} \qquad (10.7.4b)$$

Thus, if the simple measurements of $\Delta\overline{q}$ and $\Delta\overline{\theta}$ are made in the surface layer from sensors at two different heights on a mast, and if the net radiation is measured and the ground flux is measured or estimated, then we can find the surface sensible and latent heat fluxes.

The advantages of this approach are that it is simple, and the resulting fluxes balance the surface energy budget by definition. The disadvantage is that there are times (such as sunrise and especially sunset) when $\Delta\overline{q}$ or $\Delta\overline{\theta}$ are small, causing equations (10.7.4b) to blow up and give unrealistic results. At other times during the day or night, this approach appears to work well.

10.7.5 Spectral Method — for dissipation rate

Dissipation rates are difficult to measure directly, but they are easy to infer from similarity approaches. From Section 9.9.1 recall that $S(\kappa) = \alpha_k\ \varepsilon^{2/3}\ \kappa^{-5/3}$ in the inertial subrange. This can be solved for ε using the following steps:

1) Make fast response measurements of velocity to get a time series.
2) Calculate the FFT of the time series, from which spectral densities, S, can be found.
3) Plot the spectra on a log-log graph.
4) Identify the inertial subrange: that portion of the spectrum that exhibits a -5/3 slope.
5) Fit a straight line with -5/3 slope to this portion of the graph.
6) Pick any point on this line, and record the S and κ values of that point
7) Solve for ε using: $\varepsilon = 0.49\ S^{3/2}\ \kappa^{5/2}$.

The prime reason for using this method is that it is the only way to find dissipation without using structure functions. A disadvantage is that typical atmospheric spectra has considerable variability, even in the inertial subrange. This variability translates into error bars on the estimate of ε.

10.7.6 Structure Function Method — for dissipation rate

Using (8.3.1d), we can easily solve for dissipation rate: $\varepsilon = 0.35\ (c_{v2})^{3/2}$. An advantage is that measurements can be made at points in the atmosphere distant from the sensor, but the corresponding disadvantage is that the remote sensors necessary to do this are somewhat expensive. Similar methods can be used to estimate ε_θ.

10.7.7 Residual Method — for estimating terms that are difficult to measure

Some of the higher moments and pressure terms are difficult to measure directly. However, if a conservation or budget equation contains these terms and if all the other terms in the equation can be measured by other techniques, then the unknown terms can be found by making the equation balance. The advantage is that there may be no known alternative for measuring some of these terms. The disadvantage is that all of the errors from the known terms accumulate and are added into the estimate of the unknown term. Most studies that have used this method make an appropriate disclaimer that the residual values include errors as well as the term of interest.

Suppose, for example, that we wish to calculate the term $\overline{w'p'}$. After searching through our catalog of budget equations, we find the TKE equation: (5.1b). Using eddy correlation methods for terms I-V and spectral methods for term VII, we can solve for $\overline{w'p'}$ as a residual (Wyngaard and Coté, 1971).

In another example, Gal-Chen and Kropfli (1984) have used Doppler radar data to estimate pressure perturbation and buoyancy flux. This is a good trick, because the Doppler radar measures only velocities. Their approach was to use a number of budget equations together to eliminate most of the terms that could not be measured by their one sensor.   Needless to say, a number of simplifying assumptions (e.g., steady state, anelastic, isotropic, etc.) must be made.

Wilczak and Businger (1984) used the momentum equations along with assumptions of large-eddy ramp structures, stationarity, homogeneity, Boussinesq, and others to calculate pressure perturbations, using only temperature and velocity measurements from the BAO tower.

10.7.8 Similarity Methods

We have already described a variety of similarity methods, including dissipation rate estimation using spectral similarity in the inertial subrange, and the Businger-Dyer flux-profile similarity. Many other similarity relationships exist between various boundary layer variables that can be employed to solve for unknowns.

For example, we know the similarity shape for vertical velocity variance as a function

of scaled height in the mixed layer (9.6.3c): $\overline{w'^2}/w_*^2 = 1.8 \, (z/z_i)^{2/3} \, (1 - 0.8 \, z/z_i)^2$.

We could measure $\overline{w'^2}$ at any height in the mixed layer, and if we also know the mixed

layer depth z_i, then we can solve for the surface buoyancy flux $\overline{w'\theta_v'}_s$ because it appears

in the w_* scaling velocity. It's hard to believe, but by making velocity measurements 1 km above the surface we can infer surface buoyancy flux, without even measuring temperature.

An advantage is that a large number of similarity relations exist, allowing a large variety of variables to be estimated this way. The disadvantage is that the similarity approach is empirical, based on best fits of arbitrary curves to previously measured data rather than being based on conservation equations derived from first principles.

10.7.9 Examples

Problem: Given the following measurements in the surface layer:

$\overline{w'T'} = 0.2$ K m/s, $\overline{w'q'} = 0.1$ (g/kg)·(m/s), $\overline{T} = 300$ K, $-\tilde{Q}^*_s + \tilde{Q}_G = 546$ W/m^2,

$\Delta\overline{\theta} = -2$ K, $\Delta\overline{q} = -0.818$ g/kg, $\Delta\overline{U} = 5$ m/s, $c_{v2} = 0.4$ m$^{4/3}$ s^{-2},   and

$S(\kappa) = 0.015$ m^3 s^{-2} at $\kappa = 6.28$ m^{-1}.   Find:

(a) $\tilde{Q}_H$ and $\tilde{Q}_E$ in W/m^2 using the eddy correlation method.

(b) $\tilde{Q}_H$ and $\tilde{Q}_E$ in W/m^2 using the Bowen ratio method.

(c) $\overline{u'w'}$ using the profile similarity method.

(d) ε using the structure function method.

(e) ε using the inertial subrange (spectral) method.

Solution: (a) We must utilize (10.7.1c and d), to find the fluxes. We can use ρC_{pd} = 1200 (W/m²)/(K m/s) and $\rho L_v = 2760$ (W/m²)/[(g/kg)·(m/s)] to conveniently convert between dynamic and kinematic units. Remembering to first convert the moisture flux to units of (g/g)(m/s) for the sensible heat flux equation, we find:

$$\tilde{Q}_H = \overline{\rho} \, C_{pd} \left(\overline{w'T'} + 0.84 \, \overline{T} \, \overline{w'q'} \right)$$

$$\tilde{Q}_H = (1200) \cdot [0.2 + 0.025] \quad = \quad 270 \text{ W/m}^2$$

and

$$\tilde{Q}_E = (2760) \cdot [0.1] \quad = \quad 276 \text{ W/m}^2$$

(b) Using (10.7.4b) and assuming that the psychrometric constant is 0.4 (g/kg)K^{-1}:

$$\tilde{Q}_H = \frac{-\tilde{Q}_s^* + \tilde{Q}_G}{\dfrac{\Delta \overline{q}}{\gamma \, \Delta \overline{\theta}} + 1} \qquad \text{and} \qquad \tilde{Q}_E = \frac{-\tilde{Q}_s^* + \tilde{Q}_G}{\dfrac{\gamma \, \Delta \overline{\theta}}{\Delta \overline{q}} + 1}$$

$$\tilde{Q}_H = (546) / \{[-0.818/(-2 \cdot 0.4)]+1\} \quad = \quad 270 \text{ W/m}^2$$

$$\tilde{Q}_E = (546) / \{[(-2 \cdot 0.4)/(-0.818)]+1\} \quad = \quad 276 \text{ W/m}^2$$

(c) Assuming for simplicity that the ratio of eddy diffusivities = 0.74 (which is not quite true in statically unstable conditions), and using (10.7.3) we find that:

$$\overline{u'w'} = \overline{w'\theta'} \, \frac{K_m \, \Delta \overline{U}}{K_H \, \Delta \overline{\theta}} = (0.2) \, (0.74) \left(\frac{5}{-2} \right) = -0.37 \text{ m}^2\text{s}^{-2}$$

(d)

$$\varepsilon = (0.5 \, c_{v2})^{3/2} = (0.2)^{3/2} = 0.089 \text{ m}^2\text{s}^{-3}$$

(e)

$$\varepsilon = 0.49 \cdot S^{3/2} \, \kappa^{5/2} = (0.49) \, (0.015)^{3/2} \, (6.28)^{5/2} = 0.089 \text{ m}^2\text{s}^{-3}$$

Discussion: As demonstrated above, there are often a variety of methods that can be used to calculate fluxes and other variables. We recommend that field experiments utilize several methods to estimate each flux. Invariably, the different methods will give different answers, but the overall result will give the researcher a better estimate of the true answer, and will also provide an estimate of the errors.

10.8 References

André, J.-C., J.-P. Goutorbe, and A. Perrier, 1986: HAPEX-MOBILHY, a hydrologic atmospheric pilot experiment for the study of water budget and evaporation flux at the climatic scale. *Bull. Am. Meteor. Soc.*, **67**, 138-144.

André, J.-C., J.-P. Goutorbe, and A. Perrier, 1988: HAPEX-MOBILHY: first results from the special observing period. *Annales Geophysicae* (in press).

André, J.C., and P. Lacarrere, 1980: Simulation numérique détaillée de la couch limite atmosphérique, compasaison avec le situation des 2 er 3 Juilles 1977 à Voves. *La Météorologie VI*, **22**, 5-49.

AMS, 1985: Project to find causes of destructive winter storms (GALE). *Bull. Am. Meteor. Soc.*, **66**, 705-706.

Betts, A.K. and R.D. Miller, 1975: *VIMMHEX-1972 Rawinsonde Data*. Dept. of Atmos. Sci., Colorado State Univ. Fort Collins. 150pp.

Brook, R.R., 1978: The influence of water vapor fluctuations on turbulent fluxes. *Bound.-Layer Meteor.*, **15**, 481-487.

Changnon, S.A. (Ed.), 1981: *METROMEX: A Review and Summary*. Meteor. Monographs, 18, No.40. Am. Meteor. Soc. 181pp.

Clarke, R.H., A.J. Dyer, R.R. Brook, D.G. Reid and A.J. Troup, 1971: *The Wangara Experiment: Boundary Layer Data*. Div. of Meteor. Phys. Tech. Paper No. 19. CSIRO, Melbourne. 350pp.

Clarke, R.H. and R.R. Brook (Eds.), 1979: *The Koorin Expedition, Atmospheric Boundary Layer Data over Tropical Savannah Land*. Dept. of Sci. and the Environ., Bureau of Meteor. Australian Gov. Publ. Service, Canberra. 359pp.

Cunning, J.B., 1986: The Oklahoma-Kansas Preliminary Regional Experiment for STORM-Central. *Bull. Am. Meteor. Soc.*, **67**, 1478-1486.

Deardorff, J.W., 1972: Numerical investigation of neutral and unstable planetary boundary layers. *J. Atmos. Sci.*, **29**, 91-115.

Deardorff, J.W., 1974: Three-dimensional numerical study of turbulence in an entraining mixed layer. *Bound.-Layer Meteor.*, **7**, 199-226.

Deardorff, J.W. and G.E. Willis, 1985: Further results from a laboratory model of the convective planetary boundary layer. *Bound.-Layer Meteor.*, **32**, 205-236.

Deardorff, J.W. and G.E. Willis, 1987: Turbulence within a baroclinic laboratory mixed layer above a sloping surface. *J. Atmos. Sci.*, **44**, 772-778.

Deardorff, J.W. and S.-C. Yoon, 1984: On the use of an annulus to study mixed layer entrainment. *J. Fluid Mech.*, **142**, 97-120.

DeHeer-Amissah, A., U. Högström, and A. Smedman-Högström, 1981: Calculation of sensible and latent heat fluxes, and surface resistance from profile data. *Bound.-Layer Meteor.*, **20**, 35-49.

Desbraux, G. and A. Weill, 1986: Mean turbulent properties of the stable boundary layer observed during the COAST experiment. *Atmos. Res.*, **20**, 151-164.

Gal-Chen, T. and R.A. Kropfli, 1984: Buoyancy and pressure perturbations derived from dual-Doppler radar observations of the planetary boundary layer: applications for matching models with observations. *J. Atmos. Sci.*, **41**, 3007-3020.

Gryning, S.E., 1985: The Øresund experiment - A Nordic mesoscale dispersion experiment over a land-water-land area. Bull. Am. Meteor. Soc., 66, 1403-1407.

Harris, R., 1987: Amazon boundary layer experiment. Airborne Science Newsletter, 87-2 (June). NASA Airborne science program office. 2

Hill, K, G.S. Wilson and R.E. Turner, 1979: SESAME News: NASA's participation i the AVE-SESAME '79 program. *Bull. Am. Meteor. Soc.*, **60**, 1323-1329.

Hoeber, H., 1982: KONTUR: Convection and turbulence experiment: Field phase Report. Hamburger Geophysikalische Einzelschriften, Reihe B, No. 1.

Högström, U., 1974: A field study of the turbulent fluxes of heat, water vapour and momentum at a typical agricultural site. *Quart. J. Roy. Meteor. Soc.*, **100**, 624-639.

Hooke, W.H. (Ed.), 1979: *Project PHOENIX, The September 1978 Field Operation.* NOAA/ERL Wave Propagation Lab. & NCAR. NCAR Publications Office, P.O. Box 3000, Boulder. 281pp.

Izumi, Y., 1971: *Kansas 1968 Field Program Data Report.* Air Force Cambridge Res. Lab. AFCRL-72-0041. Environ. Res. Paper No. 379, Hanscom AFB, MA . 79pp.

Izumi, Y., and J.S. Caughey, 1976: *Minnesota 1973 Atmospheric Boundary Layer Experiment Data Report.* Air Force Cambridge Research Lab. AFCRL-TR-76-0038, Environ. Res. Papers No. 547, Hanscom AFB, MA 01731. 28pp.

Katsaros, K.B., S.D. Smith and W.A. Oost, 1987: HEXOS - Humidity exchange over the sea. A program for research on water-vapor and droplet fluxes from sea to air at moderate to high wind speeds. *Bull. Am. Meteor. Soc.*, **68**, 466-476.

Keuttner, J.P. and J. Holland, 1969: The BOMEX project. *Bull. Am. Meteor. Soc.*, **50**, 394-402.

Keuttner, J.P., and D.E. Parker, 1976: GATE: Report on the field phase. *Bull. Am. Meteor. Soc.*, **57**, 11-30.

Knight, C.A. (Ed.), 1982: The cooperative convective precipitation experiment (CCOPE), 18 May-7 August 1981. *Bull. Am. Meteor. Soc.*, **63**, 386-398.

KONTUR Results, 1985: Collection of papers, *Beitr. Phys. Atmosph.*, **58**, 1-52.

Kraus, H. , 1982: PUKK - A mesoscale experiment at the German North Sea Coast. *Beitr. Phys. Atmosph.*, **55**, 370-382.

Lenschow, D.H. (Ed.), 1986: *Probing the Atmospheric Boundary Layer.* Amer. Meteor. Soc., Boston, MA. 269pp.

Lenschow, D.H., and E.M. Agee, 1976: Preliminary results from the air mass transformation experiment (AMTEX). *Bull. Am. Meteor. Soc.*, **57**, 1346-1355.

Lenschow, D.H. and B.B. Stankov, 1986: Length scales in the convective boundary layer. *J. Atmos. Sci.*, **43**, 1198-1209.

Lettau, H.H. and B. Davidson, 1957: *Exploring the Atmosphere's First Mile, Proceedings of the Great Plains Field Program 1 August to 8 September 1953, O'Neill, Nebraska. Vols I and II.* Pergamon Press, NY. 578pp.

Lilly, D., 1984: NCAR, NOAA Oklahoma University scientists gear up for field project. News and Notes. *Bull. Am. Meteor. Soc.*, **65**, 721.

Mason, P.J. and D.J. Thomson, 1987: Large eddy simulations of the neutral-static stability planetary boundary layer. *Quart. J. Roy. Meteor. Soc.*, **113**, 413-443.

Moeng, C.-H., 1984a: A large-eddy simulation model for the study of planetary boundary-layer turbulence. *J. Atmos. Sci.*, **41**, 2052-2062.

Moeng, C.-H., 1984b: Eddies in the atmosphere. *Annual Report, Fiscal Year 1983, National Center for Atmospheric Research.*, **NCAR/AR-83**, 57-58.

Moeng, C.-H., 1986: Large-eddy simulation of a stratus-topped boundary layer, Part I: Structure and budgets. *J. Atmos. Sci.*, **43**, 2886-2900.

Moeng, C.-H., 1987: Large-eddy simulation of a stratus-topped boundary layer, Part II: Implications for mixed-layer modeling. *J. Atmos. Sci.*, **44**, 1605-1614.

Monna, W.A.A., and J.G. Van der Vliet, 1987: Facilities for research and weather observations on the 213 m tower at Cabauw and at remote locations. KNMI Scientific Report WR-87-5, De Bilt, The Netherlands.

Negri, A.J., 1982: SESAME News: Cloud-top structure of tornadic storms on 10th April 1979 from rapid scan and stereo satellite observations. *Bull. Am. Meteor. Soc.*, **63**, 1151-1159.

Nieuwstadt, F.T.M., 1984: The turbulent structure of the stable, nocturnal boundary layer. *J. Atmos. Sci.*, **41**, 2002-2216.

Nieuwstadt, F.T.M. and R.A. Brost, 1986: The decay of convective turbulence. *J. Atmos. Sci.*, **43**, 532-546.

Noonkester, V.R., 1979: A technique for coding boundary layer echoes from surface-based remote sensors. *Bull. Am. Meteor. Soc.*, **60**, 20-27.

Reihl, H., G. Greenhut, and B.R. Bean, 1978: Energy transfer in the tropical subcloud layer measured with DC-6 aircraft during GATE. *Tellus*, **30**, 524-536.

Scranton, D.R. and W.R. Lindberg, 1983: An experimental study of entraining, stress-driven, stratified flow in an annulus. *Phys. Fluids*, **26**, 1198-1205.

Smedman-Högström,A., and U. Högström, 1973: The Marsta micrometeorological field project. Profile measurement system and some preliminary data. *Bound.-Layer Meteor.*, **5**, 259-273.

Stage, S.A. and R.A. Weller, 1986: The frontal air-sea interaction experiment (FASINEX); Part II: Experimental plan. *Bull. Am. Meteor. Soc.*, **67**, 16-20.

Stull, R.B., 1985: Predictability and scales of motion. *Bull. Am. Meteor. Soc.*, **66**, 432-436.

Weill, A., F. Baudon, G. Resbraux, C. Mazaudier, C. Klapisz and A.G.M. Driedonks, 1985: A mesoscale shear-convective organization and boundary-layer modification: an experimental study performed with acoustic Doppler sounders during the COAST experiment. *Proceedings of the Second Conference on Mesoscale Processes*, Am. Meteor. Soc., June 3-7, 1985, Univ. Park, PA.

Weill, A., C. Mazaudier, F. Baudin, C.Klapisz, F. Leca, M. Masmoudi, D. Vida Madjar, R. Bernard, O. Taconet, B.S. Gera, A. Sauvaget, A. Druilhet, P. Durand J.Y. Caneill, P. Mery, G. Dubosclard, A.C.M. Beljaars, W.A.A. Monna, J.G. Van der Vliet, M. Crochet, D. Thomson, T. Carlson, 1987: MESO-GERS 84 Experiment (A report to appear in *Bound.-Layer Meteor.*)

Wessels, H.R.A., 1984: Cabauw meteorological data tapes 1973-1984; description of instrumentation and data processing for the continuous measurements. KNM

Scientific Report, WR-84-6. De Bilt, The Netherlands.

Wilczak, J.M. and J.A. Businger, 1984: Large-scale eddies in the unstably stratified atmospheric surface layer. Part II: Turbulent pressure fluctuations and the budgets of heat flux, stress, and turbulent kinetic energy. J. *Atmos. Sci.*, **41**, 3551-3567.

Willis, G.E. and J.W. Deardorff, 1974: A laboratory model of the unstable planetary boundary layer. J. *Atmos. Sci.*, **31**, 1297-1307.

Wyngaard, J.C. and O.R. Coté, 1971: The budgets of turbulent kinetic energy and temperature variance in the atmospheric surface layer. *J. Atmos. Sci.*, **28**, 190-201.

10.9 Exercises

1) Suppose that an upward looking lidar sees a layer of aerosol swelling at z=1.5 km, just above the top of a 1 km deep mixed layer.
   a) Would a (microwave) radar be able to detect this 1.5 km layer? Why?
   b) Would a sodar be able to detect this layer? Why?

2) Given $\overline{w'T'} = 0.2$ Kms^{-1} and $\overline{w'q'} = 0.001$ (g$_{water}$ / g$_{air}$)·m·s^{-1}, find Q_H .

3) Which remote sensor (lidar, sodar, radar) best measures the following?
   a) Dispersion of smoke
   b) Depth of the marine boundary layer near San Francisco, California
   c) Clear-air turbulence (CAT) at the jet-stream level
   d) Top of a 2-km thick mixed layer.
   e) Temperature fluctuations in the entrainment zone
   f) Moisture fluctuations near cloud base

4) a) What instruments are required to measure the Bowen ratio in the surface layer using eddy-correlation methods?
   b) What sampling frequency is necessary to make the above measurement (i.e., how fast should the data be digitized)?
   c) Over what averaging time should measurements be taken?

5) Describe the differences and similarities between large-eddy simulation and laboratory water-tank simulation of the boundary layer. What are the advantages and disadvantages of each?

6) Let $K_m = 5$ m^{2}s^{-1} = constant with height. Calculate and plot: (a) $\overline{u'w'}$ and (b) $\overline{w'\theta_v'}$
   from z = 0 to 50 m using data from problem 26 of Chapt 5.

7) Which remote sensors could be used to measure the thickness of the mixed layer, for cases where the mixed layer is over 1 km thick?

8) Describe 5 methods for finding $\overline{w'\theta'}$ at z=10 m.

9) If upside-down "U" structures are observed with a vertically pointing radar, then what type of boundary layer can be inferred?

10) Using the similarity theory results from Section 5.2.7, find the dissipation rate, ε, at
    (a) z= 50 m   and (b) z= 600 m

    for a day and time where $\overline{w'\theta_v'} = 0.3$ Kms^{-1}, $g/\overline{\theta_v} = 0.0333$ ms^{-2}K^{-1}, and z$_i$=500 m.

11) Given the following spectrum taken in statically neutral conditions, find the value of the dissipation rate (ε), and give its units.

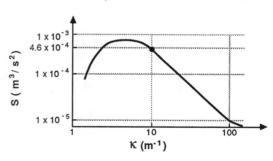

12) Given the following wind speed and temperature profile data as a function of height from the Kansas field experiment (Izumi, 1971). Use the Businger-Dyer flux profile relationships to find u_* and $\overline{w'\theta_v'}$ for each of the two different examples (a) and (b). Previous analyses suggested that $z_o = 2.4$ cm, and the zero plane displacement was about 10 cm, although it would be better if you determined your own values of these parameters from the data below. Note that both parts are difficult, but (b) is more difficult than (a).

	(a) Nocturnal (Run 25C): 0300-0315 CST, 27 Jul 68		(b) Daytime (Run 28B): 1500-1515 CST, 29 Jul 68	
z(m)	U (m/s)	θ (°C)	U (m/s)	θ (°C)
2	3.07	22.42	3.80	30.73
4	3.67	22.60	4.20	30.03
8	4.36	22.81	4.54	29.52
16	5.41	23.18	4.91	29.14
22.63	6.08	23.41	5.01	28.98
32	6.98	23.73	5.27	28.85

13) Given the following time series of w and T, use the eddy correlation method to calculate: (a) $\overline{w'T'}$, (b) $\overline{w'^3}$, and (c) $\overline{w'^2T'}$.

t (s)	w (m/s)	T (°C)
0	1	20
1	2	21
2	2	22
3	-2	19
4	0	18
5	-1	19
6	0	21
7	-1	20
8	-2	18
9	1	22

14) Identify one additional direct sensor for measuring
(a) temperature, (b) humidity, (c) wind speed
that are not listed in the sensor lists of Section 10.2. Hint, consult recent journals.

15) If you wanted to make ambient nocturnal boundary layer measurements near the surface, but did not want to have your data contaminated by drainage winds, where would you locate your sensor platform?

16) Identify new boundary layer experiments that are not listed in Table 10.1. Hint, consult the literature, especially the *Bull. Am. Meteor. Soc.*

17) Consult the meteorological literature to determine the current smallest eddy size that can be resolved with numerical simulation models. Would you classify that model as a large eddy simulation model? What order closure was used for the sub-grid parameterization?

18) Plot the variation of C_p for humidities ranging from 1 to 15 g/kg.

19) Given $\overline{w'T'} = 0.2$ Kms^{-1} at z = 6 m. If you made the following measurements with an instrumented mast, find the moisture flux $\overline{w'q'}$ at z = 6 m using the profile similarity method.

Mast Data:

z(m)	T (°C)	q (g/kg)
4	28	12
8	27	10

20) Suppose you have the mast data from problem (19), but no information on $\overline{w'T'}$ nor $\overline{w'q'}$. Use the Bowen ratio method to find $\overline{w'T'}$ and $\overline{w'q'}$, given $-\tilde{Q}^*_s + \tilde{Q}_G = 0.6$ Kms^{-1}.

21) Given a heat flux of $\tilde{Q}_H = \overline{\rho\ w'\ (C_p T)'}$ and specific heat $C_p = C_{pd}(1+0.84q)$.

Expand and simplify the expression for $\tilde{Q}_H$ in terms of kinematic fluxes of heat, moisture, etc. [The expansion given in Section 10.7.1 had deleted some terms that were small compared to the others, so it will be different than your answer here (with no terms deleted).]

11 Convective Mixed Layer

Buoyancy is the dominant mechanism driving turbulence in a convective boundary layer. Such turbulence is not completely random, but is often organized into identifiable structures such as thermals and plumes (Young, 1988). Entrainment happens at a variety of scales: lateral entrainment by small eddies into the sides of thermals, and vertical entrainment on the thermal scale into the whole mixed layer. In this chapter we examine the structure and evolution of the convective boundary layer, and study the forcings acting on it.

Three layers can be identified within the convective boundary layer (Driedonks and Tennekes, 1984) as shown in Fig 11.1:

    (1) the surface layer in the bottom 5 to 10%,

    (2) the mixed layer composing the middle 35 to 80%, and

    (3) the entrainment zone in the top 10 to 60%.

In the unstable *surface layer* there are small-scale structures such as buoyant vertical plumes, convergence lines, sheets of rising air, and dust devils. Higher in the *mixed layer* we observe larger-diameter thermals, horizontal roll vortices, and mesoscale cellular convection patterns. In the *entrainment zone* at the top of the mixed layer we find intermittent turbulence, overshooting thermals, Kelvin-Helmholtz waves, internal gravity waves, and sometimes clouds. Often the whole convective boundary layer is called the mixed layer.

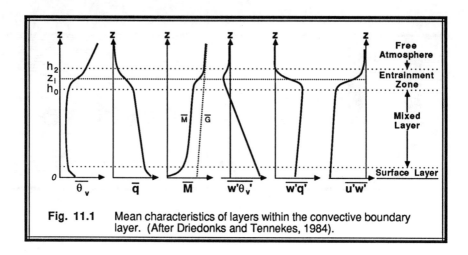

Fig. 11.1   Mean characteristics of layers within the convective boundary layer. (After Driedonks and Tennekes, 1984).

11.1 The Unstable Surface Layer

11.1.1 Mean Characteristics

The surface layer is characterized by a superadiabatic lapse rate, moisture decrease with height, and strong wind shear (see Fig 11.1) that are all well described by Monin-Obukhov similarity theory. The temperature and moisture at any height depend strongly on the recent history of the surface layer, and must be predicted using prognostic equations (Chapter 3) together with the initial and boundary conditions. Nevertheless, the shape (e.g., vertical gradients) of these profiles is quasi-steady, allowing diagnostic description using similarity theory (Chapter 9). The wind profile is not as dependent on the time history, because the wind speed is always zero at the surface, but it can be altered by a nonuniform bottom boundary (see Chapter 14). As a result, both the wind gradient and the mean wind profile itself can usually be described diagnostically by the log wind profile (Chapter 9).

The virtual potential temperature decreases very rapidly across the microlayer, from the relatively hot surface skin temperature to the warm temperatures at the bottom of the surface layer. In the remainder of the surface layer, the virtual potential temperature gradient decreases smoothly with height, becoming nearly zero at the top of the surface layer. The strong gradients in the microlayer support molecular conduction, as previously discussed, while the gradients in the remainder of the surface layer support down-gradient (small-eddy) turbulent diffusion. The gradient diffusion processes decrease in importance with height within the surface layer as large-eddy nonlocal transport and mixing take over.

Evaporation/transpiration from the surface occurs if the ground is moist and/or vegetated. The moisture gradient decreases with height in a manner similar to the temperature gradient, becoming relatively small at the top of the surface layer.

Strong convective mixing brings higher momentum air to the ground, where drag at the surface acts as a momentum sink. The mixing process is vigorous enough to maintain substantial wind speeds at the typical anemometer height of 10 m, in agreement with our casual observations that the surface winds are stronger during the day than at night over land. Superimposed on this mean wind profile is a gustiness associated with the passage of structures such as convergence lines and plumes.

In the next subsections we will explore some of the details of surface-layer structures. The structures examined here are those generated by heating at the bottom of the boundary layer, rather than by cooling at the top.

11.1.2   Plumes

Plumes are coherent vertical structures of warm rising air having diameters and depths on the order of the the surface layer depth — namely ~100 m. Above the surface layer, the plumes become more diffuse and appear to merge together to form the larger-diameter (order of 1 km) mixed-layer thermals. Some studies indicate that plumes occupy an average of about 42% of the horizontal area, with weak cool downdrafts between them. Fig 11.2 shows a sketch of an idealized plume and identifies its components (Kaimal and Businger, 1970; Wilczak and Tillman, 1980; Wilczak, 1984).

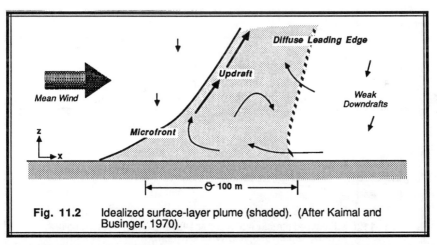

Fig. 11.2   Idealized surface-layer plume (shaded). (After Kaimal and Businger, 1970).

Even though buoyancy is the dominant forcing, the ambient mean wind modifies the plume structure so that the leading edge is diffuse and the trailing edge is sharp. The trailing edge is called the *microfront*. Plumes translate horizontally with a speed equal to the mean wind speed averaged over their depth. Thus, the plumes translate horizontally at a speed faster than the surface wind speed, but slower (70-80%) than the mean mixed layer wind.

Wind shear causes the plumes to tilt in the down shear direction at an inclination angle of about 45°. The tilt is relatively constant with time in spite of the shear, but the plume is more vertical when the shear is weak and the buoyancy strong. For weak convection and strong winds, plumes are elongated in the along-wind direction with an aspect ratio of approximately 8:1. Lengths of hundreds of meters (e.g., 500 m) and widths of tens of meters are typical in strong winds. In weak winds and strong convection, the opposite pattern occurs, with elongation in the the cross-wind direction, contraction in the along-wind direction, and a meandering propagation like a miniature front.

When a plume moves past a fixed mast or tower, the resulting temperature trace shows a characteristic **ramp structure** or sawtooth shape close to the surface, as sketched in Fig 11.3. The diffuse leading edge marks the start of the warming temperatures in the ramp, while passage of the trailing microfront results in a sharp 1-2°C drop in temperature back to the surrounding environmental values. Higher in the surface layer (above about 30 m), the microfront becomes more diffuse and the temperature excess in the plume becomes smaller.

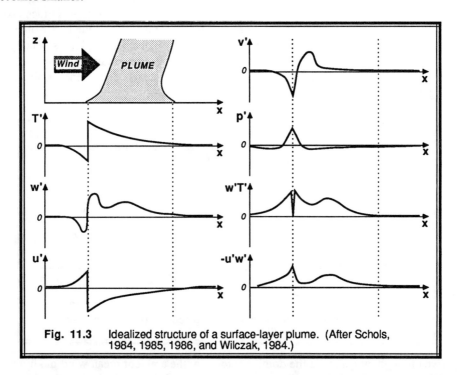

Fig. 11.3   Idealized structure of a surface-layer plume. (After Schols, 1984, 1985, 1986, and Wilczak, 1984.)

Air within a plume accelerates upward due to its buoyancy, leading to average vertical velocities of about 1 m/s. The vertical acceleration helps prevent the plume from increasing its tilt within the sheared environment of the surface layer. Air in the ramp is more turbulent than the surrounding descending air, resulting in vertical velocity fluctuations of on the order of 5 m/s in the ramp. In fact, in some random spots within the ramp the vertical velocity is downward.

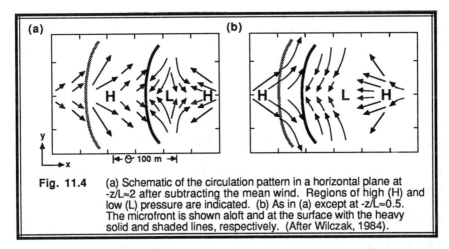

Fig. 11.4   (a) Schematic of the circulation pattern in a horizontal plane at -z/L≈2 after subtracting the mean wind. Regions of high (H) and low (L) pressure are indicated. (b) As in (a) except at -z/L≈0.5. The microfront is shown aloft and at the surface with the heavy solid and shaded lines, respectively. (After Wilczak, 1984).

Idealized horizontal and vertical circulation patterns, ensemble averaged over many ramp structures, are sketched in Fig 11.4. High perturbation pressures at the surface below the downdraft regions are associated with divergent horizontal circulations, while beneath the plume updraft there is convergence and a surface low.

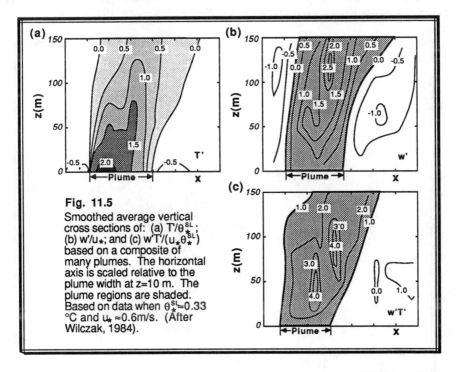

Fig. 11.5

Smoothed average vertical cross sections of: (a) T'/θ_*^{SL}; (b) w'/u_*; and (c) $\overline{w'T'}/(u_*\theta_*^{SL})$ based on a composite of many plumes. The horizontal axis is scaled relative to the plume width at z=10 m. The plume regions are shaded. Based on data when $\theta_*^{SL}\approx 0.33$ °C and $u_* \approx 0.6$m/s. (After Wilczak, 1984).

We can envision the plume as analogous to a vacuum cleaner that propagates horizontally faster than the mean surface wind, and "peels up" warm surface-layer air into the trailing edge of the plume. Similar processes have been suggested for the inflow and ramp structure in small cumulus clouds (Telford, 1986).   Most of the turbulent heat flux happens near the trailing edge of the plume. Fig 11.5 shows highly smoothed vertical cross sections of the temperature, velocity, pressure, and fluxes within a plume.

The picture imagined above applies only to convection over a uniform surface. Surface inhomogeneities that are hotter than average (asphalt parking lots or dark plowed fields) or taller than average (hills, tree lines, hedge/fence lines, dikes) can trigger plumes. Glider pilots look for these favored areas for enhanced lift.

One last caution about the word "plume". For air pollution and diffusion the word "plume" usually refers to a quasi-horizontal smoke plume downwind of an emission source. This can get confusing, because the dispersion of smoke plumes in the surface layer depend on the quasi-vertical buoyant plume structures just defined.

11.1.3 Surface Convergence Bands and Updraft Curtains

The unaided eye can detect microfronts and surface convergence bands from "cats paws" on lakes, observed variations in precipitation densities, steam fog patterns, blowing leaves and dust, and wavy fields of grain.   Quantitative measurements of surface convergence bands have been made with Doppler radar (Kropfli, 1979).

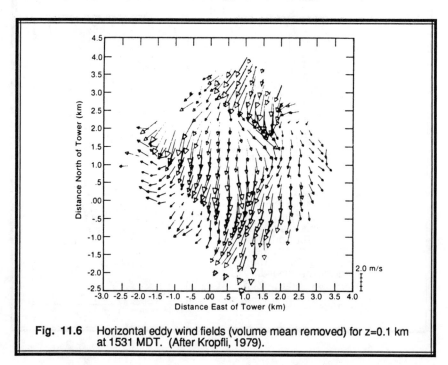

Fig. 11.6   Horizontal eddy wind fields (volume mean removed) for z=0.1 km
            at 1531 MDT. (After Kropfli, 1979).

Fig 11.6 shows one example, where the volume-mean wind has been subtracted. Horizontal convergence patterns resembling "fish nets" or "honeycombs" can be seen in the flow. The scale of the cell size in the net is about 2 km for this case, suggesting that these surface patterns are artifacts of mixed-layer thermals rather than surface-layer plumes. Other examples not shown here suggest surface layer scales, with pattern cell dimensions on the order of hundreds of meters.

In these fish net patterns we can see lines of enhanced convergence. Above these lines it is likely that there exist sheets or *updraft curtains* of rising buoyant air. These updraft curtains might be one realization of the elongated plumes described earlier. At the intersection of these curtains (i.e., at the knots in the fish net) we would expect to find enhanced buoyancy and rising motion. Although it is difficult to study directly, we can infer that a number of these curtains merge to form the larger-diameter thermals in the mixed layer.

Young (1988) points out the similarity of surface layer structures to the corresponding entrainment structures at the top of the mixed layer. Namely, updraft curtains in the surface layer merge to form larger more diffuse updrafts, while entrained downdraft curtains in the entrainment zone become more diffuse mixed-layer downdrafts. Together, these structures can be envisioned as sketched in Fig 11.7.

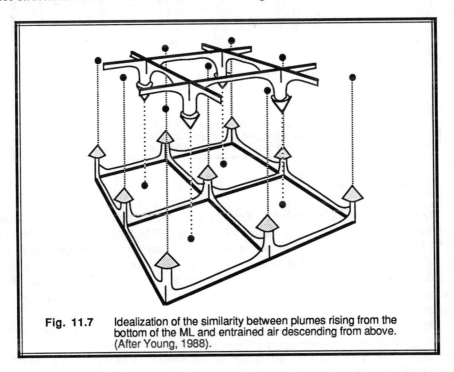

Fig. 11.7 Idealization of the similarity between plumes rising from the bottom of the ML and entrained air descending from above. (After Young, 1988).

Convergence and divergence can be measured using the inflow rate across the boundaries of known areas. For example, laser scintillation sensors were arranged in a triangle that was 450 m on a side, with each sensor 4 m above the ground during the PHOENIX experiment (Fritz and Wang, 1979). This "optical" triangle, centered on the BAO tower, measured convergence as shown in Fig 11.8a. The short-period oscillations in convergence are probably correlated with plume structures, and agree fairly well with the vertical velocities measured at the 100 m height on the tower (Fig 11.8b).

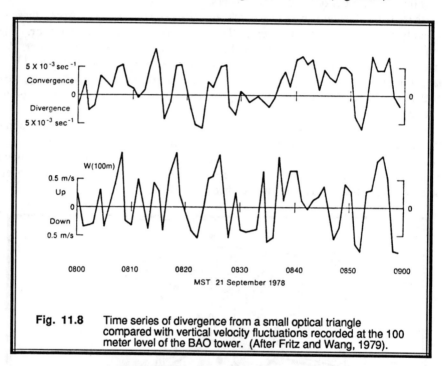

Fig. 11.8   Time series of divergence from a small optical triangle compared with vertical velocity fluctuations recorded at the 100 meter level of the BAO tower. (After Fritz and Wang, 1979).

11.1.4   Dust Devils

Dust devils are rotating updrafts of buoyant air that frequently form in fair weather conditions (Fitzjarrald, 1973; Golden, 1974; Idso, 1975a,b,c; Kaimal and Businger, 1970, News and Notes, 1976; Sinclair, et al., 1977). They are not necessarily associated with clouds, and are not related to tornadoes or water spouts. Tangential velocities in dust devils are on the order of 10 m/s, compared to 50 m/s for water spouts and 100 m/s for tornadoes. Diameters of the order of tens of meters, and depths of about 100 m are typical.

Dust devils are visible to the naked eye because the tangential velocities are sufficient to pick up dust from the ground, and vertical velocities on the order of 4 m/s can carry the finer dust aloft. Close cousins are *steam devils*, formed in steam fog regions of cold air advection over warm bodies of water or other moist surfaces, and *ash devils*, formed over ground covered by volcanic ash. There is evidence that dust devils form almost as often over nondusty ground, although the absence of dust makes them invisible to the eye. Dust devils form in the same general conditions as plumes and coexist with them, but are only about 10% as numerous.

Fig 11.9 shows the circulations structure of an idealized dust devil (Sinclair, et al., 1977). A weak downdraft with vertical velocities on the order of 2 m/s in the center of the dust devil is usually less dusty than the surrounding vortex, and is often visible to the eye. The dust devil is approximately cylindrical, and is tilted with the ambient shear. Like the plume, it translates at a speed equal to the wind speed averaged over its depth, and thus moves faster than the surface wind.   Although it occurs less frequently than the plume, it transports about ten times the vertical heat flux (Kaimal and Businger, 1970).

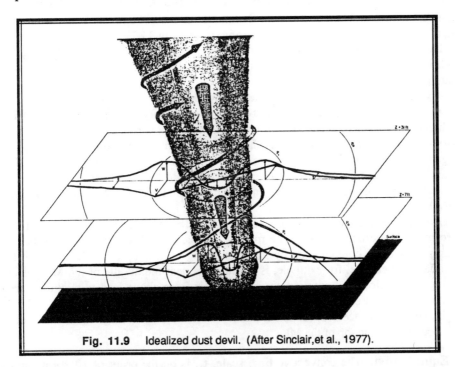

Fig. 11.9   Idealized dust devil. (After Sinclair,et al., 1977).

It has been suggested that a certain critical tangential velocity is needed to allow the dust devil to persist against the disruptive action of turbulence. These critical vorticities can sometimes be achieved by eddies generated by the mean wind flowing around obstacles. It is also possible that chance occurrences of vorticity in the mean flow are enhanced by vortex line stretching in plume updrafts. No favored direction of rotation in dust devils has been noted.

11.2 The Mixed Layer

11.2.1 Mean Characteristics

Profile Shapes. The mixed layer (ML) is so named because intense vertical mixing tends to leave conserved variables such as potential temperature and humidity nearly constant with height (see Fig 11.1). Even wind speed and direction are nearly constant over the bulk of the mixed layer. Sometimes the mixed layer is called the *well-mixed layer*.

The top of the whole convective mixed layer, z_i, is often defined as the level of most negative heat flux. This level is near the middle of the entrainment zone, often at the height where the capping inversion is strongest (see Fig 11.1). The capping inversion acts like an interface between the ML and the FA. Another measure of the average ML depth is the height at which an undiluted air parcel rising from the surface becomes neutrally buoyant.

Mixing can be generated mechanically by shears, or convectively by buoyancy. Buoyantly generated MLs tend to be more uniformly mixed than ones driven mechanically, because anisotropy in convection favors vertical motions, while shear anisotropy favors horizontal motions. Shears near the ground are usually more important for generating mixing than shears across the top of the ML, for atmospheric situations. Shears at the ML top, however, can cause a separate layer to form. A mixed layer dominated by buoyant turbulence generation is called a *convective boundary layer (CBL)* or *convective mixed layer*.

Looking more closely at the conserved variables, we find that they are not quite uniform in the vertical, because the mixing process is not instantaneous and because mixing is partially counteracted by forcings acting on the top and bottom of the ML. Potential temperature, for example, is a minimum near the middle of the ML, because heating from below and entrainment of warm air from above lead to slightly warmer potential temperatures in those regions. Moisture, however, often decreases slightly with height, because surface evaporation is adding moisture below, while entrainment of dry air is occurring at the top of the ML.

Equilibrium. The convective time scale, t_*, is on the order of 10-20 minutes in many cases. This is the typical time period for air to circulate between the surface and the top of the ML. Thus, changes in surface heat flux and other surface forcings can be communicated to the rest of the ML in a relatively short time — about 15 minutes.

As discussed in the next subsection, the ML depth changes relatively slowly in the early morning and in the afternoon. During these periods the ML is in a state of *quasi-equilibrium*, because of the short convective time scale. Similarity theory works well for these cases.

Conservation Equations. Let the angle brackets < > denote an average of any quantity, ξ, over the depth of the mixed layer:

$$<\xi> \equiv \frac{1}{z_i} \int_{z=0}^{z_i} \xi \, dz \qquad (11.2.1a)$$

One can integrate the conservation equations (3.5.3) to yield:

$$z_i \frac{d<\overline{\theta}>}{dt} = \overline{w'\theta'}_s - \overline{w'\theta'}_{z_i} \qquad (11.2.1b)$$

$$z_i \frac{d<\overline{q}>}{dt} = \overline{w'q'}_s - \overline{w'q'}_{z_i} \qquad (11.2.1c)$$

$$z_i \frac{d<\overline{U}>}{dt} = \overline{w'u'}_s - \overline{w'u'}_{z_i} - f_c <\overline{V}_g - \overline{V}> z_i \qquad (11.2.1d)$$

$$z_i \frac{d<\overline{V}>}{dt} = \overline{w'v'}_s - \overline{w'v'}_{z_i} + f_c <\overline{U}_g - \overline{U}> z_i \qquad (11.2.1e)$$

where body source terms have been neglected. Note that the only fluxes that are important are those at the top and bottom of the ML. The flux at the bottom is usually specified as a boundary condition, while the flux at the top can be found from $\overline{w'\xi'} = -w_e(\overline{\xi}_{z_i^+} - <\overline{\xi}>)$ if the entrainment velocity is known. To complete this set of equations we need the continuity equation, which depends on the evolution of the ML described next.

11.2.2 Evolution

Phases. Growth of the ML depth during a diurnal cycle is often a 4-phase process:
        (1) Formation of a shallow ML, which slowly deepens
        (2) Rapid ML growth
        (3) Deep ML of nearly constant thickness
        (4) Decay of turbulence
Fig 11.10 shows an example of the first three phases as measured by a ground-based lidar (Wilde, et al., 1985).

During the early morning the mixed layer is shallow, starting with a depth on the order of tens of meters for calm situations to depths of a couple hundred meters for windier situations. Its depth increases slowly at first because of the strong nocturnal stable layer that caps the young ML. This first phase is sometimes referred to as the **burning off** of the nocturnal inversion.

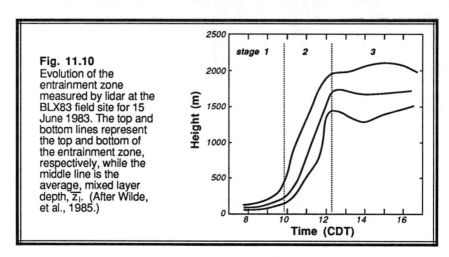

Fig. 11.10
Evolution of the entrainment zone measured by lidar at the BLX83 field site for 15 June 1983. The top and bottom lines represent the top and bottom of the entrainment zone, respectively, while the middle line is the average, mixed layer depth, $\overline{z_i}$. (After Wilde, et al., 1985.)

By late morning, for many cases, the cool nocturnal air has been warmed to a temperature near that of the residual layer, and the top of the ML has moved up to the residual-layer base. Since there is virtually no stable layer capping the ML at this point, the thermals penetrate rapidly upward during the second phase, allowing the top of the mixed layer to rise at rates of up to 1 km per 15 minutes.

When the thermals reach the capping inversion at the top of the residual layer, they meet resistance to vertical motion again and the ML growth rate rapidly decreases. During this third phase the ML depth is relatively constant during most of the afternoon. Slow depth changes are related to the balance between entrainment and subsidence. These final depths vary widely from place to place, depending on synoptic and mesoscale conditions. Final depths of 400 m have been observed over some tropical ocean regions, while depths in some desert areas reach 5 km. Typical depths over land in mid-latitudes are on the order of 1 to 2 km.

As the sun sets, the generation rate of convective turbulence decreases to the point where turbulence can not be maintained against dissipation (Nieuwstadt and Brost, 1986). In the absence of mechanical forcings, turbulence in the ML decays completely, causing us to reclassify that layer as a residual layer. Temperature fluctuations decay the fastest, while TKE decays more slowly. During this decay process the last few weak thermals may still be rising in the upper part of the ML and can still cause entrainment, while the surface layer has already become stably stratified (Stull and Driedonks, 1987). Thermals and other eddies formed as the surface heating approaches zero appear to scale to the time scale, t_*, that existed at the time of thermal creation, resulting in a possible decoupling of large and small scales (e.g., old thermals vs. new shear eddies) and a failure of similarity theory.

Turbulent Entrainment and Mixed-Layer Growth. During free-convection, buoyant thermals from the surface layer gain momentum as they rise through the ML. Upon reaching the warmer free atmosphere they find themselves negatively buoyant, but overshoot a short distance because of their momentum (Fig 11.11). This overshooting is called *penetrative convection* (Deardorff, et al., 1969). The tops of the overshooting thermals form dome or hummock-like structures. There is little ambient turbulence in the FA, and hence no way to disperse air from the overshooting thermal into the rest of the free atmosphere. The negatively-buoyant thermal sinks back down into the ML mostly intact. Any pollutants from the ML return to, and are trapped within, the ML.

During the overshoot into the inversion, wisps or sheets or curtains of warm FA air are pushed into the ML. These curtains become rapidly mixed down into the ML because of the strong turbulence there, and do not return up to the capping stable layer in spite of their positive buoyancy. The net result is entrainment of FA air into the ML. Thus, the ML grows in thickness due to a *one-way entrainment process: less turbulent air is entrained into more turbulent air.* Thus, the ML erodes into the FA. The ML can never become shallower by entrainment.

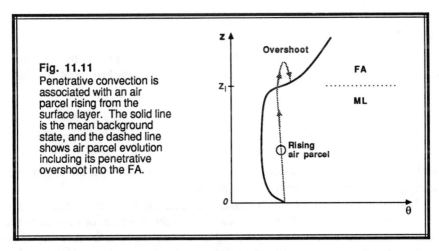

Fig. 11.11
Penetrative convection is associated with an air parcel rising from the surface layer. The solid line is the mean background state, and the dashed line shows air parcel evolution including its penetrative overshoot into the FA.

The volume of air entrained into the top of the ML per unit horizontal area per unit time (i.e., the volume flux) has the same units as velocity, and is called the entrainment velocity, w_e. This velocity is governed by the turbulence intensity and the strength of the capping inversion. Several parameterizations for the entrainment velocity are given in Section 11.4.

Continuity Equation. To first order, air density in the mixed layer can be assumed constant, allowing us to use volume conservation in place of mass conservation. In a column of ML air over a given area, A, on the earth, the volume is $A{\cdot}z_i$. If η represents the net volumetric flow rate into the volume, then volume conservation yields:

$$A\frac{dz_i}{dt} = \eta$$

Inflow can occur in the vertical because of entrainment at the top of the ML, and in the horizontal because of convergence within the ML:

$$\eta = w_e A - \int_{z=0}^{z_i} \iint_A \text{Div } dx\, dy\, dz$$

where Div is the horizontal divergence.

The last term on the right can be integrated and rewritten in terms of the mean large-scale vertical motion, w_L, acting at the top of the ML (i.e., subsidence). Upon combining the above two equations and dividing by A, we find:

$$\frac{dz_i}{dt} = w_e + w_L \qquad (11.2.2a)$$

where w_L is negative for subsidence. When active clouds are present that vent air out of the top of the ML, the volume conservation equation can be rewritten as (Stull, 1985):

$$\frac{dz_i}{dt} = (1 - \sigma_A) w_e - \sigma_A w_c + w_L \qquad (11.2.2b)$$

where σ_A is the fraction of sky covered by active clouds, and w_c is the average vertical velocity (positive upward) within the clouds at height z_i.

When there are no clouds and no subsidence, (11.2.2a) shows that the mixed layer top rises at a rate equal to w_e. Subsidence can reduce the rise rate, or even push the ML down. Subsidence, however, can never inject air into the ML, because entrainment is controlled by w_e. In convergent situations, the ML top can rise much faster than w_e. Also, the total derivatives on the left side of both equations above can be split into local derivative and advective parts. Thus, advection of higher or lower ML depths into a region contribute to the local change, and should not be neglected.

Thermodynamic (Encroachment) Mixed-Layer Growth. One of the simplest ways to forecast ML depth and temperature is to focus only on the thermodynamics, and neglect the dynamics of turbulent entrainment. For example, consider the early morning potential temperature sounding sketched in Fig 11.12a. If later in the morning the temperature reaches a value θ_1, then the mixed layer depth at that time is z_{i1}, assuming that the potential temperature within the whole mixed layer is constant with height. The amount of heat needed to reach this state is indicated by the shading in Fig 11.12a, and is given in units of K·m. Obviously more heat is required to reach a deeper ML depth with a warmer ML temperature.

Given a sensible heat flux curve such as shown in Fig 11.12b, we know that the amount of heat supplied to the air by time t_1 is shown by the area shaded under that curve. Thus, a simple graphical method for estimating ML depth at time t_1 is (1) find the area under the heat flux curve up to that time; (2) estimate what adiabat under the early morning sounding corresponds to the same amount of heating; and (3) determine the ML depth as the height where the adiabat intercepts the morning sounding. This approach for estimating the ML depth is called the ***thermodynamic method***, and it neglects factors such as advection or subsidence which might alter the sounding during the day.

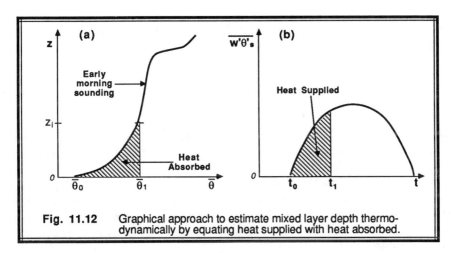

Fig. 11.12   Graphical approach to estimate mixed layer depth thermo-dynamically by equating heat supplied with heat absorbed.

The same result is obtained by neglecting advection, radiation, and latent heating, and then integrating the heat conservation equation (3.5.3f) over height and time, and using the chain rule:

$$\int_{t=0}^{t_1} \overline{w'\theta'}_s(t) \ dt \ = \ \int_{\theta=\theta_o}^{\theta_1} z(\theta) \ d\theta \qquad (11.2.2c)$$

where $z(\theta)$ represents the early morning sounding. Knowing t_1, this equation can be integrated and solved for θ_1, which is then used in the sounding to estimate the ML depth: $z_{i1} = z(\theta_1)$.

An equally-valid approach utilizes the local lapse rate of the morning sounding above the current top of the ML:

$$\frac{\partial z_i}{\partial t} \ = \ \frac{1}{\gamma} \frac{\partial \langle \overline{\theta} \rangle}{\partial t} \qquad (11.2.2d)$$

where γ is the local value of $\partial\bar{\theta}/\partial z$ just above the top of the ML, and where $\langle\bar{\theta}\rangle$ is the potential temperature vertically averaged over the ML depth. When this is combined with (11.2.1b), the result is a prognostic equation for z_i:

$$\frac{\partial z_i}{\partial t} = \frac{\overline{w'\theta'}_s - \overline{w'\theta'}_{z_i}}{\gamma \ z_i} \qquad (11.2.2e)$$

For the special case of constant heat flux with time, and constant lapse rate with height, (11.2.2e) can be integrated to yield:

$$z_i^2 - z_{i_0}^2 = \frac{2}{\gamma}\left[\overline{w'\theta'}_s - \overline{w'\theta'}_{z_i}\right] \cdot (t - t_0) \qquad (11.2.2f)$$

which shows that the ML depth could increase with the square root of time for this special case. Sometimes, a complex sounding can be broken into segments and integrated piece by piece using (11.2.2f).

If one assumes that heating from the surface is the sole source of warming of the ML, then the heat flux at z_i is assumed to be zero in (11.2.2e-f). This special case is called *encroachment*, because the top of the ML is never higher than the intercept of the ML potential temperature with the early morning sounding. The ML encroaches upward only as the ML warms.

Even though the thermodynamic approach neglects turbulent entrainment, it explains roughly 80-90% of the observed variation of the ML depth (Stull, 1976; Boers, et al., 1984). One reason is that if turbulent entrainment "gets ahead" of the thermodynamic support, the temperature inversion at the top of the ML will intensify and limit the entrainment rate (see Section 11.4).

11.2.3 Models

Bulk Model. The simplest representation of the ML assumes that mean variables are constant with height within the ML, with a sharp discontinuity between the ML and the FA. This idealization is sometimes called a *bulk or slab model* because the ML is represented by a uniform slab of air, or a *jump model* because of the discontinuity or step across the top. Another name used in the literature is *integral model*. Fluxes within the ML are assumed to be linear with height, with a jump at the top (Fig 11.13). The entrainment zone is assumed to be infinitesimally thin.

Since the actual values of mean variables in the ML are identically equal to their vertical averages, the conservation equations assume a particularly simple form:

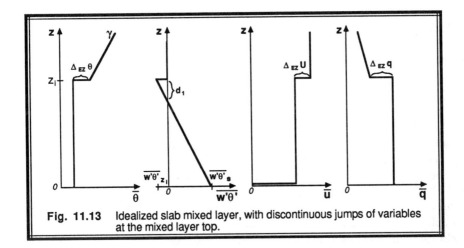

Fig. 11.13   Idealized slab mixed layer, with discontinuous jumps of variables at the mixed layer top.

$$z_i \frac{d\overline{\theta}}{dt} = \overline{w'\theta'}_s - \overline{w'\theta'}_{z_i} \tag{11.2.3a}$$

with similar equations for humidity and winds.   Above the ML the mean variables are assumed to change only by horizontal advection and subsidence.

For the sake of completeness, two other governing equations are rewritten here; the continuity equation (11.2.2a) and the equation for transport across an interface (7.4.3a):

$$\frac{dz_i}{dt} = w_e + w_L \tag{11.2.3b}$$

$$\overline{w'\theta'}_{z_i} = -w_e \Delta_{EZ}\overline{\theta} \tag{11.2.3c}$$

where $\Delta_{EZ}\overline{\theta} = \overline{\theta}_{z_{i+}} - \overline{\theta}_{ML}$ is the temperature jump across the entrainment zone at the top of the ML. Looking at the geometry of the idealized temperature profile, we expect the temperature jump to decrease as the ML warms, and to increase as entrainment eats upward into the warmer air (Tennekes, 1973):

$$\frac{d\Delta_{EZ}\overline{\theta}}{dt} = \gamma w_e - \frac{\partial\overline{\theta}}{\partial t} \tag{11.2.3d}$$

If γ, w_L, and $\overline{w'\theta'}_s$ are specified as boundary conditions, then the above four equations contain five unknowns: z_i, $\overline{\theta}$, $\overline{w'\theta'}_{z_i}$, $\Delta_{EZ}\overline{\theta}$, and w_e. To close this set of equations, we must make a closure assumption or parameterization for one of the unknowns. As will be described in Section 11.4, some investigators make assumptions for w_e, while others make them for $\overline{w'\theta'}_{z_i}$. Recall that this approach can be categorized as half-order closure, because the shape of the mean profiles are fixed in advance, and only one integrated value for each variable is forecast in the ML.

The slab ML model has been used in oceanography and meteorology for a long time (Ball, 1960; Kraus and Turner, 1967; Lilly, 1968; Kraus, 1972; Stull, 1973; Betts, 1973; Tennekes, 1973), and continues to be a popular approach. Kraus and Leslie (1982), for example, have used coupled ocean and atmospheric slab mixed layers to study air-mass modification and stratus formation. Brutsaert (1987) studied ML drying associated with entrainment of dry air.

Higher-Order Local Closure. Higher-order local closure models (one-and-a-half through third order) have been very successful. Advantages of these higher-order models include being able to forecast TKE, variances, and fluxes (if using the second and higher order models). Examples and figures of ML forecasts made with these models have already been shown in Chapter 6.

Top down / bottom up (TDBU) Diffusion. As discussed earlier, K-theory (first-order local closure) often has difficulties in the mixed layer. Specifically, infinite values of the eddy diffusivity are required to maintain fluxes in the absence of mean gradients, and negative diffusivities are needed for counter-gradient fluxes.

One attempt to circumvent this dilemma while continuing to use first-order local closure is to study the mixing upward from the ground separately from the mixing downward from the top of the ML (Wyngaard and Brost, 1984; Moeng and Wyngaard, 1984; Wyngaard, 1987; Fairall, 1987; Young, 1988). For example, if one injects a red tracer into the bottom of the ML and a green tracer into the top, we would expect the red to disperse upward (down the red gradient) independent of the green concentration, and the green to disperse downward (down the green gradient) independent of the red concentration. The superposition of the red and green should not affect the the dispersion of either one.

If we instead inject a red tracer at both the top and bottom of the ML, the dispersion should be the same as the previous case. Thus, some of the red dispersing upward will do so counter to the gradient of red diffusing down, and vise versa. The strict formulation of the TDBU model requires the use of artificial gradient functions to allow separate down-gradient diffusion from both the top and bottom of the ML. As Young (1988) points out, however, there are still problems because the gradient functions cross zero and become negative, implying counter-gradient diffusion.

Nevertheless, we can use the results of TDBU to help explain the difference in profile shape of any mean ML variable, $\overline{\xi}$, from its idealized slab shape:

$$\overline{\xi}(z) \;=\; 0.8\,\frac{\overline{w'\xi'}_s}{w_*}\left(\frac{z}{z_i}\right)^{-1/2} - \; 2.0\,\frac{\overline{w'\xi'}_{z_i}}{w_*}\left(1-\frac{z}{z_i}\right)^{-1/2} + \text{const.} \qquad (11.2.3e)$$

The vertical gradient of this function was already presented in Section 9.6.

The shape of this profile depends on both the surface flux and the entrainment flux (Fig 11.14). When the entrained flux is negative and the surface flux is positive, as it is for potential temperature, the mean profile of potential temperature is slightly concave to the right, with a stronger negative gradient near the surface and positive gradient near the ML top. For variables such as humidity that often have positive fluxes at both the top and bottom of the ML, the mean humidity profile exhibits a tilt with enhanced curvature at the top and bottom. The ratio of top to bottom fluxes, A_R, provides one measure of the overall shape (see Fig 11.14).

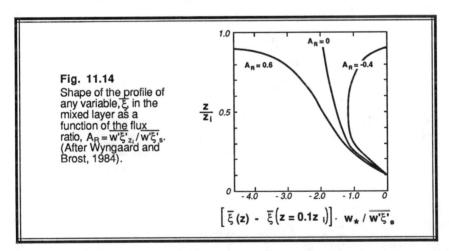

Fig. 11.14
Shape of the profile of any variable, $\overline{\xi}$, in the mixed layer as a function of the flux ratio, $A_R = \overline{w'\xi'}_{z_i}/\overline{w'\xi'}_s$. (After Wyngaard and Brost, 1984).

The mean gradients across the whole ML can also be related to the relative contributions of top and bottom fluxes (Mahrt and André, 1983; Driedonks and Tennekes, 1984):

$$\overline{\xi}_{z_i} - \overline{\xi}_s \;=\; -\,1.7\,\frac{\overline{w'\xi'}_s}{w_*} \;-\; 5\,\frac{\overline{w'\xi'}_{z_i}}{w_*} \qquad (11.2.3f)$$

Although the equation above gives approximate agreement with observed profiles, there is quite a lot of scatter in the data and uncertainty in the precise values of the two constants.

Transilient Models. Fiedler (1984) demonstrated that the TDBU model can be considered to be a special case of the non-local transilient closure model. Since the transilient approach explicitly models eddies mixing across finite distances, it can easily model the mixing associated with thermals rising in the ML. Local gradients are not used. Instead, differences between moving parcels and the surrounding environment determine the amount and direction of mixing.

Fig 11.15 shows a simulation of ML evolution observed near the Cabauw tower (Stull and Driedonks, 1987). The subset of grid points that were turbulent in this 1-D transilient model are shaded in the figure, and indicate the whole turbulent domain including the entrainment zone. Other measures of ML depth, namely the height at which a parcel rising from the surface would first become neutrally buoyant, and the height of most negative heat flux, are both indicated. The four phases of ML evolution are evident in this figure.

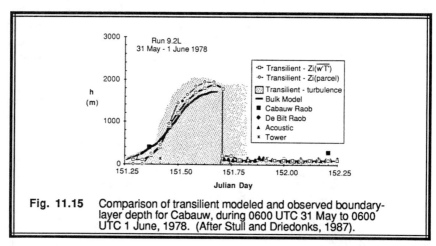

Fig. 11.15   Comparison of transilient modeled and observed boundary-layer depth for Cabauw, during 0600 UTC 31 May to 0600 UTC 1 June, 1978. (After Stull and Driedonks, 1987).

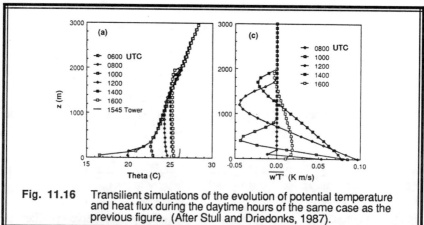

Fig. 11.16   Transilient simulations of the evolution of potential temperature and heat flux during the daytime hours of the same case as the previous figure. (After Stull and Driedonks, 1987).

The corresponding potential temperature and heat flux profiles are shown in Fig 11.16. Evident are the curvature of the potential temperature profiles, and the nearly linear heat flux profiles. On this day there were significant mechanical contributions to turbulence in addition to the dominant buoyant contributions. Heating of the lowest grid point by surface fluxes generated buoyant instabilities, to which the transilient model responded by generating mixing and producing a ML, as shown in the figures here and discussed in Chapter 6.

11.2.4 Thermal Structures

Thermals are large columns of rising buoyant air in the convective mixed layer (Fig 11.17). The convective circulation, including both the thermal updraft and associated downdraft, have horizontal scales of roughly 1.5 z_i (Caughey and Palmer, 1979; Young, 1988). Thus, early morning thermals are on the order of 100 m in diameter, while by late afternoon when the mixed layer is deeper thermals can be 1 to 2 km in diameter. The vertical extent of a thermal is also roughly equal to the ML depth.

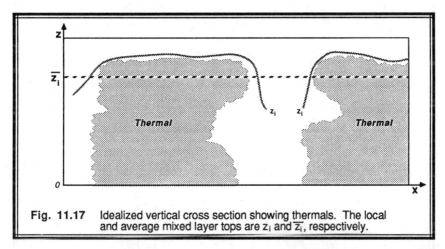

Fig. 11.17 Idealized vertical cross section showing thermals. The local and average mixed layer tops are z_i and $\overline{z_i}$, respectively.

Vertical velocities in thermals can reach 5 m/s or more, although weaker updrafts of 1 to 2 m/s are more common. Velocities scale to the convective velocity scale, w_*. Samples of vertical velocity (Caughey, et al., 1983; Deardorff and Willis, 1985) measured both in and between thermals typically show a negatively skewed distribution in the bulk of the ML (Fig 11.18), although the distributions tend to become more symmetric in the entrainment zone. There are a large number of weak interthermal downdrafts, and a small frequency of strong thermal updrafts. In a very idealized sense, this is pictured (Fig 11.19) as broad regions of gentle downdraft surrounding smaller regions of strong updraft.

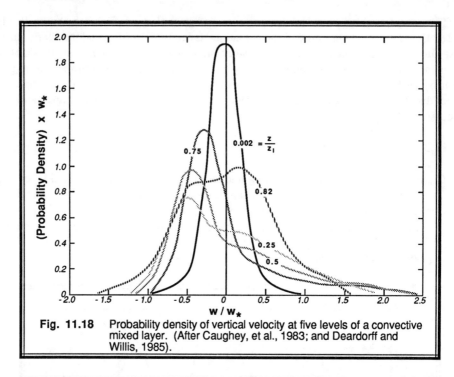

Fig. 11.18   Probability density of vertical velocity at five levels of a convective mixed layer. (After Caughey, et al., 1983; and Deardorff and Willis, 1985).

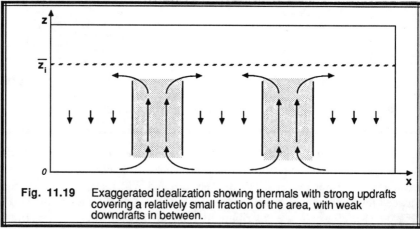

Fig. 11.19   Exaggerated idealization showing thermals with strong updrafts covering a relatively small fraction of the area, with weak downdrafts in between.

Although not plotted here, temperature histograms show negatively skewed frequency distributions in the ML, which change to symmetric distributions in the entrainment zone, similar to those of vertical velocity.

Joint probability distributions showing the relative frequency of vertical velocities and temperature fluctuations are shown in Fig 11.20 (Mahrt and Paumier, 1984; Deardorff and

Willis, 1985). In the bottom 2/3 of the ML (Fig 11.20b) there is a predominance of cool downdrafts. At the top of the ML, however, we see a peak associated with cool updrafts (Fig 11.20c). In fact, there is a broad peak associated with both updrafts and downdrafts that are cool. These are the tops of thermals overshooting into the entrainment zone and then sinking back down.

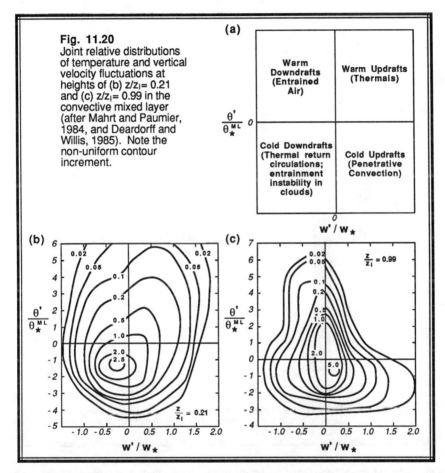

Fig. 11.20
Joint relative distributions of temperature and vertical velocity fluctuations at heights of (b) z/z_i= 0.21 and (c) z/z_i= 0.99 in the convective mixed layer (after Mahrt and Paumier, 1984, and Deardorff and Willis, 1985). Note the non-uniform contour increment.

Associated with the updrafts and downdrafts are corresponding horizontal convergence zones under thermals, and divergence above. The whole process is a circulation that moves air up and down in the mixed layer with a time period on the order of $t_* = z_i/w_*$, namely about 5 to 15 min.

When the tops of the thermals rise into the statically stable air of the entrainment zone, they become negatively buoyant. This causes a deceleration of the rising thermal, and eventually leads to its sinking back down into the mixed layer. This process is sometimes called penetrative convection (Scorer, 1957; Deardorff, et al., 1969, Stull, 1973).

Thermals are trapped within the mixed layer, which means that moisture and pollutants transported from the surface by thermals are also trapped.

The penetrative convection process and overshooting of thermals causes entrainment of large (horizontal and vertical thicknesses on the order of z_i) blobs or curtains of free atmosphere air down into the mixed layer between thermals, resulting in growth of the mixed layer thickness. Although portions of some of these entrained blobs have been observed to occasionally move down to the surface layer, most of the entrained air is mixed with surrounding ML air before reaching the surface. Using moisture as an indicator of source region of air, Crum and Stull (1987) found that undiluted surface layer air frequently reaches the top of the ML, presumedly in undiluted cores of thermals, while entrained FA air did not reach below $0.5\ z_i$ for the cases studied (Fig. 11.21).

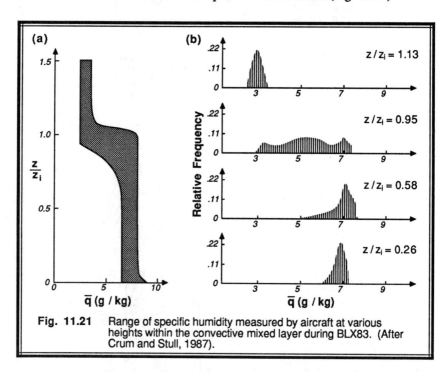

Fig. 11.21   Range of specific humidity measured by aircraft at various heights within the convective mixed layer during BLX83. (After Crum and Stull, 1987).

On a smaller scale, small eddies cause mixing into the top and sides of the thermal. Mixing into the top of individual thermals has been related to enhanced wind shear across the overshooting tops, resulting in the generation of turbulence via the Kelvin-Helmholtz process (Rayment and Readings, 1974). In spite of this process, little if any thermal air is mixed out of the thermal and deposited into the free atmosphere.

Lateral entrainment, or *intromission*, appears to be a process where blobs of environmental air from outside the thermal are mixed horizontally into the thermal (Crum, et al., 1987). In the center of the thermal is a relatively undiluted core, carrying surface layer air up to the top of the boundary layer. Around this core is a thick contorted layer of varying thickness in which some of the entrained environmental air and thermal air have mixed. Thus, there are no well defined edges to the typical thermal (Fig 11.22).

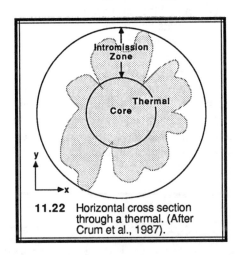

11.22 Horizontal cross section through a thermal. (After Crum et al., 1987).

This amorphous boundary between the updraft and downdraft region results in a variety of interpretations of convective structures as visualized using both *in-situ* and remote sensors. *Conditional sampling* is the technique of sorting data into thermal or non-thermal categories according to some selection criteria, called an *indicator function*. The three most used indicators of thermals based on direct measurements are vertical velocity, humidity, and/or turbulence.

Usually, some minimum *threshold amplitude* of the indicator function must be met to determine the category of a data value. Within thermals, the fluctuations of these variables are sometimes greater than their respective conditionally sampled means. Thus, some sampling strategies also require that the indicator variable maintain a certain value over specified time *durations* (corresponding to a minimum size of the structure). Using both amplitude and duration, many data points can be sorted into the *updraft (thermal) or downdraft categories*. Those remaining data points in the time series that satisfy neither the criteria for thermal updraft nor between-thermal downdraft are sometimes grouped into a third category called *environment*.

Conditional sampling of aircraft data suggests that well-defined updrafts cover about 15 to 43% of the horizontal area, while well-defined downdrafts cover about 20 to 55% (Lenschow and Stephens, 1980, 1982; Greenhut and Khalsa, 1982, 1987; and Khalsa and Greenhut, 1985, 1987; Crum, et al., 1987; Young, 1988a-c). The remaining percentage are weaker circulations of mixed environmental air. Godowitch (1986) observed 36 to 40% updrafts and 49 to 52% downdrafts over both rural and urban sites, but thermals at the urban sites had greater vertical velocities.

Thermal diameters, as well as cumulus diameters, occur with a frequency that is approximately lognormally distributed. On some field experiments, a range of one to three thermal centers are encountered per horizontal flight distance of z_i. This suggests that updrafts cluster, because the dominant wavelength of the vertical velocity spectra in MLs is at a wavelength of 1.5 z_i (Young, 1988)

Also, the number of well-defined thermal events decreases with height within the ML, suggesting that some of the thermals do not succeed in rising to the ML top. However, the diameter of thermals increases with height, suggesting that some thermals merge as they rise (Greenhut and Khalsa, 1987).

Conditionally sampled average values of vertical velocities and temperatures in updrafts and downdrafts are plotted in Fig 11.23. As expected, the virtual temperature excess in thermals is greatest at the ground, and can become negative at the top of the ML. Downdrafts at the ML top consist of warm entrained air, but lower in the ML the downdrafts are increasingly colder than the average ML temperature. The wide range of values indicated in this figure is related, in part, to differing definitions for indicator functions used by different investigators.

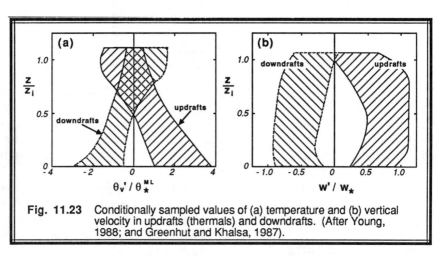

Fig. 11.23   Conditionally sampled values of (a) temperature and (b) vertical velocity in updrafts (thermals) and downdrafts. (After Young, 1988; and Greenhut and Khalsa, 1987).

Part of the difficulty in defining thermal boundaries is that although thermals start rising in the bottom third of the boundary layer as elements that are warmer than their environment, they are found to be cooler than their environment in the entrainment zone region. Some thermals gain most of the buoyancy from their moisture content, allowing the top half to be cooler than the environment even though the middle third might still be positively buoyant (using virtual potential temperature).

Lidar observations based on aerosol backscatter above some threshold (Hooper and Eloranta, 1986) tend to show the diameters of thermals decreasing with height (Fig 11.24), while Doppler sodar based on upward velocity above some threshold sometimes shows constant or increasing diameters (Taconet and Weill, 1983; Coulter, et al., 1985).

Almost all of the observations indicate that the thermals are not like bubbles, but are more like finite length columns that persist for some time. This suggests that the best model might be the "wurst" model — namely, the idealized thermal shape is like that of a sausage or wurst. Real thermals are not perfect columns of rising air, but twist and meander horizontally and bifurcate and merge as they rise. Nevertheless, thermals are anisotropic, with most of their energy in the vertical.

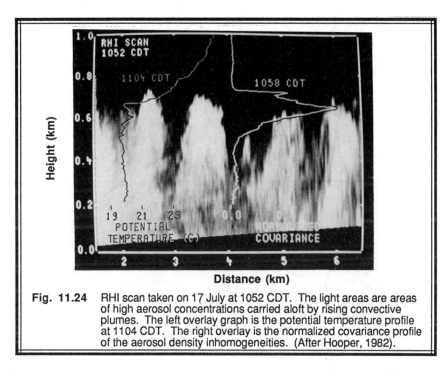

Fig. 11.24   RHI scan taken on 17 July at 1052 CDT. The light areas are areas of high aerosol concentrations carried aloft by rising convective plumes. The left overlay graph is the potential temperature profile at 1104 CDT. The right overlay is the normalized covariance profile of the aerosol density inhomogeneities. (After Hooper, 1982).

When hot spots exist on land surfaces, thermals predominantly form there (Smolarkiezicz and Clark, 1985). Glider pilots look for such hot spots as indicators of persistent lift (Reichmann, 1978). Over moist or vegetated surfaces, thermals are often moister than their environment. They are also usually more turbulent. Thermals are observed over oceans as well as land surfaces, suggesting that a surface hot spot is not necessarily needed as a triggering mechanism. In the absence of hot spots, thermals can be swept into lines or rings by weak mesoscale motions called *secondary circulations*. These patterns are visible from satellite by the cloud streets and open/closed cells.

The "ultimate" hot spot was created in a classical field experiment in France (Bénech, et al., 1986; Noilhan, et al., 1986; Noilhan and Bénech, 1986) where an array of 105 oil burners generating a total heat output of 1000 MW distributed over 15000 m^2 of the surface were used to artificially create a convective thermal. Upwind of the warm updraft a cool downdraft was observed, while downwind a cool updraft was found. A pressure deficit of about 0.1 kPa (1 mb) was generated near the surface under the updraft, which induced horizontal convergence.

11.2.5 Horizontal Roll Vortices and Mesoscale Cellular Convection

During conditions of combined surface heating and strong winds, weak horizontal helical circulations can form in the boundary layer (Kuettner, 1959; Brown, 1970; LeMone, 1973, 1976; Kropfli and Kohn, 1978; Doviak and Berger, 1980; Mason and Sykes, 1980, 1982; Reinking, et al., 1981; Rabin and Doviak, 1982; Wilczak and Businger, 1983; Kelly, 1984; Atlas, et al., 1986). These circulations are called *horizontal roll vortices* or *rolls,* and consist of clockwise and counterclockwise pairs of helices with their major axis aligned almost parallel with the mean wind. Some studies suggest that the roll axis should be roughly 18° to the left of the geostrophic wind direction for nearly neutral conditions, and that the angle decreases as the mixed layer becomes more statically unstable. The depth of these rolls equals the ML depth, and the ratio of lateral to vertical dimensions for a roll pair is about 3:1 (see Fig 11.25).

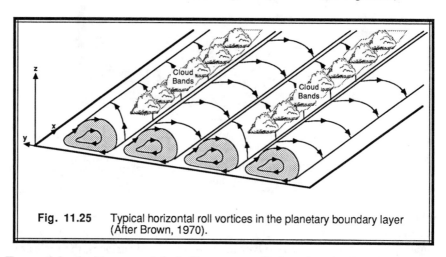

Fig. 11.25   Typical horizontal roll vortices in the planetary boundary layer (After Brown, 1970).

Tangential velocities around the helices are usually less than 1 m/s and are difficult to measure directly, although faster velocities occasionally occur. However, these velocities sweep other convective plumes and thermals into rows, creating along-wind rows of strong updraft. If sufficient moisture is present, *cloud streets* can form along these updraft rows. Such cloud streets are readily apparent from satellite imagery, particularly over ocean regions where there are no surface hot spots to destroy the organization.

Rolls are frequently observed during cold-air advection over warmer bodies of water, and are strongly associated with air-mass modification. Rolls are also common in the low-level jet ahead of cold fronts, and can occur between pairs of closed isobars of warm season anticyclones. Forest fires are also modulated by rolls, as is evident by long rows of unburned tree crowns in the middle of burned areas (Haines, 1982).

Theories for roll formation include thermal instabilities and inertial instabilities. Thermally, we would expect there to be less friction on the rising thermals if they align in rows, because thermals would have neighbors that are also updrafts. Thus, alignment into rows can be buoyantly more efficient, and the alignment provides protection from the ambient wind shear. Other studies have suggested that secondary circulations such as rolls develop whenever an inflection point occurs in the mean wind profile. For example, the Ekman spiral solution always has an inflection point near the top of the Ekman layer.

Over the ocean, roll patterns and cloud streets gradually change to cellular patterns further downwind in a cold air advection situation. These are evident as honeycomb cloud patterns visible by satellite. This pattern is caused by *mesoscale cellular convection* (*MCC*, not to be confused with mesoscale convective complex). Open cells consists of hexagonal rings of updraft and clouds around clear central areas of downdraft, while closed cells are clear rings of descending air around mesoscale cloud clusters and updrafts. Cells typically have diameters on the order of 10 to 100 km, but are nevertheless boundary layer phenomena. Cells have depths on the order of 2 to 3 km, yielding aspect ratios of about 10:1 to 30:1 (Agee, et al., 1973; Rosmond, 1973; Lenschow and Agee, 1976; Sheu and Agee, 1977; Agee, 1984; Rothermel and Agee, 1980, 1986; Ray 1986).

The cellular pattern is reminiscent of laboratory *Rayleigh-Bénard convection* cells, except that the laboratory convection has an aspect ratio of about 1:1. These laboratory cells are well explained by the *Rayleigh number*, Ra:

$$\text{Ra} \quad = \quad \frac{g \, \overline{\Delta\theta} \, h^3}{\overline{\theta} \, \nu \, \nu_\theta} \qquad (11.2.5a)$$

for nonturbulent flow, where $\Delta\theta$ is the potential temperature difference across the convective layer of depth h, and where molecular viscosity, ν, and thermal diffusivity, ν_θ, dominate.

Many attempts have been made to define a *turbulent Rayleigh number* (using eddy diffusivities in place of the molecular ones) that would work for the atmosphere (Rothermel and Agee, 1986; Chang and Shirer, 1984; Fiedler, 1984). Although investigators have been able to numerically simulate some of the characteristics of MCC, they were often forced to make rather arbitrary choices for some of the model parameters. Frequently, the parameterization requires a rather large value of eddy diffusivity: in the range of 20 to 2000 $m^2 s^{-1}$. Ray (1986) has suggested that horizontal variability of K is necessary to explain MCC.

11.2.6 Dispersion

When pollutants are emitted at the bottom of the mixed layer they are quickly drawn by the convergence zones into the rising updraft, causing the center of mass of the tracer to initially move slightly above $0.5 z_i$ as it drifts downwind (Fig 11.26a). Pollutants emitted above the surface layer move downward on the average (Fig 11.26b) because downdrafts comprise a larger area than updrafts, and because the updraft air is quickly recycled back down (Willis and Deardorff, 1976, 1978, 1981). Eventually, the pollutants become

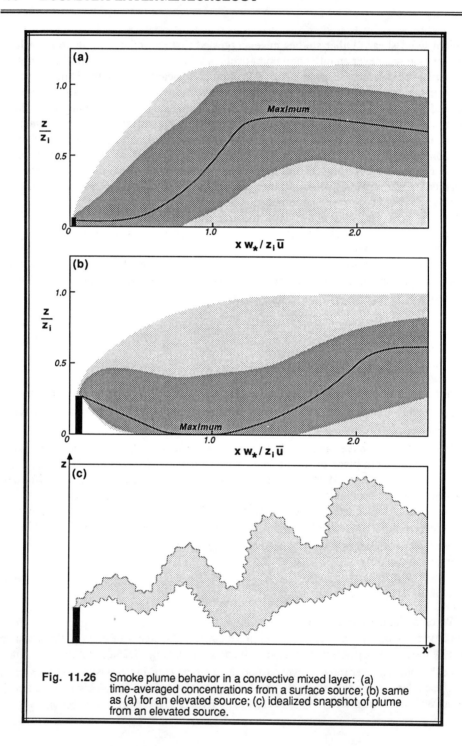

Fig. 11.26   Smoke plume behavior in a convective mixed layer: (a) time-averaged concentrations from a surface source; (b) same as (a) for an elevated source; (c) idealized snapshot of plume from an elevated source.

uniformly mixed in the vertical regardless of their source altitude, as expected in a well-mixed layer (Nieuwstadt and van Dop, 1982).   This time-averaged statistical behavior should not be confused with the instantaneous *looping* of smoke plumes (Fig 11.26c).

Such dispersion within MLs has been observed during the Condors field program (Kaimal, et al., 1986). It has also been simulated with LES models (Deardorff, 1972; Lamb, 1978), and parameterized using asymmetric matrices in transilient turbulence theory (Stull, 1988).

As mentioned before, pollutants transported up from the surface by thermals are trapped beneath the capping inversion.   Certain types of cumulus clouds (see Chapter 13) can break through the stable entrainment zone to *vent* pollutants and moisture out of the boundary layer into the free atmosphere.

11.2.7 Examples

Problem 1: Given a cloud-free ML with constant $w_e = 0.1$ m/s, and a constant divergence of Div $= 5 \times 10^{-5}$ s^{-1}, find the ML depth versus time. The initial conditions are $z_i = 0$ at $t = t_o = 0$.

Solution: We can integrate the continuity equation from the surface up to z_i to find the subsidence velocity at the ML top:

$$w_L\big|_{z_i} = - \text{Div} \cdot z_i \tag{11.2.7a}$$

When this is used in (11.2.2a), we get the prognostic equation:

$$\frac{dz_i}{dt} = w_e - \text{Div} \cdot z_i \tag{11.2.7b}$$

Upon collecting terms, we can rewrite this equation as:

$$\frac{dz_i}{w_e - \text{Div } z_i} = dt$$

which can be easily integrated to yield:

$$z_i = \frac{w_e}{\text{Div}} - \left[\frac{w_e}{\text{Div}} - z_i(t_o) \right] \cdot e^{-\text{Div} \cdot (t - t_o)}$$

The resulting values for ML depth as a function of time are:

t (h):	0	1	2	3	4	5	10	20
z_i (m):	0	329	605	834	1026	1187	1669	1945

Discussion: We see that in a ML with constant divergence, the subsidence velocity increases with height, making ML growth more difficult. In fact, for our particular example, the ML top can not grow past 2 km, because at that height the subsidence is strong enough to completely counteract entrainment (Fig 11.27). As we will discover in section 11.4, the entrainment velocity also becomes smaller as the ML deepens. These two factors combine to limit the depth of the ML. Cases have been observed in the field where subsidence becomes so strong as to push the ML top downward, even while entrainment is happening.

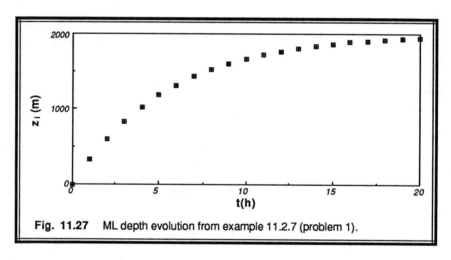

Fig. 11.27   ML depth evolution from example 11.2.7 (problem 1).

Problem 2: Suppose that the initial potential temperature sounding is $\theta = \theta_o + \gamma \cdot z$, where $\theta_o = 300$ K and $\gamma = 0.01$ K/m. Also assume that the surface heat flux is constant with time: $\overline{w'\theta'}_s = 0.2$ K m/s. If $z_i = 0$ at $t = 0$, use the thermodynamic method to find z_i and θ_{ML} at $t = 4$ h.

Solution: This problem can be represented graphically (see Fig 11.28). We wish to make the two shaded areas equal.

Heat Supplied $= (0.2$ K m/s$) \cdot (4$ h$) \cdot (3600$ s/h$) = 2880$ K·m,

Heat Absorbed $= 0.5 \cdot ($base$) \cdot ($height$) = 0.5 \cdot (\theta_{ML} - \theta_o) \cdot z_i = 0.5 \gamma \cdot z_i^2$.

Equating the two heats, we find that: $z_i = [2 \cdot (2880)/0.01]^{1/2} = 758.9$ m. The ML potential temperature is $\theta_{ML} = \theta_o + \gamma \cdot z_i = 300 + (0.01) \cdot (758.9) = 307.6$ K.

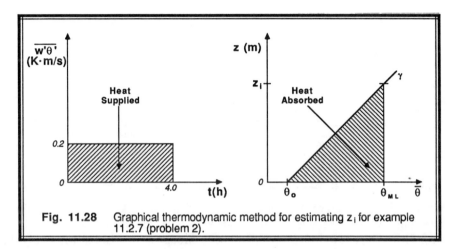

Fig. 11.28   Graphical thermodynamic method for estimating z_i for example 11.2.7 (problem 2).

Discussion: In reality, the heat flux usually varies with time, and the initial sounding is much more complex, making the solution more difficult, but not impossible.

11.3 The Entrainment Zone

11.3.1 Characteristics

The entrainment zone (EZ) is the region of statically stable air at the top of the ML, where there is entrainment of FA air downward and overshooting thermals upward. The EZ can be quite thick — averaging about 40% of the depth of the ML.

The top of the entrainment zone is defined as the altitude, h_2, of the top of the highest thermal within a region. The bottom, h_o, is more difficult to define because there is no sharp demarcation. The bottom is usually taken as that altitude where about 5 to 10 % (Deardorff, et al., 1980; Wilde, et al., 1985) of the air on a horizontal plane has FA characteristics (Fig 11.29a). An alternative definition of the EZ is that region where the

buoyancy flux, $\overline{w'\theta_v'}$, is negative (see Fig 11.1). The entrainment zone is quite apparent and easily measurable using remote sensors such as lidar (Boers, et al., 1984; Crum and Stull, 1987).

The altitude of the *local ML top*, shown by the solid line in Fig 11.29a, varies significantly between the EZ top and the middle of the ML. Its shape looks like a series of inverted "U"s marking the tops of penetrating thermals. The *average ML depth*, which we traditionally call z_i, is the altitude where 50% of the air has FA characteristics on a horizontal average. Thus, z_i is just below the midpoint between h_2 and h_o.

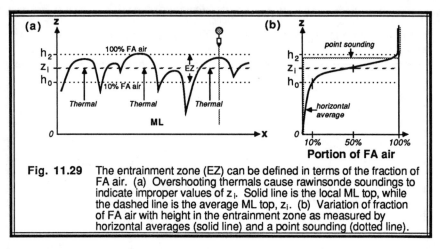

Fig. 11.29   The entrainment zone (EZ) can be defined in terms of the fraction of FA air. (a) Overshooting thermals cause rawinsonde soundings to indicate improper values of z_i. Solid line is the local ML top, while the dashed line is the average ML top, z_i. (b) Variation of fraction of FA air with height in the entrainment zone as measured by horizontal averages (solid line) and a point sounding (dotted line).

When horizontal averages of temperature or moisture are taken by aircraft in the EZ, the resulting profile of percentage of FA air vs height exhibits a smooth profile (Fig 11.29b). However, a rawinsonde sounding made through a single point will likely show a sharp jump (Mahrt, 1979) in temperature, moisture and wind (Fig 11.29b) when it encounters the local ML top. Since this local top can be at a significantly different altitude than the average top depending on which location is penetrated by the sensor, we recommend that single rawinsonde soundings NOT be used to estimate z_i. Rawinsonde z_i errors can easily be as large as 0.4 z_i.

11.3.2 Evolution and Models

When the ML is shallow in the morning over land, the EZ is proportionally shallow. As the ML grows, so does the EZ thickness (see Fig 11.10). Thin EZs are expected for large temperature changes across the ML top, because thermals will not penetrate as far and entrainment will be slow. Thick EZs are expected with more intense ML turbulence where convection is vigorous.

Deardorff, et al. (1980) have suggested that the ratio of EZ thickness (Δh) to depth of the ML (using h_o as one measure), is:

$$\frac{\Delta h}{h_o} = 0.21 + 1.31 \, (Ri^*)^{-1} \tag{11.3.2a}$$

where Ri^* is a *convective Richardson number*, defined by:

$$Ri^* = \frac{g \, \Delta_{EZ} \overline{\theta}_v \, z_i}{\overline{\theta}_v \, w_*^2} \tag{11.3.2b}$$

The -1 power law dependence is still debated in the literature, with other suggestions ranging from 0 to -2.

In the evening over land, when ML turbulence decays, we are left with the statically stable *capping inversion* in place of the EZ. In fact, the EZ is often called the capping inversion during the day because it acts as a lid to convection. In the absence of turbulence, there is little change to the capping inversion at night, except by subsidence and advection.

We can also model the distribution of FA air within the EZ. As mentioned before, the amount of FA air decreases with distance down from the top of the EZ. Also present in the EZ are some thermal cores that have transported up rather undiluted surface layer (SL) air. Smaller eddies mixing these two extremes creates a third category of air that we will call mixture air.

The relative fractions of these three air categories is plotted in Fig 11.30 based on the tank measurements of Deardorff, et al. (1980) and the aircraft field observations of Crum, et al. (1987). Sometimes mixture air and SL air are grouped together and called mixed layer air (Deardorff, et al., 1980). Wilde, et al. (1985) found that the distribution of ML air with height is fairly-well described by the integral of a truncated double-exponential function, with the peak centered at z_i.

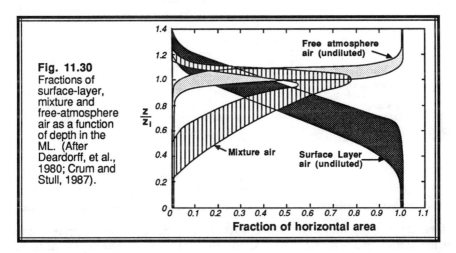

Fig. 11.30 Fractions of surface-layer, mixture and free-atmosphere air as a function of depth in the ML. (After Deardorff, et al., 1980; Crum and Stull, 1987).

11.3.3 Structures

Overshooting and Intermittancy. The entrainment zone essentially consists of turbulent thermals imbedded within non-turbulent FA air. An aircraft flying a horizontal flight leg within the entrainment zone will experience intermittent periods of turbulence.

A variety of smaller scales of motion are present in the entrainment zone in addition to the large overshooting thermals (Crum and Stull, 1987). In spite of this smaller-scale turbulence, there is very little (almost negligible) ML air dispersed out of the thermals into the FA.

K-H Waves. Kelvin-Helmholtz (K-H) waves can also exist on a variety of scales in the entrainment zone. On the largest scale, we recognize that there is wind shear between the sub-geostrophic ML air and the nearly geostrophic FA air. This shear across the statically stable entrainment zone is ideal for the formation of K-H waves, as discussed in Chapter 5. The waves formed in this manner often have a wavelength proportional to the depth of the EZ (i.e., hundreds of meters), and evolve rather slowly. They break, become turbulent, and can contribute to the entrainment of FA air into the ML.

On a smaller scale, K-H waves can form along the top boundary of overshooting thermals (Fig 11.31, after Rayment and Readings, 1974). The thermals are carrying up the cooler and slower ML air, and their tops are marked by a sharp interface with the FA air. As a result, there can be very strong shears and statically stable lapse rates across a distance of two to ten meters. This situation is ideal for the formation of short-wavelength, rapidly evolving K-H waves that form, break, and decay into turbulence during the few minutes that the thermal is penetrating into the EZ. The thin layer of turbulence that results along the top helps dilute the thermal, but contributes little to the overall entrainment into the ML.

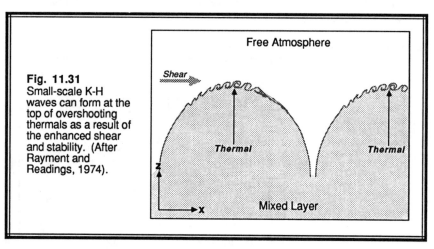

Fig. 11.31 Small-scale K-H waves can form at the top of overshooting thermals as a result of the enhanced shear and stability. (After Rayment and Readings, 1974).

Gravity Waves and Resonance with Convection. When thermals penetrate into the statically stable EZ and FA, *internal gravity (buoyancy) waves* can be excited in the FA (Fig 11.32a). These waves have been observed in laboratory tanks (Deardorff, et al., 1969) and simulated numerically (Deardorff, 1974; Carruthers and Moeng, 1987) with LES models. Depending on the static stability, these waves may propagate vertically and horizontally away from the thermals, and thus drain kinetic energy and momentum from those thermals (Deardorff, 1969). However, it appears that the portion of ML energy drained by internal waves is relatively small, and can usually be neglected (Stull, 1976).

Penetrating thermals and associated clouds are often moving at a different speed (usually slower) than the air aloft. As a result, the cloud acts like an obstruction to the flow, and forces air to flow over its top, analogous to air flow over a mountain (Fig 11.32b). The result can be the same; namely, mountain-type waves can be generated. Such waves have be used by glider pilots, who have named them *thermal waves or convection waves* (Clarke, et al., 1986; Kuettner, et al., 1987). These convection waves have a horizontal wavelength of 5 to 15 km, a vertical motion amplitude of 1 to 4 m/s, and can extend upwards throughout the troposphere if the static stability is favorable. A 90° direction change in the wind shear vector near z_i supports thermal wave generation. Climatologically, thermals waves are often found in pre- and postfrontal zones, and the waves are often stronger over cloud streets than over isolated cumulus.

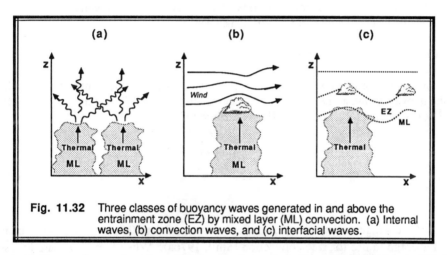

Fig. 11.32   Three classes of buoyancy waves generated in and above the entrainment zone (EZ) by mixed layer (ML) convection. (a) Internal waves, (b) convection waves, and (c) interfacial waves.

A third type of wave is an interfacial wave formed on (or trapped within) the EZ (Fig 11.32c). These are analogous to the surface waves excited when a pebble is tossed into a pond, and do not require a mean wind or wind shear in order to form. Wavelengths of the diameter of thermals and smaller experience very strong nonlinear interaction with ML turbulence, and are quickly absorbed into the ML (Weinstock, 1987). Longer wavelengths, however, can persist and modulate the organization of convection within the ML. As a result, resonance can occur where ML convection and clouds are enhanced at the same wavelength as the waves (Clark, et al., 1986; Noilhan, et al., 1986).

11.4 Entrainment Velocity and Its Parameterization

11.4.1 Typical Values

There are obvious bounds on the entrainment velocity based on the observed rise of the ML top. By definition, it can never be negative. If there is entrainment happening in both directions across a stable interface, then we can use separate (positive) velocities for the entrainment in each direction. It can be zero whenever there is no turbulence. Large values on the order of $w_e = 1$ m/s are possible during the rapid-rise phase of the late morning ML, when the ML turbulence is strong and the capping inversion is weak. Typical values, however, are often in the range of $w_e = 0.01$ to 0.20 m/s.

As mentioned earlier, we must make an assumption to close the ML equations. This assumption can be in the form of a parameterization for the entrainment velocity, or a parameterization for any of the other unknowns in the set of equations (11.2.3a-d). Obviously, if assumptions are made for the other unknowns, then the entrainment velocity can be calculated from the other knowns. The following sections describe some of these closure assumptions. The thermodynamic method (Section 11.2.2) is not described here because it is a direct calculation of ML depth, without really using entrainment velocities. Of course, the thermodynamic methods works only during free convection, and fails whenever mechanical generation of turbulence dominates.

11.4.2 Flux-Ratio Method

For the case of free convection, the turbulence causing entrainment is directly related to the buoyancy flux at the surface. This causes the buoyancy flux at the top of the ML to be a nearly constant fraction of the flux at the bottom (Ball, 1960):

$$\frac{-\overline{w'\theta'_v}_{z_i}}{\overline{w'\theta'_v}_s} = A_R \tag{11.4.2a}$$

Most of the suggested values of A_R range between 0.1 to 0.3, with $A_R = 0.2$ being a good average to use for free convection (Stull, 1976a). When humidities are low such that potential temperature and virtual potential temperature are approximately equal, (11.4.2a) is often written in terms of the heat flux.

When (11.4.2a) is used in an expression analogous to (11.2.3c) but for virtual potential temperature, we find that the entrainment velocity is:

$$w_e = \frac{A_R \overline{w'\theta'_v}_s}{\Delta_{EZ}\overline{\theta}_v} \tag{11.4.2b}$$

Almost all of the entrainment parameterizations have this same fundamental form: entrainment is proportional to some measure of the turbulence or external turbulence forcing, and is inversely proportional to the virtual potential temperature jump across the top of the ML. Stronger capping inversions reduce entrainment, while stronger surface buoyancy flux increases entrainment.

11.4.3 Energetics Method

In situations where wind shear generates turbulence, the previous method fails. Quite a few alternative parameterizations have been proposed in the literature that include mechanical as well as thermal generation of turbulence. One class of entrainment closure, described here, uses the TKE equation.

In order for warm air to be entrained into the cooler ML, it must be forced down against the restoring force of gravity. In the process of lowering the buoyant entrained air, the potential energy of the ML/FA system is increased. The rate of change of potential energy with time equals the integral over height of the negative portions of buoyancy flux. Some of the TKE of the ML is expended to do the work necessary to bring the entrained air down. The net result is that some TKE has been converted to potential energy.

To get an entrainment velocity from this approach, start by integrating the TKE equation (5.1) over the total depth of the boundary layer (Stull, 1976a):

$$\frac{d}{dt}\int \bar{e}\, dz = \frac{g}{\theta_v}\int \overline{w'\theta_v'}\, dz - \int \overline{u'w'}\frac{\partial \bar{U}}{\partial z}\, dz - \frac{1}{\rho}\overline{w'p'}\Big|_{z_i} - \int \varepsilon\, dz \qquad (11.4.3a)$$

$$\text{St} \qquad\qquad \text{B}_{PN} \qquad\qquad \text{MP} \qquad\qquad \text{GW} \qquad\qquad \text{Dis}$$

where St is the integrated storage term, B_{PN} is buoyancy, MP is the mechanical production, GW is a pressure-velocity correlation related to gravity waves draining energy from the top of the ML, and Dis is viscous dissipation. Note that the turbulent transport term (Term V in equation 5.1b) integrates to zero and does not appear above because there is no turbulent transport into the nonturbulent FA or into the ground. As mentioned before, the GW term is relatively small, and can usually be neglected.

Suppose that the buoyancy flux at any height can be split into positive and negative contributions, identified as production and consumption respectively (Stage and Businger, 1981):

$$\overline{w'\theta_v'} = \overline{w'\theta_v'}\Big|_{production} + \overline{w'\theta_v'}\Big|_{consumption}$$

The integral of buoyancy flux in term B_{PN} could thus be split into two parts, $B_{PN} = B_P + B_N$:

$$B_P = \frac{g}{\theta_v}\int \overline{w'\theta_v'}\Big|_{production}\, dz \qquad \text{and} \qquad B_N = \frac{g}{\theta_v}\int \overline{w'\theta_v'}\Big|_{consumption}\, dz \qquad (11.4.3b)$$

where B_P represents the *production* of TKE via buoyancy, and B_N represents the *consumption* of TKE as it is converted into potential energy. Although B_P and B_N

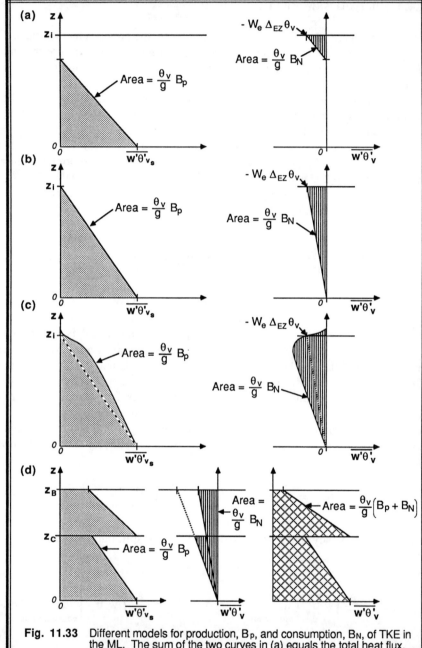

Fig. 11.33   Different models for production, B_P, and consumption, B_N, of TKE in the ML. The sum of the two curves in (a) equals the total heat flux, as plotted in Fig. 11.13. The curves in (b) and (c) also sum to the same total. Fig. (d) show the corresponding curves for an idealized stratocumulus-topped ML. (After Stage and Businger, 1981).

appear to have the same definition, only the subset of buoyancy flux that is negative is used in B_N, and the remainder is used in B_P.

The critical question is how to partition the buoyancy flux profile into its production and consumption components. One way is to focus on just the positive and negative portions of the total buoyancy flux profile (Fig 11.33a). Randall (1984) has called this the *Eulerian partitioning* approach. Alternative geometries have also been suggested (Fig 11.33b&c) based on *process partitioning* of updrafts vs. downdrafts (Randall, 1984), and some parameterizations can be used for cloud topped MLs (Fig 11.33d). The correct answer has yet to be pinned down, but luckily most partitioning schemes yield similar answers for the entrainment velocity.

To close this equation and solve for entrainment velocity, we must make a few additional assumptions or parameterizations. Quasi-steady state is often assumed, allowing us to set St = 0. The integrated TKE equation now becomes:

$$-B_N = B_P - MP - Dis \qquad (11.4.3c)$$

where MP and B_N are usually negative. The buoyant production term, B_P, is known from the partitioning described above, and is equal to $(g/\overline{\theta_v})$ times the "production area" as sketched in Fig 11.33. When clouds are present, this production area should include the effects of radiation divergence (Stage and Businger, 1981). The mechanical production term, -MP, is often parameterized in terms of forcings, such as the shear or u_* at the surface and the shear ΔU across the top of the ML. The dissipation rate, Dis, is often assumed to be proportional to the total production rate (B_P - MP). Between 80% to 99% of the TKE produced is dissipated (Stull, 1976; Stage and Businger, 1981). At this point, we know all of the terms on the right of (11.4.3c), and can solve for the value of B_N.

Knowing B_N, the "consumption area" of the buoyancy flux profile (Fig 11.33) can be found from: consumption area = $B_N / (g/\overline{\theta_v})$. Using the geometry assumed earlier in the partitioning scheme, the consumption area can be related to the buoyancy flux at the top of the ML: $\overline{w'\theta_v'}_{z_i}$. The entrainment velocity can then be found from the relationship between temperature jump and entrained heat flux:

$$w_e = \frac{-\overline{w'\theta_v'}_{z_i}}{\overline{\Delta_{EZ}\theta_v}} \qquad (11.4.3d)$$

The advantage of this production/consumption approach is that it can be applied to a variety of situations: free convection, forced convection, mixed convection, and cloud-topped MLs.

As an example, when the energetics method is applied to a cloud-free ML with shear at both the surface and at the top of the ML, the following relationship can be derived (Stull, 1976a) assuming an Eulerian partitioning geometry as sketched in Fig 11.33a:

$$w_e = \frac{2\,\overline{\theta}_v}{g\,d_1\,\Delta_{EZ}\overline{\theta}_v} \left[c_1\,w_*^3 + c_2\,u_*^3 + c_3\,(\Delta_{EZ}\overline{U})^3 \right] \qquad (11.4.3e)$$

where the three parameters are $c_1 = 0.0167$, $c_2 = 0.5$, and $c_3 = 0.0006$. The first term in square brackets parameterizes buoyant production, the second surface mechanical production, and the last mechanical production at the top of the ML. The distance, d_1, is the distance between the top of the ML and the height where the heat flux profile crosses zero:

$$\frac{d_1}{z_i} = \frac{-\overline{w'\theta'_v}_{z_i}}{-\overline{w'\theta'_v}_{z_i} + \overline{w'\theta'_v}_s}$$

It can easily be shown that this energetics parameterization reduces to the flux-ratio parameterization for the special case of free convection.

11.4.4 Other Methods

There are many other methods that have appeared in the literature, and which won't be listed here. Most are variations of the above methods. Some, such as proposed by Deardorff (1979), offer generalized approaches that do not depend on a jump or slab type of model. Most approaches have been tested against observed data and give realistic results within the uncertainties of the data. Contributing to some of the uncertainty is subsidence, which is not easily measurable and can be as large as the entrainment velocity. Section 11.5 examines some of the characteristics of boundary layer subsidence.

There were some additional methods to forecast ML depth that were tested, but which did not work well. One method assumed that the ML depth was proportional to the Ekman layer depth. Another assumed that a bulk Richardson number could be formed using the wind and temperature difference across the whole ML depth, and that z_i was found using the requirement that the bulk Richardson number equal some constant critical value.

11.4.5 Example

Problem: Given: $\overline{w'\theta'_v} = 0.2$ K m/s, $u_* = 0.2$ m/s, $g/\overline{\theta}_v = 0.0333$ m s^{-2} K^{-1},

$z_i = 1$ km, $\Delta_{EZ}\overline{\theta}_v = 2$ K. Find w_e using the:

a) Flux ratio method;   b) Energetics method (special case equation 11.4.3e).

Solution:

a) Flux ratio method:

$$w_e = A_R \frac{\overline{w'\theta_v'}_s}{\Delta_{EZ}\overline{\theta_v}} = \frac{0.2 \cdot 0.2}{2} = 0.0200 \text{ m/s}$$

b) Energetics method:

$$w_e = \frac{2}{(g/\overline{\theta_v}) \, d_1 \, \Delta_{EZ}\overline{\theta_v}} \left[c_1 w_*^3 + c_2 u_*^3 \right] = \frac{30.03}{d_1} \left[0.1112 + 0.004 \right] = \frac{3.46}{d_1}$$

But

$$d_1 = 1000 \left[\frac{2 \, w_e}{0.2 + 2 \, w_e} \right]$$

Combining these two equations gives:

$$w_e^2 = 0.00173 \, [\, 2 \, w_e + 0.2 \,]$$

which can be solved to yield $w_e = 0.0204$ m/s.

Discussion: We see that the addition of small values of surface stress have little effect on the entrainment rate in a free convection situation, and that the flux ratio method gives essentially the same answer with much fewer computations. For forced convection, however, the flux ratio method fails completely, but the energetics method can be used in the form (neglecting shear at the ML top, and using $d_1 = z_i$):

$$w_e = \frac{2 \, c_2 \, u_*^3}{(g/\overline{\theta_v}) \, z_i \, \Delta_{EZ}\overline{\theta_v}}$$

11.5 Subsidence and Advection

11.5.1 Advection

Even with the large vertical fluxes and vigorous turbulence in a convective ML, horizontal advection of state characteristics by the mean wind can have as large an effect as turbulence. The equations of Chapters 3 to 5 include the mean advection terms, and will not be discussed in more detail here. Neglect of advection is unwarranted for most simulations of real boundary layer.

One measure of the relative importance of turbulence vs. the mean wind is the dimensionless convective distance, X^{ML} (Willis and Deardorff, 1976):

$$X^{ML} = \frac{x \; w_*}{z_i \; \overline{M}} \tag{11.5.1a}$$

This can be interpreted as the ratio of measured horizontal distance to the theoretical distance traveled during one convective circulation (updraft and downdraft). For large X^{ML} (>>1) turbulent mixing dominates over the mean wind, while for small X^{ML} (<<1) the turbulent mixing is less important.

In addition to horizontal advection of momentum, moisture, heat, and pollutants, we must be concerned about advection of z_i. In essence, the latter is a measure of the advection of volume within the ML. For example, a slowly rising, shallow ML over a moist irrigated region can grow rapidly if a deeper ML advects into the area. The local change of z_i is described by:

$$\frac{\partial z_i}{\partial t} = -U_j \frac{\partial z_i}{\partial x_j} + w_e + w_L \tag{11.5.1b}$$

The slope of z_i can be significant, and its neglect can cause forecast errors in ML depth tendency as great the magnitude of the entrainment velocity (Lenschow, 1973).

For strong winds and abrupt changes in surface conditions, advection can dominate to prevent growth of the local ML with time. The resulting thermal internal boundary layer, which is a function of downwind distance x, is discussed in Chapter 14.

11.5.2 Subsidence and Divergence

Mean vertical velocities ranging from -0.22 m/s (subsidence) to 0.27 m/s (upward motion) have been observed in a limited case study based on BLX83 field experiment data (Vachalek, et al., 1988). These magnitudes are very large compared to the entrainment velocity, and can not be neglected.

Unfortunately, subsidence at the top of the ML is very difficult to measure. Vertical velocity measurements from aircraft often have a mean bias that is greater than the true subsidence, and therefore cannot be used. The mean vertical velocity magnitudes are also below the noise level of most Doppler remote sensors. The downward movement of elevated smoke, moisture, or stable layers can be tracked, but usually horizontal advection affects this movement and is frequently not known. Also, the tracking of elevated layers can be applied only if the layers are in the FA above the mixed layer, so as not to be influenced by ML growth.

Alternately, mean vertical motion at z_i can be estimated if divergence is known as a function of height within the ML, using:

$$w_L(z_i) = -\int_{z=0}^{z_i} \text{Div}(z) \, dz \qquad (11.5.2a)$$

For the special case of constant divergence with height, (11.5.2a) reduces to

$$w_L(z_i) = -\text{Div} \cdot z_i \qquad (11.5.2b)$$

This latter expression is frequently used due to lack of better divergence data.

Horizontal divergence is not trivial to measure. Theoretically, we must measure the normal velocities out of a specified horizontal area, and we must make these measurements at every point on the perimeter of the area. Some remote sensors, such as Doppler radar and Doppler lidar, come closest to satisfying this requirement by measuring radial velocity at any specified range as a function of azimuth. A plot of this data is called a *velocity-azimuth display (VAD)*. Unfortunately, ground clutter for the radar can contaminate the velocity statistics.

Surface mesonetwork stations placed close together along the perimeter of the area can be used to estimate divergence near the surface, assuming that surface terrain features do not contaminate the velocities. As stations are spaced further apart along the perimeter, divergence accuracy drops. If the diameter of the network is too small, then the horizontal velocity difference across the network could be too small to resolve.

Finally, a network of rawinsonde launch sites can be used to find divergence using the *Bellamy method*; however, sonde accuracy, large site separations, and sonde-to-sonde calibration errors can contaminate divergence calculations. Vachalek, et al. (1988) found that the rawinsonde divergence integrated over the ML depth, and surface divergence from mesonetwork stations yielded the best results.

Divergence fluctuations on a wide range of horizontal and temporal scales are usually superimposed on each other. Smaller diameter features appear to have greater magnitudes (by factors of up to 100) and shorter durations than the large diameter features. For example, a region of diameter 5 km was found to have peak divergence magnitudes in the range of 10^{-4} to 10^{-5} s^{-1}, while regions of diameter 100 km had peak divergences of 10^{-5} to 10^{-7}. A comparison of the relative frequency of events of different divergence magnitude and horizontal scale is presented in Fig 11.34.

The short duration (1 h) divergence events occur about 10 times more frequently than long duration events, and about 95% of all divergence events had durations shorter than 8 h. This implies that divergence and subsidence estimated from a large-diameter network will not show large-magnitude short-period variations, and thus may be useless for estimating subsidence over a fixed point at any specific time.

Vachalek, et al. (1988) studied how the divergence variations were associated with meso- and synoptic-scale features, and developed the following idealized divergence models:

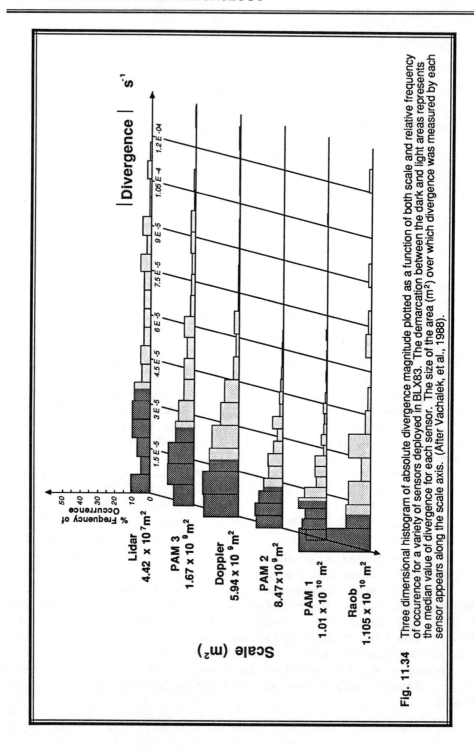

Fig. 11.34   Three dimensional histogram of absolute divergence magnitude plotted as a function of both scale and relative frequency of occurrence for a variety of sensors deployed in BLX83. The demarcation between the dark and light areas represents the median value of divergence for each sensor. The size of the area (m^2) over which divergence was measured by each sensor appears along the scale axis. (After Vachalek, et al., 1988).

(1) 2-D divergence from high (convergence into low) pressure centers,
(2) 1-D divergence from ridges (convergence into troughs),
(3) 1-D step-like changes in velocity normal to fronts, associated with very large
    convergence across a very small zone, and
(4) convergence/divergence couplets associated with thunderstorm passage.
These models are sketched in Fig 11.35.

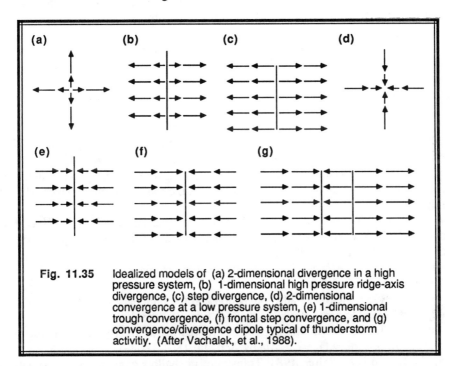

Fig. 11.35   Idealized models of (a) 2-dimensional divergence in a high
pressure system, (b) 1-dimensional high pressure ridge-axis
divergence, (c) step divergence, (d) 2-dimensional
convergence at a low pressure system, (e) 1-dimensional
trough convergence, (f) frontal step convergence, and (g)
convergence/divergence dipole typical of thunderstorm
activitiy. (After Vachalek, et al., 1988).

We learned in this chapter that the ML has a well-defined top, and is usually in quasi-equilibrium with the surface forcings. These factors allowed us to describe relatively simple models for ML characteristics and evolution. In the next chapter, however, we will learn that the stable boundary layer has a poorly-defined top, and is rarely in equilibrium with the surface. As a result, Chapter 12 reviews other techniques that can be used to describe the stable boundary layer. Clouds, which can be a part of either stable or convective boundary layers, are described in Chapter 13.

11.6 References

Agee, E.M., 1984: Observations from space and thermal convection: a historical perspective. *Bull. Am. Meteor. Soc.*, **65**, 938-949.
Agee, E.M., T.S. Chen and K.E. Dowell, 1973: A review of mesoscale cellular

convection. *J. Appl. Meteor.*, **13**, 46-53.

André, J.-C., J.-P. Goutorbe, and A. Perrier, 1986: HAPEX-MOBILHY, a hydrologic atmospheric pilot experiment for the study of water budget and evaporation flux at the climatic scale. *Bull. Am. Meteor. Soc.*, **67**, 138-144.

Arritt, R.W., 1987: Effect of water surface temperature on lake breezes and thermal internal boundary layers. *Bound.-Layer Meteor.*, **40**, 101-125.

Atlas, D., B. Walter, S.-H. Chou and P.J. Sheu, 1986: The structure of the unstable marine boundary layer viewed by lidar and aircraft observations. *J. Atmos. Sci.*, **43**, 1301-1318.

Bader, D.C., T.B. McKee and G.J. Tripoli, 1987: Mesoscale boundary layer evolution over complex terrain. Part 1. Numerical simulation of the diurnal cycle. *J. Atmos. Sci.*, **44**, 2823-2838.

Ball, F.K., 1960: Control of inversion height by surface heating. *Quart. J. Roy. Meteor. Soc.*, **86**, 483-494.

Bénech, B., J. Noilhan, A. Druilhet, J.M. Brustet and C. Charpentier, 1986: Experimental study of an artificial thermal plume in the boundary layer. Part 1. Flow characteristics near the heat source. *J. Clim. Appl. Meteor.*, **25**, 418-437.

Boers, R. and E.W. Eloranta, 1986: Lidar measurements of the atmospheric entrainment zone and the potential temperature jump across the top of the mixed layer. *Bound.-Layer Meteor.*, **34**, 357-375.

Boers, R., E.W. Eloranta and R.L. Coulter, 1984: Lidar observations of mixed layer dynamics: Tests of parameterized entrainment models of mixed layer growth rate. *J. Clim. Appl. Meteor.*, **23**, 247-266.

Brown, R.A., 1970: A secondary flow model for the planetary boundary layer. *J. Atmos. Sci.*, **27**, 742-757.

Brutsaert, W., 1987: Nearly steady convection and the boundary layer budgets of water vapor and sensible heat. *Bound.-Layer Meteor.*, **39**, 283-300.

Carruthers, D.J. and C.-H. Moeng, 1987: Waves in the overlying inversion of the convective boundary layer. *J. Atmos. Sci.*, **44**, 1801-1808.

Chang, H.-R. and H.N. Shirer, 1984: Transitions in shallow convection: an explanation for lateral cell expansion. *J. Atmos. Sci.*, **41**, 2334-2346.

Clark, T.L., T. Hauf and J.P. Kuettner, 1986: Convectively forced internal gravity waves: results from two-dimensional numerical experiments. *Quart. J. Roy. Meteor. Soc.*, **112**, 899-925.

Crum, T.D. and R.B. Stull, 1987: Field measurements of the amount of surface layer air versus height in the entrainment zone. *J. Atmos. Sci.*, **44**, 2743-2753.

Crum, T.D., R.B. Stull and E.W. Eloranta, 1987: Coincident lidar and aircraft observations of entrainment into thermals and mixed layers. *J. Clim. Appl. Meteor.*, **26**, 774-788.

Deardorff, J.W., 1969: Numerical study of heat transport by internal gravity waves above a growing unstable layer. *Phys. Fluids, Suppl. II*, **12**, 184-194.

Deardorff, J.W., 1972: Numerical investigation of neutral and unstable planetary boundary layers. *J. Atmos. Sci.*, **29**, 91-115.

Deardorff, J.W., 1974: Three-dimensional numerical study of turbulence in an entraining mixed layer. *Bound-Layer. Meteor.*, **7**, 199-226.

Deardorff, J.W., 1979: Prediction of convective mixed-layer entrainment for realistic capping inversion structure. *J. Atmos. Sci.*, **36**, 424-436.

Deardorff, J.W. and G.E. Willis, 1985: Further results from a laboratory model of the convective planetary boundary layer. *Bound.-Layer Meteor.*, **32**, 205-236.

Deardorff, J.W., G.E. Willis and D.K. Lilly, 1969: Laboratory investigation of non-steady penetrative convection. *J. Fluid Mech.*, **35**, 7-31.

Deardorff, J.W., G.E. Willis and B.H. Stockton, 1980: Laboratory studies of the entrainment zone of a convectively mixed layer. *J. Fluid Mech.*, **100**, Part 1, 41-64.

Doviak, R.J. and M. Berger, 1980: Turbulence and waves in the optically clear planetary boundary layer resolved by dual-Doppler radar. *Radio Science*, **15**, 297-317.

Driedonks, A.G.M. and H. Tennekes, 1984: Entrainment effects in the well-mixed atmospheric boundary layer. *Bound.-Layer Meteor.*, **30**, 75-105.

Fairall, C.W., 1987: A top-down and bottom-up diffusion model of c_T2 and c_q2 in the entraining convective boundary layer. *J. Atmos. Sci.*, **44**, 1009-1017.

Ferrare, R.A., 1984: Lidar observations of organized convection within the atmospheric mixed layer. M.S. Thesis. Dept. of Meteorology, Univ. of Wisconsin-Madison. 204pp.

Fiedler, B.H., 1984: The mesoscale stability of entrainment into cloud-topped mixed layers. *J. Atmos. Sci.*, **41**, 92-101.

Fitzjarrald, D.E., 1973: A field investigation of dust devils. *J. Appl. Meteor.*, **12**, 808-813.

Fritz, R.B. and T.-I. Wang, 1979: Chapt. 9. Optical systems measuring surface-level convergence during PHOENIX, *Project PHOENIX, Report No. 1.* (Hooke, Ed.), 57-73. Available from NOAA.ERL, Boulder, CO 80303.

Godowitch, J.M., 1986: Characteristics of vertical turbulence velocity in the urban convective boundary layer. *Bound.-Layer Meteor.*, **35**, 387-407.

Greenhut, G.K. and S.J.S. Khalsa, 1982: Updraft and downdraft events in the atmospheric boundary layer over the equatorial Pacific Ocean. *J. Atmos. Sci.*, **39**, 1803-1818.

Greenhut, G.K. and S.J.S. Khalsa, 1987: Convective elements in the marine atmospheric boundary layer. Part 1: Conditional sampling statistics. *J. Clim. Appl. Meteor.*, **26**, 813-822.

Haines, D.A., 1982: Horizontal roll vortices and crown fires. *J. Clim. Appl. Meteor.*, **21**, 751-763.

Hanna, S.R., 1987: An empirical formula for the height of the coastal internal boundary layer. *Bound.-Layer Meteor.*, **40**, 205-207.

Hooper, W.P., 1982: *The Diurnal Evolution of the Planetary Boundary Layer: Lidar Observations above a Flat Homogeneous Surface.* M.S. Thesis, Univ. of Wisconsin - Madison. 160pp.

Hooper, W.P. and E.W. Eloranta, 1986: Lidar measurements of wind in the planetary boundary layer: the method, accuracy and results from joint measurements with radiosonde and kytoon. *J. Clim. Appl. Meteor.*, **25**, 990-1001.

Hsu, S.A., 1986: A note on estimating the height of the convective internal boundary layer near shore. *Bound.-Layer Meteor.*, **35**, 311-316.

Idso, S.B., 1975: Tornado-like dust devils. *Weather*, **30**, 115-117.

Idso, S.B., 1975: Arizona weather watchers: past and present. *Weatherwise*, **28**, 56-60.

Idso, S.B., 1975: Whirlwinds, density currents, and topographic disturbances: a meteorological melange of intriguing interactions. *Weatherwise*, **28**, 61-65.

Kaimal, J.C. and J.A. Businger, 1970: Case studies of a convective plume and a dust devil. *J. Appl. Meteor.*, **9**, 612-620.

Kaimal, J.C., W.L. Eberhard, W.R. Moniger, J.E. Gaynor, S.W. Troxel, T. Uttal, G.A. Briggs, and G.E. Start, 1986: *Project Condors, Convective Diffusion observed by remote sensors.* NOAA/ERL, Boulder, CO 80303.

Khalsa, S.J.S. and G.K. Greenhut, 1985: Conditional sampling of updrafts and downdrafts in the marine atmospheric boundary layer. *J. Atmos. Sci.*, **42**, 2550-2562.

Khalsa, S.J.S. and G.K. Greenhut, 1987: Convective elements in the marine atmospheric boundary layer. Part 2: Entrainment at the capping inversion. *J. Clim. Appl. Meteor.*, **26**, 824-836.

Kraus, E.B., 1972: *Atmosphere-Ocean Interaction.* Cambridge Univ. Press., Oxford. 275pp.

Kraus, E.B. and L.D. Leslie, 1982: The interactive evolution of the oceanic and atmospheric boundary layers in the source regions of the trades. *J. Atmos. Sci.*, **39**, 2760-2772.

Kropfli, R.A., 1979: Chapt. 3. PHOENIX multiple Doppler radar operations. *Project PHOENIX - Report No. 1.* (Hooke, Ed.), 33-56. Available from NOAA/ERL, Boulder, CO 80303.

Kuettner, J.P., P.A. Hildebrand and T.L. Clark, 1987: Convection waves: observations of gravity wave systems over convectively active boundary layers. *Quart. J. Roy. Meteor. Soc.*, **113**, 445-468.

Lamb, R.G., 1978: A numerical simulation of dispersion from an elevated point source in the convective planetary boundary layer. *Atmos. Environ.*, **12**, 1297-1304.

LeMone, M.A., 1973: The structure and dynamics of horizontal roll vortices in the planetary boundary layer. *J. Atmos. Sci.*, **30**, 1077-1091.

LeMone, M.A., 1976: Modulation of turbulence energy by longitudinal rolls in an unstable planetary boundary layer. *J. Atmos. Sci.*, **33**, 1308-1320.

Lenschow, D.H., 1973: Two examples of planetary boundary layer modification over the great lakes. *J. Atmos. Sci.*, **30**, 568-581.

Lenschow, D.H. and E.M. Agee, 1986: Preliminary results from the Air Mass Transformation Experiment (AMTEX). *Bull. Am. Meteor. Soc.*, **57**, 1346-1355.

Lilly, D.K., 1968: Models of cloud-topped mixed layers under a strong inversion. *J. Atmos. Sci.*, **30**, 1092-1099.

Lyons, W.A., 1975: Turbulent diffusion and pollutant transport in shoreline environments. *Lectures on Air Pollution and Environmental Impact Analysis*, D.A. Haugen (Ed.), Am. Meteor. Soc. 136-208.

Mahrt, L., 1979: Penetrative convection at the top of a growing boundary layer. *Quart.*

J. Roy. Meteor. Soc., **105**, 469-485.

Mahrt, L. and J.-C. André, 1983: On the stratification of turbulent mixed layers. *J. Geophys. Res.*, **88**, 2662-2666.

Mahrt, L. and J. Paumier, 1984: Heat transport in the atmospheric boundary layer. *J. Atmos. Sci.*, **41**, 3061-3075.

Mahrt, L. and J. Paumier, 1985: Simple formulation of heat flux in the unstable atmospheric boundary layer. *Bound.-Layer Meteor.*, **33**, 61-76.

Mason, P.J. and R.I. Sykes, 1980: A 2-D numerical study of horizontal roll vortices in the neutral atmospheric boundary layer. *Quart. J. Roy. Meteor. Soc.*, **106**, 351-336.

Mason, P.J. and R.I. Sykes, 1982: A 2-D numerical study of horizontal roll vortices in an inversion-capped planetary boundary layer. *Quart. J. Roy. Meteor. Soc.*, **108**, 801-823.

Moeng, C.-H. and J.C. Wyngaard, 1984:   Statistics of conservative scalars in the convective boundary layer. *J. Atmos. Sci.*, **41**, 3161-3169.

News and Notes, 1976: Dust devil wind velocities. *Bull. Am. Meteor. Soc.*, **57**, 600.

Nieuwstadt, F.T.M. and R.A. Brost, 1986: The decay of convective turbulence. *J. Atmos. Sci.*, **43**, 532-546.

Nieuwstadt, F.T.M. and H. van Dop, 1982: *Atmospheric Turbulence and Air Pollution Modelling.* Reidel Publ. Co., Dordrecht. 358pp.

Noilhan, J. and B. Bénech, 1986: Experimental study of an artificial thermal plume in the boundary layer.   Part 3: Dynamic structure within the plume. *J. Clim. Appl. Meteor.*, **25**, 458-467.

Noilhan, J., B. Bénech, G. Letrenne, A. Druilhet and A. Saab, 1986: Experimental study of an artificial thermal plume in the boundary layer. Part 2: Some aspects of the plume thermodynamical structure. *J. Clim. Appl. Meteor.*, **25**, 439-457.

Rabin, R.M., R.J. Doviak and A. Sundara-Rajan, 1982: Doppler radar observations of momentum flux in a cloudless convective layer with rolls. *J. Atmos. Sci.*, **39**, 851-863.

Randall, D.A., 1984: Buoyant production and consumption of turbulence kinetic energy in cloud-topped mixed layers. *J. Atmos. Sci.*, **41**, 402-413.

Ray, D., 1986: Variable eddy diffusivities and atmospheric cellular convection. *Bound.-Layer Meteor.*, **36**, 117-131.

Rayment, R. and C.J. Readings, 1974: A case study of the structure and energetics of an inversion. *Quart. J. Roy. Meteor. Soc.*, **100**, 221-223.

Reichmann, H., 1975: *Cross-country Soaring (Streckensegelflug).*   Thompson Publications. P.O. Box 1175, Pacific Palisades, CA 90272. 150pp.

Reinking, R.F., R.J. Doviak and R.O. Gilmer, 1981: Clear-air roll vortices and turbulent motions as detected with an airborne gust probe and dual-Doppler radar. *J. Appl. Meteor.*, **20**, 678-685.

Rosmond, T.E., 1973: Mesoscale cellular convection. *J. Atmos. Sci.*, **30**, 1392-1409.

Rothermal, J. and E.M. Agee, 1980: Aircraft investigation of mesoscale cellular convection during AMTEX 75. *J. Atmos. Sci.*, **37**, 1027-1040.

Rothermel, J. and E.M. Agee, 1986: A numerical study of atmospheric convective scaling. *J. Atmos. Sci.*, **43**, 1185-1197.

Schols, J.L.J., 1984: The detection and measurement of turbulent structures in the atmospheric surface layer. *Bound.-Layer Meteor.*, **29**, 39-58.

Schols, J.L.J., A.E. Jansen and J.G. Krom, 1985: Characteristics of turbulent structures in the unstable atmospheric surface layer. *Bound.-Layer Meteor.*, **33**, 173-196.

Schols, J.L.J. and L. Wartena, 1986: A dynamical description of turbulent structure in the near neutral atmospheric surface layer: the role of static pressure fluctuations. *Bound.-Layer Meteor.*, **34**, 1-15.

Scorer, R.S., 1957: Experiments on convection of isolated masses of buoyant fluid. *J. Fluid Mech.*, **2**, 583-594.

Sheu, P.J. and E.M. Agee, 1977: Kinematic analysis and air-sea heat flux associated with mesoscale cellular convection during AMTEX 75. *J. Atmos. Sci.*, **34**, 793-801.

Sinclair, P.C., V.H. Leverson and R.F. Abbey, Jr., 1977: The vortex structure of dust devils, water spouts, and tornadoes. *10th AMS Conference on Severe Local Storms, Oct 18-21, Omaha, NE.* Amer. Meteor. Soc., Boston. 533pp.

Smolarkiewicz, P.K. and T.L. Clark, 1985: Numerical simulation of the evolution of a three-dimensional field of cumulus clouds. Part 1: Model description, comparison with observations and sensitivity studies. *J. Atmos. Sci.*, **42**, 502-522.

Stage, S.A. and J.A. Businger, 1981: A model for entrainment into a cloud-topped marine boundary layer. Parts 1 & 2. *J. Atmos. Sci.*, **38**, 2213-2242.

Stull, R.B., 1973: Inversion rise model based on penetrative convection. *J. Atmos. Sci.*, **30**, 1092-1099.

Stull, R.B., 1976a: The energetics of entrainment across a density interface. *J. Atmos. Sci.*, **33**, 1260-1267.

Stull, R.B., 1976b: Mixed layer depth model based on turbulent energetics. *J. Atmos. Sci.*, **33**, 1268-1278.

Stull, R.B., 1976c: Internal gravity waves generated by penetrative convection. *J. Atmos. Sci.*, **33**, 1279-1286.

Stull, R.B., 1985: A fair-weather cumulus cloud classification scheme for mixed-layer studies. *J. Clim. Appl. Meteor.*, **24**, 49-56.

Stull, R.B., 1988: Pollutant dispersion and mixed-layer modeling using asymmetric transilient matrices. *8th AMS Symposium on Turbulence and Diffusion.* San Diego, 25-29 April 1988. Amer. Meteor. Soc., Boston. 4pp.

Stull, R.B. and A.G.M. Driedonks, 1987: Applications of the transilient turbulence parameterization to atmospheric boundary-layer simulations. *Bound.-Layer Meteor.*, **40**, 209-239.

Tennekes, H., 1973: A model for the dynamics of the inversion above a convective boundary layer. *J. Atmos. Sci.*, **30**, 558-581.

Vachalek, R.E., R.B. Stull, and E.W. Eloranta, 1988: Mean vertical velocity and divergence measurements in the boundary layer. (Submitted to J. Appl. Meteor.)

Venkatram, A., 1977: A model for internal boundary layer development. *Bound.-Layer Meteor.*, **11**, 419-437.

Weinstock, J., 1987: The turbulence field generated by a linear gravity wave. *J. Atmos.*

Sci., **44**, 410-420.

Wilczak, J.M. and J.A. Businger, 1983: Thermally indirect motions in the convective atmospheric boundary layer. *J. Atmos. Sci.*, **40**, 343-358.

Wilczak, J.M. and J.E. Tillman, 1980: The three-dimensional structure of convection in the atmospheric surface layer. *J. Atmos. Sci.*, **37**, 2424-2443.

Wilde, N.P., R.B. Stull, and E.W. Eloranta, 1985: The LCL zone and cumulus onset. *J. Clim. Appl. Meteor.*, **24**, 640-657.

Willis, G.E. and J.W. Deardorff, 1976: A laboratory model of diffusion into the convective planetary boundary layer. *Quart. J. Roy. Meteor. Soc.*, **102**, 427-445.

Willis, G.E. and J.W. Deardorff, 1978: A laboratory study of dispersion from an elevated source within a modeled convective planetary boundary layer. *Atmos. Environ.*, **12**, 1305-1311.

Willis, G.E. and J.W. Deardorff, 1981: A laboratory study of dispersion from a source in the middle of the convectively mixed layer. *Atmos. Environ.*, **15**, 109-117.

Wyngaard, J.C., 1987: A physical mechanism for the asymmetry in top-down and bottom-up diffusion. *J. Atmos. Sci.*, **44**, 1083-1087.

Wyngaard, J.C. and R.A. Brost, 1984: Top-down and bottom-up diffusion of a scalar in the convective boundary layer. *J. Atmos. Sci.*, **41**, 102-112.

Young, G.S., 1988a&b: Turbulence structure of the convective boundary layer. Parts I & II. *J. Atmos. Sci.*, **44**, (in press).

Young, G.S., 1988c: Convection in the atmospheric boundary layer. *Earth Science Reviews* (in press).

11.7 Exercises

1) Given a ML that is initially 500 m thick with an average virtual potential temperature $<\overline{\theta_v}> = 280$ K. Assume that the mixed layer depth, z_i, and $<\overline{\theta_v}>$ change with time

according to: $dz_i/dt = w_e$ and $d<\overline{\theta_v}>/dt = [\overline{w'\theta_v'}_s - \overline{w'\theta_v'}_{z_i}]/z_i$

where the entrainment velocity, $w_e = 0.01$ m/s. At the top of the mixed layer is a sharp (step) inversion, above which $\overline{\theta_v} = 292$ K and is constant with height. This

mixed layer is over a warm ocean, having a sea surface temperature of $<\overline{\theta_v}>_s = 290$ K. This temperature remains constant with time, both day and night. Assume that the mean wind speed is $\overline{M} = 5$ m/s and is constant with time, and that the bulk transfer coefficient is $C_H = 0.002$.

a) Find an equation for $<\overline{\theta_v}>$ as a function of time.

b) Plot your answer over a 30 h period.

c) Assume the mixed layer remains well mixed as it grows in thickness. Plot $\overline{w'\theta_v'}$ vs. z at the initial time (t = 0).

2) Given the following sounding at sunrise: $\overline{\theta_v} = a - b \exp(-z/c_1)$, where $a = 300$ K, $b = 10$ K, and $c_1 = 100$ m. Use the thermodynamic (encroachment) approach to find and plot the ML depth, z_i, as a function of time, t, since sunrise (out to 6 h), given

$\overline{w'\theta_v'}_s = c_2 \sin(\pi t/12h)$, where $c_2 = 0.3$ K m/s. Assume no subsidence.

3) Given a slab-like ML with mean properties $< >$, capped by a discontinuity of $\Delta( )$,

where: $z_i = 1000$ m, no subsidence, $<\overline{\theta_v}> = 300$ K, $<\overline{U}> = 5$ m/s, $\Delta\overline{\theta_v} = 1$ K,

$\Delta\overline{U} = 2$ m/s, $\partial\overline{\theta_v}/\partial z = 5$ K/km above the discontinuity, $\overline{w'\theta_v'}_s = 0.2$ K m/s, and

$u_* = 0.35$ m/s.

a) Calculate the entrainment velocity, w_e, using the (1) flux-ratio approach, and
   (2) energetics approach.

b) If $\overline{w'\theta_v'}_s = $ constant with time, then use the flux ratio approach to calculate and plot

z_i, $<\overline{\theta_v}>$, and w_e as a function of time for a 3 h period.

c) If the entrainment-zone thickness, Δh equals $0.3\ z_i$, and the lifting condensation level (LCL) is at a constant height of 1500 m, then when will the first convective clouds form? Use the results from part (b).

4) Given the steady-state advection of cold air from land over warmer water, write the equation set necessary to find the change in height of the internal boundary layer (i.e., the new ML forming over the warmer water) as a function of distance downwind from the shore.   Do NOT solve these equations, just write the equations without solution, but simplified as much as possible.

Assume: The initial cold air from the land has a uniform lapse rate, γ. You can use the flux ratio approach to determine the new ML depth over water. Align the coordinate system with the mean wind. Assume lateral homogeneity (i.e., no crosswind), neglect subsidence, and neglect $\overline{u'\theta'}$.   Assume the water surface temperature, θ_G, is constant in time and space. Assume the mean wind speed and direction are constant with height. Neglect radiation and latent heating.

5) If the moisture flux at z_i is $\rho\overline{w'q'}= 0.10\ g_{water}\ m^{-2}s^{-1}$, and $\Delta\overline{q} = -5\ g_{water}/kg_{air}$, and $\Delta\overline{\theta} = 3$ K, then:
a) Find the value of entrainment velocity (w_e).
b) Find the value of heat flux (in kinematic units) at z_i.

6) Given a dry convective mixed layer with initial conditions $z_i = 500$ m, and $\Delta\overline{\theta} = 5$ K, as sketched below.   Assume the surface heat flux is $\overline{w'\theta'}_s= 0.2$ K m/s and is constant with time. Use the flux ratio approach.   Remember that $\partial(\Delta\overline{\theta})/\partial t = \gamma\ w_e - \partial<\overline{\theta}>/\partial t$, where γ is the lapse rate in the free atmosphere.
a) What must the subsidence he initially to prevent the mixed layer from rising (i.e., z_i = constant = 500 m)?
b) How must the subsidence vary with time over a 3 h period to prevent the mixed layer from rising?   Plot your result in a graph like the one shown below.

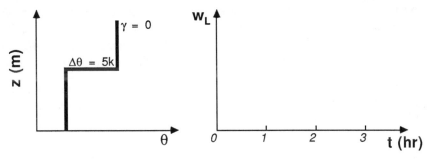

7) Given the initial sounding at $t_o = 0600$ local time as plotted below. Assume that the sensible heat flux varies with time over this land surface according to:

$$\overline{w'\theta'} = c_1 \cdot \sin[\pi \ (t\text{-}t_o)/ \ c_2 \ ], \text{ where } c_1 = 0.2 \text{ K m s}^{-1}, \text{ and } c_2 = 12 \text{ h.}$$ Use the thermodynamic (encroachment) approach to estimate the mixed layer depth (z_i) and mean potential temperature $<\overline{\theta}>$ at $t = 1800$ local time.

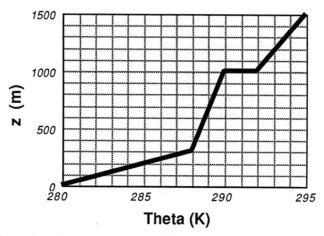

8) What do the surface-layer plume and dust devil have in common?

9) If the initial lapse rate at sunrise is linear (with $\partial\overline{\theta}/\partial z = 10$ K/km and a surface temperature of $\overline{\theta} = 10°C$), and if the total heat input from sunrise to 10 AM is 10^3 m·K, then use the thermodynamic method to estimate z_i at 10 AM.

10) List all the kinds and scales of buoyancy waves that are observed near the entrainment zone and lower FA.

11) Suppose that a slab ML has an initial depth $z_i = 100$ m and an initial temperature of $\overline{\theta}$ = 280 K. This air is advecting from land to warmer water ($\overline{\theta}$ = 290 K) with a mean wind speed of $\overline{M} = 5$ m/s. Over land, there is no flux into the bottom of the ML and no entrainment into the top. Over water, the flux at the bottom can be approximated with a bulk transfer relationship (use $C_D = C_H = 10^{-3}$ as the bulk transfer coefficient). The flux ratio method can be used to model entrainment at the top of the ML. (Neglect subsidence, and assume that the inversion capping the ML has a constant strength of $\Delta\overline{\theta} = 2$ K.) How do the thickness and temperature measured 1 km downwind from the shoreline vary with speed (your answer can be in word and/or equation form, although no numbers need be found)?

12) If an air parcel rises through the mixed layer and penetrates into the entrainment zone with an initial vertical velocity at the base of the inversion of 5 m/s and initial temperature equal to the ML temperature of $\overline{\theta}$ = 300 K, how far will the parcel overshoot, if:
   a) the temperature step at the top of the ML is 3 K and the lapse rate in the FA is adiabatic?
   b) there is no temperature step at the Ml top, but there is a constant lapse rate of

   $\partial\overline{\theta}/\partial z$ = 5 K/km?

13) How do surface layer plumes relate to updraft curtains and convergence lines?

14) Suppose that the FA is filled with high concentrations of a tracer called BSP (blue sky pigment). This tracer is entrained into the top of the ML with flux $\overline{w'c'}_{z_i}$ = - 0.1 g m/s. This tracer is lost from the bottom of the ML by being absorbed into lakes (causing blue lakes) at the same rate: $\overline{w'c'}_s$ = - 0.1 g m/s. Use top-down bottom-up methods to calculate and plot the shape of the mean concentration, $\overline{C}$, of BSP as a function of height, z/z_i, in the ML. What is the mean gradient across the ML? Do the calculations for: a) $w_* = 2$ m/s.   b) $w_* = 0.2$ m/s.

15) How would smoke disperse from a 200 m high smoke stack if the ML were 800 m thick, and if the vertical velocity frequency distribution were positively skewed (instead of the negatively skewed distribution that is observed in nature)?

16) What are the similarities and differences between the entrainment zone and the intromission zone?

17) For a 1 km thick boundary layer with a 3 K potential temperature difference across it, what is the value of
   a) the Rayleigh number for air?
   b) the turbulent Rayleigh number for air (and state assumptions for K)?

18) Find and plot Δh vs. time for a 3 h period. Assume for simplicity that the average ML top, z_i, occurs in the middle of the entrainment zone. Suppose $z_i = a \cdot t^{1/2}$, and $\Delta\overline{\theta}$ = $b \cdot (t+c)^{-1/2}$, where a = 8.333 m s$^{-1/2}$, b = 180 K s$^{1/2}$, and c = 1 h. Assume the air is dry, and the surface flux is constant at $\overline{w'\theta'}$ = 0.2 K m/s, and $g/\overline{\theta}$ = 0.03333 m s^{-2} K^{-1}.

19) If a ML grows approximately 1 km during a 12 h period, calculate the average entrainment velocity. Is it possible to have great enough subsidence to oppose the ML growth? If so, which horizontal scale would be most likely to support subsidence of that magnitude?

12

Stable
Boundary Layer

The boundary layer can become stably stratified whenever the surface is cooler than the air. This stable boundary layer (SBL) often forms at night over land, where it is known as a nocturnal boundary layer (NBL). It can also form by advection of warmer air over a cooler surface.

The balance between mechanical generation of turbulence and damping by stability varies from case to case, creating stable boundary layers that range from being well mixed to nonturbulent. Sometimes the SBL turbulence is sporadic and patchy, allowing the upper portions of the boundary layer to decouple from surface forcings. As a result of this complexity, the SBL is difficult to describe and model, and some of the concepts presented in this chapter will change with future research.

12.1 Mean Characteristics

12.1.1 Vertical Profiles

Fig 12.1 shows typical profiles of mean variables in the SBL for the case of weak turbulent mixing. The greatest static stability is near the ground, with stability decreasing smoothly toward neutral with height. If stabilities are great enough near the surface to cause temperatures to increase with height, then that portion of the SBL is classified as a *temperature inversion*. In fact, sometimes the whole SBL is loosely called a *nocturnal inversion*.

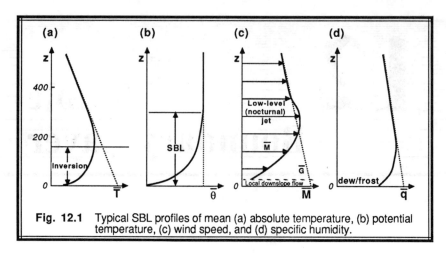

Fig. 12.1   Typical SBL profiles of mean (a) absolute temperature, (b) potential temperature, (c) wind speed, and (d) specific humidity.

In urban areas, a ML can continue throughout the night because of the large heat capacity of buildings and streets, and because of heat released from transportation and space heating. Early evening shallow SBLs in the neighboring rural countryside are not observed in the city. Later at night, when the rural stable layer is deeper than the height of urban buildings, a shallow layer of air within the city can remain well mixed, but it is capped by a stable layer (Godowitch, et al., 1985).

SBL winds can have very complex characteristics. In the lowest 2 to 10 m, cold air will drain down hill. Wind direction in this layer is determined by local topography; wind speed is governed by buoyancy, friction, and entrainment. In flat areas or at the bottom of valleys or topographic depressions, the wind can become calm.

Higher in the SBL synoptic and mesoscale forcings become important. The wind speed can increase with height, reaching a maximum near the top of the stable layer. The layer of peak wind speed is sometimes greater than the geostrophic speed, and is called the **nocturnal jet**. Wind directions often **veer** (turn clockwise) with height. Above the jet, the wind speed and direction smoothly change to geostrophic. The wind profile is often not in steady-state, and evolves with time during the night.

Humidity is also difficult to classify, because sometimes evaporation from the surface continues at night, while at other locations or times condensation results in dew or frost formation. In cases of reduced turbulence, strong moisture gradients can occur near the surface. The delicate balance between turbulence and stability is apparent in the formation of fog vs. dew, and in the fog processes of dissipation, thickening, or lifting. Fog processes will be discussed in Chapter 13.

Pollutants or other scalar tracers that are emitted into the SBL spread out horizontally in thin layers. This process is called **fanning** (see Fig 1.10). Even without tracer emission from a single source at night, the strong wind shears and wind direction changes advect in air of differing tracer concentrations at different altitudes, resulting in a "layer-cake" pattern of tracer concentration (see Fig 12.2). This layered fine-scale vertical structure in aerosols is also apparent in temperature, humidity, and turbulence (Gossard, et al., 1985). Since there is little vertical mixing, photosensitive constituents already present in the layers of air can chemically react during the night in the absence of solar radiation.

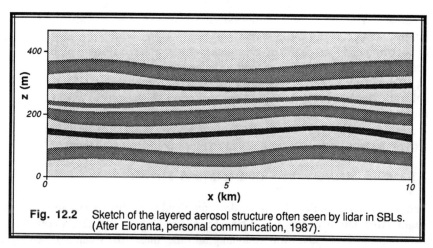

Fig. 12.2   Sketch of the layered aerosol structure often seen by lidar in SBLs. (After Eloranta, personal communication, 1987).

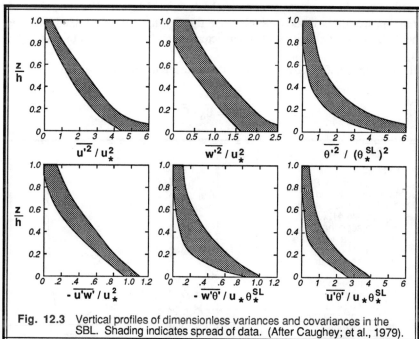

Fig. 12.3   Vertical profiles of dimensionless variances and covariances in the SBL. Shading indicates spread of data. (After Caughey; et al., 1979).

Fig 12.3 shows an idealization of vertical profiles of some turbulent quantities for the weakly turbulent case, based on the previous data from Chapters 3 to 5. Turbulence,when averaged over long times, decreases smoothly with height. At the top of the SBL, the turbulence blends into the residual layer turbulence, or can become nonturbulent. Vertical turbulent motions are suppressed by stability, but buoyant oscillations can occur as *gravity waves*.

An instantaneous snapshot of turbulence gives a far different picture. Patchy turbulence can occur in thin layers, and turbulent bursting can cause sporadic vertical mixing. Turbulence at height z is governed by local shears and stability at that height rather than by forcings at the surface. For some cases, this results in local *z-less* scaling of turbulence, and *decoupling* of the flow state from the ground state. Turbulence can be suppressed or enhanced by the passage of gravity waves, can ride up and down hundreds of meters on these waves, and can nonlinearly interact with them.

12.1.2 Bulk Measures of SBL Depth and Strength

Even if turbulence in the SBL is weak, patchy and sporadic, cases can occur where flow-state information is eventually transferred throughout the SBL depth, given a sufficient time interval. Over a long averaging time, these SBLs can act like a single entity, rather than like a group of completely decoupled layers. For the subset of SBL cases ranging from this weakly-coupled one through the strongly-mixed SBL, we can treat the SBL as a single entity with a characteristic bulk strength, depth, and turbulence intensity.

SBL *strength* is defined by the near-surface potential-temperature difference:

$$\Delta\overline{\theta}_s = \overline{\theta}_o - \overline{\theta}_s \qquad (12.1.2a)$$

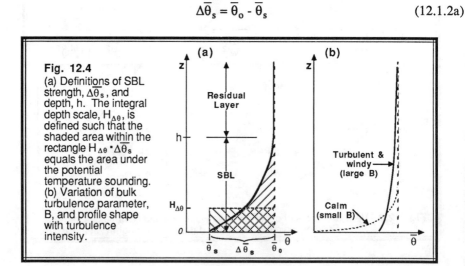

Fig. 12.4
(a) Definitions of SBL strength, $\Delta\overline{\theta}_s$, and depth, h. The integral depth scale, $H_{\Delta\theta}$, is defined such that the shaded area within the rectangle $H_{\Delta\theta} \cdot \Delta\overline{\theta}_s$ equals the area under the potential temperature sounding.
(b) Variation of bulk turbulence parameter, B, and profile shape with turbulence intensity.

between the residual layer air, $\overline{\theta}_o$, and the near-surface air, $\overline{\theta}_s$. This is a measure of the amount of cooling that has occurred since SBL formation (i.e., since *transition*), because the potential temperature of the residual layer is close to the initial near-surface air temperature at transition time (Fig 12.4). Typical magnitudes for SBL strength range from zero at transition to values on the order of 15°C by morning, depending on the turbulence intensity and cloud cover.

The height, h, of the top of the SBL (i.e., the SBL *depth*) is more difficult to quantify, because in many cases the SBL blends smoothly into the residual layer (RL) aloft without a strong demarcation at its top. Thus, many of the definitions of SBL depth that appeared in the literature are based on relative comparisons of SBL state aloft to near-surface state. For example, h can be defined as the lowest height where:

- $\partial\overline{\theta}/\partial z = 0$ (stable layer top; i.e., the height where the lapse rate is adiabatic)

- $\partial\overline{T}/\partial z = 0$ (inversion top; i.e., the height where the lapse rate is isothermal)
- TKE $= 0$ (top of the turbulent layer, or mixed layer if one exists)
- TKE $= 0.05$ TKE   (height where turbulence is 5% of its surface value)

- $\overline{u'w'} = 0$ (top of the stress layer)

- $\overline{u'w'} = 0.05\ \overline{u'w'}_s$ (height where stress is 5% of its surface value)

- $\overline{M}$ is maximum   (the nocturnal jet level)

- $\overline{M} = \overline{G}$ (bottom of free atmosphere, where winds are geostrophic)
- sodar returns disappear   (top of the layer with temperature fluctuations)

Typical magnitudes for height of the top of the SBL range from near zero at transition time to over 1000 m. Usually, however, SBLs grow to depths of about 100 to 500 m.

Those definitions requiring a gradient or flux to equal zero are extremely difficult to use, because those quantities often approach, but never exactly equal zero. Also, different investigators might use the same criteria with the same data and still select different SBL depths, given errors in the experimental data. It has been difficult to compare SBL models and normalized data, because different investigators have used different depth definitions. Mahrt and Heald (1979) have shown that for some cases, these different definitions are poorly correlated with each other, even for the same SBL.

An alternative measure of the SBL depth is the *integral depth scale*, H_ξ, where:

$$H_\xi = \frac{\displaystyle\int_z \xi\,dz}{\xi_s} \tag{12.1.2b}$$

and ξ is any appropriate variable. For example, ξ can be TKE, $\overline{u'w'}$, or it can be $\Delta\overline{\theta}$,

where $\Delta\overline{\theta}(z) = \overline{\theta}_o - \overline{\theta}(z)$. This definition for a depth scale can be applied even if there is no sharp demarcation at the SBL top (Fig 12.4). Typical magnitudes for $H_{\Delta\theta}$ range from 0 to about 150 m.

Rearranging (12.1.2b) and substituting $\Delta\theta$ for ξ shows that the **accumulated cooling** within the SBL is:

$$\int_0^h \Delta\theta \, dz \; = \; \Delta\theta_s \cdot H_{\Delta\theta} \tag{12.1.2c}$$

A **bulk turbulence scale**, B, can be defined from the bulk depth scale, which for temperature becomes:

$$B \; = \; \frac{H_{\Delta\theta}}{\overline{\Delta\theta}_s} \tag{12.1.2d}$$

Large values of B correspond to deep SBLs with small surface temperature change, while small values correspond to shallow $H_{\Delta\theta}$ and large surface cooling. Hence, B is a measure of the overall effect of external forcings on the SBL. Typical magnitudes range from 3 m/K for light turbulence through 15 m/K for strong turbulence.

12.1.3 Idealized Models for the Potential Temperature Profile

The wide variety of observed flow states has resulted in a variety of idealized models for the long-time averaged picture of SBLs. No single model works for all situations, so the best model should be chosen on a case-by-case basis.

Stable Mixed Layer. At the one extreme, strong winds, strong turbulence and weak surface cooling can lead to well-mixed layers (Zeman, 1979) that behave as described in Chapter 11.

$$\Delta\overline{\theta}(z) \; = \; \begin{cases} 0 & \text{for } z > h \\ \text{constant} & \text{for } z < h \end{cases} \tag{12.1.3a}$$

Although buoyancy acts like a sink rather than a source, there is enough mechanical mixing by the wind shears to create a mixed layer (Fig 12.5a). This layer can grow in depth by entrainment, although entrainment rates generally decrease during the night as the stability across the entrainment zone increases. For a well-mixed layer, it is easy to show that the integral depth scale equals the actual SBL depth: $h = H_{\Delta\theta}$.

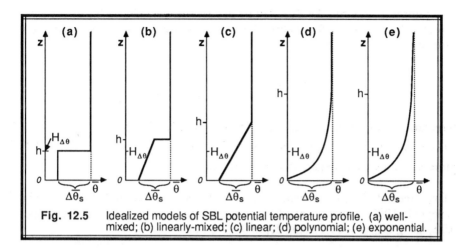

Fig. 12.5   Idealized models of SBL potential temperature profile. (a) well-mixed; (b) linearly-mixed; (c) linear; (d) polynomial; (e) exponential.

The well-mixed SBL can also occur in fog or stratocumulus situations, where strong radiative cooling at cloud top creates convective mixing. For this case, buoyancy is a source of turbulence.

Mixed-Linear. Another idealization allows a potential temperature increase (often linearly) with height, but retains a strong temperature step, $\Delta\overline{\theta}_h$, at the top of the SBL (Fig 12.5b). This can occur with moderately strong winds and turbulence.

$$\Delta\overline{\theta}(z) = \left(1 - \frac{z}{h} \right) \Delta\overline{\theta}_s \;+\; \left(\frac{z}{h} \right) \Delta\overline{\theta}_h \qquad (12.1.3b)$$

Above h it is assumed that $\Delta\overline{\theta} = 0$. This not-so-well-mixed layer occurs with slightly weaker mechanical forcings than the previous case. The relationship between top of the SBL and the integral depth scale is $h = 2 \left[\Delta\overline{\theta}_s / (\Delta\overline{\theta}_s + \Delta\overline{\theta}_h) \right] H_{\Delta\theta}$.

Linear. A third idealization (Estournel, et al., 1985) describes a linear increase of potential temperature from a finite value near the surface to zero at the top of the SBL (Fig 12.5c).

$$\Delta\overline{\theta}(z) = \left(1 - \frac{z}{h} \right) \Delta\overline{\theta}_s \qquad (12.1.3c)$$

Above h it is again assumed that $\Delta\overline{\theta} = 0$. For this case, $h = 2H_{\Delta\theta}$.

Polynomial. Fourth, the SBL can be described (Yamada, 1979) by a polynomial that passes through the surface value of potential temperature, and has a zero gradient at the top of the SBL (Fig 12.5d).

$$\overline{\Delta\theta}(z) = \left(1 - \frac{z}{h} \right)^{\alpha} \overline{\Delta\theta}_s \qquad (12.1.3d)$$

where parameter α is usually taken as 2 or 3. Above h it is assumed that $\overline{\Delta\theta} = 0$. The relationship between integral depth and SBL top is $h = (1 + \alpha) H_{\Delta\theta}$. All four parameterizations above require that h be specified.

Exponential. The last scheme (Stull, 1983; Carlson and Stull, 1986; Surridge and Swanepoel, 1987; Wong, et al., 1987) assumes an exponential potential temperature profile, using $H_{\Delta\theta}$ as the e-folding depth (Fig 12.5e).

$$\overline{\Delta\theta}(z) = \overline{\Delta\theta}_s \, e^{-\frac{z}{H_{\Delta\theta}}} \qquad (12.1.3e)$$

Since there is no well-defined SBL top for this case, we must make an arbitrary assumption. If h is the height where $\overline{\Delta\theta}$ is 5% of it surface value, then $h = 3H_{\Delta\theta}$.

Instead, if h is the height where $\overline{\Delta\theta}$ is 2% of it surface value, then $h = 4H_{\Delta\theta}$. This demonstrates the possible magnitudes of h-errors for a profile with no well-defined top.

12.2 Processes

Forcings that act on the SBL include radiation, conduction, turbulence, subsidence and advection. An additional forcing, local terrain slope, will be discussed separately in Section 12.7. It will be shown that none of these forcings can be neglected.

12.2.1 Radiation

Unsaturated conditions. The absence of solar radiation in nocturnal boundary layers allows one to focus on only the longwave radiation budget (i.e., $Q^* = I^*$). Strong radiative cooling of the ground surface occurs at night, with net upward radiative fluxes of about 100 W/m^2 (0.8 K m/s), as already discussed in Chapter 7. In addition, radiation divergence within the air causes small, but significant, cooling in the boundary layer above the ground. This in-situ radiative cooling was neglected for the ML, but can not be neglected in the SBL because many of the other flux terms are small too.

Water vapor is the most important constituent controlling the infrared radiation budget, with carbon dioxide playing a smaller role. In spite of the importance of water vapor, the radiation budget is rather insensitive to the exact amount of moisture present, unless the air becomes saturated and fog or stratocumulus forms.

Typical values of net longwave radiative flux (positive upward) are about 100 W/m^2 in the boundary layer. The radiative flux increases slightly with height, as shown in Fig 12.6a, resulting in flux divergence and radiative cooling at every height. Greater flux divergence often occurs in the lowest tens of meters. The flux divergence and radiative cooling are related by:

$$\left. \frac{\partial \overline{\theta}}{\partial t} \right|_{\text{radiative}} = - \frac{\partial I^*}{\partial z} \qquad (12.2.1a)$$

which is a simplification of (3.5.3f).

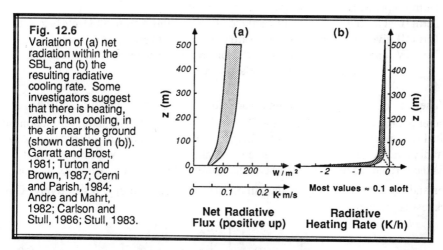

Fig. 12.6 Variation of (a) net radiation within the SBL, and (b) the resulting radiative cooling rate. Some investigators suggest that there is heating, rather than cooling, in the air near the ground (shown dashed in (b)). Garratt and Brost, 1981; Turton and Brown, 1987; Cerni and Parish, 1984; Andre and Mahrt, 1982; Carlson and Stull, 1986; Stull, 1983.

Many studies (Garratt and Brost, 1981; André and Mahrt, 1982; Stull, 1983; Cerni and Parish, 1984; Carlson and Stull, 1986; Turton and Brown, 1987) have indicated a constant cooling rate of 0.1 K/h at heights above 500 m, with cooling rates increasing slowly to about 0.2 K/h at 50 m (see Fig 12.6b). Within a few meters of the ground the cooling rate increases to 1 to 3 K/h in many cases. Temperature inversions near the ground, however, can occasionally lead to radiative heating from the warmer air aloft.

Models. It is beyond the scope of this book to go into the details of radiative transfer models, other than to briefly discuss some of the general approaches (Cerni and Parish, 1984). (1) The *line-by-line* approach uses spectral-line information to calculate absorption and emission as a function of wavelength. This is the most accurate method, but is also the most computationally expensive. (2) The *band* method uses average absorption and emission over a smaller number of wavelength bands, and is less expensive and less accurate. (3) The *flux-emissivity*, or *graybody*, method assumes that each atmospheric layer can be represented by a single transmissivity or emissivity averaged over all wavelengths. This approach is the least expensive to use, but is sufficiently accurate for many applications. Mahrt (personal communication) notes that any of the three models can give poor results if the vertical resolution near the surface is too coarse.

Using the flux-emissivity method, the net upward longwave flux, I^*, at any height z is described by:

$$I^*(z) = \int_0^{\varepsilon_{FB}} \sigma_{SB} T^4 d\varepsilon_F - \int_0^{\varepsilon_{FT}} \sigma_{SB} T^4 d\varepsilon_F + (1 - \varepsilon_{FB}) \sigma_{SB} T_G^4 \quad (12.2.1b)$$

$$\ \ \ \text{I} \qquad\qquad \text{II} \qquad\qquad\quad \text{III} \qquad\qquad \text{IV}$$

Term I represents the net longwave flux upward
Term II gives the amount of upward radiation reaching z from all levels below z
Term III gives the amount of downward radiation from all levels above z
Term IV represents the amount of radiation emitted from the surface that is not
        absorbed by the time it gets to level z.

The flux emissivity, ε_F, is found from tables (e.g., Staley and Jurica, 1970), graphs or other parameterizations (Cerni and Parish, 1984; Chou, 1984), and is a function of the optical thickness, u(z,Z), between the height of interest, z, and any other height Z where the radiation was emitted. The absolute temperature, T, is also evaluated at height Z; and T_G is the surface radiative skin temperature. Emissivities from the bottom of the boundary layer and the (effective) top of the atmosphere are ε_{FB} and ε_{FT}, respectively. Optical thickness is given by:

$$u(z,Z) = \left| \int_{z'=z}^{z'=Z} q\, dz' \right| \quad (12.2.1c)$$

Equation (12.2.1b) must be solved numerically, using the observed temperature soundings and the parameterizations for flux emissivity (Cerni and Parish, 1984). See Section 13.2 for more discussion on radiation.

The bulk heat flux, Q_R, associated with radiation divergence across the whole SBL is defined as the difference in longwave radiative flux between the bottom and top of the SBL, and is negative when radiative cooling occurs:

$$Q_R = I^*_s - I^*_h \quad (12.2.1d)$$

12.2.2 Conduction

As in other boundary layers, conduction is small enough to be negligible everywhere except within a few millimeters above the ground. Within these few millimeters, however, the molecular flux dominates over the turbulent flux, and is responsible for the cooling of air in these lowest layers. Above this height, turbulence can transport the colder air upward into the remainder of the SBL. As before, the *effective turbulent flux* at ground level, Q_H, is composed mostly of the molecular flux, because the true turbulent flux goes to zero at the surface.

Within these few millimeters, molecular diffusion and conduction control dew and frost formation at night. Dew forms primarily at the tips (i.e., points) of blades of grass, because diffusion of water vapor from the air towards the tip can come from the largest range of angles. Also, the tips can radiate over a large range of angles towards the sky, without experiencing as much return radiation from other objects (like other blades of grass). The reader is directed to any of the cloud microphysics texts for more details of droplet and ice crystal formation.

12.2.3 Turbulence

Turbulence Classification. A variety of types of turbulence can occur in the SBL (Mahrt and Gamage, 1987). In some situations, the SBL is continuously and strongly turbulent over the whole depth of the SBL. In other situations, turbulence might be patchy, weak and intermittent, but when time averaged over a couple hours the average turbulence flux might appear to be acting over the whole SBL depth. Finally, some observations show turbulence aloft that is relatively disconnected with processes at the surface.

In the absence of fog or stratocumulus, turbulence is generated only mechanically in the SBL, usually by wind shears. Wind shear can be created near the ground by friction acting on the ambient flow. It can be generated aloft by variations in the geostrophic wind speed with height. When a low-level jet forms, shears can be enhanced both below and above the nose of the jet. Over sloping terrain, shears can develop at the top of the layer of drainage flow (Mahrt, 1985). Also, shears can be enhanced or reduced at different locations relative to the phase of buoyancy waves. Finally, flow around obstacles such as trees and buildings can generate wake turbulence.

Time scales. Brost and Wyngaard (1978) suggested that a timescale for the effects of turbulence on the whole SBL is approximately:

$$\tau_R = \frac{h}{0.01\, u_*} \qquad (12.2.3a)$$

This scale represents the time it takes for surface information to be communicated across the depth of the SBL via turbulence. Using typical ranges of values for u_* and h for weak turbulence, we find that τ_R is on the order of 7 to 30 h.   Thus, changes in surface forcings might not be felt near the top of the SBL until many hours later if the turbulence is weak.

Compared with the 10 to 15 min time scales for convective ML turbulence, we see that the weakly turbulent SBL is essentially not in equilibrium with its surface forcings (i.e., with surface heat flux or roughness). The SBL is trying to react to surface changes, but is always lagging behind by several hours. Therefore, we can not expect similarity theory to work if our key variables are the surface forcing variables, and we anticipate that statistically robust fluxes and variances might be difficult to measure. Another way to look at it is that since typical mid-latitude nights last about 12 h, the weakly turbulent NBL cannot approach equilibrium before the night ends and the NBL disappears.

Another time scale for motions in the SBL is the inverse of the Brunt-Väisälä frequency, $1/N_{BV}$. This is related to the period of oscillation of a parcel in a statically stable environment, and is discussed in the section on buoyancy waves.

Length Scales. One length scale for turbulence within the SBL is given by:

$$l_B = \frac{\sigma_w}{N_{BV}} \tag{12.2.3b}$$

which indicates the degree of suppression of vertical motions by the static stability. This scale is called the **buoyancy length scale**, and it can range from a couple hundred meters for weak stratification, to less than a meter for weak turbulence in strong stratification.

Mahrt and Gamage (1987) defined a length scale based on the structure functions for vertical velocity [$D_{ww}(r)$] and potential temperature [$D_{\theta\theta}(r)$], for separation distance r:

$$l_{SB}(r) = \frac{D_{ww}(r)}{(g/\theta_v)\,[\,D_{\theta\theta}(r)\,]^{1/2}} \tag{12.2.3c}$$

For SBLs with strong stability and weak turbulence, this **structure buoyancy length scale** is on the order of a meter, while for weak stability and strong turbulence, the scale can be 50 to 100 m. Mahrt and Gamage also used structure functions to study the anisotropy of turbulence, and found that the ratio of vertical to horizontal scales of large turbulent eddies ranged from 0.1 for strong stratification to about 0.5 for weaker stratification.

A **dissipation length scale** (Hunt, et al., 1985) can be defined by:

$$l_\varepsilon = 0.4\,\frac{\sigma_w^3}{\varepsilon} \tag{12.2.3d}$$

which typically ranges from 10 to 100 m.

The *Ozmidov scale* (Hunt, et al., 1985) is given by:

$$l_O = \left(\frac{\varepsilon}{N_{BV}^3} \right)^{1/2}$$

(12.2.3e)

Typical values range from 10 to 100 m.

The inverse of the wavelength at the peak in the spectrum of $f \cdot S_w(f)$ can also be used to define a length scale (Hunt, et al., 1985). Typical values are 10 to 50 m.

Another length scale, already reviewed in Chapter 9, is the local Obukhov length, L_L (Nieuwstadt, 1984). Typical values range from zero to 50 m; it usually approaches zero as height increases. In the case of strong contiguous turbulence, the Obukhov length based on surface fluxes can be used instead.

A variety of mixing-length scales have be used successfully to model the SBL, because of the absence of large eddies (Estournel and Guedalia, 1987; Lacser and Arya, 1986; and Delage, 1974). For example, Delage's (1974) suggestion was:

$$\frac{1}{l} = \frac{1}{k\,z} + \frac{f_c}{0.0004\,G} + \frac{\beta}{k\,L_L}$$

(12.2.3f)

where β is an empirical parameter. Additional parameterizations are a function of the Richardson number.

Continuous/Contiguous Turbulence. Cooling rates and heat fluxes for a couple of cases are shown in Fig 12.7. Unlike the corresponding radiation figure, the variation of heat flux and cooling rate from case to case varies so widely that it is useless to show the whole possible range on one figure. Given a turbulent heat flux into the bottom of the

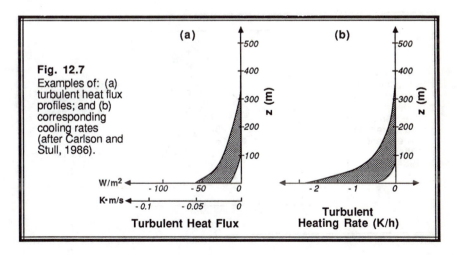

Fig. 12.7
Examples of: (a) turbulent heat flux profiles; and (b) corresponding cooling rates (after Carlson and Stull, 1986).

SBL, and no turbulence at the top, the bulk turbulent heat flux is equal to the surface flux, Q_H.

Similarity theory is useful for the contiguously turbulent SBL, where surface forcings can be used as the key scaling variables. Such parameterizations were described in Chapter 9, and were used to scale the variables in Fig 12.3. Also, K-theory, higher-order closure models, and transilient theory have also been successfully applied to the SBL.

For the continuously and contiguously turbulent situation that is in equilibrium with the surface, it is reasonable to use the bulk measures of turbulence described earlier. We expect greater turbulence to occur in stronger winds and over rougher ground. However, weaker turbulence is anticipated during cases of strong cooling at the surface. These influences can be incorporated into a parameterization (Stull, 1983) for the bulk turbulence:

$$ B = \frac{\left(\left| f_c G \right| Z_s \right)^{3/2}}{(- g Q_H)} \tag{12.2.3g} $$

where Z_s is a measure of the surface roughness averaged over many kilometers upwind. The roughness factor, Z_s, is not like a roughness length, but has values on the order of 2 to 10 km. This roughness factor must be determined aerodynamically for each wind direction at each site based on observed SBL development, because no direct description for it has appeared in the literature.

Instead of using wind speed directly in the above parameterization the pressure gradient forcing $f_c G$ is used because it drives the winds (Estournel and Guedalia, 1985). Note that the definition of G includes a factor of $1 / f_c$, thus making the product $f_c G$ independent of f_c, as required for these schemes to work in tropical regions. The variation of B with geostrophic speed and direction for Wangara is shown in Fig 12.8.

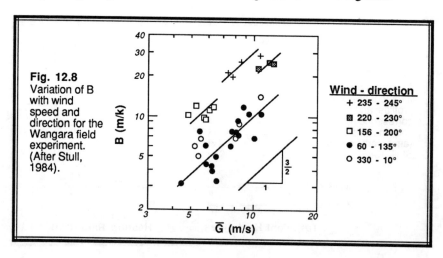

Fig. 12.8 Variation of B with wind speed and direction for the Wangara field experiment. (After Stull, 1984).

Wind - direction
+ 235 - 245°
▨ 220 - 230°
□ 156 - 200°
● 60 - 135°
○ 330 - 10°

Patchy Turbulence in Equilibrium with the Surface. We have already seen in Fig 12.3 an example of time-averaged turbulence variables, during which the SBL might not have been turbulent at all times or at all heights during the averaging period. Nevertheless, we can often utilize the continuous/contiguous turbulence parameterizations described above to model this time-averaged SBL.

One phenomenon sometimes seen in the SBL is an occasional bursting of turbulence. It is speculated that in nonturbulent SBLs, wind shears can increase because of the dynamic forcings and the lack of mixing. Eventually, the shears are great enough to trigger turbulence (i.e., the Richardson number is low enough). This burst of turbulence causes vertical mixing of both heat and momentum, causing shears to decrease and the Richardson number to increase. Eventually, the shears are too weak to continue to support turbulence, and turbulence ceases. During the resulting quiescent period, shears can again build to the point of another turbulent burst. Such a scenario can also occur with discontinuous turbulence, described next.

Discontinuous Turbulence. In situations where turbulence aloft is disconnected with the surface, it has been shown in Chapter 9 that local *z-less* similarity scaling can be used (Wyngaard, 1973; Nieuwstadt, 1984). Turbulence variables are consistent with, and scale to, each other, but do not scale with surface forcings. In order to use this approach, one must first measure a subset of the turbulence variables aloft, against which the remaining variables can be related. Although this is a useful diagnostic tool to relate variables within the turbulent layer, it precludes forecasts or parameterizations. The assumption of constant Richardson number is often made with z-less scaling, although as shown below the Richardson number is rarely constant with time or location.

The Richardson number is one of the indicators for the location of turbulence within the SBL. Fig 12.9 (André and Mahrt, 1982) shows the potential temperature and wind profiles for one case study, and also shows the corresponding vertical profile of Richardson number. Regions of Richardson number less than 0.5 are shaded, and

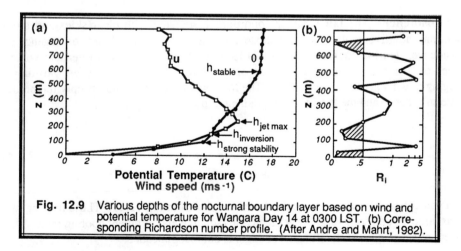

Fig. 12.9 Various depths of the nocturnal boundary layer based on wind and potential temperature for Wangara Day 14 at 0300 LST. (b) Corresponding Richardson number profile. (After Andre and Mahrt, 1982).

indicate layers where turbulence is likely. During the night, these turbulent layers can form and dissipate, and reform at other heights. Mahrt, et al. (1979) found that the Richardson number is large at altitudes corresponding to the LLJ maximum, but they found smaller Richardson numbers and more intense turbulence just above the jet where shears are stronger and static stability is weak.

Forecasting the Richardson number for the SBL can be tenuous because of its sensitivity to relatively small changes in potential temperature and wind profile gradient (see Fig 12.10, for a different case, after Gossard and Frisch, 1987). Also, coarse measurement of the bulk Richardson number across a thick layer might yield a large value (i.e., nonturbulent), while measurements made at higher resolution over the same region might indicate the possibility of thin layers of low Richardson number (i.e., turbulent).

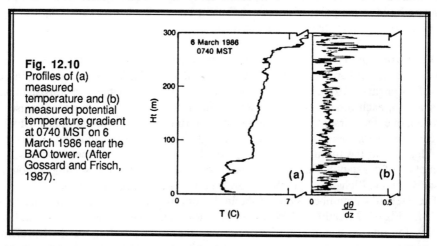

Fig. 12.10 Profiles of (a) measured temperature and (b) measured potential temperature gradient at 0740 MST on 6 March 1986 near the BAO tower. (After Gossard and Frisch, 1987).

12.2.4 Subsidence and Advection

Subsidence. The fair weather conditions associated with clear skies and strong nocturnal radiative cooling are also those associated with anticyclones, divergence and subsidence. During such conditions, Carlson and Stull (1986) found vertical velocity values of -0.1 to -0.5 m/s near the top of the SBL. Subsidence causes warming by bringing down warmer air from aloft. The local warming rate is thus simply the product of the subsidence and the local vertical gradient of potential temperature:

$$\left.\frac{\partial \overline{\theta}}{\partial t}\right|_{\text{subsidence}} = -\,w\,\frac{\partial \overline{\theta}}{\partial z} \qquad (12.2.4a)$$

Typical heating rates due to subsidence are shown in Fig 12.11. These heating rates are as large as, and sometimes larger than, the radiative and turbulent cooling rates. Thus, neglect of subsidence in the SBL heat budget is not recommended.

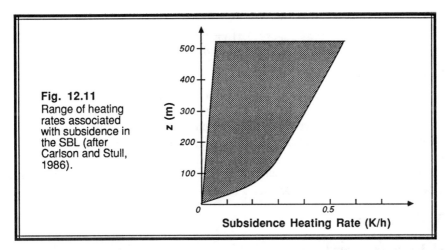

Fig. 12.11 Range of heating rates associated with subsidence in the SBL (after Carlson and Stull, 1986).

Another way to look at the effects of subsidence is to recognize that the associated divergence, Div, is removing some of the chilled air from the SBL. Using $\Delta\theta_s \cdot H_{\Delta\theta}$ as a measure of the accumulated cold air within the SBL, the effective bulk heat flux into the whole SBL, Q_w, that would cause heating equal to the amount of cold air lost is:

$$Q_w = \text{Div} \cdot \Delta\theta_s H_{\Delta\theta} \qquad (12.2.4b)$$

Typical magnitudes of this flux are 0.01 K m/s.

It is possible to have BL cooling and SBL formation in regions of convergence, where warm air advection over a colder surface can provide the cooling in the absence of radiative cooling. Convergence sweeps together the surrounding cold air to make a deeper and/or colder SBL than would have otherwise occurred via local processes. The equations (12.2.4a & b) can be used, recognizing that Div is negative and w is positive.

Advection. Often the largest term in the heat budget of the SBL is advection, yet it is the one least often considered or measured. Schaller and Wichmann (1985) and Carlson and Stull (1986) have demonstrated that advective heating rates can easily range between ± 2 K/h, where the heating rate is given by:

$$\left. \frac{\partial\overline{\theta}}{\partial t} \right|_{\text{advection}} = -\overline{U}_j \frac{\partial\overline{\theta}}{\partial x_j} \qquad (12.2.4c)$$

for j = 1 and 2.

Furthermore, advection can be a strong heat source in one part of the SBL, and a heat sink in other parts. The effective bulk heat flux, Q_A, associated with horizontal advection into the whole SBL is:

$$Q_A = -\int_0^h \overline{U}_j \frac{\partial \overline{\theta}}{\partial x_j} \, dz \qquad (12.2.4d)$$

for $j = 1$ and 2.

Traditional means of measuring advection must be used with care, because advection will tend to follow isentropic surface, except where the surfaces touch the ground or where there is mixing in the vertical. The synoptic rawinsonde network might not offer sufficient resolution in the vertical nor horizontal to adequately measure the advection at night.

There is no simple way to parameterize this term. It must be calculated directly using a 3-D (mesoscale) forecast model, or it must be measured directly.

12.3 Evolution

12.3.1 Bulk Growth

The combined effects of turbulence, radiation, subsidence and advection cause the SBL to evolve as sketched in Fig 12.12.   Shortly after surface cooling starts (e.g., evening), a shallow weak SBL forms, leaving a deep residual layer above it. Gradually, the SBL becomes deeper by the cooling of the lower residual layer. The strength also increases until such time as the surface cooling stops (e.g., morning). The example shown in Fig 12.12 corresponds to one of relatively weak contiguous turbulence.

The evolution of accumulated cooling, $\Delta\theta_s H_{\Delta\theta}$, within the continuously/contiguously turbulent SBL is given by a heat balance of all the forcings:

$$-\frac{\partial (\Delta\theta_s \cdot H_{\Delta\theta})}{\partial t} = Q_H + Q_R + Q_w + Q_A = Q_T \qquad (12.3.1a)$$

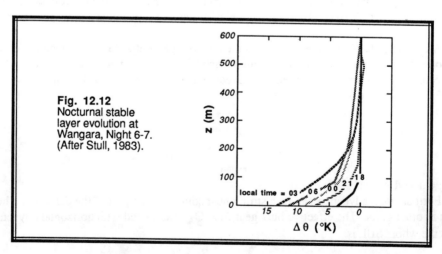

Fig. 12.12
Nocturnal stable
layer evolution at
Wangara, Night 6-7.
(After Stull, 1983).

where Q_T , the total (net) heat flux acting on the bulk SBL, is negative. If Q_T were positive, then there would be net heating causing the SBL to be eliminated.

During most nights Q_T varies significantly with time, requiring (12.3.1) to be integrated numerically. For the special case of constant Q_T, we can employ the definition for bulk turbulence together with (12.3a) to analytically solve for the integral depth and strength of the SBL:

$$H_{\Delta\theta} = \left(-Q_T\, t\, B \right)^{1/2} \tag{12.3.1b}$$

$$\Delta\theta_s = \left(\frac{-Q_T\, t}{B} \right)^{1/2} \tag{12.3.1c}$$

These equations show that the depth and strength increase with the square root of time, as sketched in Fig 12.13. They also show that a more-turbulent SBL (i.e., one with larger B) will be deeper and have less surface cooling than a less-turbulent SBL. The square-root of time growth has been observed by Brunt (1939), Brost and Wyngaard (1978), Stull (1983a,b), Brook, 1985), and Surridge and Swanepoel (1987).

With forecasts of $\Delta\theta_s$ and $H_{\Delta\theta}$, the profile models of Section 12.1.3 can be used to forecast the shape of the potential temperature profile as it evolves throughout the night.

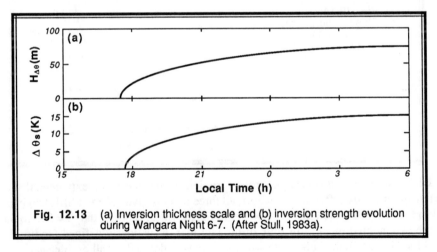

Fig. 12.13   (a) Inversion thickness scale and (b) inversion strength evolution during Wangara Night 6-7. (After Stull, 1983a).

12.3.2   Example

Problem. Given a clear night with brisk geostrophic winds (G = 10 m/s) and strong bulk cooling (Q_T = -0.03 K m/s) at a location where $f_c = 10^{-4}$ s^{-1} and Z_s = 4 km, find the bulk depth and strength of the SBL after 6 h. Also plot the resulting potential temperature profiles for the mixed, linear, and exponential models.

Solution. First, find the bulk turbulence parameter, B:

$$B = \frac{\left(\left| f_c \overline{G} \right| z_s \right)^{3/2}}{- Q_T \, g} = \frac{[\, (10^{-4}) \, (10) \, (4000) \,]^{3/2}}{(0.03) \, (9.8)} = 27.2 \text{ m/K}$$

Next, use B to find $H_{\Delta\theta}$ and $\Delta\theta_s$ at t = 6 h:

$$H_{\Delta\theta} = \left(- Q_T \, t \, B \right)^{1/2} = [\, (0.03) \, (6) \, (3600) \, (27.2) \,]^{1/2} = 132.8 \text{ m}$$

$$\Delta\theta_s = \left(- Q_T \, t \, / \, B \right)^{1/2} = [\, (0.03) \, (6) \, (3600) \, / \, (27.2) \,]^{1/2} = 4.88 \text{ K}$$

Plots of the profiles are shown in Fig 12.14.

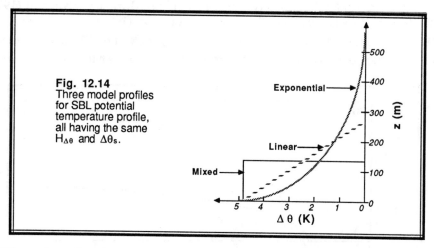

Fig. 12.14
Three model profiles for SBL potential temperature profile, all having the same $H_{\Delta\theta}$ and $\Delta\theta_s$.

Discussion. For this case of strong turbulence, the linear and exponential models are not significantly different. However, all three models give different SBL heights: h = 133 m for the well-mixed model, h = 266 m for the linear model, and h = 398 to 531 m for the exponential model. In spite of the fact that h is poorly defined (and probably should not be used), the bulk depth scale, $H_{\Delta\theta}$, is well defined for all profile shapes.

12.4 Other Depth Models

In some of the earlier research on SBLs, it was hoped that a *diagnostic equation* could be found for an equilibrium SBL depth. Later, when the time evolution was understood better, *rate equations* were devised to allow the SBL depth to slowly approach an equilibrium depth. Neither of these approaches yielded the best results, and are only briefly reviewed here.

12.4.1 Diagnostic Models

The following models, sans empirical constants, have been proposed at various times for *equilibrium SBL depths*, h_{eq}: u_*/f_c, $u_*^2/(f_c G \sin \alpha_{GM})$, $[\, u_* L / |f_c| \,]^{1/2}$, L, G/N_{BV}, $h_{tropopause}$, $G^2 \theta / (g \, \Delta\theta_s)$, and combinations of these, where α_{GM} is the angle between the geostrophic and surface winds. The last model listed above is based on the assumption that the bulk Richardson number for the whole SBL is constant. None of these diagnostic models were found to be satisfactory, because the SBL is rarely in equilibrium. It evolves with time as cold air is accumulated and as forcings change.

12.4.2 Rate Equations

After diagnostic models were found to be unsatisfactory, an improvement was suggested where the actual depth tends to adjust toward an equilibrium depth, with a response time of τ_R. As a result, the so-called rate equations were born:

$$\frac{dh}{dt} = \frac{h_e - h}{\tau} \tag{12.4.2}$$

For constant h_{eq}, the equation above describes an exponential approach of h toward h_{eq}, with an e-folding time of τ_R.

Over ten versions of this rate equation are in the literature. Although most are better than the diagnostic approach, they still have some drawbacks. Part of the problem is the assumption that there exists some appropriate equilibrium depth, h_{eq}. Many of the same models of equilibrium depth described in the diagnostic section are also used in the rate equations. A variety of models for response time have been proposed, including: h/u_*, h/G, h/M, $1/f_c$, and $\Delta\theta_s/(d\theta_s/dt)$.

12.4.3 Other Prognostic Models

A few other approaches are suggested in the literature, some of which are of the form: $dh/dt \propto u_* L / h$. Although this yields the appropriate square-root time dependence, it is not a function of the accumulated cooling. This scheme also has difficulties.

Surridge and Swanepoel (1987) found that the square-root time dependence occurs on many occasions at many sites, but not at all sites. For some of the sites, the SBL depth and strength approach equilibrium values with an error-function time dependency:

$$H = H_{eq} \ \text{erf}\left(\frac{t}{\tau_R}\right) \qquad (12.4.3a)$$

$$\Delta\theta_s = \Delta\theta_{seq} \ \text{erf}\left(\frac{t}{\tau_R}\right) \qquad (12.4.3b)$$

where the response time, τ_R, is on the order of 1 to 2 h for the cases studied. An exponential approach to equilibrium, such as predicted using the rate equations, would have probably worked as well. Their observations of equilibrium values for strength and depth ($\Delta\theta_{seq}$ and H_{eq}, respectively) suggest that radiative cooling must be balanced by other processes such as advection or subsidence for these situations.

12.5 Low-Level (Nocturnal) Jet

12.5.1 Characteristics

As described in Chapter 1, the low-level jet (LLJ) is a thin stream of fast moving air, with maximum wind speeds of 10 to 20 m/s usually located 100 to 300 m above the ground. Peak speeds up to 30 m/s have been reported, and altitudes of the peak were occasionally as high as 900 m above ground. The LLJ can have a width of hundreds of kilometers and a length of a thousand kilometers, making it more like a sheet than a narrow ribbon, in some cases.

The LLJ has been observed in Europe (Sladkovic and Kanter, 1977; Kraus, et al., 1985), Africa (Anderson, 1976; Hart, et al., 1978), North and South America (Blackadar, 1957; Bonner, 1968; Lettau, 1967), and Australia (Malcher and Kraus, 1983;

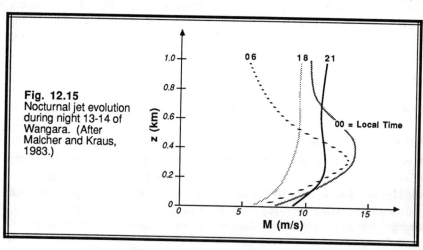

Fig. 12.15
Nocturnal jet evolution during night 13-14 of Wangara. (After Malcher and Kraus, 1983.)

Brook, 1985; Garratt, 1985). In many cases, the LLJ forms during the night and reaches its peak during the predawn hours (Figs 12.15 and 12.16). These LLJs are often called *nocturnal jets*. Examples include:
- Koorin nocturnal jet (Australia)
- Great Plains nocturnal jet (USA)

In other cases, the jet forms during the daytime:
- Paracas LLJ (Peru)
- Southerly Buster LLJ (East-Australia)

Chile's Atacama boundary layer wind peaks twice, once from the southwest during the afternoon, and then from the northeast at night (Lettau, personal communication). The East African (Somali) jet lasts day and night for many days.

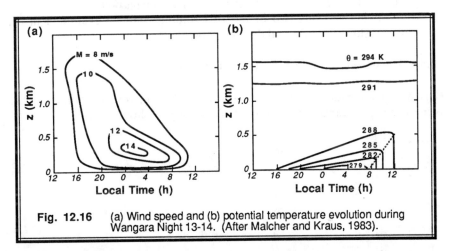

Fig. 12.16   (a) Wind speed and (b) potential temperature evolution during Wangara Night 13-14. (After Malcher and Kraus, 1983).

Different investigators have used different criteria for identifying LLJs. Some have required wind speeds greater than a specific speed (e.g.,12, 16, or 20 m/s) below a specified height (e.g., 1000, 1500, or 2500 m), while others require the speed to be supergeostrophic (Bonner, 1968; Brook, 1985). We will pragmatically define the LLJ as occurring whenever there is a relative wind speed maximum that is more than 2 m/s faster than wind speeds above it within the lowest 1500 m of the atmosphere.

Investigations have shown that there are many possible causes for the LLJ (Kraus, et al., 1985), including:
- synoptic-scale baroclinicity associated with weather patterns
- baroclinicity associated with sloping terrain
- fronts
- advective accelerations
- splitting, ducting and confluence around mountain barriers
- land and sea breezes
- mountain and valley winds
- inertial oscillations

In some situations, more than one of the above factors contribute to jet formation (Garratt, 1985). Thus, there are many different types of LLJs with different characteristics. The first three factors listed above can generate LLJs with subgeostrophic wind speeds. The others can create supergeostrophic winds.

LLJs are not rare phenomena. Brook reports nocturnal jets on 19% of the winter nights in parts of Australia, with peak speeds reached between midnight and 0500 local time. Bonner (1968) observed North American LLJs occurring most frequently in the central plains, particularly in the Kansas and Oklahoma region, where 30% of all rawinsonde soundings had LLJs. Many of these were associated with southerly or southwesterly wind components, and the majority occurred in the late night or early morning. A distinct jet core was observed from Texas to Minnesota. The Somali jet lasts many days, and reoccurs almost every year with the Indian monsoon.

In the following sections two of the LLJ forcings will be examined in detail: baroclinicity over sloping terrain, and the inertial oscillation.

12.5.2 Baroclinicity Over Sloping Terrain

Horizontal temperature gradients cause changes of the geostrophic wind with height, as given by the approximate *thermal wind* relationship (Holton, 1972):

$$\frac{\partial U_g}{\partial z} = - \frac{g}{f_c T} \frac{\partial T}{\partial y} \tag{12.5.2a}$$

$$\frac{\partial V_g}{\partial z} = + \frac{g}{f_c T} \frac{\partial T}{\partial x} \tag{12.5.2b}$$

For example, given a southerly geostrophic wind at the surface (i.e., V_g is positive), V_g will decrease with height if the air is warmer to the west in the northern hemisphere (Figs 12.17a & b).

If the actual wind is geostrophic everywhere except near the ground (where friction reduces it below geostrophic), then we see in Fig 12.17b that a LLJ is formed given the same situation of southerly geostrophic wind. This is an example of a subgeostrophic LLJ that can occur day or night, although it is less likely to occur during the day because vigorous mixing in the ML would tend to mix the fast LLJ air with the slower air at adjacent levels, leaving a well-mixed wind profile.

The thermal wind associated with baroclinicity can be caused by a variety of forcings, including synoptic-scale temperature gradients, mesoscale land-sea temperature gradients, and gradients across fronts. In regions such as the central USA, *sloping terrain* can also generate horizontal temperature gradients that change sign with the diurnal cycle (Lettau, 1967; Holton, 1968). This can lead to a diurnal oscillation in the strength of the LLJ, as shown below.

The scenario of Fig 12.17a corresponds to a late afternoon situation where the ground has been warmed by the sun, and where the temperature decreases adiabatically with height within a deep ML. In any plane at constant height above sea level, the higher terrain in the west causes the temperature there to be warmer than the temperature in the east. This creates the horizontal temperature gradient of the previous example, and causes negative geostrophic wind gradients at all heights within the ML (indicated by the thermal wind vectors labeled T.W. in Fig 12.17a).

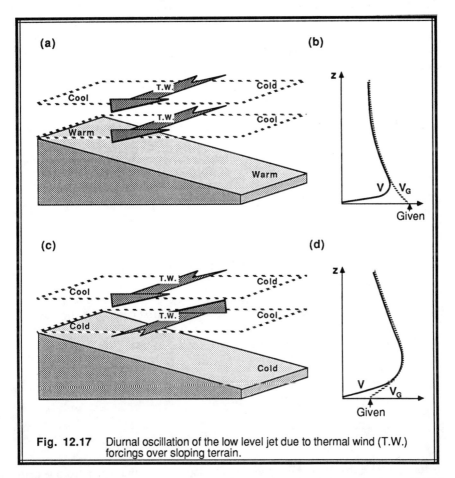

Fig. 12.17   Diurnal oscillation of the low level jet due to thermal wind (T.W.) forcings over sloping terrain.

At night, the ground cools the adjacent air, but not the air higher in the residual layer. This reverses the thermal wind at low altitudes in the SBL, but leaves the thermal wind unchanged higher in the residual layer (Fig 12.17c). The resulting geostrophic wind has a peak at low levels (Fig 12.17d), assuming a southerly surface geostrophic wind. The actual wind has an even more predominant jet, resulting from drag at the surface (Fig 12.17d).

The southerly geostrophic winds often occur in the warm sectors of approaching cyclones in the central USA. These sectors are also sometimes conditionally unstable. Raymond (1978) and Uccellini and Johnson (1979) have shown that thunderstorms can be triggered by LLJs in these situations.

12.5.3 Inertial Oscillation

During the daytime, winds in the ML are subgeostrophic because of strong frictional drag at the ground. At sunset when ML turbulence ceases, pressure gradients tend to accelerate the winds back toward geostrophic. However, the Coriolis force induces an *inertial oscillation* in the wind, causing it to become supergeostrophic later at night (Blackadar, 1957). Details of this oscillation are now discussed.

The starting point in this analysis are momentum equations (3.5.3c & d) for the boundary layer. For simplicity, choose a coordinate system such that $V_g = 0$, and abbreviate the Reynolds stress divergence (friction) terms by $\partial \overline{u'w'}/\partial z \equiv f_c \cdot F_u$, and $\partial \overline{v'w'}/\partial z \equiv f_c \cdot F_v$, where F_u and F_v have units of velocity. The equations become:

$$\frac{d\overline{U}}{dt} = + f_c \ \overline{V} \qquad - f_c \, F_u \qquad\qquad (12.5.3a)$$

$$\frac{d\overline{V}}{dt} = f_c (\overline{U}_g - \overline{U}) - f_c \, F_v \qquad\qquad (12.5.3b)$$

Initially, the winds are subgeostrophic, so we must first determine the daytime ML winds to be used as initial conditions for the nocturnal case. Assuming steady state during the day, the above equations can be easily solved for the winds:

$$\overline{U}_{day} = \overline{U}_g - F_{v\,day}$$

$$\overline{V}_{day} = F_{u\,day} \qquad\qquad (12.5.3c)$$

In this form, we see that F_u and F_v represent the departure of the winds from geostrophic (i.e., the *geostrophic departure*).

Next, assume that friction suddenly disappears above the surface layer at sunset, and that friction remains zero throughout the night. The nocturnal winds are expected to evolve with time, so we cannot assume steady state. Combine (12.5.3 a & b) into one equation by taking the time derivative of the first equation, and then substituting in the second equation:

$$\frac{d^2 \overline{U}}{dt^2} = -f_c (\overline{U} - \overline{U}_g)$$

The solution will be of the form:

$$\overline{U} - \overline{U}_g = A \sin(f_c t) + B \cos(f_c t)$$

The parameters A and B are then determined from the initial conditions (12.5 3c), yielding $A = F_{u\,day}$ and $B = -F_{v\,day}$. The final result is then:

$$\overline{U}_{night} = \overline{U}_g + F_{u\,day} \cdot \sin(f_c t) - F_{v\,day} \cdot \cos(f_c t)$$

$$\overline{V}_{night} = F_{u\,day} \cdot \cos(f_c t) + F_{v\,day} \cdot \sin(f_c t) \qquad (12.5.3d)$$

We see that the winds oscillate about the geostrophic value, but never converge to the geostrophic value in this idealized scenario. The period of oscillation, called the *inertial period*, is $2\pi/f_c$. At midlatitudes, the inertial period is about 17 h. The magnitude of the oscillation at night depends on the amount of geostrophic departure at the end of the day. Typical geostrophic departures are on the order of 2 to 5 m/s at the end of the day, leading to nocturnal jet maxima that can be 2 to 5 m/s faster than geostrophic (Garratt, 1985; and Kraus, et al., 1985).

12.5.4 Example

Problem. Frictional drag during the day results in geostrophic departures of $F_u = F_v$ = 3 m/s at sunset. Calculate and plot the resulting nocturnal winds for every hour over a full inertial period. Also determine the time of occurrence and maximum speed of the nocturnal jet. Assume $f_c = 10^{-4}$ s^{-1}, $U_g = 10$ m/s, and $V_g = 0$.

Solution. Solving (12.5.3d) using the conditions above gives the winds listed in the Table 12-1. These are plotted as a hodograph in Fig 12.18. We see that the wind vectors describe a circle about the geostrophic wind, with a radius of 4.24 m/s. A maximum wind speed of 14.24 m/s is reached about 6.5 h after sunset. In fact, the wind speeds are supergeostrophic for about a 9 h period.

Discussion. Initially, the winds are subgeostrophic and cross the isobars toward low pressure, as expected with friction. Shortly after sunset, the winds continue to turn toward low pressure (Mahrt, 1981). However, between 7 and 15 h after sunset the winds cross the isobars toward high pressure during a portion of the inertial oscillation. Such ageostrophic winds can lead to convergence regions that can trigger thunderstorms.

Also, midlatitude nights last only 8 to 16 h, depending on the season and latitude. Thus, the full cycle of the oscillation might not be realized before daytime mixing destroys the nocturnal jet.

Table 12-1. Example of an inertial oscillation.

t (h)	U - U_g (m/s)	V_g (m/s)		t (h)	U - U_g (m/s)	V_g (m/s)
0	-3.00	3.00		10	1.36	-4.02
1	-1.75	3.86		11	-0.14	-4.24
2	-0.28	4.23		12	-1.62	-3.92
3	1.23	4.06		13	-2.90	-3.10
4	2.58	3.37		14	-3.81	-1.88
5	3.60	2.24		15	-4.22	-0.41
6	4.16	0.83		16	-4.10	1.10
7	4.19	-0.69		17	-3.45	2.47
8	3.67	-2.12		17.45	-3.00	3.00
9	2.69	-3.28				

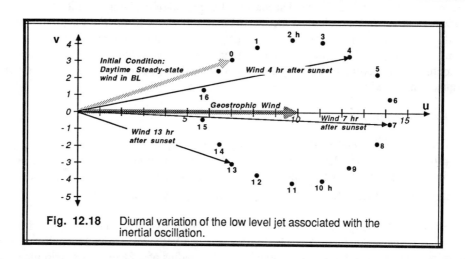

Fig. 12.18   Diurnal variation of the low level jet associated with the inertial oscillation.

12.6 Buoyancy (Gravity) Waves

Statically stable environments support buoyancy waves, so we expect to find waves within the SBL (Gossard and Hooke, 1975). Weakly turbulent SBLs are usually filled with waves. Fig 12.19 shows a sodar record of waves near the Boulder Atmospheric Observatory tower (Hooke and Jones, 1986). This figure clearly exhibits the ubiquitous superposition of waves of various wavelengths and amplitudes.

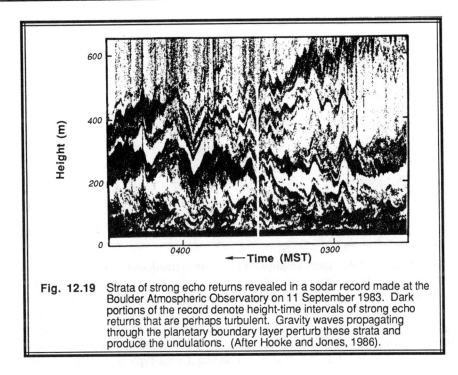

Fig. 12.19   Strata of strong echo returns revealed in a sodar record made at the Boulder Atmospheric Observatory on 11 September 1983. Dark portions of the record denote height-time intervals of strong echo returns that are perhaps turbulent. Gravity waves propagating through the planetary boundary layer perturb these strata and produce the undulations. (After Hooke and Jones, 1986).

12.6.1 Characteristics

Wave frequencies for internal waves at any altitude must be less than the Brunt-Väisälä frequency, N_{BV}, at that height, where:

$$N_{BV}^2 = \frac{g}{\theta_v} \frac{\partial \overline{\theta_v}}{\partial z} \qquad (12.6.1a)$$

Fig 12.20 shows the variation of Brunt-Väisälä frequency as a function of height for the 0300 local time potential temperature sounding plotted in Fig 12.12. Higher stabilities near the surface support a larger range of frequencies than those higher in the SBL. In a residual layer of neutral stratification, vertically propagating waves are not supported, and their amplitude is damped with height within the RL. Waves that propagate upward within the SBL eventually reach a level where their frequency matches the ambient Brunt-Väisälä frequency, at which point they reflect back down toward the ground. Waves are thus trapped between the ground and the neutral layers aloft, resulting in horizontally propagating waves and modal oscillations only, as shown in Figure 12.19. Wave periods (2π/frequency) ranging from less than a minute to 40 min have been reported, depending on the stability. Fig 12.19 shows waves with periods between 1 and 10 minutes.

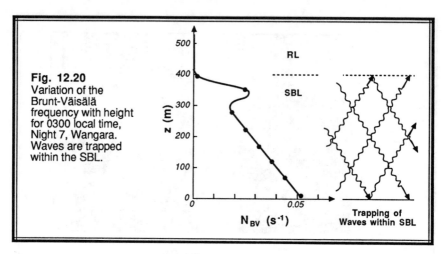

Fig. 12.20
Variation of the
Brunt-Väisälä
frequency with height
for 0300 local time,
Night 7, Wangara.
Waves are trapped
within the SBL.

Vertical wave displacement amplitudes range from fractions of meters to about 200 m (Nai-Ping, et al., 1983; Hooke and Jones, 1986; Finnigan, et al., 1984). Fig 12.19 shows waves with displacement amplitudes between 10 and 100 m. Large amplitude waves can induce surface pressure fluctuations as large as 10 Pa (0.1 mb), although most waves create pressure fluctuations smaller than 5 Pa (0.05 mb). These can be easily measured with surface-based microbarographs. Wave-induced pressure variations can be large compared to the 1 Pa fluctuations associated with typical turbulent wind gusts (Gedzelman, 1983).

Because wave amplitudes can be large compared to the depth of the SBL, we might expect the state of the SBL to be strongly modulated by such waves. This is indeed the case, as is shown for N_{BV}^2, the shear squared, gradient Richardson number, and TKE in Fig 12.21 for one wave period (Finnigan, et al., 1984). For this case, the Richardson

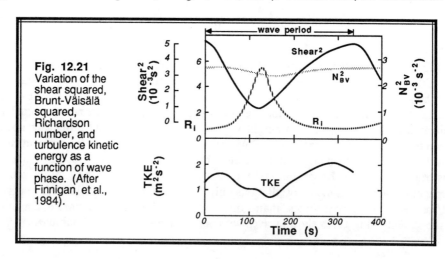

Fig. 12.21
Variation of the
shear squared,
Brunt-Väisälä
squared,
Richardson
number, and
turbulence kinetic
energy as a
function of wave
phase. (After
Finnigan, et al.,
1984).

number was lowered, dynamic instability induced, and turbulence was generated during portions of the wave cycle, while during other portions turbulence was suppressed. Weinstock (1987) also suggested that some waves can modify the SBL to the point where periods of static instability occur, leading to bursts of convective turbulence during portions of the wave cycle.

Waves can be generated by *wind shear* and by *impulses* (Gedzelman, 1983). At heights where the Richardson number is small, wind shear can generate K-H waves. These heights are called the *critical levels*. The K-H waves excite other internal gravity waves that propagate elsewhere in the SBL. Normally, such shear-induced waves have a phase speed equal to the mean wind speed at their critical level. It is possible for multiple critical levels to form, generating a variety of waves that coexist and interact within the SBL. Any shear waves that reflect off the ground back to their respective critical level are absorbed there.

Finnigan, et al. (1984) observed dramatic sequences of large amplitude shear waves at the BAO tower, and computed critical levels at 550 and 650 m for those cases. Hooke and Jones (1986) found that wave-related turbulence can form in thin sheets, perhaps where the Richardson number is small.

Impulse-generated waves have a high coherency, are virtually nondispersive, and are associated with some specific event such as an explosion, thunderstorm, or flow over an obstacle (Gedzelman, 1983). Waves propagate away from this event analogous to water waves propagating away from a pebble thrown into a pond. Often these waves have a long wavelength and large period. As these wave propagate through the SBL, they can induce large displacements of SBL air that modify the shear and stability.

For SBLs that are strongly turbulent, waves are subjected to strong nonlinear interactions with the turbulence. It is an academic question whether such flows should be classified as stably-stratified turbulence, or strongly-nonlinear waves. Any waves generated elsewhere that propagate into the region of strong turbulence will be absorbed by the turbulence, and the wave energy will be converted into TKE.

12.6.2   Analysis Techniques and Governing Equations

We will start with a brief review of linear wave theory, and progress into wave-turbulence interaction. For more details, see Finnigan, et al. (1984), Holton (1972) or Gossard and Hooke (1975).

Linear Wave Theory. Starting with the basic conservation equations described in Chapter 3, we can: (1) expand each dependent variable into a basic state ($\bar{\xi}$, usually assumed as a constant reference background state) and a perturbation wave part (ξ''); (2) assume that the perturbations are small; and (3) eliminate terms consisting of products of perturbation variables because scale analysis shows they are smaller than the other terms.

This process is known as *linearizing the governing equations*, because the only terms that are left contain basic state variables, or products of the basic state and a single linear perturbation variable. For many special cases, wave solutions can be found for these equations, resulting in *linear wave theory*. Some real atmospheric waves appear to have many of the characteristics described by linear wave theory, and are called *linear waves*. Many others do not, and are sometimes called *nonlinear waves*.

If κ_x, κ_y and κ_z are the wavenumbers in the x, y, and z directions, respectively, and f is the local wave frequency observed from a fixed point on the ground, then *linear internal waves* are described by (Stull, 1976):

$$w'' = A \cdot \sin(J)$$

$$\eta'' = \frac{A}{f_i} \cos(J)$$

$$u'' = -\left(\frac{\kappa_x \kappa_z A}{\kappa_H^2} \right) \sin(J)$$

$$v'' = -\left(\frac{\kappa_y \kappa_z A}{\kappa_H^2} \right) \sin(J)$$

$$p'' = -\left(\frac{\kappa_z \bar{\rho} f_i A}{\kappa_H^2} \right) \sin(J)$$

$$\theta'' = -\left(\frac{A}{f_i} \frac{\partial \bar{\theta}}{\partial z} \right) \cos(J) \qquad (12.6.2a)$$

where the double prime denotes the wave perturbation, and:

   η'' is wave vertical displacement,

   A   is an amplitude parameter,

   κ_H is the *horizontal wavenumber* defined by $\kappa_H^2 = \kappa_x^2 + \kappa_y^2$,

   f_i is the *intrinsic wave frequency* defined by $f_i = f - \kappa_x \bar{U} - \kappa_y \bar{V}$, and

   J = $\kappa_x x + \kappa_y y + \kappa_z z - f t$.

Also, the following frequency relation must be satisfied:

$$\frac{f_i^2}{N_{BV}^2} = \frac{\kappa_H^2}{\kappa_H^2 + \kappa_z^2} \qquad (12.6.2b)$$

Equations (12.6.2) show that linear waves are deterministic; we can solve for the wave perturbation displacement, vertical velocity, etc. at any location at any time. This is unlike turbulence, where equations (3.5.3) forecast only the statistics of the perturbations. Furthermore, simple sine and cosine functions describe the waves. By looking in time series records for simple sinusoidal oscillations, we can identify linear waves.

Linear waves can transport momentum and kinetic energy vertically, but they can not transport heat or other scalars. For example, (12.6.2a) shows that θ'' and w'' are 90° out of phase, because one is a sine function and the other is a cosine function. When the product $\theta'' w''$ is averaged over an integer number of wave periods, the result (i.e., the heat flux) is identically zero.

Nonlinear Waves and Turbulence. Finnigan, et al. (1984) show how analyses of time series data can be extended to include both waves and turbulence. Split any variable, ξ, into three parts: the mean, wave, and turbulence components:

$$\xi = \bar{\xi} + \xi'' + \xi'$$
(12.6.2c)

The mean is defined by the usual time average (2.4.1a). A phase averaging operator $\backslash\ \backslash$ can be defined as:

$$\backslash\xi\backslash = \lim_{N\to\infty} \left(\frac{1}{N} \sum_{k=1}^{N} \xi(t + k\,\mathbb{P}) \right)$$
(12.6.2d)

where $\mathbb{P}$ is the time period of the wave of interest. The wave component can then be found from:

$$\xi'' = \backslash\xi\backslash - \bar{\xi}$$
(12.6.2e)

Knowing both the mean and wave components, these can be subtracted from the instantaneous value to yield the remaining turbulent part from (12.6.2c). The above technique can be used to analyze the wave and turbulence components of an observed time series.

The set of averaging rules for this situation is an extension of the Reynolds averaging rules (Finnigan, et al., 1984):

$$\backslash\xi\backslash = 0 \qquad\qquad \overline{\xi''} = 0$$

$$\overline{\xi'} = 0 \qquad\qquad \overline{a\,b} = \bar{a}\,\bar{b}$$

$$\backslash a''\,b\backslash = a''\,\backslash b\backslash \qquad\qquad \backslash\bar{a}\,b\backslash = \bar{a}\,\backslash b\backslash$$

$$\backslash\bar{\xi}\backslash = \bar{\xi} \qquad\qquad \overline{\backslash\xi\backslash} = \bar{\xi}$$

$$\overline{a''b'} = 0 \qquad\qquad \overline{\backslash a''b\backslash} = 0 \qquad\qquad (12.6.2f)$$

Nonlinear waves can transport heat and scalars in addition to energy and momentum. (Hunt, et al., 1985). When (12.6.2c) is used in the basic conservation equations, one can derive averaged conservation equations for turbulence kinetic energy, wave kinetic energy, turbulent heat flux, wave heat flux, etc. using methods analogous to those used in Chapters 2 to 5. A detailed derivation of the averaged equations is beyond the scope of this book. However, the TKE equation for turbulence shown below gives a taste the interaction between waves and turbulence (Finnigan, et al., 1984):

$$\frac{\partial \bar{e}}{\partial t} = \frac{g}{\bar{\theta}_v}\overline{w'\theta_v'} - \overline{u'w'}\frac{\partial \overline{U}}{\partial z} - \frac{\partial \overline{w'e}}{\partial z} - \frac{1}{\bar{\rho}}\frac{\partial \overline{w'p'}}{\partial z}$$

$$- \overline{r_{ij}''\frac{\partial u_i''}{\partial x_j}} - \overline{u_j''\frac{\partial (r_{ii}''/2)}{\partial x_j}} - \varepsilon \qquad (12.6.2g)$$

where

$$r_{ij}'' = \backslash u_i'u_j'\backslash - \overline{u_i'u_j'} \qquad (12.6.2h)$$

which can be viewed as wavelike fluctuations in the background Reynolds stress due to the presence of the wave. The terms in the first line of (12.6.2g) are the same as those described earlier in (5.1b). The terms in the second line are a result of the wave-turbulence interaction. Also, the viscous dissipation terms are different than those described in Chapter 5.

12.6.3 Spectra

Both waves and turbulence can be present in the SBL. Finnigan, et al. (1984) found that about 20% of the total variance for their case study was explained by coherent wave motion, the remainder being turbulence. The portion of any spectrum composed of idealized linear gravity waves can be identified using cospectral techniques, because vertical motions and temperature will be 90° out of phase. Portions of the spectrum where phase differences are 0 or 180° usually correspond to turbulence. Difficulties can arise because many of the real waves might be highly nonlinear.

For the special case of waves excited at only one wavenumber, we would expect nonlinear interactions to remove the wave energy from the excited wavenumber (*source*) and "spread" it to both higher and lower wavenumbers. The spectrum of a mixture of waves and turbulence can often be split into three subranges (Fig 12.22):

(1) At wavenumbers lower than the source (i.e., large eddies), similarity arguments give a *-1 power law* spectral shape [$S(\kappa) \propto \kappa^{-1}$, or $\kappa \cdot S(\kappa) \propto \kappa^0$ as in Fig 12.22].

(2) Above the source wavenumber, in the *buoyancy subrange* of middle-size eddies, eddies are quasi-two-dimensional because of the suppression of vertical motions by stability. For quasi-2-D turbulence, TKE is not transferred across the spectrum, but *enstrophy* (mean-square vorticity) is. Similarity arguments give a *-3 power law*

spectral shape [$S(\kappa) \propto \kappa^{-3}$, or $\kappa \cdot S(\kappa) \propto \kappa^{-2}$ as in Fig 12.22] for these middle-size eddies.

(3) The inverse of the Ozmidov scale defines the *buoyancy wavenumber*, which separates the buoyancy subrange (middle wavenumbers) from the inertial subrange (high wavenumbers). At wavenumbers higher than the buoyancy wavenumber, eddies do not feel the static stability directly, are three-dimensional, and obey *-5/3 power law* inertial-subrange scaling [$\kappa \cdot S(\kappa) \propto \kappa^{-2/3}$ as in Fig 12.22].

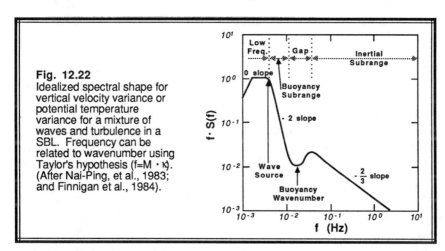

Fig. 12.22
Idealized spectral shape for vertical velocity variance or potential temperature variance for a mixture of waves and turbulence in a SBL. Frequency can be related to wavenumber using Taylor's hypothesis (f=M · κ). (After Nai-Ping, et al., 1983; and Finnigan et al., 1984).

Within the buoyancy subrange we might expect the only relevant variables to be S, N_{BV}, and κ. Dimensional analysis (Turner, 1973) gives:

$$S(\kappa) \propto N_{BV}^2 \; \kappa^{-3} \tag{12.6.3}$$

This is the same -3 power law dependence expected using enstrophy arguments.

Between the buoyancy and inertial subranges there is evidence of a gap. This gap is even more apparent in the potential temperature spectrum (Nai-Ping, 1983). One possible explanation (Finnigan, et al., 1984) for the gap is that the energy transfer in the inertial subrange towards lower wavenumbers is blocked at the -3 power law region, leading to a build-up of energy near the buoyancy wavenumber.

Many real atmospheric spectra might be affected by waves at more than one wavelength. The resulting spectrum would be very complex, and would be a function of the relative strengths of the different waves. In addition, Mahrt and Gamage (1987) have found that mesoscale forcings can dominate at low wavenumbers.

12.6.4   Bores

In northern Australia a propagating hydraulic jump phenomenon has frequently been observed during the night and early morning. This undular bore is locally known as the "morning glory", because it is visible on humid mornings as a spectacular-looking, very low-altitude, rapidly propagating (10-15 m/s), rolling, wave-cloud band (Smith, et al., 1982).

A *bore* is a rapidly propagating increase in fluid depth, usually associated with tidal signals propagating up estuaries or canals (Lamb, 1932). In the atmosphere, it is believed to be associated with a propagating increase in SBL depth, triggered by a sea breeze or a katabatic flow, and enhanced by the presence of a nocturnal low-level jet (Noonan and Smith, 1986). Bore passage is characterized by a wind squall (surface wind speeds increase from 0.5 m/s to 6 m/s in just a few minutes; wind directions change to point in the direction of bore movement), a pressure increase (0.1 kPa in 10 to 30 min), and the rolling cloud band.

The morning glory phenomenon additionally exhibits a series of 3 to 5 oscillations (of pressure, wind speed, and wind direction) and the corresponding number of cloud bands over a 1 h period at the leading edge of the bore. Because of these oscillations, the phenomenon is classified as an *undular bore*.

12.7 Terrain Slope and Drainage Winds

12.7.1 Characteristics

Winds formed when cold dense air is accelerated downslope by gravity are called *drainage winds, gravity flows, slope flows*, or *katabatic winds*. Brost and Wyngaard (1978) and Mahrt (1981) showed that even gentle slopes of $\Delta z/\Delta x = 0.001$ to 0.01 over a large area can cause drainage winds of 1 to 2 m/s. Slopes at the Wangara field site in Australia were on the order of 0.007, while the average slope of the Great Plains in the USA is roughly 0.001. Both of these sites look flat to the eye, yet slopes are great enough to create drainage winds. Thus, we should expect to have drainage winds just about everywhere within SBLs over land. Kottmeier (1986) observed gravity flows over the Ekström ice shelf.

Large lakes and oceans are about the only places where we can ignore drainage winds. Even the smaller lakes and dry lake beds could be influenced by drainage winds, as cold air from the neighboring slopes drain into the low areas and accumulate as a pool of cold air. Since most weather stations are located in valleys, near rivers, or near lakes, we must expect that the reported winds and temperature are influenced by cold air drainage on radiatively clear nights, and are not representative of the rest of the boundary layer nor of the synoptic flow (assuming light synoptic winds).

Fig 12.23 shows typical profiles of wind in a slope flow. Near the ground, drag reduces the wind speed. Just above the ground where friction is less but the air is still relatively cold, there is a wind maximum or nose. Higher in the profile, shears across the top of the gravity flow in the presence of weaker stability allow significant turbulence,

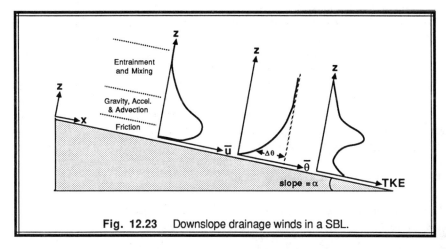

Fig. 12.23   Downslope drainage winds in a SBL.

mixing, and entrainment. Fig 12.23 also shows the typical coordinate system used for modeling studies of slope flows.

As the air flows down the slope, the depth of the drainage flow can grow, and the peak speed can increase. Fig 12.24 shows the slope flow evolution down a 10° average slope covered by grass and short brush (Doran and Horst, 1983; Horst and Doran, 1986). For this case, a very shallow flow with nose at about 50 m over the initial 21° slope evolves into a deeper flow with nose at about 300 m further down the hill where the slope is only 8°.

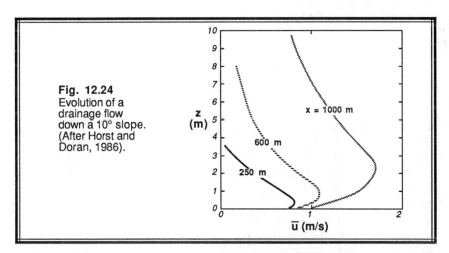

Fig. 12.24
Evolution of a
drainage flow
down a 10° slope.
(After Horst and
Doran, 1986).

In the absence of ambient mean flow, drainage wind speeds range between 0.5 to 3.5 m/s. When ambient mean winds are light, the drainage flows in the same direction as the mean wind are enhanced, while those opposing the mean wind are reduced (Yamada,

1983; Heilman and Dobosy, 1985; and Wong, et al., 1987). For moderate and strong mean winds, opposed slope flows are eliminated. Similar interesting interactions between slope flows and sea breezes can occur when mountains are located near the shore. Fitzjarrald (1984, 1986) found that the diurnal cycle of slope flows and land/sea breezes can cause a reversal of the local winds at Veracruz, Mexico.

When drainage winds flow out of a valley into a flatter plain, the leading edge behaves like a weak thunderstorm gust front, with a well-defined nose and K-H waves along the top of the flow. In measurements of such flows past the BAO tower, Blumen (1984) observed the flow to be about 150 m thick, with peak velocities of 4 m/s found below 50 m. The propagation speed of the drainage flow front was estimated to be 2.5 to 3.5 m/s.

12.7.2 Governing Equations

Using the coordinate system shown in Fig 12.23, Mahrt (1982) has written the equation of motion for downslope component of wind as:

$$\frac{\partial U}{\partial t} + U\frac{\partial U}{\partial x} + V\frac{\partial U}{\partial y} + W\frac{\partial U}{\partial z} = g\frac{\Delta\theta}{\theta}\sin(\alpha) - \cos(\alpha)\frac{g}{\theta}\frac{\partial\overline{\overline{\theta}}\,h_d}{\partial x} + f_c V - \frac{\partial\overline{u'w'}}{\partial z}$$

$$\text{(12.7.2a)}$$

$$\quad\text{I}\qquad\text{II}\qquad\text{III}\qquad\text{IV}\qquad\text{V}\qquad\qquad\text{VI}\qquad\qquad\text{VII}\qquad\text{VIII}$$

where α is the slope angle, h_d is the depth of the flow, $\Delta\theta(z)$ is the potential temperature difference between the ambient air and the colder slope flow, and the average potential temperature depression in the flow is:

$$\overline{\overline{\theta}}(z) = \frac{1}{h_d}\int_{z'=z}^{h_d}\Delta\theta\,dz'$$

$$\text{(12.7.2b)}$$

The terms in the above equation have the following interpretation:

Term I	Acceleration
Term II	Downslope advection
Term III	Cross slope advection
Term IV	Vertical advection
Term V	Buoyancy
Term VI	Thermal wind
Term VII	Coriolis
Term VIII	Stress divergence

Obviously different terms are important for different slopes with different ambient flow conditions. Mahrt (1982) discussed 8 different categories of flow based on a scale analysis: *nonstationary, advective-gravity, near equilibrium, shooting, combination, combination with friction, tranquil, and non-gravity*, only two of which will be reviewed here.

For *advective-gravity flows*, only terms II and V are important. This flow is constant with time, but is a function of downslope distance. The average wind speed over the depth of the flow [using averaging like (12.7.2), except integrated from 0 to h_d] is:

$$\overline{\overline{U}} = \left[\frac{g\,\overline{\overline{\theta}}}{\theta} \cdot x \cdot \sin(\alpha) \right]^{1/2} \tag{12.7.2c}$$

where x is downslope distance. Briggs (1979) suggested a similar equation, but using the surface heat flux instead of the average potential temperature departure:

$$U = 2.15 \left[\sin(\alpha) \right]^{2/9} \left[\frac{g}{\theta} (-\overline{w'\theta'}_s) \, x \right]^{1/3} \tag{12.7.2d}$$

He also suggested that the depth of the flow is given by:

$$h_d = 0.037 \left[\sin(\alpha) \right]^{2/3} x \tag{12.7.2e}$$

For *equilibrium flows*, the pull of gravity is balanced by frictional drag at the top and bottom of the flow (i.e., terms V and VIII are important). These flows are constant in both space and time. The equilibrium average wind speed is:

$$\overline{\overline{U}}_{eg} = \left[\frac{g\,(\overline{\overline{\theta}}/\theta)\,\sin(\alpha)\,h_d}{C_D + C_{Dh}} \right]^{1/2} \tag{12.7.2f}$$

where C_{Dh} is like a drag coefficient, except at the top of the flow: $C_{Dh} = \overline{u'w'}_h / U_{eq}$.

12.7.3   Interactions of SBL Processes

At the start of this chapter, we explored some of the problems associated with SBL modeling. These included factors such as long time scales, turbulence bursting, z-less scaling, inertial oscillations, and now drainage flows. Most of these processes were examined in an idealized scenario in which the other processes were neglected. In the real atmosphere, however, many of these processes can occur simultaneously, making the SBL behavior very difficult to describe.

For example, the accumulated cooling method of examining SBL growth assumes that none of the chilled air is drained away downslope. The downslope flows neglected inertial oscillations. The relationship between fog formation and drainage flow could be important. Turbulence, mixing, and SBL growth are modulated by buoyancy waves. We conclude that a model of any arbitrary SBL must be designed to include all of the physics described in this chapter in order for it to succeed in all situations. Such a model might be impractical.

12.8 References

Anderson, D.L.T., 1976: The low-level jet as a western boundary current. *Mon. Wea. Rev.*, **104**, 907-921.

Blackadar, A.K., 1957: Boundary layer wind maxima and their significance for the growth of nocturnal inversions. *Bull. Amer. Meteor. Soc.*, **38**, 283-290.

Blumen, W., 1984: An observational study of instability and turbulence in nighttime drainage winds. *Bound.-Layer Meteor.*, **28**, 245-269.

Bonner, W.D., 1968: Climatology of the low level jet. *Mon. Wea. Rev.*, **96**, 833-850.

Briggs, G.A., 1979: Analytic modeling of drainage flows. Draft document, Atmospheric Turbulence and Diffusion Laboratory, NOAA. (Available from EPA, Environmental Sciences Research Lab., Research Triangle Park, NC 27722).

Brook, R.R., 1985: Koorin nocturnal low-level jet. *Bound.-Layer Meteor.*, **32**, 133-154.

Brost, R.A. and J.C. Wyngaard, 1978: A model study of the stably stratified planetary boundary layer. *J. Atmos. Sci.*, **35**, 1427-1440.

Brunt, D., 1939: *Physical and Dynamical Meteorology*, 2nd Ed. Cambridge Univ. Press, Cambridge, GB. 428pp.

Carlson, M.A. and R.B. Stull, 1986: Subsidence in the nocturnal boundary layer. *J. Clim. Appl. Meteor.*, **25**, 1088-1099.

Cerni, T.A. and T.R. Parish, 1984: A radiative model of the stable nocturnal boundary layer with application to the polar night. *J. Clim. Appl. Meteor.*, **23**, 1563-1572.

Chou, M.-D., 1984: Broadband water vapor transmission functions for atmospheric IR flux computations. *J. Atmos. Sci.*, **41**, 1775-1778.

Delage, Y., 1974: A numerical study of the nocturnal atmospheric boundary layer. *Quart. J. Roy. Meteor. Soc.*, **100**, 351-364.

Doran, J.C. and T.W. Horst, 1983: Observations and models of simple nocturnal slope flows. *J. Atmos. Sci.*, **40**, 708-717.

Estournel, C. and D. Guedalia, 1985: Influence of geostrophic wind on atmospheric nocturnal cooling. *J. Atmos. Sci.*, **42**, 2695-2698.

Estournel, C. and D. Guedalia, 1987: A new parameterization of eddy diffusivities for nocturnal boundary layer modeling. *Bound.-Layer Meteor.*, **39**, 191-203.

Estournel, C., R. Vehil and D. Guedalia, 1986: An observational study of radiative and turbulent cooling in the nocturnal boundary layer (ECLATS experiment). *Bound.-Layer Meteor.*, **34**, 55-62.

Finnigan, J.J., F. Einaudi and D. Fua, 1984: The interaction between an internal gravity wave and turbulence in the stably-stratified nocturnal boundary layer. *J. Atmos. Sci.*, **41**, 2409-2436.

Fitzjarrald, D.R., 1984: Katabatic wind in opposing flow. *J. Atmos. Sci.*, **41**, 1143-1158.

Fitzjarrald, D.R., 1986: Slope winds in Veracruz. *J. Clim. Appl. Meteor.*, 25, 133-144.

Garratt, J.R., 1983: Surface influence upon vertical profiles in the nocturnal boundary layer. *Bound.-Layer Meteor.*, **26**, 69-80.

Garratt, J.R., 1985: Inland boundary layer at low latitudes. Part 1, the nocturnal jet. *Bound.-Layer Meteor.*, **32**, 307-327.

Garratt, J.R. and R.A. Brost, 1981: Radiative cooling effects within and above the nocturnal boundary layer. *J. Atmos. Sci.*, **38**, 2730-2746.

Gedzelman, S.D., 1983: Short-period atmospheric gravity waves: a study of their statistical properties and source mechanisms. *Mon. Wea. Rev.*, **111**, 1293-1299.

Godowich, J.M., J.K.S. Ching and J.F. Clark, 1985: Evolution of the nocturnal inversion layer at an urban and non-urban location. *J. Appl. Meteor.*, **24**, 791-804.

Gossard, E.E., J.E. Gaynor, R.J. Zamora, and W.D. Neff, 1985: Fine structure of elevated stable layers observed by sounder and in situ tower measurements. *J. Atmos. Sci.*, **42**, 2156-2169.

Gossard, E.E. and W.H. Hooke, 1975: *Waves in the Atmosphere, Atmospheric Infrasound and Gravity Waves — their Generation and Propagation.* Elsevier Scientific Publ. Co., NY. 456pp.

Hart, J.E., G.V. Rao, H. van de Boogaard, J.A. Young, and J. Findlater, 1978: Aerial observations of the E. African low-level jet stream. *Mon. Wea. Rev.*, **106**, 1714-1724.

Heilman, W. and R. Dobosy, 1985: A nocturnal atmospheric drainage flow simulation investigating the application of one-dimensional modeling and current turbulence schemes. *J. Appl. Meteor.*, **24**, 924-936.

Holton, J.R., 1967: The diurnal boundary layer wind oscillation above sloping terrain. *Tellus*, **19**, 199-205.

Holton, J.R., 1972: *An Introduction to Dynamic Meteorology.* Academic Press, NY. 319pp.

Hooke, W.H. and R.M. Jones, 1986: Dissipative waves excited by gravity wave encounters with the stably stratified planetary boundary layer. *J. Atmos. Sci.*, **43**, 2048-2060.

Horst, T.W. and J.C. Doran, 1986: Nocturnal drainage flow on simple slopes. *Bound.-Layer Meteor.*, **34**, 263-286.

Hunt, J.C.R., J.C. Kaimal and J.E. Gaynor, 1985: Some observations of turbulence structure in stable layers. *Quart. J. Roy. Meteor. Soc.*, **111**, 793-815.

Kottmeier, C., 1986: Shallow gravity flows over the Ekström ice shelf. *Bound.-Layer Meteor.*, **35**, 1-20.

Kraus, H., J. Malcher and E. Schaller, 1985: Nocturnal low-level jet during PUKK. *Bound.-Layer Meteor.*, **31**, 187-195.

Lacser, A. and S.P.S. Arya, 1986: A comparitive assessment of mixing length parameterizations in the stably stratified nocturnal boundary layer (NBL). *Bound.-Layer Meteor.*, **38**, 1-22.

Lamb, H., 1935: *Hydrodynamics* (6th Ed.). Dover, New York. 738pp.

Lettau, H.H., 1967: Small to large scale features of boundary layer structure over mountain slopes. *Proc. Symp. on Mountain Meteor., Colorado State Univ.*, Ft. Collins, Part 2. 1-74.

Mahrt, L., 1981: The early evening boundary layer transition. *Quart. J. Roy. Meteor. Soc.*, **107**, 329-343.

Mahrt, L., 1982: Momentum balance of gravity flows. *J. Atmos. Sci.*, **39**, 2701-2711.

Mahrt, L., 1985: Vertical structure and turbulence in the very stable boundary layer. *J. Atmos. Sci.*, **42**, 2333-2349.

Mahrt, L. and N. Gamage, 1987: Observations of turbulence in stratified flow. *J. Atmos. Sci.*, **44**, 1106-1121.

Mahrt, L. and R.C. Heald, 1979: Comments on "Determining height of the nocturnal boundary layer. *J. Appl. Meteor.*, **36**, 383.

Mahrt, L., R.C. Heald, D.H. Lenschow, B.B. Stankov and IB Troen, 1979: An observational study of the sturcture of the nocturnal boundary layer. *Bound.-Layer Meteor.*, **17**, 247-264.

Malcher, J. and H. Kraus, 1983: Low-level jet phenomena described by an integrated dynamic PBL model. *Bound.-Layer Meteor.*, **27**, 327-343.

Nai-Ping, L., W.D. Neff and J.C. Kaimal, 1983: Wave and turbulence structure in a disturbed nocturnal inversion. *Bound.-Layer Meteor.*, **26**, 141-155.

Nieuwstadt, F.T.M., 1984: The turbulent structure of the stable, nocturnal boundary layer. *J. Atmos. Sci.*, **41**, 2202-2216.

Nieuwstadt, F.T.M., 1984: Some aspects of the turbulent stable boundary layer. *Bound.-Layer Meteor.*, **30**, 31-55.

Noonan, J.A. and R.K. Smith, 1986: Sea-breeze circulations over Cape York Peninsula and the generation of Gulf of Carpentaria cloud line disturbances. *J. Atmos. Sci.*, **43**, 1679-1693.

Raymond, D.J., 1978: Instability of the low-level jet and severe storm formation. *J. Atmos. Sci.*, **35**, 2274-2280.

Sladkovic, R. and H.J. Kantor, 1977: Low-level jet in the Bavarian pre-alpine regime. *Arch. Met. Geoph. Biokl.*, Ser. A., **25**, 343-355.

Smith, R.K., N. Crook and G. Roff, 1982: The Morning Glory: an extraordinary atmospheric undular bore. *Quart. J. Roy. Meteor. Soc.*, **108**, 937-956.

Staley, D.O. and G.M Jurica, 1970: Flux emissivity tables for water vapor carbon dioxide and ozone. *J. Appl. Meteor.*, **9**, 365-372.

Stull, R.B., 1976: Internal gravity waves generated by penetrative convection. *J. Atmos. Sci.*, **33**, 1279-1286.

Stull, R.B., 1983a: Integral scales for the nocturnal boundary layer. Part 1: Empirical depth relationships. *J. Clim. Appl. Meteor.*, **22**, 673-686.

Stull, R.B., 1983b: Integral scales for the nocturnal boundary layer. Part 2: Heat budget, transport and energy implications. *J. Clim. Appl. Meteor.*, **22**, 1932-1941.

Surridge, A.D. and D.J. Swanepoel, 1987: On the evolution of the height and temperature difference across the nocturnal stable boundary layer. *Bound.-Layer Meteor.*, **40**, 87-98.

Turner, J.S., 1973: *Buoyancy Effects in Fluids.* Cambridge Univ. Press, Cambridge, GB. 367pp.

Turton, J.D. and R. Brown, 1987: A comparison of a numerical model of radiation fog with detailed observations. *Quart. J. Roy. Meteor. Soc.*, **113**, 37-54.

Uccellini, L.W. and D.R. Johnson, 1979: The coupling of upper and lower tropospheric jet streams and implications for the development of severe convective storms. *Mon. Wea. Rev.*, **107**, 682-703.

Weinstock, J., 1987: The turbulence field generated by a linear gravity wave. *J. Atmos. Sci.*, **44**, 410-420.

Wong, R.K.W., K.D. Hage and Leslie D. Phillips, 1987: The numerical simulation of drainage winds in a small urban valley under conditions with supercritical Richardson numbers. *J. Clim. Appl. Meteor.*, **26**, 1447-1463.

Wyngaard, J.C., 1973: On surface layer turbulence. *Workshop on Micrometeorology* (Ed. by D. Haugen). Amer. Meteor. Soc., Boston. 101-150.

Xing-Sheng, L., J.E. Gaynor and J.C. Kaimal, 1983: A study of multiple stable layers in the nocturnal lower atmosphere. *Bound.-Layer Meteor.*, **26**, 157-168.

Yamada, T., 1983: Simulations of nocturnal drainage flows by a $q^2 l$ turbulence closure model. *J. Atmos. Sci.*, **40**, 91-106.

12.9 Exercises

1) Assume the following PBL initial conditions: time is 1800 local, temperature sounding is adiabatic ($\bar{\theta}$=300 K) up to z = 1 km, above which the lapse rate is $\gamma = \Delta\bar{\theta}/\Delta z =$ 0.01 K/m. Also, assume the following boundary conditions: B = 10 m/K (bulk NBL turbulence parameter), no subsidence, no winds during daytime, and a surface heat flux vs time as plotted below:

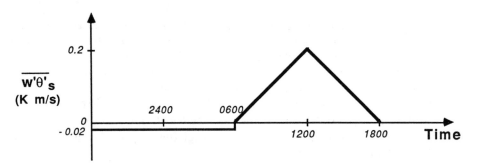

To perform the following exercises, use the thermodynamic (encroachment) method during the day to find the rise of z_i with time, and neglect $\overline{w'\theta'}$ at z_i:

(a) Forecast and plot the depth of the PBL (RL, NBL, and ML) as a function of time for the full 24 h period, starting at 1800 local time on Day 1, and ending at 1800 local time on Day 2.

(b) Also plot the potential temperature profiles at 0600 and 1800 local times for Day 2.

2) What are typical response times of the stable (nocturnal) boundary layer and the mixed layer, and how does the flow vary with these response times?

3) Given: $<\bar{\theta}>$ = 5 C at sunset (1800 local time) = constant with height. The following are constant with time during the night:

P_s = 100 kPa   (surface pressure)

$\overline{U}_g$ = 5 m/s   (geostrophic wind speed)

Z_s = 5000 m   (roughness scale)

Q_H = -0.02 K m/s (surface kinematic heat flux)

f_c = 10-4 s-1   (Coriolis parameter)

An orchard owner hires you to forecast the near surface air temperature, because she is concerned about freeze damage to the fruit. Will the near-surface air temperature reach freezing on this night? If so, when? (Hint: Neglect moisture and direct radiative cooling of the trees).

4) Assume a nocturnal boundary layer with Q_H = - 0.02 K m/s (constant throughout the night), Z_s = 3 km.   Also assume that the transition to negative heat flux happens at 8 PM local time (call this time t = 0), and that the heat flux becomes positive at 6 AM.

Forecast the depth (h) and strength ($\Delta\bar{\theta}_s$) of the NBL as a function of time for (a) $\overline{U}_g$ = 10 m/s, and (b) $\overline{U}_g$ = 3 m/s.

5) Given the previous answer, plot vertical profiles of $\Delta\bar{\theta}$ corresponding to a local time of 3 AM for both geostrophic-wind cases.

6) Neglecting thermal wind effects, forecast the boundary layer winds in Minneapolis at 3, 6, and 9 h after sunset (assume turbulence ceases at sunset).   Given: $\overline{U}_g$ = 8 m/s, $\overline{V}_g$ = 0. Assume that the initial conditions at sunset are $\overline{U}$ = 6 m/s and $\overline{V}$ = 3 m/s. Use the inertial oscillation method to answer this question.

7) Weather stations are often located in the bottom of valleys. Will such a station give surface winds that are representative of the synoptic-scale flow? Why?

8) Name two theories for nocturnal jet formation.

9) Given initial conditions at 1800 local time during Day 1 with an adiabatic temperature profile ($\bar{\theta}$ = 290 K) up to z = z_i = 2 km,   no temperature step at the top of the ML, but a constant lapse rate of $\gamma = \partial\bar{\theta}/\partial z$ = 0.01 K/m. Also given the same surface heat flux plotted in question (1), except that the constant nighttime flux is -0.1 K m/s. When in the morning of Day 2 will an elevated thin layer of smoke 500 m above the ground be fumigated down to the ground?   Hint: use an exponential-shaped SBL potential temperature profile with B = 10 m/K at night, and use the thermodynamic (encroachment) method for the ML during the day.

10) Calculate the average speed as a function of downslope distance for an advective-gravity flow, where the slope is 20° and the average temperature departure (ambient temperature minus drainage flow temperature) is 3 K.

11) Derive using similarity arguments the spectral power law for enstrophy cascade in a SBL.

12) Prove the averaging rules of (12.6.2f).

13) Derive the linearized equations of motion for which (12.6.2a & b) are solutions.

14) Derive an expression for $\overline{u'w'}$, $\overline{w'\theta'}$ and TKE for a linear wave, using (12.6.2a). Assume the average applies over an integer multiple of wavelengths.

13 Boundary Layer Clouds

Clouds can form at the top of mixed layers, and at the bottom of stable boundary layers. The amount and distribution of short and long-wave radiative flux divergence in the boundary layer are altered by clouds, and these effects are emerging as important aspects of the climate-change problem. In addition, the radiative effects combine with latent heating to modulate BL dynamics, turbulence generation, and evolution. This chapter provides a brief review of cloud thermodynamics, radiative processes, the role of entrainment, and descriptions of fogs, cumulus and stratocumulus clouds.

13.1 Thermodynamics

13.1.1 Variables

Table 13-1 lists many of the thermodynamic variables that have been used in the literature to describe the state of the cloudy boundary layer. The *static energies* (s, s_v, s_e, s_L, s_{es}) are similar to the respective potential temperatures (θ, θ_v, θ_e, θ_L, θ_{es}). However, static energies are based on the assumption that any kinetic energy is locally dissipated into heat, while the potential temperatures do not utilize that assumption.

Static energies have units of m^2/s^2, or equivalently J/kg. Typical magnitudes for static energies in the boundary layer are 3×10^5 J/kg. By expressing static energies in units of kJ/kg, the resulting values are on the order of 300 kJ/kg. This helps to reinforce its analogy with potential temperature, which is typically on the order of 300 K.

Table 13-1. Thermodynamic variables useful for cloud studies. Many of the expressions are approximate.

Water Variables:

Mixing ratio (for water vapor)

r $r = r_{sat}$ when the air is saturated

Saturated mixing ratio (for water vapor)

r_{sat} calculated based on T and P

Liquid water mixing ratio

r_L

Total water mixing ratio

$r_T = r + r_L$

Temperature Variables:

Absolute temperature

T

Virtual temperature

$T_v = T \cdot (1 + 0.61 r - r_L)$

Potential Temperature Variables:

Potential temperature

$$\theta = T \left(\frac{P_0}{P} \right)^{0.286} \cong T + \frac{g z}{C_p}$$

Virtual potential temperature

$$\theta_v = \theta \cdot (1 + 0.61 r - r_L)$$

Equivalent potential temperature

$$\theta_e = \theta + \left(\frac{L_v \theta}{C_p T} \right) \cdot r$$ for unsaturated air, T is the parcel's temperature at the LCL

Liquid water potential temperature

$$\theta_L = \theta - \left(\frac{L_v \theta}{C_p T} \right) \cdot r_L$$

Saturation equivalent potential temperature

$$\theta_{es} = \theta + \left(\frac{L_v \theta}{C_p T} \right) \cdot r_{sat}$$ use calculated value of r_{sat} based on T and P

Static Energies:

Dry static energy (also known as the Montgomery stream function)
$$s = C_p T + g z$$
Virtual dry static energy
$$s_v = C_p T_v + g z$$
Moist static energy
$$s_e = C_p T + g z + L_v r$$
Liquid water static energy
$$s_L = C_p T + g z - L_v r_L$$
Saturation static energy
$$s_{es} = C_p T + g z + L_v r_{sat} \qquad \text{use } r_{sat} \text{ calculated from T and P}$$

Saturation Points:

Saturation point pressure
$$P_{SP}$$
Saturation point temperature
$$T_{SP}$$

Some of the characteristics of these variables are:

$s_L = s$	for unsaturated air
$s = s_{es}$	for saturated air
s_{es}	is conserved during *moist* adiabatic ascent/descent (no mixing)
s	is conserved during *dry* adiabatic ascent/descent (no mixing)
s_L and s_e	are conserved for both *dry and moist* adiabatic ascent/descent
$\partial s/\partial z$	determines *dry static stability* (stable if gradient is positive)
$\partial s_{es}/\partial z$	determines *moist static stability* (stable if gradient is positive)
r_T	is conserved for both *dry and moist* ascent/descent (no mixing and no precipitation).

Air is *conditionally unstable* if $\partial s/\partial z > 0$ and $\partial s_{es}/\partial z < 0$. Similar characteristics apply to the corresponding potential temperatures.

For air at a pressure level P, the *saturation level* is found by dry (moist) adiabatic ascent (descent) of an unsaturated air parcel (a saturated cloud parcel) to the pressure level, P_{SP}, where the parcel is just saturated with no cloud liquid water (Betts, 1982a&b, 1985). This level is also known as the *lifting condensation level* (LCL) for rising unsaturated air parcels, and often defines *cloud base*. The temperature of an air parcel when moved to this level is symbolized as T_{SP}. On a thermodynamic diagram, the point represented by (P_{SP}, T_{SP}) is called the *saturation point* (SP).

13.1.2 Conserved Variables

Variables such as potential temperature and water-vapor mixing ratio are *not conserved* in a cloud, because of latent heat release/absorption and condensation/evaporation. We would rather use variables that are conserved under adiabatic processes regardless of the state of saturation of the air parcel, because then we can use those variables to study how diabatic processes and external forcings affect the evolution of the cloudy boundary layer.

The following variables *are conserved* for both dry and moist adiabatic processes with no precipitation, and change in proportion to the relative amounts of air undergoing isobaric mixing:

equivalent potential temperature,	θ_e
liquid-water potential temperature,	θ_L
moist static energy,	s_e
liquid water static energy,	s_L
total water mixing ratio,	r_T
saturation point pressure,	P_{SP}
saturation point temperature,	$T_{SP}.$

Any two of these conserved variables, together with the actual pressure or height of the air parcel, can be use to completely define the *thermodynamic state and water content* of the air (Betts and Albrecht, 1987). The most popular sets are (P, θ_e, r_T), (P, θ_L, r_T), (P, s_e, r_T), (P, s_L, r_T), or (P, P_{SP}, T_{SP}). Frequently, z is used in place of P in the above sets. Using any one set, we could calculate all of the variables listed in Table 13-1 (see Example 13.1.6).

By plotting one conserved variable against the other, we can create a *conserved variable diagram* that is useful for diagnostic studies of cloud and boundary layer processes (Hanson, 1984; Betts, 1985; Betts and Albrecht, 1987). Fig 13.1 shows a

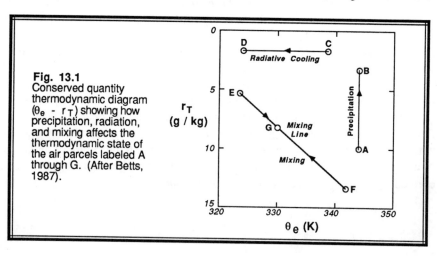

Fig. 13.1
Conserved quantity thermodynamic diagram (θ_e - r_T) showing how precipitation, radiation, and mixing affects the thermodynamic state of the air parcels labeled A through G. (After Betts, 1987).

θ_e - r_T diagram, along with some of the processes that can be represented. Precipitation leaving the parcel reduces r_T, but not θ_e. Radiative cooling affects θ_e, but not r_T. Mixing between two different air parcels results in a mixture parcel that falls on the *mixing line*, a straight line connecting the thermodynamic states of the two original parcels. The portion of each original parcel in the final mixture is represented by the relative position along the mixing line.

Fig 13.2 shows a θ_e - r_T diagram for data from the CCOPE experiment in Montana. It is clear that a deep layer between about 85 kPa (850 mb) and almost 60 kPa (600 mb) is undergoing mixing, because the data falls along a mixing line. This mixing includes both the subcloud boundary layer and the vigorous cumulus convection above. At cloud top levels (60 to 50 kPa) radiative cooling dominates.

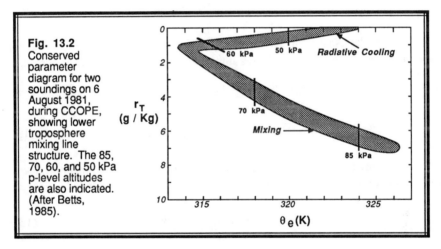

Fig. 13.2 Conserved parameter diagram for two soundings on 6 August 1981, during CCOPE, showing lower troposphere mixing line structure. The 85, 70, 60, and 50 kPa p-level altitudes are also indicated. (After Betts, 1985).

13.1.3 Conservation Equations for Turbulent Flow

Conservation of total water (3.4.4b) has already been discussed in Chapter 3. It is rewritten here in terms of a total water mixing ratio, and molecular diffusion is neglected:

$$\frac{\partial \overline{r_T}}{\partial t} + \overline{U}_j \frac{\partial \overline{r_T}}{\partial x_j} = \frac{S_{rT}}{\overline{\rho}_{air}} - \frac{\partial \overline{u_j' r_T'}}{\partial x_j} \qquad (13.1.3a)$$

where S_{rT} is a body source term that is negative for net precipitation leaving the air parcel, and is positive for net precipitation falling into the parcel. Change of water phase within the parcel does not alter the total water conservation.

Conservation of heat is rewritten in terms of s_L, although s_e or the corresponding potential temperatures can be used instead:

$$\frac{\partial \overline{s_L}}{\partial t} + \overline{U_j} \frac{\partial \overline{s_L}}{\partial x_j} = -\frac{1}{\overline{\rho} \, C_p} \frac{\partial \overline{Q_j^*}}{\partial x_j} - \frac{\partial \overline{u_j' s_L}}{\partial x_j} \qquad (13.1.3b)$$

where molecular conduction has been neglected. The evaporation term in (3.4.5b) does not appear because phase change is already incorporated into the definition of the liquid water static energy.

The fluxes at the top of a ML are found using the entrainment parameterizations:

$$\overline{w' r_T'}_{z_i} = -w_e \, \Delta_{EZ} \overline{r_T} \qquad (13.1.3c)$$

$$\overline{w' s_L'}_{z_i} = -w_e \, \Delta_{EZ} \overline{s_L} \qquad (13.1.3d)$$

where $\Delta_{EZ}(\ )$ is the usual jump across the top of the ML. Rogers, et al. (1985a) suggest the following parameterization for fluxes at the surface (in this case, over an ocean):

$$\overline{w' r_T'}_s = C_E \, \overline{M} \, [\, r_{sat}(T_s) - r_T \,] \qquad (13.1.3e)$$

$$\overline{w' s_L'}_s = C_H \, \overline{M} \, [\, s_s - \overline{s_L} \,] \qquad (13.1.3f)$$

The buoyancy flux is needed to calculate TKE production, entrainment production vs. consumption, or to find scaling variables such as w_*. Moeng and Randall (1984) suggest that:

$$\overline{w' s_v'} = \alpha_1 \, \overline{w' s_L'} + \alpha_2 \, \overline{w' r_T'} \qquad (13.1.3g)$$

where $\alpha_1 = 1$ and $\alpha_2 = 0$ for unsaturated air. For cloudy air:

$$\alpha_1 = \alpha_3 \left[1 + 1.61 \, T \frac{\partial r_{sat}}{\partial T} \right]$$

$$\alpha_2 = \alpha_3 \left[L_v \cdot \left(1 + 0.61 \, T \frac{\partial r_{sat}}{\partial T} \right) \right]$$

$$\alpha_3 = \left[1 + \frac{L_v}{C_p} \frac{\partial r_{sat}}{\partial T} + 1.61 \, T \frac{\partial r_{sat}}{\partial T} \right]^{-1}$$

Buoyancy variables such as s_v can be computed (using iteration) from the variables listed in Table 13-1 (see Example 13.1.6).

13.1.4   Saturation Point and the Lifting Condensation Level

To use many of the variables in Table 13-1 for a saturated air parcel, we must be able to determine the saturation mixing ratio. Empirical fits to the saturation curve have been reviewed by Buck (1981), Bolton (1980), Lowe (1977), Wexler (1976), and Stackpole (1967). Bolton suggests that a variation of *Tetens' formula* (1930) is sufficiently accurate to determine the saturation (with respect to liquid) water-vapor pressure (in units of kPa) for typical BL temperatures:

$$e_{sat} = (0.61078 \text{ kPa}) \cdot \exp\left[\frac{17.2694 \cdot (T - 273.16)}{T - 35.86} \right] \qquad (13.1.4a)$$

for absolute temperature in (K).

The saturation mixing ratio is then found from:

$$r_{sat} = 0.622 \frac{e_{sat}}{P - e_{sat}} \qquad (13.1.4b)$$

for pressure P in the same units as e_{sat}. Sometimes the variation of saturation mixing ratio with temperature is needed. The *Clausius-Clapeyron equation* can be written as:

$$T \frac{dr_{sat}}{dT} = 0.622 \frac{L_v \, r_{sat}}{\Re \, T} \qquad (13.1.4c)$$

The *saturation point* (SP) temperature and pressure are found from the following approximation:

$$T_{SP} = T_{LCL} = \frac{2840}{3.5 \ln(T) - \ln\left(\dfrac{P \cdot r}{0.622 + r} \right) - 7.108} + 55. \quad (13.1.4d)$$

and

$$P_{SP} = P_{LCL} = P \cdot \left(\frac{T_{SP}}{T} \right)^{3.5} \qquad (13.1.4e)$$

for P in kPa and T in K (Bolton, 1980; Wilde, et al., 1985). The lifting condensation level is the saturation level for an unsaturated parcel lifted dry adiabatically. As was shown in Chapter 11, many convective thermals have undiluted cores, allowing the actual *cloud base* to be very close to the LCL calculated from surface-layer air.

13.1.5   Convection and Cloud Available Potential Energies

Not all clouds are positively buoyant. The buoyancy of a rising air parcel can be written as: $g \cdot (\Delta \theta_v / \theta_v)$, or approximately as $g \cdot (\Delta s_v / s_v)$, where $\Delta \theta_v = \theta_{v \, cloud} - \theta_{v \, environ}$.

and $\Delta s_v = s_{v\,cloud} - s_{v\,environ}$. Fig 13.3 shows four key levels of a convective cloud:
- cloud base (near the LCL)
- level of free convection (LFC)
- limit of convection (LOC)
- cloud top

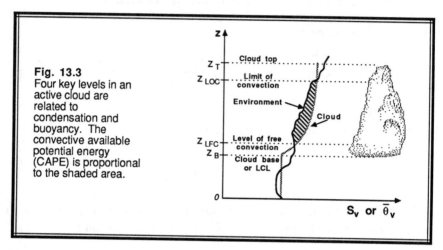

Fig. 13.3
Four key levels in an active cloud are related to condensation and buoyancy. The convective available potential energy (CAPE) is proportional to the shaded area.

Although condensation starts to occur in the parcel at *cloud base*, this condensation often forms in the negatively buoyant overshoot of a ML thermal into its capping inversion. Latent heat is released as the parcel rises. If the parcel has sufficient inertia to overshoot high enough, its potential temperature might rise to the point where it is again warmer than the environment. This point, where the cloudy parcel is first positively buoyant, is called the *level of free convection* (LFC). A cloud that reaches its LFC is classified as an *active cloud*.

The cloud parcel continues to rise due to its own buoyancy while $\Delta\theta_v$ or Δs_v is positive. The cloud temperature can be estimated by following a moist adiabat. Eventually, the rising parcel reaches a level where it is cooler than the environment. This point is called the *limit of convection*. The cloud parcel might overshoot beyond the limit of convection because of its inertia, eventually stopping its overshoot at *cloud top*.

A variety of available potential energies can be defined from cloud properties. The vertical integral of the parcel buoyancy between the LFC and LOC defines the *convective available potential energy* (CAPE):

$$\text{CAPE} = \int_{z=z_{LFC}}^{z_{LOC}} \frac{g}{\theta}\,\Delta\theta_v(z)\,dz \tag{13.1.5a}$$

and is indicated as the shaded area in Fig 13.3. A velocity scale, w_{CAPE}, can be defined by assuming that all the potential energy is converted into kinetic energy (Rogers, et al., 1985):

$$W_{CAPE} = (2 \cdot CAPE)^{1/2} \tag{13.1.5b}$$

This scale can be used to model the overshoot of the cloud top above the LOC.

The *evaporative available potential energy* (EAPE) can be defined similarly as (13.1.5a), except for the negative buoyancy associated with the evaporation of liquid water within a sinking cloudy parcel (Betts, 1985). This can be used to study cloud-top entrainment.

13.1.6 Example

Problem: Assume that the following measurements of (p, s_L, r_T) were obtained in a cloud and the surrounding environment, respectively: (80 kPa, 304.1 kJ/kg, 16.5 g/kg)$_{cloud}$ and (80 kPa, 307.0 kJ/kg, 10 g/kg)$_{env}$. Is the cloudy parcel positively or negatively buoyant?

Solution: To determine buoyancy, we must calculate s_v for both the cloud and environment. First, we can use the equations in Table 13-1 to write the expression for s_v:

$$s_v = C_p T \cdot [1 + 0.61r - r_L] + gz \tag{13.1.6a}$$

Next, we can use the *hypsometric equation* to find gz:

$$gz = \Re \overline{T_v} \ln(p_o/p)$$
$$= \Re \overline{T} \ln(p_o/p) \cdot [1 + 0.61r - r_L] \tag{13.1.6b}$$

where the overbar represents an average over the layer between p_o and p. Since we have no information about the temperature between those two levels, we will use the local value of T and r_T at p for simplicity. We will assume a reference level of $z = 0$ at $p_o = 100$ kPa.

Environment: In the unsaturated environment, $r_L = 0$ and $r = r_T$. Thus:

$$s_L = s = C_p T + gz \tag{13.1.6c}$$

Combining this with the previous equation gives:

$$gz = \frac{s_L}{1 + \left[\dfrac{\Re}{C_p} \cdot \ln\left(\dfrac{p_o}{p}\right) \cdot (1 + 0.61\, r_T)\right]^{-1}} \tag{13.1.6d}$$

Using $\Re/C_p = 0.286$, $p_o/p = 100/80$, and $r_T = 0.01$ g/g, the term in square brackets becomes 0.0642. This yields $g\,z = 18{,}523$ m^{2}s^{-2}.

Combining (13.1.6a) and (13.1.6c) yields an expression for the environmental virtual static energy in terms of the knowns: s_L, r_T, and $g\,z$:

$$s_v = s_L + 0.61\, r_T\, (s_L - gz)$$

$$= 307000 + (0.61){\cdot}(0.01){\cdot}(307000 - 18523)$$

$$= 308760 \text{ J/kg}$$

$$= 308.8 \text{ kJ/kg}$$

Cloud: In a saturated environment: $r_T = r_{sat} + r_L$. Thus, the liquid water static energy becomes:

$$s_L = C_p T + g\,z - L_v r_T + L_v r_{sat} \qquad (13.1.6e)$$

If we assume for simplicity that $gz_{cloud} = gz_{environ.}$, then all of the terms in the above equation are known except T and r_{sat}. But since r_{sat} is a function of T (and p) via Teten's formula, we can iteratively solve for the temperature that satisfies (13.1.6e). The solution is T = 289.25 K and r_{sat} = 14.54 g/kg.

The virtual static energy in the cloud is thus:

$$s_v = C_p T \cdot [\,1 + 1.61\, r_{sat} - r_T\,] + g\,z$$

$$= (1004){\cdot}(289.25){\cdot}[\,1 + (1.61){\cdot}(0.01454) - 0.0165\,] + 18523$$

$$= 310937 \text{ J/kg}$$

$$= 310.9 \text{ kJ/kg}$$

Buoyancy: The cloud is more buoyant than the environment, because $s_{v\ cloud}$ > $s_{v\ environ}$.

Discussion: We could have made the problem much simpler by defining our coordinate system such that z = 0 at p = 80 kPa. Since both the cloudy and environmental air parcels are at the same height for this problem, the gz terms would have disappeared. Although this approach is acceptable for determining relative buoyancy, the magnitudes of the resulting virtual static energies would differ from those usually obtained with a reference height of 100 kPa.

Although it was stated earlier that any two of the conserved variables plus the pressure or height are sufficient to solve for any other conserved variable, we see in this example that the solution can be far from trivial. The problem lies in the nonlinear dependency of saturation mixing ratio on temperature. Computerized solutions can be designed to iterate the equations to solve for temperature. Alternately, linearized simplifications can be used (Betts, 1982) to give an approximate solution.

13.2 Radiation

The physics relevant for radiation within and between clouds requires knowledge of liquid water content, cloud droplet size distribution, cloud temperature, cloud surface shape, cloud cover, solar zenith angle, and many more (Welch and Wielicki, 1984; Schmetz and Beniston, 1986). It is a very complex problem that is beyond the scope of this book. We will focus on only the simplest parameterizations here for continuous stratus/stratocumulus cloud decks. By neclecting radiation from cloud sides and weighting the radiation budget by the cloud cover, the stratocumulus radiation parameterization below can be extended in crude form to scattered cumulus clouds.

Fig 13.4 shows profiles of upward and downward short and longwave radiation for a stratocumulus-topped marine boundary layer during the day (Nicholls, 1984). Solar radiation is absorbed over a depth on the order of hundreds of meters in the top of this cloud, while infrared radiative divergence occurs within about 30 m of the cloud top and bottom. These are discussed in more detail next.

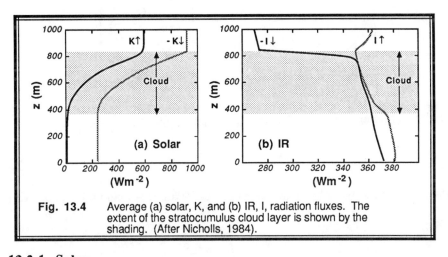

Fig. 13.4   Average (a) solar, K, and (b) IR, I, radiation fluxes. The extent of the stratocumulus cloud layer is shown by the shading. (After Nicholls, 1984).

13.2.1 Solar

Fig 13.5 shows the vertical profile of net solar radiation ($K^* = K\!\downarrow + K\!\uparrow$) for an idealized 500 m thick cloud within a 1000 m thick BL (Hanson, 1987). The exponential decrease in net radiation with depth below cloud top causes warming within the top 100 to 200 m of the cloud. Lower in the cloud there is less absorption and less heating. There is no noticeable solar flux divergence at cloud base. The cloud shades the ground, reducing

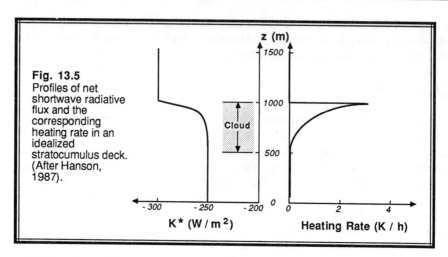

Fig. 13.5 Profiles of net shortwave radiative flux and the corresponding heating rate in an idealized stratocumulus deck. (After Hanson, 1987).

the surface heating and the associated convection during the daylight hours.

Hanson and Derr (1987) have proposed the following simplified parameterization for solar radiation absorption. It is assumed that the heights of cloud top and base are known (z_T and z_B, respectively).

$$K^*(z) = K_T^* - (K_T^* - K_B^*) \cdot \frac{1 - \exp\left[\dfrac{-(z_T - z)}{\lambda_{sol}}\right]}{1 - \exp\left[\dfrac{-(z_T - z_B)}{\lambda_{sol}}\right]}$$

(13.2.1a)

where the e-folding solar decay length, λ_{sol}, is approximated by:

$$\lambda_{sol} \cong 15 \cdot W_p^{0.335}$$

(13.2.1b)

The decay length is typically in the range of 50 to 150 m. The liquid water path, W_p, is:

$$W_p \equiv \int_{z_B}^{z_T} \rho_{air} \, r_L \, dz$$

$$\cong \frac{(z_T - z_B)^2}{880}$$

(13.2.1c)

where the last approximation gives W_p in units of g/m^2, for cloud heights in units of m. The liquid water mixing ratio is assumed to increase linearly with height from cloud base to cloud top, associated with moist adiabatic ascent. This latter assumption appears realistic for the midlatitude marine stratocumulus studied by Nicholls and Leighton (1986), but less realistic for the arctic stratus clouds studied by Curry (1986). Typical liquid water paths vary between about 10 and 1000 g/m^2. Fravalo, et al. (1981) find that the flux profiles are particularly sensitive to the liquid water content distribution.

The bulk cloud albedo, a_c, and absorption, b_c, are used to find the net shortwave fluxes at the top and bottom of the cloud (K_T^* and K_B^*, respectively, positive for upward flux):

$$K_T^* = (1 - a_c) K{\downarrow}$$

$$K_B^* = (1 - a_c - b_c) K{\downarrow} \qquad (13.2.1d)$$

The second equation above describes the transmission of downward solar radiation through the cloud. The downward solar flux at the top of the cloud deck, $K{\downarrow}$, is always negative, and can be found from (7.3.1b). The albedo and absorption are a function of the liquid water path and the solar zenith angle, as shown in Fig 13.6 (Stephens, 1978). Stephens also presents a parameterization for these curves. One must be careful using either the figures or the parameterization to ensure that the sum of the albedo and absorption do not exceed one (a problem for large liquid water paths and small zenith angles).

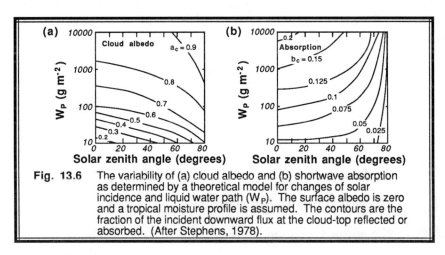

Fig. 13.6   The variability of (a) cloud albedo and (b) shortwave absorption as determined by a theoretical model for changes of solar incidence and liquid water path (W_P). The surface albedo is zero and a tropical moisture profile is assumed. The contours are the fraction of the incident downward flux at the cloud-top reflected or absorbed. (After Stephens, 1978).

13.2.2 Infrared

Fig 13.7 shows an example of the longwave radiative flux profile through a stratocumulus deck, and the associated heating rate. Water droplets are such efficient absorbers/emitters of IR radiation that the cloud essentially behaves as a black body. The changes in radiative flux associated with the cloud occur with an e-folding length of only about 30 m, causing a sharp cooling spike at cloud top and heating spike at cloud base.

Although there is some debate as to exactly where these heating and cooling spikes actually occur, we will assume that they occur just within the cloud boundaries.

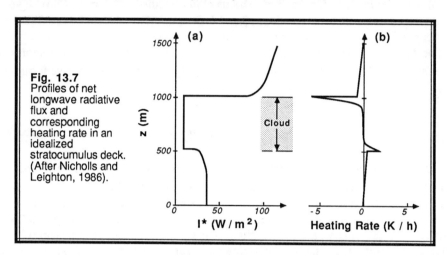

Fig. 13.7 Profiles of net longwave radiative flux and corresponding heating rate in an idealized stratocumulus deck. (After Nicholls and Leighton, 1986).

If fine resolution modeling of the flux divergence is required, a parameterization such as proposed by Hanson and Derr (1987) and Hanson (1987) could be used. Otherwise, a simpler approach allows all of the flux divergence to occur across a small fixed distance such as 30 m. Rogers, et al. (1985a) modeled the change in net longwave flux across the top (ΔI^*_T) and bottom (ΔI^*_B) of the cloud as:

$$\Delta I^*_T = \frac{\sigma_{SB}}{\rho\, C_p} \left(T^4_{cloud\ top} - T^4_{sky} \right) \tag{13.2.2a}$$

$$\Delta I^*_B = \frac{\sigma_{SB}}{\rho\, C_p} \left(T^4_{surface} - T^4_{cloud\ base} \right) \tag{13.2.2b}$$

where ($I^* = I{\downarrow} + I{\uparrow}$) is the net longwave radiation, positive upward.

13.3   Cloud Entrainment Mechanisms

13.3.1   Cloud-top Entrainment Instability in Stratocumulus

Lilly (1968) first suggested that the warm air entrained into the top of a stratocumulus cloud might cool and sink if it were initially dry enough to support considerable evaporative cooling of the neighboring cloud droplets. His criterion for this unstable entrainment state was $\Delta_{EZ}\theta_e < 0$, for $\Delta_{EZ}\theta_e = \theta_{e\ just\ above\ cloud\ top} - \theta_{e\ just\ below\ cloud\ top}$. Randall (1980) and Deardorff (1980) recognized that the virtual potential temperature was a better measure of buoyancy and instability, and suggested a minor modification to Lilly's criterion:

Entrainment is unstable if: $\quad \Delta_{EZ}\theta_e < \Delta_{EZ}\theta_e \text{ critical}$ (13.3.1a)

where the critical equivalent potential temperature jump across cloud top is (Rogers, et al., 1985a):

$$\Delta_{EZ}\theta_{e \text{ critical}} = \frac{\theta \cdot \Delta_{EZ}r_T}{\alpha_4}$$ (13.3.1b)

The factor α_4 is:

$$\alpha_4 = \frac{1 + 0.609\, r_{sat} + 1.609\left(T\dfrac{dr_{sat}}{dT}\right)}{1 + \dfrac{L_v}{C_p T}\left(T\dfrac{dr_{sat}}{dT}\right)}$$ (13.3.1c)

Negatively buoyant downdrafts formed from the entrained air produce additional TKE that can enhance mixing and entrainment. The newly entrained air can then also become unstable and sink, resulting in even more TKE and more entrainment. This positive feedback process can cause a cloud to entrain large amounts of dry air, resulting in the rapid breakup and evaporation of the cloud.

A number of investigators have suggested that (13.3.1a) is a necessary but not sufficient condition for unstable entrainment (Hanson, 1984). Rogers and Telford (1986) learned that some additional trigger or initial disturbance is required to start the whole process. Albrecht, et al. (1985) pointed out that even if the criteria above are satisfied, the cloud might not break up if the mixture between the entrained and cloudy air is unsaturated.

13.3.2 Entrainment into Cumulus

Although lateral entrainment might be the dominant mixing mechanism for unsaturated thermals, it appears to play a minor role for many of the cumulus clouds studied. Entrainment through the top of the cloud is more important. In this process, entrained air at cloud top partially mixes with some of the cloudy air, and the mixture cools because of the evaporation of some of the cloud droplets, just like stratocumulus cloud-top entrainment instability. The result is a negatively buoyant downdraft (see Fig 13.8) that sinks through the cloud and mixes with other cloudy air along the way (Pontikis, et al., 1987).

Although Squires (1958) had originally proposed this mechanism, it wasn't until the mixing analysis by Paluch (1979) that the *cloud top entrainment* mechanism was confirmed and better understood. Her analysis, modified here using the techniques of Betts (1985), uses conserved variables to diagnose the origins of air measured at various levels within a cloud.

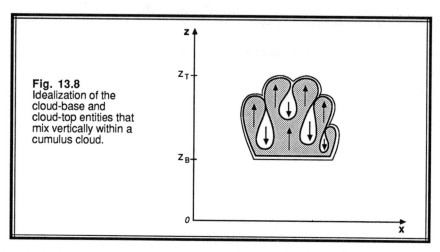

Fig. 13.8
Idealization of the cloud-base and cloud-top entities that mix vertically within a cumulus cloud.

Fig 13.9 shows an idealized conserved variable plot of an environmental sounding. Measurements within clouds, plotted as the data points in Fig 13.9, lie on, or close to, a straight line connecting a point on the environmental sounding corresponding to cloud-base or subcloud layer air, and another point on the environmental sounding at the height of cloud top.   The fact that most of the measurements fall on the same straight *mixing line* indicates that air originating from cloud base and cloud top are mixed together in various proportions and left within the cloud.

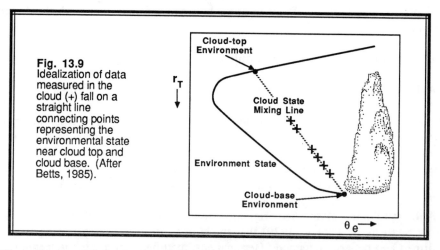

Fig. 13.9
Idealization of data measured in the cloud (+) fall on a straight line connecting points representing the environmental state near cloud top and cloud base. (After Betts, 1985).

If environmental air from other levels between cloud base and cloud top were entrained into the cloud, then data points would lie between the environmental line and the mixing line drawn in Fig 13.9. Since few such data points are usually observed, we must conclude that most of the air within the cloud came from cloud base and cloud top (Blyth and Latham, 1985; Jensen, et al., 1985). A further indication of the importance of cloud-top entrainment is that the spread of cloud droplet sizes is smaller than would be observed if lateral entrainment was a more prominent process. (Raymond and Blyth, 1986; Kitchen and Caughey, 1981).

Many of Telford's observations show a bimodal or multimodal droplet size distribution that further supports the cloud top entrainment mechanism (Rogers, et al., 1985b). This prompted him to propose an *entity-type entrainment mixing* model (Telford and Chai, 1980; Telford, et al., 1984), where there are various different but individually well-mixed entities or blobs that move to their respective levels of neutral buoyancy, and which together form the cloud as sketched in Fig 13.8. Raymond and Blyth (1986) extended this approach to allow a variety of mixing proportions between cloud top and cloud base air.

Betts (1986) and Betts and Miller (1986) have used mixing line results to develop a convective adjustment parameterization for use in large-scale models. Betts and Albrecht (1987) further noted that evaporation of falling precipitation into parts of the cloud can cause penetrative downdrafts that result in kinks in the mixing line.

In addition to thermodynamic mixing-line analyses of mixing, there have been a variety of direct observations of eddy motion fields. Kitchen and Caughey's (1981) tethered-balloon observations showed a circulation pattern shaped like a "?" or a backwards "P" (see Fig 13.10). Eymard (1984) observed a greater vertical velocity variance immediately under clouds than at the same altitude between clouds. Brümmer and Wendel (1987) confirmed that there are often downdrafts along and just outside of the lateral edges of cumulus clouds. Stith, et al. (1986) found that the updrafts at middle levels in the cloud have diameters smaller than the cloud, but that near cloud top the updraft is more diffuse and wider, implying significant mixing at cloud top.

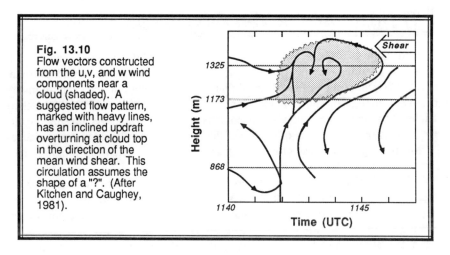

Fig. 13.10
Flow vectors constructed from the u,v, and w wind components near a cloud (shaded). A suggested flow pattern, marked with heavy lines, has an inclined updraft overturning at cloud top in the direction of the mean wind shear. This circulation assumes the shape of a "?". (After Kitchen and Caughey, 1981).

13.4 Fair-weather Cumulus

13.4.1 Cloud Classification

Fair-weather cumulus clouds can be divided into three classes based on their dynamics (see Fig 13.11): *forced*, *active*, and *passive* (Stull, 1985). This classification is different than the morphological categorization based on cloud shape and appearance. The total cumulus cloud cover (for low-altitude clouds, σ_{C_L}) consists of the sum of the covers of forced (σ_{C_F}), active (σ_{C_A}), and passive (σ_{C_P}) cloud covers:

$$\sigma_{C_L} = \sigma_{C_F} + \sigma_{C_A} + \sigma_{C_P} \tag{13.4.1}$$

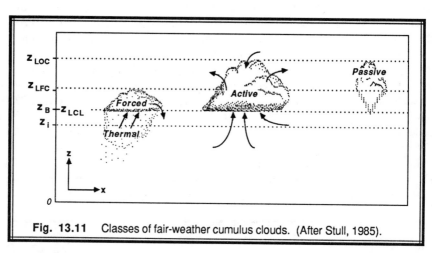

Fig. 13.11   Classes of fair-weather cumulus clouds. (After Stull, 1985).

Forced Clouds. These clouds form in the tops of ML thermals, and exist only while there is continued forcing from the parent thermal. Often, these clouds form in the negatively buoyant portion of the thermal that is overshooting into the capping inversion (entrainment zone). In spite of the latent heat release during condensation, there is insufficient heating for these clouds to become positively buoyant. As a result, the clouds behave as quasi-passive tracers of the top of the thermal. The cloud top never reaches its LFC. Morphologically, these clouds are very shallow and often flat looking, and are usually classified as cumulus humilis.

All of the air rising in the thermal up through the cloud base continues circulating through the cloud and remains within the ML (i.e., there is *no venting* of ML air out of the ML). In conditions of light wind shear, air in the cloud diverges from the center toward the lateral edges, where descending return flow into the ML is associated with droplet evaporation. In stronger wind shear, the cloud often appears as a breaking wave, with updrafts on the upshear side, and the return circulation and downdrafts on the downshear side.

Active Clouds. These clouds are also triggered by ML thermals, but at some point a portion of the updraft reaches its LFC and the clouds become positively buoyant. The rising updraft then induces its own pressure perturbations that affect its evolution and draw more air in through its cloud base. The lifetime of this cloud is now controlled by its cloud dynamics and its interaction with the environment. It may persist longer than the ML thermal that first triggered it.

These clouds can *vent* ML air out into the FA. Their vertical dimensions are often on the same order, or slightly larger than their horizontal dimensions. Morphologically, they are the cumulus mediocris.

Passive Clouds. When active clouds cease withdrawing air from the ML, we classify them as dynamically passive. The tops of the passive clouds might still be positively buoyant and may even be growing, but they no longer are venting ML air. The bottoms of these clouds are diffuse as the droplets evaporate and mix with the environment. As a result, the original cloud base disappears, leaving the remaining portion of the cloud totally above the ML and EZ where it is not dynamically interactive with the ML.

13.4.2 Feedback from the Clouds to the Mixed Layer

Radiative Feedback. All classes of boundary-layer clouds (σ_{C_L}) shade the surface. Over a land surface this results in negative feedback, because less solar heating of the ground will trigger fewer or weaker thermals and will cause the ML to grow more slowly, resulting in fewer new cumulus clouds being triggered. Thus on days over land where solar heating is the primary driving force for free convection (rather than cold air advection, ground thermal inertia, or forced mechanical convection), fair-weather cumulus clouds will tend to reach an equilibrium cloud cover that is scattered (0.1 to 0.5 coverage) or broken (0.6 to 0.9), but not overcast.

Dynamic Feedback. Active clouds withdraw some of the ML air, causing the ML to grow more slowly or even not grow at all. This negative feedback limits the number of new thermals that can penetrate high enough to trigger new active clouds. The continuity equation (11.2.2b) describes how active clouds can modify ML growth. Given typical values of entrainment velocity, subsidence, cloud base average updraft velocity, and ML growth rate (0.05, -0.01, 1.0, and 0.02 m/s, respectively), (11.2.2b) yields an active cloud cover of 2%. Thus, active clouds rarely cover more than a few percent of the area, even when there are many forced and passive clouds present.

Brümmer and Wendel (1987) observed that these few active clouds have such vigorous vertical motion that their vertical heat and momentum fluxes can be on the order of 100 W/m^2 and 1 N/m^2, respectively. When horizontally averaged over the remainder of the cloud layer, these few clouds were responsible for most of the vertical transport.

Environmental Feedbacks. There are two positive feedbacks that can be easily discussed. Active clouds tend to vent moisture from the ML and deposit it in the FA. Passive clouds tend to evaporate more slowly in a moist FA than in a dry one. Thus, the number of passive and active clouds can increase in the FA with time as new clouds develop, but older passive and active clouds linger because of slower evaporation rates.

Nonprecipitating active clouds also vent some of the cooler ML air (in a potential temperature sense) into the cloud layer. Between clouds, subsidence brings warmer air down to the ML top, where the ML can entrain it. As a result, the ML warms and the FA cools, resulting in a destabilization of the whole ML/FA system and the increase in cloud cover.

13.4.3   Cumulus Onset Time and Cloud Cover

There is a thermal-to-thermal variability in temperature and moisture associated with the variability in land use over which thermals form. Thus, thermals penetrate to a variety of heights, and have a variety of LCLs. The range of penetration heights has already been defined as the entrainment zone. The range of LCLs defines the *LCL zone* (Wilde, et al., 1985).

The first cumulus clouds of the day will form when the top of the entrainment zone reaches the bottom of the LCL zone, because at that time some of the thermals are reaching their LCLs. As more of the entrainment zone overlaps and moves above the LCL zone, the cloud cover increases (Fig 13.12). If the entrainment zone were to move completely above the LCL zone, then the sky would be overcast (> 95% cloud cover) because all of the thermals are above their respective LCLs, leaving only the unmixed entrained air as cloud-free breaks in the overcast.

The probability density function (pdf) of finding the top of a thermal at any height within the entrainment zone is well described by a double exponential function, such as is plotted in Fig 13.13. The peak in the function occurs at the mean ML height, z_i, and the width of the function covers the thickness of the entrainment zone. This function indicates

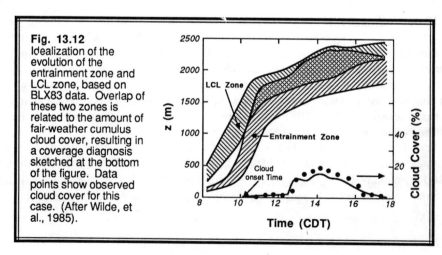

Fig. 13.12 Idealization of the evolution of the entrainment zone and LCL zone, based on BLX83 data. Overlap of these two zones is related to the amount of fair-weather cumulus cloud cover, resulting in a coverage diagnosis sketched at the bottom of the figure. Data points show observed cloud cover for this case. (After Wilde, et al., 1985).

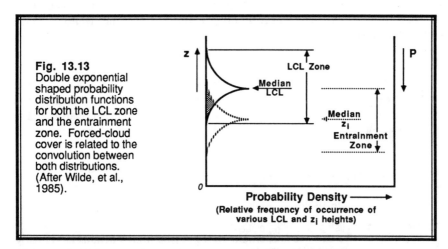

Fig. 13.13
Double exponential shaped probability distribution functions for both the LCL zone and the entrainment zone. Forced-cloud cover is related to the convolution between both distributions. (After Wilde, et al., 1985).

that the tops of thermals are more likely to be found near z_i, and less likely further away. A similar double exponential pdf describes the distribution of LCLs.

By integrating over both pdfs for the entrainment zone and the LCL zone, we can determine the cloud cover of forced clouds from the total probability of finding thermals above their LCLs (Wilde, et al., 1985):

$$\sigma_{C_F} = \int_{z_i = 0}^{\infty} pdf(z_i) \int_{z_{LCL} = 0}^{z_i} pdf(z_{LCL}) \ dz_{LCL} \ dz_i \qquad (13.3.3)$$

To forecast coverage of active and passive clouds, one must develop prognostic equations such as those by Rogers, et al. (1985a).

13.4.4   Cloud Size Distribution

Lopez (1977) demonstrated that many cloud attributes, such as diameter and depth, are distributed lognormally. Examples of observed cloud size distributions are shown in Fig 13.14, along with the best fit lognormal curve (Stull, 1984). We see that the diameters and depths of most clouds cluster together, but there is a small percentage of clouds that are much larger.

The expression for a lognormal distribution is given by:

$$pdf(X) = \frac{1}{(2\pi)^{1/2} \ X \ S_X} \exp\left[-0.5 \left(\frac{\ln(X/L_X)}{S_X} \right)^2 \right] \qquad (13.4.4)$$

where L_X and S_X are the location and shape parameters, and X represents diameter or depth. Integrating (13.3.4) over all X yields 1.0, which means that 100% of the clouds present are explained.

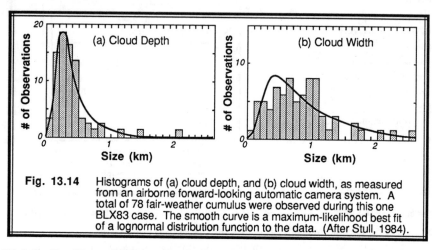

Fig. 13.14   Histograms of (a) cloud depth, and (b) cloud width, as measured from an airborne forward-looking automatic camera system. A total of 78 fair-weather cumulus were observed during this one BLX83 case. The smooth curve is a maximum-likelihood best fit of a lognormal distribution function to the data. (After Stull, 1984).

13.4.5   Profiles of Mean Variables and Fluxes

Examples of soundings in a variety of cloudy boundary layers over oceans and continents for different thermodynamic variables are shown in Figs 13.15 to 13.17. Bougeault (1982) produced idealized profiles of liquid water potential temperature and total water mixing ratio (Fig 13.15) using data from the Puerto Rico, Voves, and GATE field experiments. The Puerto Rican data was obtained over the western tropical Atlantic (Pennell and LeMone, 1974), where there was a strong capping *trade-wind inversion* at 1.5 km, strong (15 m/s) BL winds, and vigorous turbulence generation (primarily by shear in the lower third of the BL). The suppressed cumulus clouds in this region are called *trade-wind cumuli*.

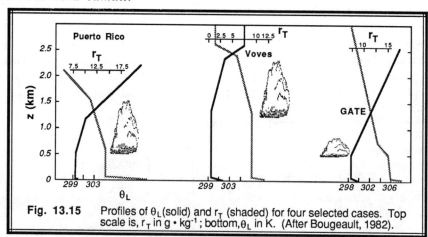

Fig. 13.15   Profiles of θ_L (solid) and r_T (shaded) for four selected cases. Top scale is, r_T in g • kg^{-1}; bottom, θ_L in K. (After Bougeault, 1982).

The GATE data was obtained over the eastern tropical Atlantic (Nicholls and LeMone, 1980), where light winds (3 to 5 m/s) and cooler sea surface temperatures resulted in weaker mixing of the BL. The trade-wind inversion was at lower altitudes in the GATE area (e.g., at about 750 m), and longwave radiative cooling of the BL approximately balanced convective heating. Both the Puerto Rican and GATE data exhibited mesoscale variability in cloudiness, and both had suppressed, nonprecipitating cumulus clouds. The Voves data was taken over central France during the summer, where stronger convection and weaker environmental stability resulted in a deeper boundary layer with taller cumulus clouds.

Betts and Albrecht (1987) show a selected set of soundings from over the equatorial Pacific Ocean obtained during the 1979 FGGE experiment (Fig 13.16). This diagram represents an average of 84 individual soundings, where only those soundings having active clouds capped by an inversion below 60 kPa (600 mb) are included. The vertical line rising from the θ_e curve represents the adiabatic rise of a parcel starting from 98 kPa (980 mb), and indicates positive buoyancy (where it is warmer than the θ_{es} curve) between about 95 and 85 kPa (950 and 850 mb). The capping inversion is enhanced by subsidence, but is also modified by radiative cooling.

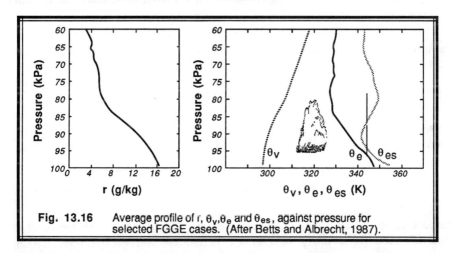

Fig. 13.16   Average profile of r, θ_v, θ_e and θ_{es}, against pressure for selected FGGE cases. (After Betts and Albrecht, 1987).

Finally, Raymond and Blyth (1986) show potential temperature, equivalent potential temperature, and parcel buoyancy ($g \cdot \Delta\theta_v / \theta_v$) for nonprecipitating towering cumulus clouds observed over the Magdalena Mountains of New Mexico (Fig 13.17). Weak cold frontal passage earlier in the day reduced the instability of the lower atmosphere, preventing thunderstorm development.

Examples of fluxes and variances averaged horizontally are shown in Fig 13.18, based on the analysis of Nicholls, et al. (1982). We find significantly large variances, particularly of temperature, within the cloud layer. Nicholls and LeMone (1980) found that clouds affect the distribution of heat and moisture within the subcloud layer, but had

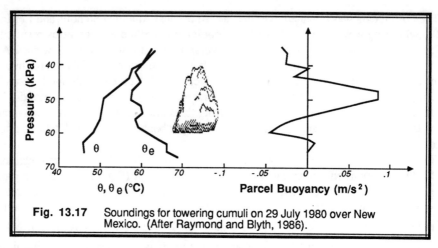

Fig. 13.17   Soundings for towering cumuli on 29 July 1980 over New Mexico. (After Raymond and Blyth, 1986).

little effect on the buoyancy profile. The greatest coupling between the cloud and subcloud layers was via large wavelengths rather than small. Soong and Ogura (1980) determined that cumulus clouds tend to respond rapidly to changes in large-scale forcings, and can be assumed to be in a state of quasi-equilibrium.

Betts (1973) suggests that vertical fluxes associated with cumulus convection can be parameterized using the concept of a *convective mass flux*, ω^*, where $\omega^* = \sigma_{C_A} \cdot w_{up}$, and where w_{up} is the average vertical velocity through the cloud bases of all active clouds. The fluxes of any conserved variable, ξ, can be written as:

$$\overline{w'\xi'} = \frac{\omega^* (\xi_{up} - \bar{\xi})}{1 - \sigma_{C_A}} \tag{13.4.5a}$$

$$\overline{w'\xi'} = \omega^* (\xi_{up} - \xi_{down}) \tag{13.4.5b}$$

where subscripts up and down represent the updraft and downdraft portions of the cloud circulation, and the overbar represents a horizontal average over both cloud and environmental air. Penc and Albrecht (1987) find that this convective mass flux approach also works fairly well for stratocumulus clouds, where they estimate $\omega^* = 0.1$ m/s near the base of the cloud, decreasing slightly with height for the cases they examined.

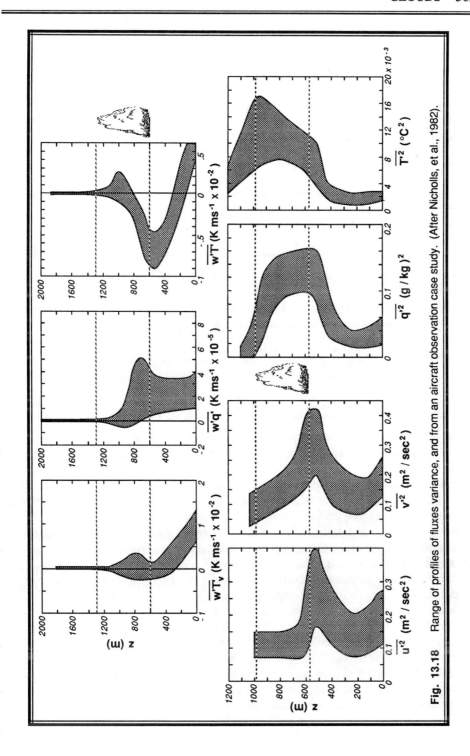

Fig. 13.18   Range of profiles of fluxes variance, and from an aircraft observation case study.   (After Nicholls, et al., 1982).

13.5 Stratocumulus

13.5.1   Vertical Profiles of Mean Variables and Fluxes

Fig 13.19 shows an idealized composite of vertical profiles of mean variables through a stratocumulus topped mixed layer, in which the cloud and subcloud layers are fully turbulently coupled (Turton and Nicholls, 1987; Rogers and Telford, 1986; Albrecht, et al., 1985; Roach, et al., 1982; Caughey, et al., 1982; Slingo, et al., 1982; Brost, et al., 1982a; and Stage and Businger, 1981a).

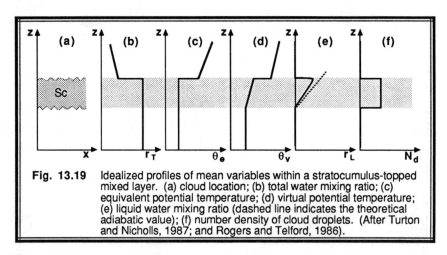

Fig. 13.19   Idealized profiles of mean variables within a stratocumulus-topped mixed layer. (a) cloud location; (b) total water mixing ratio; (c) equivalent potential temperature; (d) virtual potential temperature; (e) liquid water mixing ratio (dashed line indicates the theoretical adiabatic value); (f) number density of cloud droplets. (After Turton and Nicholls, 1987; and Rogers and Telford, 1986).

The equivalent potential temperature and total water mixing ratio are constant with height, supporting the view that the stratocumulus clouds are imbedded within, and are an integral part of, the ML. The virtual potential temperature follows the moist adiabat within the cloud layer, and the liquid water content increases almost linearly with height above cloud base as would be expected if there were no precipitation. Variations in liquid water content (or mixing ratio) are expected due to entrainment of drier air from above. The top of the cloud often has significantly less liquid water than expected, for the same reason. The number density of drops is roughly constant, however, suggesting that the droplets are larger near the top of the cloud.

Turbulent flux responses to various imposed forcings are shown in Fig 13.20 (a-e), where the solid lines indicate the response for a fully coupled cloud and subcloud BL, and the dashed lines show the response for a cloud layer just recently decoupled from the subcloud layer at cloud base. Figs 13.20 a & b show the individual effects of surface heating and entrainment, while Figs 13.20 c-e show individual responses to both IR and solar radiation. These individual curves are analogous to the process-partitioning results for TKE production and consumption. For the entrainment curve, it is assumed for the decoupled case that the cloud and subcloud layers are independently turbulent, with entrainment in both directions across the interface at cloud base.

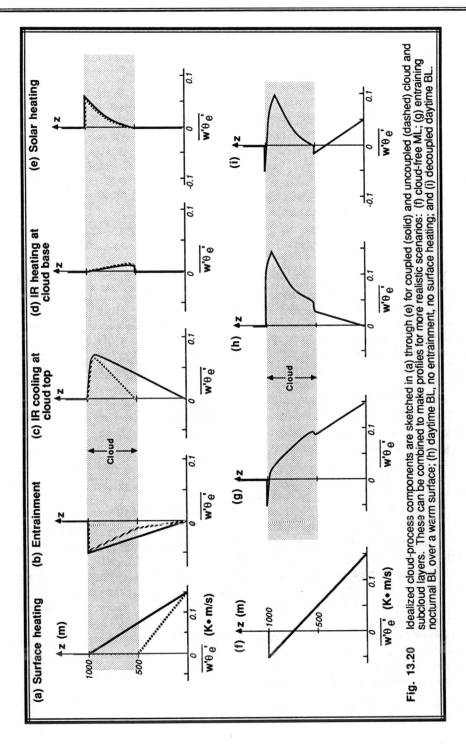

Fig. 13.20   Idealized cloud-process components are sketched in (a) through (e) for coupled (solid) and uncoupled (dashed) cloud and subcloud layers. These can be combined to make profiles for more realistic scenarios: (f) cloud-free ML; (g) entraining nocturnal BL over a warm surface; (h) daytime BL, no entrainment, no surface heating; and (i) decoupled daytime BL.

These five figures can be used as building blocks to construct the turbulence response for more complex cases. For example, adding (a) and (b) yields the usual linear ML heat flux profile (Fig 13.20f). Adding (a) thru (d) gives a profile for an entraining nocturnal stratocumulus-topped BL over a warm surface (Fig 13.20g) that looks similar to the data from Stage and Businger (1981). Adding only the radiative contributions yields the response shown in Fig 13.20h, which is similar to the transilient simulations of Stull and Driedonks (1987). Finally, the sum of all components (a-e) for the decoupled case yields a curve (Fig 13.20i) resembling the data of Turton and Nicholls (1987). Similar constructions can be made to simulate the data of Nicholls and Leighton (1986), Albrecht, et al. (1985), Brost, et al. (1982b), and Deardorff (1980). For all of these cases there are noticable changes in many of the turbulent fluxes near cloud top and cloud base, corresponding to changes in the radiative forcing.

13.5.2 Mixing Processes

There are seven processes that can generate the mixing required to maintain stratocumulus clouds: surface-based (free) convection, differential cold-air advection, surface-layer shear (forced convection), cloud-top radiative cooling, cloud-base radiative heating, cloud-layer shear, and cloud-top entrainment instability. All of these are processes that can be maintained in the absence of strong solar heating of the ground, because we are assuming a broken to overcast stratocumulus deck that shades the ground and attenuates the incoming solar radiation. Also, many of the processes can operate simultaneously.

Surface-based free convection occurs when cold air is advecting over a warmer surface, such as immediately following a cold front passage over land, or cold air masses moving over warmer lakes or ocean currents. The first situation is usually temporary, and disappears as either drier air advects in, or as the heat stored in the ground is lost. The second situation is sometimes seasonal, and is associated with semi-stationary fog and stratocumulus features such as the stratocumulus clouds off of southern California, or the snow squalls downwind of the Great Lakes.

Differential cold-air advection. Sometimes after a cold frontal passage, colder air is advected into an area aloft than is advected in at lower levels. The result is a static destabilization of the layer of air, causing convection and cloud formation. Such destabilization can occur even when the surface is colder than the overlying air.

Shear-generated mechanical turbulence associated with strong BL winds can also cause sufficient mixing (forced convection) to maintain the stratocumulus topped ML. Strong winds and the associated surface-layer shear are also found near fronts and low pressure centers.

Cloud-top radiative cooling creates upside-down "thermals" of cold air that sink from cloud top. This is another form of free convection, and can be important when there are not other higher cloud decks to reduce the infrared cooling from the lowest deck. This process can operate both day and night.

Cloud-base radiative heating is usually much weaker than cloud top cooling, because of the small absolute temperature difference between the cloud-base and the surface. This process differs from the preceeding ones because it destabilizes the cloud layer, but stabilizes the subcloud layer. If conditions are right, cloud-base heating can contribute to the decoupling of the cloud and subcloud layers.

Cloud-layer shear generates mechanical turbulence and mixing. Usually, the shear is found near cloud top, although it is less frequently found near cloud base. The turbulence generated by cloud-top shear generates very localized mixing and entrainment, and can also contribute to decoupling of the cloud layer from the subcloud layer unless there are other turbulent processes that can mix the excess TKE throughout the whole ML.

Cloud-top entrainment instability was descibed earlier, and is associated with entrainment that leads to free convection of cold descending air parcels. Given the proper ambient conditions at cloud top and the appropriate trigger, this process usually dries and warms the ML, causing the cloud-base height to rise and the cloud deck to break up.

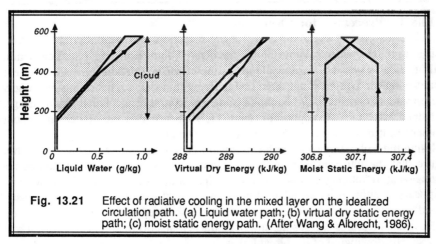

Fig. 13.21   Effect of radiative cooling in the mixed layer on the idealized circulation path. (a) Liquid water path; (b) virtual dry static energy path; (c) moist static energy path. (After Wang & Albrecht, 1986).

Both the Eulerian and process partitioning of the TKE budget have been used to study mixing in stratocumulus clouds. Some of the Eulerian methods are discussed in the next subsection. Wang and Albrecht (1986) used a process approach to model the updraft and downdraft portions of the mixing circulations directly. A sample of their modeled updraft and downdraft paths is shown in Fig 13.21, where they found that the thermodynamic differences between the two paths is much less than the differences between the cloud and its environment.

Moeng (1987) analyzed her large-eddy simulation model by grouping the warm rising and cold sinking air parcels into a process-production component of TKE, and the warm sinking and cold rising parcels into the process-consumption component. She found that consumption is a larger fraction (0.22) of the total production (buoyant and shear) of TKE for cloud-topped MLs than for cloud-free MLs, where the fraction was only 0.15 (see Fig 13.22). Moeng also confirmed that most (85%) of the longwave radiative cooling occurs near the top edge of the cloud, in the entrainment zone portion of the BL.

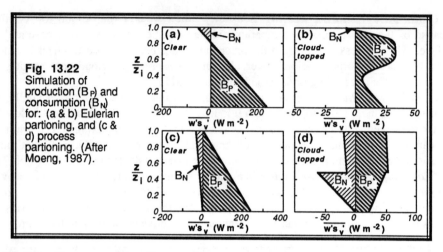

Fig. 13.22 Simulation of production (B_P) and consumption (B_N) for: (a & b) Eulerian partioning, and (c & d) process partioning. (After Moeng, 1987).

13.5.3   Processes for Decoupling

During the day when solar heating of the cloud is significant, if the surface heat flux is small, then it is possible for turbulence in the cloud layer to become decoupled from turbulence in the bottom of the subcloud layer, with a stable transition layer in between (Turton and Nicholls, 1987; Nicholls and Leighton, 1986). For this situation, TKE production is too weak to support the large TKE consumption associated with downward mixing of heated cloud-layer air into the subcloud layer.

Using the Eulerian definition for production and consumption, Turton and Nicholls require that the net consumption below cloud base be less than a fraction of the net production above cloud top, where z_L is the height of the bottom of the cloudy ML:

$$\int_{z_L}^{z_B} \overline{w'\theta_v'}\,dz \;<\; -0.4 \int_{z_B}^{z_T} \overline{w'\theta_v'}\,dz \tag{13.5.3a}$$

It this condition is satisfied for $z_L = 0$, then the cloud and subcloud layers are fully coupled. Otherwise, decoupling is suggested with $z_L > 0$ in order to satisfy (13.5.3a).

Fig 13.23 shows examples of fully coupled and uncoupled layers. A fully coupled mixed layer can become decoupled after sunrise when there is sufficient heating within the cloud layer. The two layers can stay decoupled if there is sufficient cloud heating, or can become recoupled near sunset when the cloud heating is removed (see Fig 13.24).

Other mechanisms can also cause decoupling. If cloud-top entrainment instability entrains sufficient warm air into the top of the cloud layer, then decoupling can occur (Rogers, et al., 1985a). Decoupling is also possible when shear generation of turbulence at the top of the cloud layer is large enough (Wai, 1987).

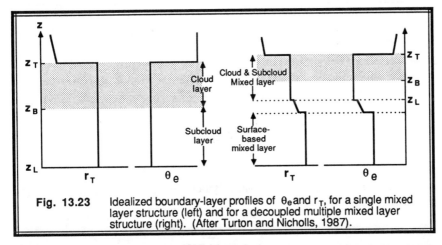

Fig. 13.23 Idealized boundary-layer profiles of θ_e and r_T, for a single mixed layer structure (left) and for a decoupled multiple mixed layer structure (right). (After Turton and Nicholls, 1987).

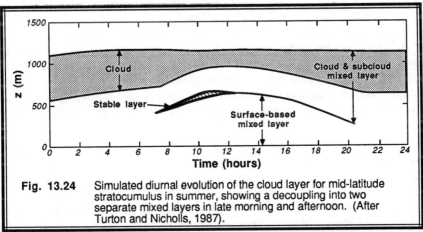

Fig. 13.24 Simulated diurnal evolution of the cloud layer for mid-latitude stratocumulus in summer, showing a decoupling into two separate mixed layers in late morning and afternoon. (After Turton and Nicholls, 1987).

13.5.4 Entrainment

Most of the cloud-top entrainment relationships are written as a balance between the production, dissipation, and consumption of TKE (Stage and Businger, 1981). The production mechanisms include surface heating, cloud-top cooling, cloud-top entrainment instability, and wind shear. Consumption is associated with warm entrained air moving down and cold air moving up. The basic premise, as discussed in Section 11.4.3, is that the portion of produced TKE which is not dissipated or transported away can be expended by consumption during the entrainment process.

Randall (1984) recognized that entrainment and parcel motion is a very nonlocal process that is best described by Lagrangian or nonlocal turbulence closure approaches. Parameterizations of the Lagrangian concept lead to the Eulerian and process partitioning approximations described earlier, but variations in the particular choice of parameterization can lead to estimates of entrainment differing by over a factor of two. It is clear that more work needs to be done to clarify the cloud-top entrainment process in stratocumulus (Hanson, 1984).

Some specialized entrainment models have also been proposed based on a limited set of production mechanisms. For example, Rogers, et al. (1985a) suggested the following entrainment velocity for cloud-top entrainment instability :

$$w_e \cong \left[\frac{g}{\theta} \left| \Delta_{EZ} \theta_e \right| l_c \right]^{1/2} \tag{13.5.4a}$$

where l_c is the distance traveled by negatively buoyant entrained elements. When cloud-top radiative cooling dominates:

$$w_e \cong \frac{\Delta I^*}{\Delta_{EZ} \theta_v} \tag{13.5.4b}$$

where ΔI^* is the net longwave radiative flux divergence near cloud top, expressed in kinematic units (K m/s). Other buoyancy flux production factors have been used in the numerator of the above equation for processes other than radiation.

Finally, many entrainment relationships have been proposed relating a dimensionless entrainment velocity to an inverse Richardson number (Nicholls and Turton, 1986):

$$\frac{w_e}{w_* \text{ or } u_*} \propto Ri^{*-1} \tag{13.5.4c}$$

where the convective Richardson number, Ri^*, can be defined in terms of w_*, u_*, or other turbulence production scales. Deardorff (1980; and Albrecht, et al., 1985) suggest that critical values of the entrainment velocity can be related to stratocumulus breakup and dissipation. Tag and Payne (1987) note that stratocumulus decks can break up slower in stronger wind shears at cloud top, given that the cloud-top entrainment instability criterion is satisfied.

13.6 Fog

The following fog types can form in the boundary layer: radiation, advection, precipitation (frontal), steam, and upslope. Only the first two will be discussed in more detail here.

13.6.1 Radiation Fog

Two types of radiation fog have been observed: (1) a mist that is most dense near the ground but becomes diffuse with height; and (2) a well-mixed fog layer with a sharp top that is similar to a stratocumulus cloud (Welch and Wielicki, 1986; and Gerber, 1981). The following scenario shows how these fogs develop, and how the mist can become a well-mixed fog (Wessels, personal communication).

As the air close to the ground cools further below the dew point temperature than air higher in the SBL, a mist can form that sometimes has greater liquid water content at the bottom than the top (Gerber, 1981). This fog has a diffuse top and is often very shallow, with depths from 1 m to about 5 m (Fig 13.25a). If the sun rises before this fog develops further, then short-wave radiation can penetrate the fog, heat the surface, and evaporate the fog. SBL scaling and dynamics can be used to describe this fog.

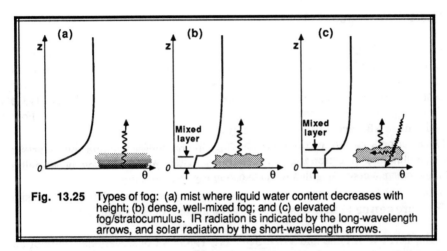

Fig. 13.25 Types of fog: (a) mist where liquid water content decreases with height; (b) dense, well-mixed fog; and (c) elevated fog/stratocumulus. IR radiation is indicated by the long-wavelength arrows, and solar radiation by the short-wavelength arrows.

Instead, if the mist thickens and liquid water content increases during the night as radiative cooling continues, then there is less net radiative heat loss from the ground and more from the dense parts of the fog. Eventually, the fog becomes optically thick enough that there is greater radiative flux divergence near the top of the fog than the bottom. At this point, cooling at fog top generates cold thermals that sink and begin to convectively mix the fog layer. Very quickly, the fog becomes more uniform in the vertical, with a well-defined top edge (Fig 13.25b). This sharp top also concentrates the radiative divergence closer to that region, which reinforces convective mixing in the fog layer.

The well-mixed fog behaves like a stratocumulus cloud, and can consists of cellular circulation patterns or patches with a horizontal scale on the order of 2 to 3 times the fog depth (Welch and Ravichandran, 1985; and Welch and Wielicki, 1986). Both the well-mixed fog and the stratocumulus cloud appear to follow convective ML scaling (using turbulent fluxes at the top of the ML where appropriate) rather than SBL scaling. Many of the parameterizations described earlier in this chapter for stratocumulus clouds can be applied to the well-mixed fog.

Well-mixed fog can persist well into the morning, because much of the solar radiation is reflected from the top, and radiative cooling of the top continues during the day. The combined effects of absorption of a little solar radiation in the interior of the fog and longwave cooling at cloud top can cause the boundary layer to warm and the fog base to *lift* from the ground. This fog is then reclassified as a stratocumulus cloud (Fig 13.25c). Sometimes the patchiness of the fog is enhanced by solar heating of the ground between patches, causing the fog to break up into cumulus clouds. A wind-speed increase can also enhance mixing of the fog with drier, warmer air aloft to trigger fog lifting or break-up.

The onset time of fog, and the transition from a mist to a well-mixed fog, are sensitive to the balance between all processes (Turton and Brown, 1987). Small errors in any one of the processes can lead to large errors in fog forecasts and models. As a result, precise onset and dissipation times for fog are often difficult to forecast.

13.6.2 Advection Fog

Advection of warm humid air over a cooler surface can result in advection fog. Such fogs are frequently found off the west coast of California, and the west coasts of South America. With persistent wind fields and sea surface temperature, it is relatively easy to forecast the onset location and mean properties of this fog. Advection fogs are also found during spring when warm, moist air advects over snow and cold lake surfaces. Forecast methods utilize SBL and TIBL evolution physics.

Once formed, radiative cooling of the fog top can enhance the mixing, creating the stratocumulus-like advection fog with a sharp top. Light winds can also induce sufficient mixing to create a cool ML filled with advection fog.

13.7 References

Albrecht, B.A., R.S. Penc and W.H. Schubert, 1985: An observational study of cloud-topped mixed layers. *J. Atmos. Sci.*, **42**, 800-822.

Anderson, D.L.T., 1976: The low-level jet as a western boundary current. *Mon. Wea. Rev.*, **104**, 907-921.

Betts, A.K., 1973: Non-precipitating cumulus convection and its parameterization. *Quart. J. Roy. Meteor. Soc.*, **99**, 178-196.

Betts, A.K., 1982a: Saturation point analysis of moist convective overturning. *J. Atmos. Sci.*, **39**, 1484-1505.

Betts, A.K., 1982b: Cloud thermodynamic models in saturation point coordinates. *J. Atmos. Sci.*, **39**, 2182-2191.

Betts, A.K., 1983: Thermodynamics of mixed stratocumulus layers: saturation point budgets. *J. Atmos. Sci.*, **40**, 2655-2670.

Betts, A.K., 1985: Mixing line analysis of clouds and cloudy boundary layers. *J. Atmos. Sci.*, **42**, 2751-2763.

Betts, A.K., 1986: A new convective adjustment scheme. I: observations and theoretical basis. *Quart. J. Roy. Meteor. Soc.*, **112**, 677-692.

Betts, A.K. and B.A. Albrecht, 1987: Conserved variable analysis of the convective boundary layer thermodynamic structure over tropical oceans. *J. Atmos. Sci.*, **44**, 83-99.

Betts, A.K. and M.J. Miller, 1986: A new convective adjustment scheme. II: single column test using GATE wave, BOMEX, and arctic air-mass data sets. *Quart. J. Roy. Meteor. Soc.*, **112**, 693-710.

Blyth, A.M. and J. Latham, 1985: An airborne study of vertical structure and microphysical variability within small cumulus. *Quart. J. Roy. Meteor. Soc.*, **111**, 773-792.

Boatman, J.F. and A.H. Auer, Jr., 1983: The role of cloud-top entrainment in cumulus clouds. *J. Atmos. Sci.*, **40**, 1517-1534.

Bolton, D., 1980: The computation of equivalent potential temperature. *Mon. Wea. Rev.*, **108**, 1046-1053.

Bougeault, P., 1982: Cloud-ensemble relations based on the gamma probability distribution for the higher-order models of the planetary boundary layer. *J. Atmos. Sci.*, **39**, 2691-2700.

Bougeault, P., 1985: The diurnal cycle of the marine stratocumulus layer: a higher-order model study. *J. Atmos. Sci.*, **42**, 2826-2843.

Brümmer, B. and M. Wendel, 1987: Observations of intermittent cumulus convection in the boundary layer. *Quart. J. Roy. Meteor. Soc.*, **113**, 19-36.

Buck, A.L., 1981: New equations for computing vapor pressure and enhancement factor. *J. Appl. Meteor.*, **20**, 1527-1532.

Caughey, S.J., B.A. Crease, and W.T. Roach, 1982: A field study of nocturnal stratocumulus: II. turbulence structure and entrainment. *Quart. J. Roy. Meteor. Soc.*, **108**, 125-144.

Curry, J.A., 1986: Interactions among turbulence, radiation and microphysics in arctic stratus clouds. *J. Atmos. Sci.*, **43**, 90-106.

Deardorff, J.W., 1976: On the entrainment rate of a stratocumulus-topped mixed layer. *Quart. J. Roy. Meteor. Soc.*, **102**, 563-582.

Deardorff, J.W., 1980: Cloud-top entrainment instability. *J. Atmos. Sci.*, **37**, 131-147.

Eymard, L., 1984: Radar analysis of a tropical convective boundary layer with shallow cumulus clouds. *J. Atmos. Sci.*, **41**, 1380-1393.

Fravalo, C., Y. Fouquart and R. Rosset, 1981: The sensitivity of a model of low stratiform clouds to radiation. *J. Atmos. Sci.*, **38**, 1049-1062.

Gerber, H.E., 1981: Microstructure of a radiation fog. *J. Atmos. Sci.*, **38**, 454-458.

Hignett, P., 1987: A study of the short-wave radiative properties of marine stratus: aircraft measurements and model comparisons. *Quart. J. Roy. Meteor. Soc.*, **113**, 1011-1024.

Hanson, H.P., 1984: On mixed layer modeling of the stratocumulus-topped marine boundary layer. *J. Atmos. Sci.*, **41**, 1226-1234.

Hanson, H.P., 1987: Radiative/turbulent transfer interactions in layer clouds. *J. Atmos. Sci.*, **44**, 1287-1295.

Hanson, H.P. and V.E. Derr, 1987: Parameterization of radiative flux profiles within layer clouds. *J. Clim. Appl. Meteor.*, **26**, 1511-1521.

Jensen, J.B., P.H. Austin, M.B. Baker and A.M. Blyth, 1985: Turbulent mixing, spectral evolution and dynamics in a warm cumulus cloud. *J. Atmos. Sci.*, **42**, 173-192.

Kitchen, M. and S.J. Caughey, 1981: Tethered balloon observations of the structure of small cumulus clouds. *Quart. J. Roy. Meteor. Soc.*, **107**, 853-874.

Lilly, D.K., 1968: Models of cloud-topped mixed layers under a strong inversion. *Quart. J. Roy. Meteor. Soc.*, **94**, 292-309.

Lopez, R.E., 1976: The lognormal distributions and cumulus cloud population. *Mon. Wea. Rev.*, **105**, 1865-1872.

Lowe, P.R., 1977: An approximating polynomial for the computation of saturation vapor pressure. *J. Appl. Meteor.*, **16**, 100-103.

Moeng, C.-H., 1987: Large-eddy simulation of a stratus-topped boundary layer. Part II. Implications for mixed-layer modeling. *J. Atmos. Sci.*, **44**, 1605-1614.

Moeng, C.-H. and D.A. Randall, 1984: Problems in simulating the stratocumulus-topped boundary layer with a third-order closure model. *J. Atmos. Sci.*, **41**, 1588-1600.

Nicholls, S., 1984: The dyamics of stratocumulus: aircraft observations and comparisons with a mixed layer model. *Quart. J. Roy. Meteor. Soc.*, **110**, 783-820.

Nicholls, S. and J. Leighton, 1986: An observational study of the structure of stratiform cloud sheets. Part I. Structure. *Quart. J. Roy. Meteor. Soc.*, **112**, 431-460.

Nicholls, S. and M.A. LeMone, 1980: The fair-weather boundary layer in GATE: The relationship of subcloud fluxes and structure to the distribution and enhancement of cumulus clouds. *J. Atmos. Sci.*, **37**, 2051-2067.

Nicholls, S., M.A. LeMone and G. Sommeria, 1982: The simulation of a fair-weather marine boundary layer in GATE using a three-dimensional model. *Quart. J. Roy. Meteor. Soc.*, **108**, 167-190.

Nicholls, S. and J.D. Turton, 1986: An observational study of the structure of stratiform cloud sheets. Part II. Entrainment. *Quart. J. Roy. Meteor. Soc.*, **112**, 461-480.

Paluch, I.R., 1979: The entrainment mechanism in Colorado cumuli. *J. Atmos. Sci.*, **36**, 2467-2478.

Penc, R.S. and B.A. Albrecht, 1987: Parameteric representation of heat and moisture fluxes in cloud-topped mixed layers. *Bound.-Layer Meteor.*, **38**, 225-248.

Pennell, W.T. and M.A. LeMone, 1974: An experimental study of turbulence structure in the fair-weather trade wind boundary layer. *J. Atmos. Sci.*, **31**, 1308-1323.

Pontikis, C., A. Rigaud and E. Hicks, 1987: Entrainment and mixing as related to the microphsical properties of shallow warm cumulus clouds. *J. Atmos. Sci.*, **44**, 2150-2165.

Randall, D.A., 1980: Conditional instability of the first kind upside down. *J. Atmos. Sci.*, **37**, 125-130.

Randall, D.A., 1984: Buoyant production and consumption of turbulence kinetic energy in cloud-topped mixed layers. *J. Atmos. Sci.*, **41**, 402-413.

Randall, D.A., J.A. Coakley, Jr., C.W. Fairall, R.A. Kropfli, and D.H. Lenschow, 1984: Outlook for research on subtropical marine stratiform clouds. *Bull. Amer. Meteor. Soc.*, **65**, 1290-1301.

Randall, D.A. and G.J. Huffman, 1982: Entrainment and detrainment in a simple cumulus cloud model. *J. Atmos. Sci.*, **39**, 2793-2806.

Raymond, D.J. and A.M. Blyth, 1986: A stochastic mixing model for nonprecipitating cumulus clouds. *J. Atmos. Sci.*, **43**, 2708-2718.

Raymond, D.J. and M.H. Wilkening, 1982: Flow and mixing in New Mexico mountain cumuli. *J. Atmos. Sci.*, **39**, 2211-2228.

Roach, W.T., R. Brown, S.J. Caughey, B.A. Crease and A. Slingo, 1982: A field study of nocturnal stratocumulus. I. Mean structure and budgets. *Quart. J. Roy. Meteor. Soc.*, **108**, 103-124.

Rogers, D.P., J.A. Businger, and H. Charnock, 1985a: A numerical investigation of the JASIN atmospheric boundary layer. *Bound.-Layer Meteor.*, **32**, 373-399.

Rogers, D.P. and J.W. Telford, 1986: Metastable stratus tops. *Quart. J. Roy. Meteor. Soc.*, **112**, 481-500.

Rogers, D.P., J.W. Telford and S.K. Chai, 1985b: Entrainment and the temporal development of the microphysics of convective clouds. *J. Atmos. Sci.*, **42**, 1846-1858.

Schmetz, J. and M. Beniston, 1986: Relative effects of solar and infrared radiation forcing in a mesoscale model. *Bound.-Layer Meteor.*, **34**, 137-155.

Slingo, A., R. Brown and C.L. Wrench, 1982: A field study of nocturnal stratocumulus: III. High resolution radiative and microphysical observations. *Quart. J. Roy. Meteor. Soc.*, **108**, 145-166.

Soong, S.-T. and Y.Ogura, 1980: Response of tradewind cumuli to large-scale processes. *J. Atmos. Sci.*, **37**, 2035-2050.

Squires, P., 1958: Penetrative downdraughts in cumuli. *Tellus*, **10**, 382-389.

Stackpole, J.D., 1967: Numerical analysis of atmospheric soundings. *J. Appl. Meteor.*, **6**, 464-467.

Stage, S.A. and J.A. Businger, 1981a: A model for entrainment into a cloud-topped marine boundary layer. Part I: model description and application to a cold-air outbreak episode. *J. Atmos. Sci.*, **38**, 2213-2229.

Stage, S.A. and J.A. Businger, 1981b: A model for entrainment into a cloud-topped marine boundary layer. Part II: discussion of model behavior and comparison with other models. *J. Atmos. Sci.*, **38**, 2230-2242.

Stephens, G.L., 1978: Radiation profiles in extended water clouds. II: parameterization schemes. *J. Atmos. Sci.*, **35**, 2123-2132.

Stith, J.L., D.A. Griffith, R.L. Rose, J.A. Flueck, J.R.Miller, Jr., and P.L. Smith, 1986: Aircraft observations of transport and diffusion in cumulus clouds. *J. Clim. Appl. Meteor.*, **25**, 1959-1970.

Stull, R.B., 1984: Models and measurements of the interaction between the mixed layer and fair-weather cumulus clouds: Part 2. Some preliminary measurements. *Transactions of An APCA Speciality Conference on Environmental Impact of Natural Emissions* (V.P. Aneja, Ed.), Air Pollution Control Assoc., 326-337.

Stull, R.B., 1985: A fair-weather cumulus cloud classification scheme for mixed layer studies. *J. Clim. Appl. Meteor.*, **24**, 49-56.

Stull, R.B., 1987: Applications of the transilient turbulence parameterization to atmospheric boundary-layer simulations. *Bound.-Layer Meteor.*, **40**, 209-239.

Tag, P.M. and S.W. Payne, 1987: An examination of the breakup of marine stratus: a three dimensional numerical investigation. *J. Atmos. Sci.*, **44**, 208-223.

Telford, J.W., 1985: Comments on 'Outlook for research on subtropical marine stratiform clouds'. *Bull. Amer. Meteor. Soc.*, **66**, 850-852.

Telford, J.W. and S.K. Chai, 1980: A new aspect of condensation theory. *Pure Appl. Geophys.*, **118**, 720-742.

Telford, J.W. T.S. Keck and S.K. Chai, 1984: Entrainment at cloud tops and the droplet spectra. *J. Atmos. Sci.*, **41**, 3170-3179.

Tetens, O., 1930: Über einige meteorologische Vegriffe. *Z. Geophys.*, **6**, 297-309.

Turton, J.D. and R. Brown, 1987: A comparison of a numerical model of radiation fog with detailed observations. *Quart. J. Roy. Meteor. Soc.*, **113**, 37-54.

Turton, J.D., and S. Nicholls, 1987: A study of the diurnal variation of stratocumulus using a multiple mixed layer model. *Quart. J. Roy. Meteor. Soc.*, **113**, 969-1010.

Wang, S. and B.A. Albrecht, 1986: A stratocumulus model with an internal circulation. *J. Atmos. Sci.*, **43**, 2374-2391.

Wai, M.M.-K., 1987: A numerical study of the marine stratocumulus cloud layer. *Bound.-Layer Meteor.*, **40**, 241-267.

Welch, R.M. and M.G. Ravichandran, 1985: Prediction of quasi-periodic oscillations in mature radiation fogs. *J. Atmos. Sci.*, **42**, 2888-2897.

Welch, R.M. and B.A. Wielicki, 1984: Stratocumulus cloud field reflected fluxes: the effect of cloud shape. *J. Atmos. Sci.*, **41**, 3085-3103.

Welch, R.M. and B.A. Wielicki, 1985: A radiative parameterization of stratocumulus cloud fields. *J. Atmos. Sci.*, **42**, 2888-2897.

Welch, R.M. and B.A. Wielicki, 1986: The stratocumulus nature of fog. *J. Clim. Appl. Meteor.*, **25**, 101-111.

Wexler, A., 1976: Vapor pressure formulation for water in range 0 to 100°C. A revision. *J. Res. Nat. Bur. Stand.*, **80A**, 775-785.

Wilde, N.P., R.B. Stull, and E.W. Eloranta, 1985: The LCL zone and cumulus onset. *J. Clim. Appl. Meteor.*, **24**, 640-657.

Yamada, T. and C.-Y.J.Kao, 1986: A modeling study on the fair weather marine boundary layer of the GATE. *J. Atmos. Sci.*, **43**, 3186-3199.

Yuen, C.-W., 1985: Dynamical modeling of flow in cumulus-filled boundary layers. *J. Atmos. Sci.*, **42**, 113-134.

13.8 Exercises

1) Given the following wind and θ_V profiles in a stratocumulus-topped mixed layer. Indicate in the table below the sign $(+ , - , 0)$ of each term of the simplified turbulence kinetic energy (e) equation for each region. (The shaded region represents cloud.)

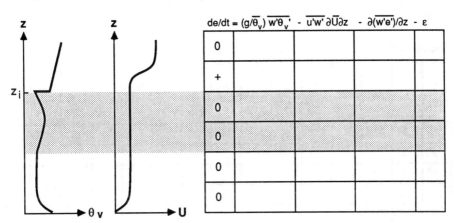

$de/dt = (g/\overline{\theta}_v) \overline{w'\theta_v'}$	$- \overline{u'w'} \partial \overline{U} \partial z$	$- \partial(\overline{w'e'})/\partial z$	$- \varepsilon$
0			
+			
0			
0			
0			
0			

2) What is the approximate relationship between static energies and potential temperatures?

3) Given the following data points of (p, θ_L, r_T) in units of (kPa, K, g/kg) representing an environmental sounding: (100, 303, 19) (95, 303, 19) (90, 303, 19) (85, 305, 13) (80, 306, 10) (70, 312, 8) (60, 322, 5) (50, 335, 3) (40, 355, 2).

   a) Plot these on a conserved-variable diagram (θ_L vs. r_T).

   b) Suppose that a cloud exists in this environment, with a top at 60 kPa and base at 90 kPa. Draw the mixing line on the diagram, and plot the points corresponding to the following mixtures of environmental air from cloud top and cloud base (top,base): (50%, 50%), (30%, 70%), (20%, 80%), and (10%, 90%).

   c) If there is additional mixing within the cloud (for example, the 50/50 mixture air mixing with the 10/90 mixture air), where do these mixtures fall on the mixing line?

4) Suppose the mixed layer includes the points at 100 to 90 kPa in the previous example. Where do these mixed layer points fall on the conserved variable diagram?

5) a) Given the atmospheric measurements in the table below, calculate the remaining thermodynamic variables to fill in the table. Assume that the cloud, for which data was supplied, is embedded within the given environment.

   b) Plot all varieties of potential temperature as a function of height on the same graph.

   c) Plot all varieties of static energies (both moist and dry) on the same graph.

   d) At what pressure does neutral buoyancy (limit of convection) occur for the cloudy air specified in the table, and for a parcel rising from cloud base?

e) Calculate the CAPE in units of m^2/s^2 of the actual cloudy air, and for an air parcel rising from cloud base.

f) Assuming no entrainment, mixing, or friction, what vertical velocity would an air parcel have at the limit of convection if it started at cloud base with an upward velocity of 1 m/s? About how far above the LOC would the air parcel overshoot (i.e., where is cloud top)?

g) What processes might be responsible for the differences between actual in-cloud measurements and the idealized moist parcel ascent?

Measurements in the environment:

p (kPa)	T (°C)	r (g/kg)	r sat (g/kg)	r_L (g/kg)	Tv (°C)	θ (K)	$θ_v$ (K)	$θ_e$ (K)	$θ_L$ (K)	$θ_{es}$ (K)	s (J/g)	s_v (J/g)	s_e (J/g)	s_L (J/g)	s_{es} (J/g)
100	30	20													
90	21	15													
80	14	10													
70	9	9													
60	5	5													
50	1	3													
40	0	2													

Measurements in the cloud:

p (kPa)	T (°C)	r (g/kg)	r sat (g/kg)	r_L (g/kg)	Tv (°C)	θ (K)	$θ_v$ (K)	$θ_e$ (K)	$θ_L$ (K)	$θ_{es}$ (K)	s (J/g)	s_v (J/g)	s_e (J/g)	s_L (J/g)	s_{es} (J/g)
90	22			0.5											
80	16			2.0											
70	11			5.0											
60	7			4.0											

6) If $\overline{w's_L'} = 0.05$ (J/g)(m/s) and $\overline{w'r_T'} = 0.05$ (g/kg)(m/s) , then calculate the value of the buoyancy flux, $\overline{w's_v'}$.

7) Find the saturation vapor pressure and mixing ratio for T = 20°C and P = 90 kPa.

8) Find the saturation-point temperature and pressure for P = 85 kPa, T = 20°C, r = 8 g/kg.

9) Find and plot the net solar radiation as a function of depth within a stratocumulus cloud, given z_T = 1000 m, z_B = 700 m, solar zenith angle = 70°, and K = -900 W/m². What are the values of the e-folding decay length, bulk cloud albedo, and bulk cloud absorption?

10) Find the change in net longwave flux across the top and bottom of a stratocumulus cloud, assuming the following temperatures: $T_{surface}$ = 30°C, $T_{cloud\ base}$ = 25°C, $T_{cloud\ top}$ = 15°C, and T_{sky} = 100 K.

11) Just above cloud top: P = 80 kPa, T = 10°C, r_T = 0.5 g/kg. Just below cloud top: P = 80 kPa, T = 5°C, r_T = 10 g/kg. Is this stratocumuls cloud unstable for cloud-top entrainment?

12) Calculate and plot the distribution of cloud diameters, assuming the lognormal location and shape parameters are 750 m and 0.75, respectively.

14 Geographic Effects

In most of the previous chapters we assumed a flat, uniform bottom boundary, but in many parts of the world the ground is neither flat nor uniform. Geographic variations can modify the boundary-layer flow, and in some cases generate circulations in conjunction with diurnal heating cycles. We have already touched on a few such flows, such as the drainage winds at night within a stable boundary layer. In this chapter we examine both geographically-generated and geographically-modified flows.

Many of the examples will be based on highly simplified and idealized hill structures and surface features, in order to make the problem tractable. Nevertheless, the results can be qualitatively applied to more realistic scenarios.

14.1 Geographically Generated Local Winds

Warming of mountain slopes by daytime sunlight or cooling by nocturnal radiation causes the air adjacent to the mountain to be warmed or cooled by conduction and turbulence. If the air near the mountain is at a different temperature than the ambient air at the same altitude over the center of the valley, then buoyant forces generate a circulation. Similar forcings occur over flat surfaces where albedo or heat capacity differences generate a horizontally inhomogeneous temperature field. In general, *local winds* are named after the source location of the wind: sea breezes come in from the sea, mountain winds come down from the mountain, etc.

In the following discussions we assume that the ambient winds are light or calm. For this situation the geographically-generated circulation can be studied without external influences. In many real situations, however, ambient synoptic or mesoscale winds can modify, or even eliminate the weak geographic circulations.

587

14.1.1   Circulations in Mountainous Regions

During the diurnal cycle in mountainous regions, three-dimensional circulations can form within and just above the valleys. We will initially examine the cross-valley-axis flow (anabatic/katabic *slope winds*) separately from the along-valley-axis flow (mountain/valley *valley winds*), and then combine the two components into a full three-dimensional picture (Geiger, 1965).

We will assume a topographic model as sketched in Fig 14.1 (Whiteman, 1982), with a river valley descending and deepening as it flows away from its source near the top of a mountain range. Along both sides of the valley and perpendicular to the mountain-range axis are ridges that also decrease in height. The valley is assumed to drain into a large open area such as a lake, ocean, plain, or a larger valley.

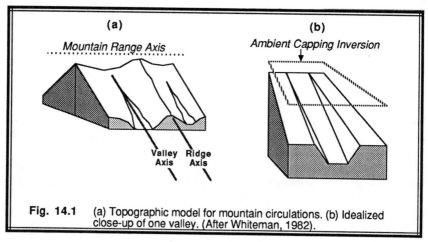

Fig. 14.1   (a) Topographic model for mountain circulations. (b) Idealized close-up of one valley. (After Whiteman, 1982).

Anabatic/Katabatic Cross-Valley Winds. We will start our diurnal cycle at sunset, where it is assumed that a deep daytime mixed layer extends well above the top of the mountain range, and that turbulence then decays to leave a neutrally stratified warm residual layer over the whole mountain and valley system (Fig 14.2a). Radiative cooling of the mountain surfaces cools the air adjacent to the surfaces, resulting in cold downslope or *katabatic winds*. These winds are very shallow (2 to 20 m), and have velocities on the order of 1 to 5 m/s. Above the valley floor, there is a gentle return circulation of upward moving air that diverges towards the ridges.

The chilled air flows into the valley and collects as a cold pool. Although some of the cold air flows down the valley axis, some can remain in the valley depending on the topography (Fig 14.2b). During the night, continued katabatic winds fill the cold air pool. Above the cold pool is the remaining warmer RL air, resulting in a temperature inversion capping the cold pool. The cold air from higher on the ridges flows downslope until it reaches an altitude where its temperature is the same as the temperature of the air in the valley. As a result, the coldest air sinks to the bottom of the pool, while the cooler air slides into the pool at higher altitudes (Fig 14.2c).

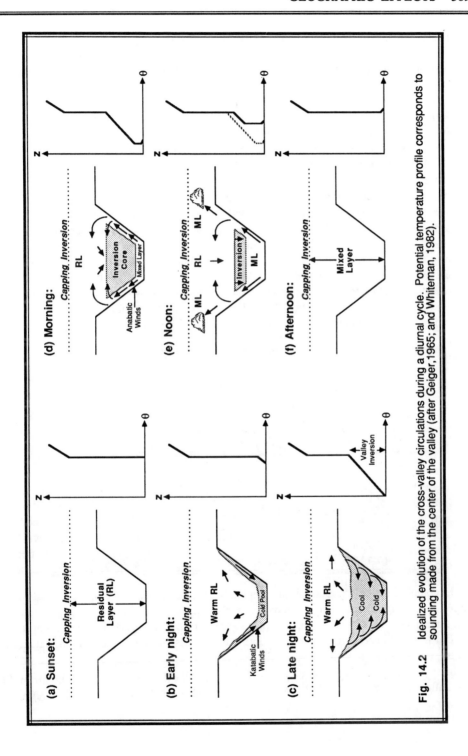

Fig. 14.2   Idealized evolution of the cross-valley circulations during a diurnal cycle. Potential temperature profile corresponds to sounding made from the center of the valley (after Geiger,1965; and Whiteman, 1982).

The resulting pool is often stably stratified throughout its depth, and is sometimes called the *valley inversion*. Pollutants emitted into this inversion can build to high concentrations because of the trapping between the valley walls, and can be hazardous to people, animals, and plant life on the slopes.

After sunrise, solar heating warms the air near the valley walls, causing warm upslope *anabatic winds* (Fig 14.2d). These gentle winds (less than 1 m/s) tend to hug the valley walls as they rise, eventually breaking away at their level of neutral buoyancy, or at the top of the ridge. Sometimes cumulus clouds known as *anabatic clouds* form in the warm air rising along the ridge axes. Above the valley inversion there is gentle convergence and subsidence.

As this warmed air leaves the valley floor, the remaining pool of cold air sinks to replace it. In addition, solar heating of the valley floor itself causes a shallow mixed layer to form. Although this ML tends to grow by the entrainment processes discussed in Chapter 11, a portion of the ML air is continuously drained away by the anabatic winds, causing the ML to rise much more slowly than would otherwise be expected over flat terrain (Fig 14.2e). The balance between ML entrainment and valley-inversion subsidence varies from situation to situation, between the extremes of no ML growth to rapid growth with little subsidence (Whiteman, 1982). For the case sketched in Fig 14.2e, the top and bottom of the valley inversion evolve as shown in Fig 14.3.

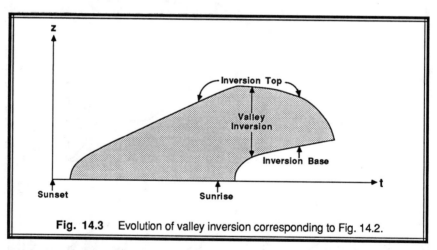

Fig. 14.3   Evolution of valley inversion corresponding to Fig. 14.2.

Eventually, the cold pool of air is completely eliminated, leaving a daytime convective mixed layer (Fig 14.2f). When this happens, the TKE can suddenly increase, sometimes to 6 times its earlier morning value (Banta, 1985). The diurnal cycle can then repeat itself.

In Figs 14.2d and e, the heating on both valley walls was idealized as being equal. In reality, the orientation of the valley walls relative to the sun position can cause one wall to be heated strongly, while the other is heated weakly or is in shadow. The resulting circulation is asymmetric, but otherwise proceeds as sketched.

Along-Valley Winds. At night, the cold winds flowing down the valley onto the plains are known as *mountain winds* or *drainage winds*. Depths range from 10 to 400 m, depending on the size and flow constrictions of the valley (Neff and King, 1987). Velocities of 1-8 m/s have been observed, and these winds are occasionally intermittent or surging. The return gentle circulation of warmer air aloft is called the *anti-mountain wind*, with velocities of about half of the mountain wind, and depths of about twice as much.

During the day, warm air gently flowing up the valley axis is known as the *valley wind*. This wind consists of a *valley-floor* component, and sometimes an *up-incline* component along the ridge tops. The cool, slow return flow aloft is called the *anti-valley wind*. Fig 14.4 shows these along-valley wind components.

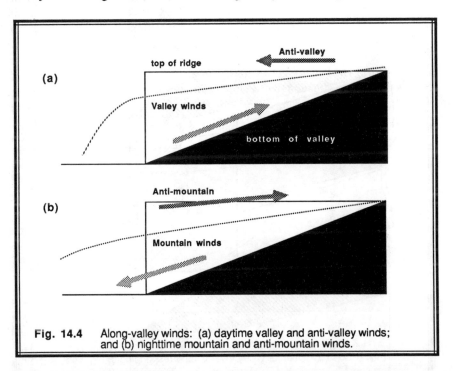

Fig. 14.4   Along-valley winds: (a) daytime valley and anti-valley winds; and (b) nighttime mountain and anti-mountain winds.

Three-dimensional Picture. At night, the 3-D circulation is fairly easy to picture, with downslope and down-valley winds converging just above the ground, and with gentle up-valley winds and divergence aloft (Fig 14.5a) (McNider and Pielke, 1984). During the morning, the picture is a bit more complicated, with the remaining elevated valley inversion continuing to slide down the valley as a stable core, riding above shallow up-valley and anabatic flows (Fig 14.5b, after Whiteman, 1982).

During the afternoon, after the stable core has disappeared, two counter-rotating helices are sometimes observed as the wind flows upslope and up-valley at low levels, and descends aloft over the center of the valley (Fig 14.5c).

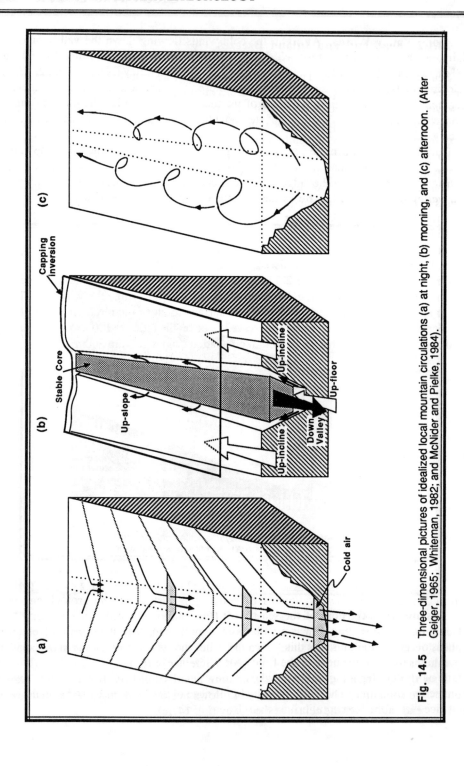

Fig. 14.5   Three-dimensional pictures of idealized local mountain circulations (a) at night, (b) morning, and (c) afternoon. (After Geiger, 1965; Whiteman, 1982; and McNider and Pielke, 1984).

14.1.2   Sea/Land and Inland Breezes

Sea and Lake Breezes. The large heat capacity of lakes and oceans reduces water-surface temperature change to near-zero values during a diurnal cycle. The land surface, however, warms and cools more dramatically because the small molecular conductivity and heat capacity in soils prevents the diurnal temperature signal from propagating rapidly away from the surface. As a result, the land is warmer than the water during the day, and cooler at night. This scenario is ideal for the formation of sea breezes (Lyons, 1975; Simpson, et al., 1977; Helmis, et al., 1987; Ogawa, et al., 1986).

During mid-morning (1000 local time) after the nocturnal SBL has been eliminated, air begins to rise over the warm land near the shoreline, and cooler air from the water flows in to replace it. This is known as the *sea-breeze* (or *lake breeze*). The inland limit of cool air progression over land is known as the *sea-breeze front*, and is marked by low-level convergence (in a band about 1 to 2 km wide), a marked temperature drop (often several °C; hence, it is a mesoscale cold front; see Fig 14.6), an increase in humidity, upward motion (of about 0.5 to 2.5 m/s), and sometimes enhanced cumulus clouds. A return circulation (the *anti-sea-breeze*) of 1 to 2 m/s aloft brings the warmer air back out to sea where it descends toward the sea surface to close the circulation (Fig 14.7). The depth of the sea breeze has been observed to be on the order of 100 to 500 m, and the total circulation depth including the return circulation can range from 500 m to 2000 m.

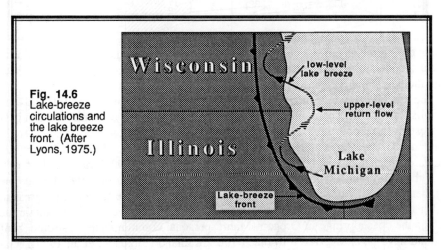

Fig. 14.6 Lake-breeze circulations and the lake breeze front. (After Lyons, 1975.)

In some cases the nose of the sea-breeze front can be about twice the depth of the trailing flow. The trailing flow is usually faster than the propagation rate of the front, with speeds of up to 5 to 7 m/s. As this flow approaches the surface front, a vortex-like pattern sometimes forms at the nose of the flow. After sea-breeze frontal passage the turbulence state of the cooler air quickly reaches a new local equilibrium, and scales to free convection similarity. The TKE intensity in the nose of the sea breeze can be on the order of twice that of the later equilibrium state (Briere, 1987).

In the absence of a background synoptic flow, the front progresses inland normal to the coastline at speeds of 1 to 5 m/s. It can easily reach 20 to 50 km inland by the end of the day. During the daytime, the sea-breeze front is driven partially by the active conversion of available potential energy to kinetic energy. After sunset, the front can continue to progress inland in the form of a gravity or density current similar to the thunderstorm gust front.

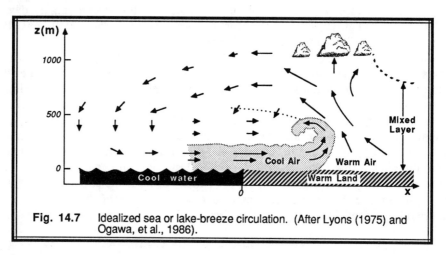

Fig. 14.7   Idealized sea or lake-breeze circulation. (After Lyons (1975) and Ogawa, et al., 1986).

Near capes and peninsulas, sea-breeze fronts from opposite shores converge and collide during the day, producing stronger upward motion. This phenomena is known to trigger thunderstorms in southern Florida (Pielke, 1974), and is believed to trigger cloud lines and possibly a bore wave in northern Australia (Noonan and Smith, 1986).

The sea-breeze wind direction at low-levels often turns under the influence of the Coriolis force and the baroclinicity between land and sea. For example, the winds along the western shore of Lake Michigan blow from east to west in the morning, and gradually rotate to blow from the southeast to northwest (Fig 14.6) by late afternoon (Lyons, 1975). Pollutants emitted into the sea breeze can thus be recirculated back further up the coastline.

If the background synoptic flow is in the same direction as the low-level sea-breeze and there are no major barriers to the flow, then the sea-breeze front can progress much further inland (over 100 km). It has been observed to flow over small mountain ranges such as the coastal range in Oregon (Mahrt, personal communication). Some fronts in England have taken over 10 h to reach 100 km inland (Simpson, et al, 1977). Garratt (1985) reported sea-breeze fronts occasionally reaching as far inland as 400 km in Australia. The speed of the front is well approximated by a linear sum of the imposed background wind component perpendicular to the front, and the speed of the front in still air (Pearson, et al., 1983).

During opposing synoptic flow, the sea-breeze can stall near the shoreline, or can be totally eliminated as the mean wind flows from land to sea. For synoptic flow parallel to the coastline, some observations indicate that the sea-breeze front becomes broader and more diffuse. In many parts of the world we find mountains bordering the sea. At these locations both the sea breeze and the valley breeze can interact and unite to form stronger flows, or can counteract and oppose each other (Fitzjarrald, 1984, 1986; Kitada, 1986; Wakimoto and McElroy, 1986; Neff and King, 1987). The occasional presence of low-level jets near these mountainous coasts can also be superimposed on the mountain and sea breezes.

Sea-breeze circulations have been modeled both analytically (Rotunno, 1983) and numerically. The simpler models are often two-dimensional, hydrostatic, nonturbulent, steady-state approximations, where crosswind uniformity and sometimes linearity are assumed . Some of the more sophisticated models have used third-order closure or three dimensions (Martin and Pielke, 1983; Briere, 1987).

Land Breeze. At night, land surfaces usually cool faster than the neighboring water bodies, reversing the temperature gradient that was present during the day. The result is a *land breeze*: cold air from land flows out to sea at low levels, warms, rises, and returns aloft toward land (*anti-land-breeze*) where it eventually descends to close the circulation. The land breeze is believed to be analogous to the sea breeze, although there have been insufficient observations of its propagation across the water to be conclusive.

Inland Sea Breeze. Some modeling studies have suggested that mesoscale variations of surface moisture over land can induce mesoscale circulations analogous to the sea breeze (Segal and Pielke, 1987; Yan and Anthes, 1987, personal communication). This type of circulation, called the *inland sea breeze*, is believed to be strongest where moist vegetated surfaces are adjacent to drier surfaces. In some cases, the circulations in these regions might be large enough to trigger or enhance precipitation.

It is believed that swaths of moist soil from a previous thunderstorm passage over an otherwise dry land surface can create a sufficiently large moisture inhomogeneity to generate the inland sea breeze. In the western plains of the United States, large farms using center-pivot irrigation systems or banded irrigated crops can also create the necessary surface inhomogeneities. Deforestation and urbanization might be similar mechanisms causing inadvertent climate modification.

14.2 Geographically Modified Flow

When air flows over a variety of surfaces, each surface characteristic affects the flow. For example, suppose that there is a flat semi-infinite plot of land that is dry, unvegetated, and smooth. Downwind of that plot is another flat plot that is vegetated, moist, and rough. A boundary layer will develop over the first field that is in equilibrium with the surface forcings. When this air crosses the border and flows over the neighboring field, the bottom of the boundary layer will be modified by the new surface features, and the depth of this modified air will increase with distance downwind of the border. Above this modified layer, the boundary layer does not "feel" the new surface, and continues to behave as it did over the upwind surface.

14.2.1 The Internal Boundary Layer and Fetch

Definitions. The air that is modified by flow over a different surface is called an *internal boundary layer* (IBL), because it forms within an existing boundary layer. When surface heat flux changes across the border between two surfaces, the modified air is called a *thermal internal boundary layer* (TIBL) (Lyons, 1975; Garratt, 1987). For a change in roughness with no change in surface heat flux or stability, the generic name internal boundary layer is usually used.

The distance, x, downwind from a change in surface features is called *fetch*. When making surface-layer measurements, it is desirable to locate the instrument mast far enough downwind of the border so that the depth of the internal boundary layer, δ, is greater that the height of the mast (see Fig 14.8). Thus, with sufficiently large fetch, the surface-layer measurements are characteristic of the local field in which the instruments are located (Gash, 1986). For this reason, many surface-layer measurements are made within large uniform farm fields, meadows, pasture land, or forests.

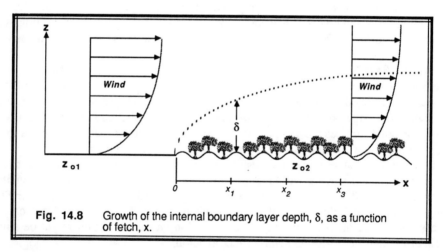

Fig. 14.8   Growth of the internal boundary layer depth, δ, as a function of fetch, x.

Change in Roughness. The IBL depth is often parameterized as a power of the fetch (Panofsky, et al., 1982; Smedman-Högström and Högström, 1978; Rao, et al., 1974):

$$\frac{\delta}{z_{o1}} = a_{IBL} \cdot \left(\frac{x}{z_{o1}} \right)^{b_{IBL}} \tag{14.2.1a}$$

where z_{o1} and z_{o2} are the aerodynamic roughness lengths upwind and downwind of the border. The power, b_{IBL}, is equal to about 0.8 for statically neutral conditions, but is slightly smaller (0.6 to 0.7) in statically stable conditions, and larger for unstable (0.8 to 1.0). Parameter a_{IBL} is in the range of 0.2 to 0.8, being large for unstable conditions and small for stable ones. Sometimes, a_{IBL} is parameterized as a function of both roughnesses:

$$a_{IBL} = 0.75 + 0.03 \ln\left(\frac{z_{o2}}{z_{o1}}\right) \qquad (14.2.1b)$$

Fig 14.9 shows the idealized evolution of the statically-neutral logarithmic wind profile as a function of fetch, for the case of flow from a smooth to a rough surface. At fetch x_1 (see Fig 14.8) only a shallow layer "feels" the increased roughness. At greater distances, deeper layers feel the increased roughness. Above the top of the IBL (i.e., above the kink in the lines), the wind maintains a profile characteristic of the upwind roughness (Beljaars, 1987).

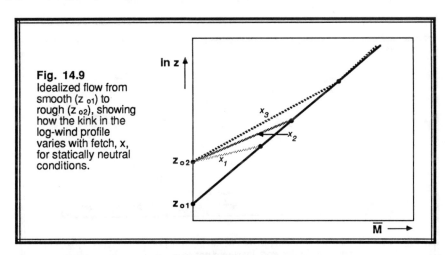

Fig. 14.9
Idealized flow from smooth (z_{o1}) to rough (z_{o2}), showing how the kink in the log-wind profile varies with fetch, x, for statically neutral conditions.

Smedman-Högström and Högström (1978) and Rao, et al. (1974) have shown that real wind profiles often exhibit a transition zone between the equilibrium IBL and the PBL aloft (Fig 14.10a). Also, Claussen (1987) suggests that the change in roughness can be felt a small distance ($x \cong 300\, z_o$) **up**stream of the boundary, resulting in a modification of the upwind profile (Fig 14.10b). Flow near the ground is slightly reduced for a change from smooth to rough, and flow aloft accelerates slightly.

In many parts of the world, there is a patchwork of surface features such as towns, farms, forests, lakes, etc. As sketched in Fig 14.11, this leads to IBLs within IBLs, with many kinks in the log-wind profile (Oke, 1978). When we combine kinks due to IBLs with curvature of the wind profile in diabatic conditions and curvature associated with a nonzero displacement distance, we find a very complicated wind profile. When hills are also considered, the result is complex indeed (Beljaars, et al., 1987).

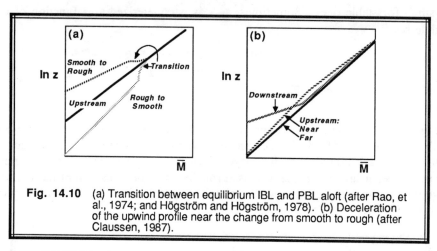

Fig. 14.10   (a) Transition between equilibrium IBL and PBL aloft (after Rao, et al., 1974; and Högström and Högström, 1978). (b) Deceleration of the upwind profile near the change from smooth to rough (after Claussen, 1987).

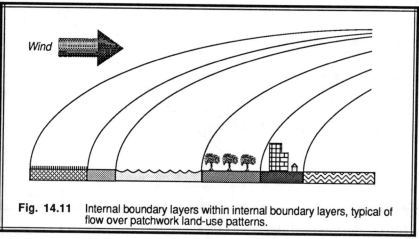

Fig. 14.11   Internal boundary layers within internal boundary layers, typical of flow over patchwork land-use patterns.

When air flows from a smooth surface to a rough surface, the air in the IBL decelerates. This results in horizontal convergence and upward motion above the boundary between smooth and rough (Fig 14.12). Similarly, flow from rough to smooth causes divergence and subsidence. These vertical motions interact with other convective motions and affect pollutant transport. A convergence/divergence dipole is often observed at the upwind/downwind borders of cities, because of the change in roughness.

Upwind of a surface change from smooth to rough the friction velocity decreases slightly, then has a step increase overshooting its final equilibrium value, and finally decays to a final equilibrium state that is larger than the friction velocity over the smooth surface (Claussen, 1987). Turbulence kinetic energy also increases in a shallow layer at the point of roughness increase, then the layer of increased TKE grows in thickness as the IBL grows, and finally the surface value decreases to a new equilibrium value that is larger than the upwind surface value. Panofsky, et al. (1982) showed that most of the increase in TKE is associated with the smaller-size eddies in the spectrum.

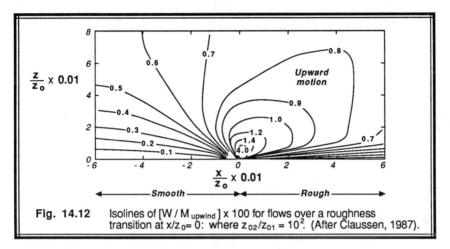

Fig. 14.12   Isolines of $[W / M_{upwind}] \times 100$ for flows over a roughness transition at $x/z_0 = 0$: where $z_{02}/z_{01} = 10^2$. (After Claussen, 1987).

14.2.2 Thermal Internal Boundary Layer (TIBL)

Shorelines are good examples of regions where surface heat flux differences between neighboring surfaces can cause TIBLs. Although we will frequently use the word "shoreline" in this subsection, one should recognize that similar physics applies to other borders as well, such as between two land surfaces with differing albedoes and surface heat fluxes.

Convective TIBL. When air flows from a cooler to a warmer surface, a steady-state convective ML (*convective TIBL*, in this case) forms and deepens with distance downwind of the shoreline (Fig 14.13).   Turbulence is vigorous over the bulk of the

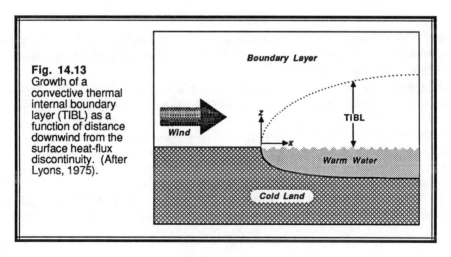

Fig. 14.13
Growth of a convective thermal internal boundary layer (TIBL) as a function of distance downwind from the surface heat-flux discontinuity. (After Lyons, 1975).

convective TIBL, and there is a relatively sharp, well-defined top. The ML depth, z_i, grows by entrainment, is suppressed by subsidence, and otherwise evolves as previously discussed for ML growth (Lyons, 1975; Venkatram, 1977; Hsu, 1986; Arritt, 1987; Hanna, 1987; Mitsuta, et al., 1986). TIBLs can occur in winter as cold continental air advects over warmer water, and in summer as air over colder water advects over warmer land.

Taylor's hypothesis can be invoked to transform the prognostic equations into gradient equations. Namely, apply (11.2.1b-e) to a column of air translating along with the mean wind. Pick a coordinate system with the x-axis perpendicular to the shoreline, and assume uniformity in the y direction. The result is:

$$\frac{\partial z_i}{\partial x} = \frac{w_e + w_L}{\overline{U}} \tag{14.2.2a}$$

$$\overline{U} \, z_i \frac{d<\overline{\theta}>}{dx} = \overline{w'\theta'}_s - \overline{w'\theta'}_{z_i} \tag{14.2.2b}$$

Similar equations can be written for moisture and winds.

Just as the ML can be shown to grow with the square root of time for special cases, the TIBL grows with the square root of distance from the shore. For example, Venkatram (1977) has suggested the following equation for TIBL depth as a function of distance, x:

$$z_i = \left[\frac{2 \, C_D \left| \overline{\theta}_{land} - \overline{\theta}_{sea} \right| x}{\gamma \, (1 - 2 \, A_R)} \right]^{1/2} \tag{14.2.2c}$$

where γ is the vertical potential-temperature gradient immediately above the TIBL, and A_R is an entrainment coefficient in the range of 0 to 0.22. Near shore, Hsu (1986) found for some cases that $z_i = 1.91 \cdot x^{1/2}$.

As the column of air advects downwind and warms, the temperature difference between the air and the ground lessens. As a result, the heat flux at the ground decreases, the mixed layer warms less rapidly, and the rise rate of the mixed layer is reduced. At large distances downwind of the shoreline, the air column is assumed to reach a state of equilibrium with (i.e., at the same temperature as) the underlying surface, with no surface heat fluxes and little or no entrainment. Mean variables approach this equilibrium state exponentially with distance, while heat fluxes approach zero exponentially.

An *air mass* is defined to be a boundary layer that has remained stationary over one location (the *source region*) long enough (days to weeks) to acquire some of the characteristics of the underlying surface. When one air mass moves away from its source region over a surface of different temperature or moisture, the process is called *air mass modification*. The TIBL is just one type of air mass modification.

Stable TIBL. For flow from a warmer to a cooler surface, a *stable TIBL* forms analogous to the SBLs discussed in Chapter 12. Immediately downwind of the surface change one finds decaying residual turbulence that is very effective at mixing some of the cooler air upward. Further downwind, the static stability suppresses turbulence, except near the surface and near other shear zones where mechanical TKE production still occurs. For these well-developed stable TIBLs, turbulence decreases with height to a poorly defined top (just as in the nonadvective SBL).

Garratt (1987) suggests that the depth of the stable TIBL increases with the square-root of distance:

$$h = 0.014 \, \overline{M} \left[\frac{x \, \overline{\theta}}{g \, \Delta\theta} \right]^{1/2} \tag{14.2.2d}$$

where $\Delta\theta$ is the initial temperature difference between the air (before growth of the TIBL) and the new cooler surface, and x is the distance from the surface change, measured in the direction of the mean wind. As the bottom of the stable TIBL cools and approaches the surface temperature, the above equation becomes invalid because the surface heat flux approaches zero. Beyond this point there is no further change in TIBL depth except for that caused by subsidence.

14.2.3 Flow Over Hills

Stable Stratification and the Froude Number. In statically stable conditions, perturbed air parcels oscillate vertically at the Brunt-Väisälä frequency, N_{BV}. When this oscillating air is imbedded in an air mass moving at mean wind speed $\overline{M}$, a wave is traced by the parcel. This oscillation has a wavelength proportional to $2\pi\overline{M}/N_{BV}$, which represents a natural wavelength of the air. If the length of the obstacle disturbing the flow is W_T, then we can write its effective wavelength as $2W_T$. The ratio of the natural wavelength of the air to the effective wavelength of the obstacle is defined as the internal *Froude number*:

$$Fr = \frac{\pi \, \overline{M}}{N_{BV} \, W_T} \tag{14.2.3a}$$

Frequently, this number is defined without the factor π. The Froude number is also sometimes described as the ratio of inertial to buoyant forces.

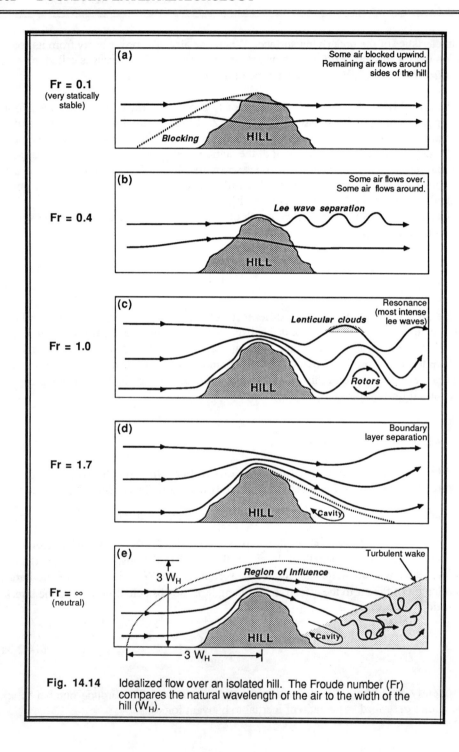

Fig. 14.14   Idealized flow over an isolated hill. The Froude number (Fr) compares the natural wavelength of the air to the width of the hill (W_H).

Fig 14.14 shows the variety of flows that are possible for different Froude numbers, for the special case of an isolated hill (Hunt, 1980). For strongly stable environments with light winds (i.e., Fr ≅ 0.1), air would rather flow around the hill than over it (Fig 14.14a). Directly upwind of the hill, some of the air is **blocked** by the hill and becomes stagnant. The combination of the blocked air and the hill acts like a larger, more streamlined obstacle, around which the remaining air must flow.

For slightly faster winds or weaker stability (i.e., Fr ≅ 0.4), some of the air flows over the top of the hill, while air at lower altitudes separates to flow around the hill (Fig 14.14b). Air that flows over the top has a natural wavelength much smaller than the size of the hill, and is perturbed by the hill to form lee waves. This **lee-wave separation** from the hill top is above the non-oscillating air that flows around the hill. For a column of air (air depth = hill height) approaching the center of the hill, the fraction of the column that flows over the top is approximately equal to Fr:

$$\frac{z_{LW}}{z_{hill}} = Fr \qquad\qquad (14.2.3b)$$

where z_{LW} is the depth of the column flowing over the hill that forms lee waves, and z_{hill} is the total depth of the column of air (equal to the height of the hill).

At a Froude number of 1.0, the stability is weaker and the winds are stronger, causing the natural wavelength of the air to match the size of the hill (Fig 14.14c). Large-amplitude **lee waves** or **mountain waves** are formed by such a natural resonance, with the possibility of rotor circulations near the ground under the wave crests. For this case, the air at the surface stagnates at periodic intervals downwind of the hill, and reverse flow can occur at the surface under the rotors. If sufficient moisture is present, **standing lenticular clouds** can form along the crests of the waves, and **rotor clouds** can form in the updraft portion of the rotor circulations.

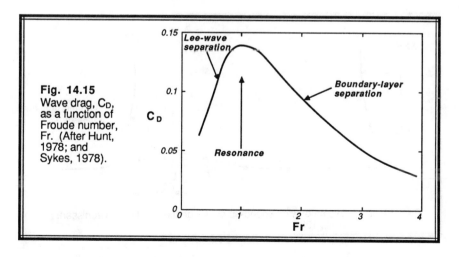

Fig. 14.15 Wave drag, C_D, as a function of Froude number, Fr. (After Hunt, 1978; and Sykes, 1978).

For even stronger winds and weaker stability (Fr $\cong$ 1.7) the natural wavelength is longer than the hill dimensions. This causes *boundary layer separation* at the lee of the hill, and creates a *cavity* with reverse surface wind direction immediately behind the hill (Fig 14.14d). Wave drag associated with flow over a hill is a maximum for the resonance state, and is less for lee-wave and boundary-layer separation (Fig 14.15).

Neutral Stratification. For strong winds and neutral stability, the Froude number approaches infinity, and is not an appropriate measure of flow dynamics. Flow of a neutral boundary layer over an isolated hill is sketched in Fig 14.14e. Streamlines are disturbed upwind and above the hill out to a distance of about three times the size of the hill. Beyond this *region of influence*, the flow does not "feel" the presence of the hill. Near the top of the hill the streamlines are packed closer together, causing a *speed-up* of the wind.

Immediately downwind of the hill in strong wind situations, there is often found a *cavity* (sometimes called a *wind shadow*) associated with boundary layer separation (Tampieri, 1987). This is the start of a turbulent *wake* behind the hill. The height of the wake is initially on the same order as the size of the hill, but it grows in size and decreases in turbulence intensity further downwind of the hill. Eventually, in the absence of other turbulence generating mechanisms, the turbulence decays completely far downwind of the hill, and the flow returns to its undisturbed state. Similar flow patterns are found near buildings (Hanna, et al., 1982). For weaker winds and smooth gentle hills, the cavity and turbulent wake sometimes do not form (Taylor, et al., 1987).

The speed-up at the crest of the hill is sketched in Fig 14.16. The fractional speed-up ratio, $\Delta\overline{M}_{hill}$ is defined as:

$$\Delta\overline{M}_{hill} = \frac{\overline{M}_x(\Delta z) - \overline{M}_A(\Delta z)}{\overline{M}_A(\Delta z)} \tag{14.2.3c}$$

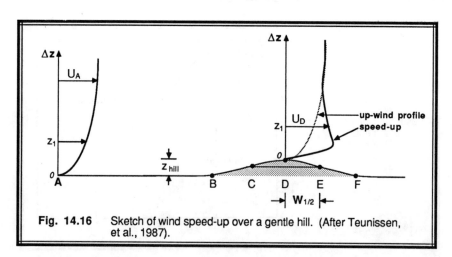

Fig. 14.16   Sketch of wind speed-up over a gentle hill. (After Teunissen, et al., 1987).

where Δz is the height above the local terrain, and subscript A denotes a point upwind of the hill where the flow is undisturbed. For gentle 2-D ridges, speed-up values of about $\Delta \overline{M}_{hill} \cong 2\ z_{hill}/W_{1/2}$ are expected, while for isolated 3-D hills the speed-up is less, $\Delta \overline{M}_{hill} \cong 1.6\ z_{hill}/W_{1/2}$ (Taylor & Teunissen, 1987; Taylor & Lee, 1984; Hunt, 1980). The half width of the hill, $W_{1/2}$, is defined as the horizontal upstream distance from the crest to the point where the elevation has decreased to half its maximum. For typical gentle hills ($z_{hill} = 100$ m, $W_{1/2} = 250$ m) the speed-up of the wind just above the crest of the hill can be 60% or more, which is important for siting wind-turbine generators.

Fig 14.17 shows the wind speed profiles at various locations on the hill (Teunissen, et al., 1987). Well upwind at point A, out of the region of influence, the wind is logarithmic with height as expected for neutral stratification. Closer to the hill, at point B, some blocking is experienced and the low-altitude winds are slower. By point C, the winds above 2 m have accelerated beyond the undisturbed upstream values, and even become faster than ambient winds aloft at point D at the crest of the hill. To the lee of the hill, the low-altitude winds rapidly decrease, and can become slower than the undisturbed upstream values, or can reverse direction as discussed earlier.

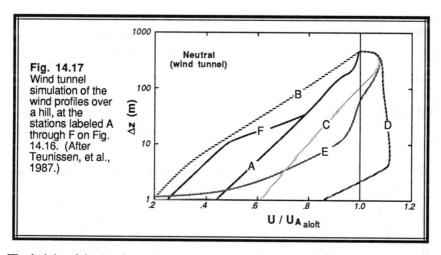

Fig. 14.17 Wind tunnel simulation of the wind profiles over a hill, at the stations labeled A through F on Fig. 14.16. (After Teunissen, et al., 1987.)

The height of the maximum speed-up can be estimated using the roughness length and mountain half width:

$$\Delta z_{max} \cong z_o\ e^{(2k^2 W_{1/2})^b} \tag{14.2.3d}$$

Below this height is an *inner layer* where friction with the ground reduces the wind speed. Above this height in an *outer layer*, the flow can be well described by inviscid *potential flow theory*.

Suggestions for the value of b range from 0.5 to 1.0 (Taylor and Teunissen, 1987; Jackson and Hunt, 1975; and Jensen, et al., 1984; Zeman and Jensen, 1987), but it appears that 0.5 gives the best agreement with observations. For many of the gentle hills reviewed by Taylor, et al. (1987), the height above the hill crest where the maximum speed-up is found ranges between 2.5 and 5 m. The inner layer for the hill of Fig 14.17, for example, is about 3.5 m thick at point D. Thus, we see that the inner layer is quite shallow.

In the outer layer, some of the changes to the turbulence field appear to be described by *rapid distortion theory* (Taylor, et al., 1987). This theory assumes that existing eddies are distorted by the flow over the hill, and that the turbulence is modified by compression and stretching of the existing vortex elements (Britter, et al., 1981; Panofsky, et al., 1982). The theory further assumes that the distortion is rapid enough that turbulence generation by the mechanical production terms is not increased. Rapid distortion theory predicts that $\overline{w'^2}$ will increase over a 2-D ridge, but $\overline{u'^2}$ will decrease in the outer layer. The vortex stretching also enhances the higher frequencies of turbulence, and reduces the lower frequencies. Observations, however, indicate that both $\overline{w'^2}$ and $\overline{u'^2}$ are reduced in the outer layer.

In the inner layer, where wind shears are enhanced and turbulence is in local equilibrium with the surface, most turbulence statistics (including stress) are larger than their upstream values (Taylor, et al., 1987). Zeman and Jensen (1987) have suggested a third layer in between the inner and outer layers. This layer occurs at roughly the height of maximum speed-up, and is characterized by a minimum in turbulent stresses as a consequence of the curvature of the mean flow streamlines. Analogous curvature effects are believed to be related in increases in turbulence statistics upstream of the hill crest.

A series of hills can modify the flow to yield a new log-wind profile that has an aerodynamic roughness length characteristic of rough terrain (Fig 14.18). Over the first

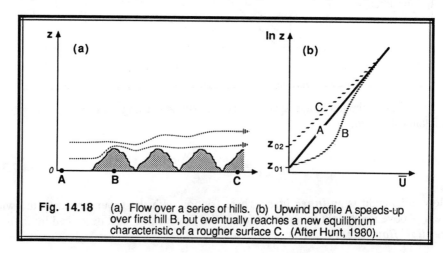

Fig. 14.18   (a) Flow over a series of hills. (b) Upwind profile A speeds-up over first hill B, but eventually reaches a new equilibrium characteristic of a rougher surface C. (After Hunt, 1980).

hill, the wind profile experiences the speed-up already discussed. Over succeeding hills, the superposition of the effects from the individual hills accumulate to generate a wind profile that is slower at low levels, characteristic of a larger roughness length (Hunt, 1980).

Mixed Layer Capped by an Inversion. Some interesting phenomena can occur when a mixed layer approaches a 2-D ridge (Fig 14.19), because the capping inversion tends to constrain the flow response to occur within the mixed layer. Define a modified Froude number, Fr*, by:

$$Fr* = \frac{\overline{M}}{N_{BV} \cdot (z_i - z_{hill})} \tag{14.2.3e}$$

For $z_i > z_{hill}$, two different types of flow can occur depending on the wind speed (Hunt, 1980; Carruthers and Choularton 1982). For relatively slow wind speeds (Fr* $\ll$ 1), acceleration over the ridge crest can locally draw down the inversion via a **Bernoulli** mechanism (Fig 14.19a). To the lee of the ridge light winds are found in the region of boundary-layer separation.

For faster winds (Fr* $\cong$ 1) with a strong capping inversion, the mixed layer accelerates down the lee side of the mountain in a very shallow high-velocity flow (Fig 14.19b) that is sometimes called a **downslope windstorm** in the Boulder, Colorado area, or **bora** in Yugoslavia. Wind gusts greater than 50 m/s have been measured in Boulder during one of these severe downslope winds, with substantial damage to houses and trees and aircraft reports of severe to extreme turbulence (Lilly, 1978). The bora can last 4 to 6 days (Smith, 1987). Downwind of the ridge there can be a **hydraulic jump**, where the boundary layer rapidly decelerates and the depth increases. The hydraulic jump can be visualized as a wave or bore that is trying to propagate upstream, in a flow that is fast enough to counteract any propagation towards the mountain. Two-layer models capture most of the important physics of this phenomenon, while three-layers offer more accuracy (Smith and Sun, 1987).

If $z_i \gg z_{hill}$, then the mixed layer tends to evolve as if the hill were not present. This is particularly true for light winds and strong convection. For a high isolated 3-D mountain and a shallow boundary layer ($z_i < z_{hill}$), the air is constrained to flow around the mountain (Fig 14.19c). This can cause a street of **von Karman vortices** (Brighton, 1978) to form downwind of the mountain (Fig 14.19d).

14.2.4 Flow Over Other Complex Terrain

Many other phenomena can occur for various valley and pass geometries (Egan, 1975). Boundary layer winds can be funneled through mountain passes, generating strong pass winds. Ambient flow crossing narrow valleys can generate both along-valley flow and cross-valley circulations within the valley (Erasmus, 1986a and b). Differential

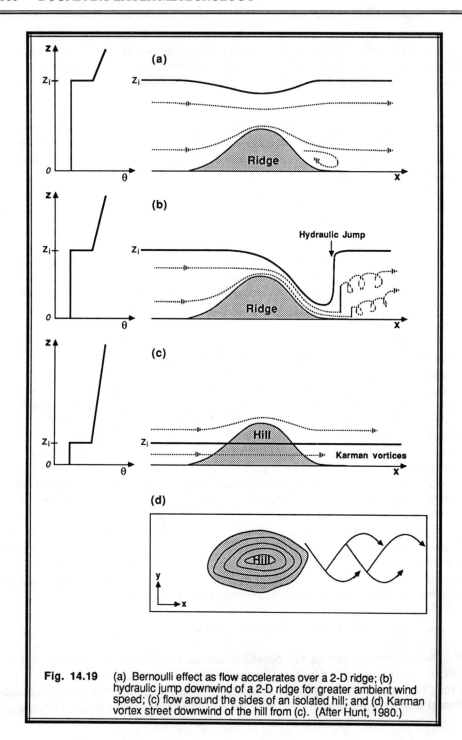

Fig. 14.19   (a) Bernoulli effect as flow accelerates over a 2-D ridge; (b) hydraulic jump downwind of a 2-D ridge for greater ambient wind speed; (c) flow around the sides of an isolated hill; and (d) Karman vortex street downwind of the hill from (c). (After Hunt, 1980.)

solar heating of one valley wall compared to the other can modify the turbulence intensity and stress (Carlson and Foster, 1986). Pollutant dispersion in complex terrain is difficult to model, and additional research is needed in this area (Egan and Schiermeier, 1986).

In large cities, tall buildings act like steep valley walls along **urban canyons** formed by the streets (Oke, 1978). As might be expected, the criss-crossing urban canyons for a very complex urban terrain system, with air being trapped in some canyons and air in others being ducted like wind-tunnels (Hosker, 1987). Not only is there differential absorption on the various canyon walls (buildings) depending on the sun position, but there can be significant multiple reflections off of the windows and emission of heat from traffic in the street and from buildings (Johnson and Watson, 1984; Steyn and Lyons, 1985). The generation of urban boundary layers is reviewed after the example below.

14.2.5 Example

Problem. Air is flowing from a corn field to a short-cut grass pasture. How far downwind from the edge of the corn field should a 10 m mast be erected so as to not "feel" the influence of the corn field (i.e., what minimum fetch within the pasture is required). Assume statically neutral conditions, and no change of surface heat flux.

Solution. First, we can estimate the roughness of the two fields using Fig 9.6: for corn, $z_{o1} \cong 0.06$; for cut grass, $z_{o2} \cong 0.006$. Using (14.2.1b) we can solve for the a_{IBL} parameter:

$$a_{IBL} = 0.75 + 0.03 \ln\left(\frac{0.006}{0.06}\right) = 0.681$$

Next, we can solve (14.2.1a) for the required fetch, knowing that δ must be at least 10 m:

$$x = z_{o1} \cdot \left(\frac{1}{a_{IBL}} \frac{\delta}{z_{o1}}\right)^{1/b_{IBL}} = 0.06 \cdot \left(\frac{1}{0.681} \frac{10}{0.06}\right)^{1/0.8} = 58 \text{ m}$$

Discussion. The mast must be located no less that 58 m from the corn field. If we anticipate a range of stabilities during our measurement, it would be best to locate the mast even further away. Note that the equation for IBL depth is not a function of wind speed, because it is assumed that turbulence is more intense and the IBL grows faster in a surface layer with greater shear.

14.3 Urban Heat Island and the Urban Plume

Most cities are anthropogenic sources of heat and pollution. In addition, the downtown areas are covered by a large percentage of asphalt and concrete, which are usually dry, water-proof surfaces with albedoes and heat capacities that convert and store incoming radiation as sensible heat better than the surrounding countryside (recall Fig

7.12). Thus, surface-layer air in cities is generally warmer than that of their surroundings (Oke, 1982, 1987; Goldreich, 1985; Bornstein, 1987). When isotherms are plotted on a surface weather map, the pattern looks like the topographic contours of an island (Fig 14.20), whence the term *heat island*.

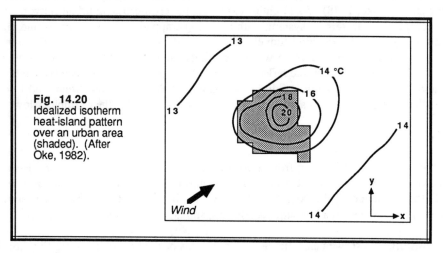

Fig. 14.20 Idealized isotherm heat-island pattern over an urban area (shaded). (After Oke, 1982).

The greatest temperature differences between urban and rural areas are usually observed during the night (Fig 14.21). In many cases, heat from the city is sufficient to maintain a shallow convective mixed layer at night, even while a substantial stable boundary layer has developed over the surrounding countryside (Fig 14.22). Cities with a population of about 1000 have been observed to have maximum temperature excesses (compared to the surrounding rural area) of 2 to 3 °C, while cities of a million or more inhabitants have been known to generate excesses of 8 to 12°C (Oke, 1982; Katsoulis and Theoharatos, 1985).

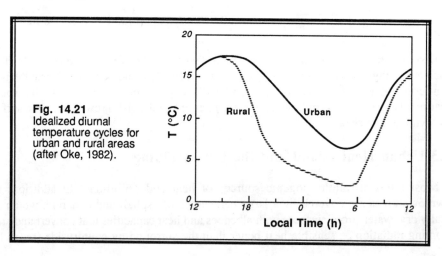

Fig. 14.21 Idealized diurnal temperature cycles for urban and rural areas (after Oke, 1982).

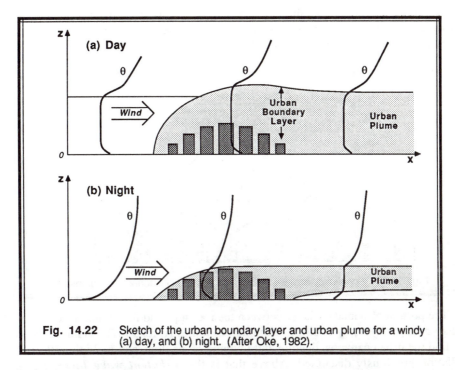

Fig. 14.22   Sketch of the urban boundary layer and urban plume for a windy (a) day, and (b) night. (After Oke, 1982).

Synoptic and topographic forcings influence the development of the urban boundary layer.   The temperature excesses at any given time are modulated by cloudiness, precipitation, and mean ambient wind speed (Ackerman, 1985; McKee, et al., 1987). Local topography such as mountains, lakes and rivers also have a large impact.   In Chicago, for example, the greatest temperature excess is often observed in early evening during the late summer.

Enhanced urban turbulence at night can create counter-rotating vortices on opposite sides of the city (Draxler, 1986). Balling and Cerveny (1987) noted an increase in wind speed over the city at night, which they suggested was due to local urban horizontal temperature gradients and enhanced vertical mixing with the faster flow aloft.

During the daytime, heat from the urban area can enhance the mixing already present in the ML, and create an urban internal boundary layer. Hildebrand and Ackerman (1984) and Clarke, et al. (1987) reported enhanced turbulent heat flux, vertical velocity variance, and entrainment rates over urban areas. The ML can sometimes be deeper over urban areas than rural ones because of greater low-level convergence, resulting in more cloud condensation nuclei, more thunderstorms and enhanced precipitation immediately downwind of the city (Changnon, 1981). The presence of large buildings increases surface drag and wake turbulence, and decreases the mean wind speed.

In the presence of a mean wind, excesses of temperature and pollutants, and deficits of humidity are carried downwind in an ***urban plume*** (Oke, 1982; Hanna, et al., 1987).

These plumes are as wide as the city, and can be transported hundreds of kilometers downstream (Fig 14.21). In the absence of wind, closed circulations analogous to the sea-breeze can form over the city (Fig 14.23).

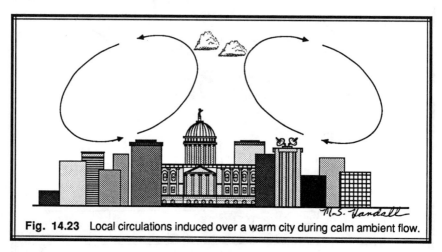

Fig. 14.23   Local circulations induced over a warm city during calm ambient flow.

The portion of boundary layer between the roof tops and the ground is known as the *urban canopy layer*, by analogy with plant canopies (Oke, 1987). Within this region we find the urban canyons, ducting and trapping of air flow, and multiple reflections of radiation previously discussed. Above that is the *turbulent wake layer*, where the wakes and IBLs from the individual buildings and surface patterns can still be discerned (Fig 14.24). Still higher is the surface layer, where the momentum and heat budgets feel the average effect of the urban area, but where individual wakes are not important. Finally, the urban mixed layer extends to the top of the urban boundary layer, which itself might be an internal boundary layer within the larger-scale flow.

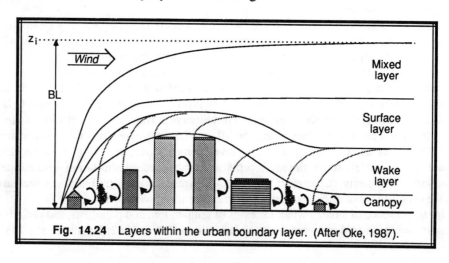

Fig. 14.24   Layers within the urban boundary layer. (After Oke, 1987).

14.4 References

Ackerman, B., 1985: Temporal march of the Chicago heat island. *J. Clim. Appl. Meteor.*, **24**, 547-554.

Arritt, R.W., 1987: Effect of water surface temperature on lake breezes and thermal internal boundary layers. *Bound.-Layer Meteor.*, **40**, 101-125.

Balling, R.C., Jr. and R.S. Cerveny, 1987: Long-term associations between wind speeds and the urban heat island of Phoenix, Arizona. *J. Clim. Appl. Meteor.*, **26**, 712-716.

Banta, R.M., 1985: Late-morning jump in TKE in the mixed layer over a mountain basin. *J. Atmos. Sci.*, **42**, 407-411.

Beljaars, A.C.M., 1987: On the memory of wind standard deviation for upstream roughness. *Bound.-Layer Meteor.*, **38**, 95-102.

Beljaars, A.C.M., J.L. Walmsley, P.A. Taylor, 1987: A mixed spectral finite difference model for neutrally stratified boundary layer flow over roughness changes and topography. *Bound. Layer Meteor.*, **38**, 273-303.

Bornstein, R.D., 1987: Mean diurnal circulation and thermodynamic evolution of urban boundary layers. *Modeling the Urban Boundary Layer.* Amer. Meteor. Soc., Boston. 53-93.

Briere, S., 1987: Energetics of daytime sea breeze circulation as determined from a two-dimensional, third-order turbulence closure model. *J. Atmos. Sci.*, **44**, 1455-1474.

Brighton, P.W.M., 1978: Strongly stratified flow past three-dimensional obstacles. *Quart. J. Roy. Meteor. Soc.*, **104**, 289-308.

Britter, R.E., J.C.R. Hunt and K.J. Richards, 1981: Airflow over a two-dimensional hill: studies of velocity speed-up, roughness effects and turbulence. *Quart. J. Roy. Meteor. Soc.*, **107**, 91-110.

Brutsaert, W. and W.P. Kustas, 1987: Surface water vapor and momentum fluxes under unstable conditions from a rugged complex area. *J. Atmos. Sci.*, **44**, 421-431.

Carlson, J.D. and M.R. Foster, 1986: Numerical study of some unstably stratified boundary-layer flows over a valley at moderate Richardson number. *J. Clim. Appl. Meteor.*, **25**, 203-213.

Carruthers, D.J. and T.W. Choularton, 1982: Airflow over hills of moderate slope. *Quart. J. Roy. Meteor. Soc.*, **108**, 603-624.

Changnon, S.A., Jr. (Ed.), 1981: *METROMEX: A Review and Summary.* Meteor. Monographs, 18, No. 40. Amer. Meteor. Soc., Boston. 181pp.

Clarke, J.F., J.K.S. Ching, J.M. Goodwitch, and F.S. Binkowski, 1987: Surface layer turbulence in an urban area. *Modeling the Urban Boundary Layer.* Amer. Meteor. Soc., Boston. 161-199.

Claussen, M., 1987: The flow in a turbulent boundary layer upstream of a change in surface roughness. *Bound.-Layer Meteor.*, **40**, 31-86.

Draxler, R.R., 1986: Simulated and observed influence of the nocturnal urban heat island on the local wind field. *J. Clim. Appl. Meteor.*, **25**, 1125-1133.

Egan, B.A. and F.A. Schiermeier, 1986: Dispersion in complex terrain: a summary of the AMS workshop held in Keystone, Colorado, 17-20 May 1983. *Bull. Amer. Meteor. Soc.*, **67**, 1240-1247.

Erasmus, D.A., 1986: A model for objective simulation of boundary-layer winds in an area of complex terrain. *J. Clim. Appl. Meteor.*, **25**, 1832-1841.

Erasmus, D.A., 1986: A comparison of simulated and observed boundary-layer winds in an area of complex terrain. *J. Clim. Appl. Meteor.*, **25**, 1842-1852.

Fitzjarrald, D.R., 1984: Katabatic wind in opposing flow. *J. Atmos. Sci.*, **41**, 1143-1158.

Fitzjarrald, D.R., 1986: Slope winds in Veracruz. *J. Clim. Appl. Meteor.*, **25**, 133-144.

Garratt, J.R., 1983: Surface influence upon vertical profiles in the nocturnal boundary layer. *Bound.-Layer Meteor.*, **26**, 69-80.

Garratt, J.R., 1987: The stably stratified internal boundary layer for steady and diurnally varying offshore flow. *Bound.-Layer Meteor.*, **38**, 369-394.

Gash, J.H.C., 1986: A note on estimating the effect of a limited fetch on micrometeorological evaporation measurements. *Bound.-Layer Meteor.*, **35**, 409-413.

Geiger, R., 1965: *The Climate Near the Ground, Revised Edition.* Harvard Univ. Press, Cambridge. 611 pp.

Goldreich, Y., 1985: The structure of the ground-level heat island in a central business district. *J. Clim. Appl. Meteor.*, **24**, 1237-1244.

Hanna, S.R., 1987: An empirical formula for the height of the coastal internal boundary layer. *Bound.-Layer Meteor.*, **40**, 205-207.

Hanna, S.R., G.A. Briggs, and R.P. Hosker, Jr., 1982: *Handbook on Atmospheric Diffusion.* Tech. Information Center, U.S. Dept. of Energy. DOE/TIC-11223. 102pp.

Hanna, S.R., J.V. Ramsdell, and H.E. Cramer, 1987: Urban Gaussian diffusion parameters. *Modeling the Urban Boundary Layer.* Amer. Meteor. Soc., Boston. 337-379.

Haugen, D.A.(Ed.), 1975: *Lectures on Air Pollution and Environmental Impact Analysis.* Amer. Meteor. Soc., Boston. 296pp.

Helmis, C.G., D.N. Asimakopoulos, D.g. Deligiorgi, and D.P. Lalas, 1987: Observations of sea-breeze fronts near the shoreline. *Bound.-Layer Meteor.*, **38**, 395-410.

Hildebrand, P.H., and B. Ackerman, 1984: Urban effects on the convective boundary layer. *J. Atmos. Sci.*, **41**, 76-91.

Hosker, R.P., Jr., 1987: Effects of buildings on local dispersion. *Modeling the Urban Boundary Layer.* Amer. Meteor. Soc., Boston. 95-160.

Hsu, S.A., 1986: A note on estimating the height of the convective internal boundary layer near shore. *Bound.-Layer Meteor.*, **35**, 311-316.

Hunt, J.C.R., 1980: Wind over hills. *Workshop on the Planetary Boundary Layer.* J.C. Wyngaard (Ed.), Amer. Meteor. Soc., Boston. 107-144.

Jackson, P.S. and J.C.R. Hunt, 1975: Turbulent wind flow over a low hill. *Quart. J. Roy. Meteor. Soc.*, **101**, 929-956.

Jensen, N.O., E.L. Petersen, and I. Troen, 1984: *Extrapolation of Mean Wind Statistics with Special Regard to Wind Energy Applications.* Report WCP-86, WMO, Geneva. 85pp.

Johnson, G.T. and I.D. Watson, 1984: The determination of view-factors in urban canyons. *J. Clim. Appl. Meteor.*, **23**, 329-335.

Johnson, G.T. and I.D. Watson, 1985: Reply. *J. Clim. Appl. Meteor.*, **24**, 386.

Katsoulis, B.D. and G.A. Theoharatos, 1985: Indications of the urban heat island in Athens, Greece. *J. Clim. Appl. Meteor.*, **24**, 1296-1302.

Kitada, T., K. Igaraschi and M. Owada, 1986: Numerical analysis of air pollution in a combined field of land/sea breeze and mountain/valley wind. *J. Clim. Appl. Meteor.*, **25**, 767-784.

Lilly, D.K., 1978: A severe downslope windstorm and aircraft turbulence event induced by a mountain wave. *J. Atmos. Sci.*, **35**, 59-77.

Lyons, W.A., 1975: Turbulent diffusion and pollutant transport in shoreline environments. *Lectures on Air Pollution and Environmental Impact Analysis*, D.A. Haugen (Ed.), Am. Meteor. Soc. 136-208.

Martin, C.L. and R.A. Pielke, 1983: The adequacy of the hydrostatic assumption in sea breeze modeling over flat terrain. *J. Atmos. Sci.*, **40**, 1472-1481.

McKee, T.B., D.C. Bader, and K. Hanson, 1987: Synoptic influence on urban circulation. *Modeling the Urban Boundary Layer*. Amer. Meteor. Soc., Boston. 201-214.

McNider, R.T. and R.A. Pielke, 1984: Numerical simulation of slope and mountain flows. *J. Clim. Appl. Meteor.*, **23**, 1441-1453.

Mitsuta, Y., N. Monji, and D.H. Lenschow, 1986: Comparisons of aircraft and tower measurements around Tarama Island during the AMTEX '75. *J. Clim. Appl. Meteor.*, **25**, 1946-1955.

Neff, W.D. and C.W. King, 1987: Observations of complex-terrain flows using acoustic sounders: experiments, topography, and winds. *Bound.-Layer Meteor.*, **40**, 363-392.

Noonan, J.A. and R.K. Smith, 1986: Sea-breeze circulations over Cape York Peninsula and the generation of Gulf of Carpentaria cloud line disturbances. *J. Atmos. Sci.*, **43**, 1679-1693.

Ogawa, Y., T. Ohara, S. Wakamatsu, P.G. Diosey, and I. Uno, 1986: Observations of lake breeze penetration and subsequent development of the thermal internal boundary layer for the Nonticoke II shoreline diffusion experiment. *Bound.-Layer Meteor.*, **35**, 207-230.

Oke, T.R., 1978: *Boundary Layer Climates*. Halsted Press, New York. 372pp.

Oke, T.R., 1982: The energetic basis of the urban heat island. *Quart. J. Roy. Meteor. Soc.*, **108**, 1-24.

Oke, T.R., 1987: The surface energy budgets of urban areas. *Modeling the Urban Boundary Layer*. Amer. Meteor. Soc., Boston. 1-52.

Panofsky, H.A., D. Larko, R. Lipschutz, G. Stone, E.F. Bradley, A.J. Bowen, J.Højstrup, 1982: Spectra of velocity components over complex terrain. *Quart. J. Roy. Meteor. Soc.*, **108**, 215-230.

Pearson, R.A., G. Carboni and G. Brusasca, 1983: The sea breeze with mean flow. *Quart. J. Roy. Meteor. Soc.*, **109**, 809-830.

Rao, K.S., J.C. Wyngaard and O.R. Coté, 1974: The structure of the two-dimensional internal boundary layer over a sudden change of surface roughness. *J. Atmos. Sci.*, **31**, 738-746.

Rotunno, R., 1983: On the linear theory of the land and sea breeze. *J. Atmos. Sci.*, **40**, 1999-2009.

Simpson, J.E., D.A. Mansfield, J.R. Milford, 1977: Inland penetration of sea-breeze fronts. *Quart. J. Roy. Meteor. Soc.*, **103**, 47-76.

Smedman-Högström, A.-S., and U. Högström, 1978: A practical method for determining wind frequency distributions for the lowest 200 m from routine meteorological data. *J. Appl. Meteor.*, **17**, 942-954.

Smith, R.B., 1987: Aerial observations of the Yugoslavian Bora. *J. Atmos. Sci.*, **44**, 269-297.

Smith, R.B. and J. Sun, 1987: Generalized hydraulic solutions pertaining to severe downslope winds. *J. Atmos. Sci.*, **44**, 2934-2939.

Steyn, D.G. and T.J. Lyons, 1985: Comments on "The determination of view-factors in urban canyons". *J. Clim. Appl. Meteor.*, **24**, 383-385.

Sykes, R.I. 1978: Stratification effects in boundary layer flow over hills. *Proc. Roy. Soc. A.*

Tampieri, F., 1987: Separation features of boundary-layer flow over valleys. *Bound.-Layer Meteor.*, **40**, 295-308.

Taylor, P.A., P.J. Mason and E.F. Bradley, 1987: Boundary-layer flow over low hills (A review). *Bound.-Layer Meteor.*, **39**, 107-132.

Taylor, P.A. and H.W. Teunissen, 1987: The Askervein Project: overview and background data. *Bound.-Layer Meteor.*, **39**, 15-39.

Teunissen, H.W., M.E. Shokr, A.J.Bowen, C.J. Wood, and D.W.R. Green, 1987: The Askervein Hill Project: wind tunnel simulations at three length scales. *Bound.-Layer Meteor.*, **40**, 1-29.

Venkatram, A., 1977: A model for internal boundary layer development. *Bound.-Layer Meteor.*, **11**, 419-437.

Wakimoto, R.M. and J.L. McElroy, 1986: Lidar observation of elevated pollution layers over Los Angeles. *J. Clim. Appl. Meteor.*, **25**, 1583-1599.

Whiteman, C.D., 1982: Breakup of temperature inversions in deep mountain valleys: part 1. observations. *J. Appl. Meteor.*, **21**, 270-289.

Wood, D.H., 1978: Calculation of the neutral wind profile following a large step change in surface roughness. *Quart. J. Roy. Meteor. Soc.*, **104**, 383-392.

Zeman, O. and N.O. Jensen, 1987: Modification of turbulence characteristics in flow over hills. *Quart. J. Roy. Meteor. Soc.*, **113**, 55-80.

14.5 Exercises

1) Define each of the layers below:

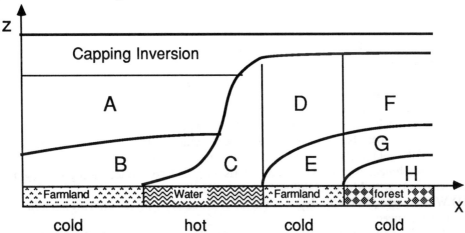

2) Given the surface-layer wind profile at right, explain all the possible reasons why the line is not straight when plotted on a semi-log graph.

3) Given the wind profile at right, determine whether the boundary layer is flowing from smooth to rough or rough to smooth, and estimate the distance upwind at which the change in roughness occurred. Assume neutral static stability and zero displacement height.

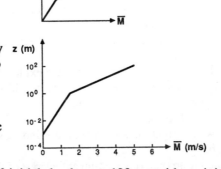

4) Assume a slab mixed layer of initial depth $z_i = 100$ m, with an initial potential temperature of 280 K. This air is advecting from land to warmer water (water surface effective potential temperature is 290 K) with a mean wind speed of 5 m/s. Over land, there is no flux into the bottom of the mixed layer, and no entrainment into the top. Over water, the flux at the bottom can be approximated with a bulk transfer relationship (use $C_D = 2 \times 10^{-3}$). The flux ratio method can be used to model the entrainment across the top. Neglect subsidence, and assume for simplicity that the strength of the capping inversion is always 2 K.

a) Calculate and plot the IBL depth and slab potential temperature as a function of downwind distance, x.

b) How does the thickness and temperature at x = 1 km downwind of the shoreline vary with wind speed.

5) Suppose that the air flowing over two hills is statically stable. The stability is such that the Froude number for the first hill is Fr = 0.5.

a) Sketch the flow over the first hill.

b) If the second hill is half the width of the first hill, what is its Froude number, and sketch the flow over it, assuming the same stability and wind speed as for the first hill.

6) What is the Froude number, and how is it used?

7) What is an anabatic cumulus cloud?

8) Given a wide valley with ridges on both sides, describe the dispersion of smoke over a full diurnal cycle from two stacks:

a) One very short stack at the side of the valley floor at the base of the ridge.

b) A tall stack (100 m) at the same location as in part (a).

9) If a fully enclosed basin fills with cold air during the night, describe the flow across the top of the cold pool of air in the morning.   State all your assumptions.

10) If a sea breeze is caused by warm air rising over land surface during the day, then what causes the low-level inflow from sea to land?

11) Why is the equation for depth of the internal boundary layer (14.2.1a) not a function of wind speed?

12) Plot the depth of the IBL as a function of distance downwind of a change in roughness, given $z_{o1} = 10^{-3}$ m and $z_{o2} = 10^{-2}$ m.   State your assumptions.

13) Suppose air is flowing from warm land over a cooler lake during the Spring.

a)If the wind speed is 7 m/s and the lake is 15°C cooler than the land surface (290 K), then plot the depth of the stable TIBL as a function of distance from the shore.

b) If the same situation occurs in the Fall, except that the land and lake temperatures are reversed, plot the depth of the convective TIBL as a function of distance downwind from the shore.

c) Compare the plots from (a) and (b), and explain why they differ.

14) Assume the mean wind speed is 5 m/s, the hill width is 200 m, and the vertical potential temperature gradient is 0.02 K/m.

a) Calculate the Froude number.

b) Above what height does flow separate and flow over the hill. Assume the hill is 100 m high.

c) Calculate the amount of speed-up at the crest of the hill, and the height above the hill at which this speed-up is found. State your assumptions.

d) What other factors besides speed-up might be important for determining whether a site could be used for wind turbine power generation?

17) Assuming an exponential change of temperature with height in the SBL (see Chapter 12) and a hill of width and height 500 m, calculate and plot the Froude number as a function of height.   Sketch a possible flow field for this situation.

18) Given an ambient wind speed of 15 m/s and a mountain height of 1000 m, plot a curve of lapse rate vs. depth of the mixed layer that delineates the threshold for the onset of severe downslope windstorms.

19) Describe the nature of the urban boundary layer for your own town.

Appendices

Contents

Appendix A

Scaling Variables and Dimensionless Groups

Scaling Variables

Scaling variables are used in similarity theory to nondimensionalize other variables. The following tables summarize length, velocity, time, temperature and humidity scales. Detailed discussions of these variables and their applications are presented in the body of this book. See Appendix B for definitions of notation.

Dimensionless Groups

Sometimes, the ratio of terms from an equation has significance concerning the nature of turbulence in the boundary layer. For example, if one term represents the suppression of turbulence and another term represents the generation of turbulence, then the ratio of these two terms might give an indication of the existence of turbulence or its intensity. The ratio of such terms defines a dimensionless group.

The following tables present many of the dimensionless groups that are used in boundary layer meteorology.

Length Scales:

Symbol	Name	Dimensions	Definition	Characteristics & Use
z_i	Inversion height or ML depth	m	Average mixed layer depth	Convective BL
L	Obukhov length	m	$\dfrac{-\overline{\theta}_v \, u_*^3}{k\,g\,\overline{w'\theta'_{vs}}}$	Surface layer scale
λ_{max}	Wavelength of peak in TKE spectrum	m	$1/\kappa_{max}$ or $2\pi/\kappa_{max}$	Dominant eddy size
z_o	Aerodynamic roughness length	m	Height where $\bar{u}=0$ in a log wind profile	Measure of the effective earth-surface roughness (surface layer)
$H_{\Delta\theta}$	Integral depth scale	m	$\displaystyle\int_{t_{sunset}}^{t} \overline{w'\theta'}_s \; d\tau / \Delta\bar{\theta}_s$	Stable (nocturnal) BL
η	Kolmogorov viscous length scale	m	$(\nu^3/\varepsilon)^{1/4}$	The largest eddies that feel the effects of viscosity
h_e	Ekman layer depth scale	m	u_*/f_c	Related to the depth of the idealized Ekman BL
L_L	Local Obukhov length	m	$\dfrac{-\bar{\theta}_v \,(\overline{u'w'}^2 + \overline{v'w'}^2)^{3/4}}{k\,g\,\overline{w'\theta'_v}}$	z - less scaling in the stable BL
W_H	Hill width	m	Hill width	Flow over hills and obstacles

Velocity Scales:

Symbol	Name	Dimensions	Definition	Characteristics & Use
w_*	Free-convection scaling velocity	m/s	$\left[\dfrac{g\, z_i}{\bar{\theta}_v} (\overline{w'\theta'_v})_s \right]^{1/3}$	Convection BL (used only when $(\overline{w'\theta'_v})_s \geq 0$)
u_*	Friction velocity	m/s	$\left(\overline{u'w'}_s^2 + \overline{v'w'}_s^2 \right)^{1/4}$	Surface layer and forced convection
$e^{1/2}$	TKE velocity scale	m/s	$(\overline{u'^2} + \overline{v'^2} + \overline{w'^2})^{1/2}$	Eddy velocity
G	Geostrophic wind	m/s	$-\dfrac{1}{\rho} \dfrac{\partial \bar{p}}{f_c \, \partial y}$	A measure of the pressure - gradient force
σ_u	Standard deviation of u	m/s	$(\overline{u'^2})^{1/2}$	Eddy velocity in x-direction
u_ε	Dissipation velocity scale	m/s	$(k\, z\, \varepsilon)^{2/3}$	Related to flow of TKE from large to small wave numbers
w_{Lf}	Local free convection velocity scale	m/s	$\left[\dfrac{g}{\theta_v} z\, \overline{w'\theta'_v} \right]^{1/3}$	Bottom of the mixed layer
u_L	Local friction velocity scale	m/s	$\left(\overline{u'w'}^2 + \overline{v'w'}^2 \right)^{1/4}$	Top of the stable boundary layer
u_*^{ML}	Convective stress velocity scale	m/s	u_*^2 / w_*	Stress in a mixed layer

Time Scales:

Symbol	Name	Dimensions	Definition	Characteristics & Use
t_*	Free convection time scale	s	z_i / w_*	Time for a thermal to rise in convective BL
$1/N_{BV}$	Oscillation time period (inverse Brunt - Väisälä frequency)	s	$\left(\dfrac{g}{\overline{\theta}_v} \dfrac{\partial \overline{\theta}_v}{\partial z} \right)^{-1/2}$	Buoyant oscillation period in statically stable environment. (used when $\partial \overline{\theta}_v / \partial z \geq 0$)
$1/f_c$	Inertial period	s	$\left[(1.46 \times 10^{-4}\, s^{-1}) \sin(\phi) \right]^{-1}$	Related to earth's rotation. (Not used much any more in BL studies.)

Miscellaneous Scales:

Symbol	Name	Dimensions	Definition	Characteristics & Use
θ_*^{SL}	Surface-layer temperature scale	K	$-\dfrac{(\overline{w'\theta'_v})_s}{u_*}$	Eddy temperature fluctuations in surface layer
θ_*^{ML}	Mixed-layer temperature scale	K	$\dfrac{(\overline{w'\theta'_v})_s}{w_*}$	Eddy temperature fluctuations in convective BL
θ_L	Local surface-layer temperature scale	K	$-\dfrac{(\overline{w'\theta'_v})}{u_L}$	Top of stable surface layer
θ_{Lf}	Local free convection temperature scale	K	$\dfrac{(\overline{w'\theta'_v})}{w_{Lf}}$	Bottom of convective boundary layer
$\Delta\theta$	Temperature jump	K	$\theta_o - \theta_s$ or $\theta(z_i^+) - \theta(z_i^-)$	Stable boundary layer strength or capping inversion strength
q_*^{SL}	Surface-layer humidity scale	$\dfrac{g\ water}{g\ air}$	$-\dfrac{(\overline{w'q'})_s}{u_*}$	Eddy moisture fluctuations in surface layer
q_*^{ML}	Mixed-layer humidity scale	$\dfrac{g\ water}{g\ air}$	$\dfrac{(\overline{w'q'})_s}{w_*}$	Eddy moisture fluctuations in mixed layer
q_L	Local surface layer humidity scale	$\dfrac{g\ water}{g\ air}$	$-\dfrac{(\overline{w'q'})}{u_L}$	Top of stable surface layer
q_{Lf}	Local free convection humidity scale	$\dfrac{g\ water}{g\ air}$	$\dfrac{(\overline{w'q'})}{w_{Lf}}$	Bottom of convective boundary layer

Dimensionless Groups:

Symbol	Name	Definition	Characteristics & Use
Fr	Froude number	$\dfrac{M}{N_{BV}\,W}$	Flow characteristic past an obstacle of size W. (inertia / buoyancy) or (natural wavelength / physical wavelength)
Ra	Rayleigh number	$\dfrac{g\,(\Delta T)\,D^3}{\rho\,\nu\,\nu_\theta}$	Benard convection between 2 plates of separation D and temperature difference ΔT (not used in BL because Ra >> 0).
Re	Reynolds number	$\dfrac{MD}{\nu}$	Ratio of inertial to viscous forces. Indicates when statically neutral flow becomes turbulent. (not used in BL because Re >> 0).
Ri	Richardson number (gradient)	$\dfrac{(g/\overline{\theta}_v)\ \partial\overline{\theta}_v/\partial z}{(\partial\overline{u}/\partial z)^2 + (\partial\overline{v}/\partial z)^2}$	Dynamic stability parameter. Indicates when laminar flow becomes turbulent.
R_f	Flux Richardson number	$\dfrac{(g/\overline{\theta}_v)\,\overline{w'\theta'_v}}{\overline{u'w'}\,\partial\overline{U}/\partial z + \overline{v'w'}\,\partial\overline{V}/\partial z}$	Dynamic stability parameter. Indicates when turbulent flow becomes laminar. Ratio of consumption to generation terms of TKE.
R_B	Bulk Richardson number	$\dfrac{g}{\overline{\theta}_v}\,\dfrac{\Delta\theta_v\ \Delta z}{(\Delta U)^2\ (\Delta V)^2}$	Approximation to the gradient Richardson number based on differences across distance Δz.
Ro	Surface Rossby number	$\dfrac{G}{f_c\,z_o}$	Ratio of pressure-gradient to friction (roughness). Used to estimate drag coefficients.

Dimensionless Groups:

Symbol	Name	Definition	Characteristics & Use
ϕ_M	Dimensionless wind shear	$\dfrac{kz}{u_*}\dfrac{\partial \bar{M}}{\partial z}$	Flux-profile relationships in the surface layer.
ϕ_H	Dimensionless lapse rate	$\dfrac{kz}{\theta_*}\dfrac{\partial \bar{\theta}_v}{\partial z}$	Flux-profile relationships in the surface layer.
μ	μ (Ekman)	$\dfrac{k u_*}{f_c L}$	An internal stability parameter.
μ_i	μ_i (Mixed Layer)	$\dfrac{k z_i}{L}$	An internal stability parameter.
C_D	Drag coefficient	$u_*^{\,2}/M^2$	Relates frictional drag to mean wind.
k	von Karman constant	$\lim\limits_{z \to z_o}\left[\dfrac{-u_*\,\partial^2\bar{u}/\partial z^2}{(\partial\bar{u}/\partial z)^2}\right]$	A "universal" constant related to turbulent flow near a surface ($k\approx0.35$ to 0.41).
ζ	Zeta	z/L	An internal stability parameter (used in surface layer). ($\zeta \approx R_i$ in unstable conditions).
z/z_i	Fraction of the mixed layer height	z/z_i	Height within the mixed layer.
X_M	Convective horizontal advection distance	$\dfrac{x\,w_*}{z_i\,\bar{M}}$	Actual horizontal distance traveled relative to one convective timescale distance.

Appendix B

Notation

A Note on Notation. Some notational compromises were required to avoid the conflicting definitions of symbols that appeared in the boundary-layer literature. For example, h is used for both moist static energy and boundary-layer depth, f for frequency and Coriolis parameter, and U for wind speed and wind component in the x-direction.

For some cases we resolved the conflicts by adding subscripts (e.g., f_c for Coriolis parameter). In other situations we extended existing classes of notation (e.g., s for both dry and moist static energies). Finally, in a few cases we had to define new symbols (e.g., M for wind magnitude).

A

a	albedo; various constants and parameters
a_c	bulk cloud albedo
a_{FR}	conductivity between air and ground in force-restore model
a_{IBL}	internal boundary layer parameter
agl	above ground level
A	various variables and constants; fundamental dimension of electric current; amplitude
A_R	an entrainment coefficient
A(k)	amplitude
A(t,x)	inverse Fourier transform
ABL	atmospheric boundary layer
ABLE	Amazon Boundary Layer Experiment
Am	amplitude spectrum
AMTEX	Air Mass Transformation Experiment

B

b	various constants and parameters
b_c	bulk cloud absorption
b_{IBL}	internal boundary layer parameter
B	various variables and constants; bulk measure of turbulence in the SBL
B_N	buoyant consumption of TKE
B_o	parameter for spectral diffusivity
B_P	buoyant production of TKE
B_{PN}	buoyancy
BAO	Boulder Atmospheric Observatory
BL	boundary layer
BLS77	Boundary Layer Structure 1977
BLX83	Boundary Layer Experiment 1983
BOMEX	Barbados Oceanography and Meteorology Experiment

C

c	pollutant or tracer concentration; generic constant
c_m	drag coefficient for individual roughness elements
$c_{n_{ref}^2}$	structure function parameter for index of refraction
c_{q^2}	structure function parameter for moisture
c_{rh^2}	structure function parameter for relative humidity
c_{T^2}	structure function parameter for temperature
c_{V^2}	structure function parameter for velocity
c_{A^2}	structure function parameter for variable A

629

C_D bulk momentum transfer coefficient (drag coefficient)

C_{Dh} drag coefficient at top of drainage flow

C_{DN} bulk momentum transfer coefficient for neutral conditions

C_E bulk moisture transfer coefficient

C_{EN} bulk moisture transfer coefficient for neutral conditions

C_g soil heat capacity

C_{GA} soil heat capacity per unit area

C_H bulk heat transfer coefficient (Stanton number)

C_{HN} bulk heat transfer coefficient for neutral conditions

C_p specific heat at constant pressure for moist air

C_{pd} specific heat at constant pressure for dry air

C_s speed of sound in air

C_* dimensionless canopy density

CAPE   convective available potential energy

CAT   clear air turbulence

CBL   convective boundary layer

CCOPE   Cooperative Convective Precipitation Experiment

CDT   Central Daylight Time (CDT = UTC - 5 h)

CL   cloud layer

Co   cospectrum

Coh   coherence spectrum

CPU   central processing unit

D

d day; displacement distance

d_r day of the summer solstice

d_s depth of upper slab in force-restore model

d_y average number of days in a year

d_1 distance between the top of the ML and the height where the heat flux profile is zero

D nonlocal dissipation parameter; structure function; pipe diameter

$D(L)$ structure function

$D_{AA}(r)$ structure function for variable A and itself

DFT   Discrete Fourier Transform

Dis   dissipation rate of TKE

Div   horizontal divergence

DYCOMS   Dynamics and Chemistry of Marine Stratocumulus

E

e instantaneous turbulence kinetic energy; water vapor pressure

e_{sat} saturation vapor pressure

E phase change rate (usually evaporation rate)

$E_A(n)$ discrete spectral intensity or energy of variable A

EAPE   evaporative available potential energy

EZ   entrainment zone

F

f frequency (often used for circular frequency in radians/s)

f_c Coriolis parameter

f_i intrinsic wave frequency

f_{max} frequency corresponding to peak in the spectrum

F momentum flux; discrete Fourier transform

F_{Ar} real part of the Fourier transform of variable A

F_{Ai} imaginary part of the Fourier transform of variable A

$F_A(n)$ discrete Fourier transform of variable A

F_u departure of the u-wind speed from geostrophic

F_v departure of the v-wind speed from geostrophic

F_w correction term in the Penman-Monteith Method

F_x generic flux

F_A^* complex conjugate of F_A

$F(t,\kappa)$ forward Fourier transform

FA   free atmosphere

FASINEX   Frontal Air Sea Interaction Experiment

FIFE   First ISLSCP Field Experiment

FFT   Fast Fourier Transform

Fr   Froude number

Fr^* modified Froude number

G

g acceleration due to gravity

G geostrophic wind speed

G_A unfolded spectral energy of F_A

G_{AB} cross spectrum between variables A and B

G_s geostrophic wind speed at the surface

G_{z_i} geostrophic wind speed at the top of the boundary layer

G_2 wind scale within the upper boundary layer

GALE   Genesis of Atlantic Low Experiment

GARP   Global Atmospheric Research Program

GATE   GARP Atlantic Tropical Experiment

GCM   general circulation model

GW   pressure-velocity correlation (and gravity waves)

H

h hour;
boundary layer depth;
depth of the stable boundary layer

h_d depth of the drainage flow

h_e Ekman layer depth

h_{eq} equilibrium SBL depth

h_i height of roughness element i

h_* average height of the canopy

h_o height of the bottom of the entrainment zone

h_2 height scale within the upper boundary layer;
height of the top of the entrainment zone

h^* average vertical extent of roughness elements

H SBL integral length scale (heat-flux-history scale);
height of obstacle

H_{eq} equilibrium SBL integral length scale

$H_{\Delta\theta}$ SBL integral depth scale

HAPEX   Hydrologic-Atmosphere Pilot Experiment

HEXOS   Humidity Exchange Over the Sea

I

I dimensionless turbulence intensity;
longwave radiation;
fundamental dimension of luminous intensity

I^* net longwave radiation

I^*_B net longwave radiative flux at the bottom of the cloud

I^*_h longwave radiative flux at the SBL top

I^*_s longwave radiative flux at the SBL bottom

I^*_T net longwave radiative flux at the top of the cloud

IBL   internal boundary layer

IR   infrared radiation

ISLSCP   International Satellite Land Surface Climatology

J

J argument for wave function

K

k von Karman constant

k_g thermal molecular conductivity of the soil

K generic eddy diffusivity;
shortwave radiation;
fundamental temperature dimension

K^* net shortwave radiative flux

K_B^* net shortwave radiative flux at the bottom of the cloud

K_T^* net shortwave radiative flux at the top of the cloud

K_E eddy diffusivity for moisture

K_H eddy diffusivity for heat

K_m eddy diffusivity for momentum (eddy viscosity)

$K(\kappa)$ spectral turbulent diffusivity

K-H   Kelvin-Helmholtz (waves)

KONTUR   Convection and Turbulence over Sea

L

l mixing length

l_B buoyancy length scale

l_c distance travelled by negatively buoyant entrained elements

l_O Ozmidov scale

l_ε dissipation length scale

$l(r)$ structure buoyancy length scale

L Obukhov length;
time lag;
fundamental length dimension

L_f latent heat of fusion of water

L_L local Obukhov length

L_p latent heat associated with some phase change

L_s latent heat of sublimation of water

L_T length of straight line travelled in summing over i roughness elements

L_v latent heat of vaporization of water

L_X location parameter for lognormal
     distribution
L transilient mixing distance
$\mathbb{L}$ length scale
LCL   lifting condensation level
LES   large eddy simulation
LFC   level of free convection
LLJ   low-level jet
LOC   limit of convection

M

m   mass
msl   (above) mean sea level
M   magnitude of wind (wind speed);
    mass flux;
    fundamental mass dimension
M_s wind speed at the surface
M_{z_i} wind speed at the top of the boundary layer
MCC   mesoscale cellular convection
METROMEX   Metropolitan Meteorology
    Experiment
MKE   kinetic energy associated with mean wind
ML   mixed layer
MP   mechanical production of TKE

N

n   frequency (cycles per time period $\mathbb{P}$)
n_f Nyquist frequency
n_{ref} index of refraction
ñ   frequency (cycles per second)
N   number count
N_d cloud droplet number density
N_{BV} Brunt-Väisälä frequency
NBL   nocturnal boundary layer

P

p   pressure
p_0 reference pressure
pdf   probability density function
P   pressure
P_0 reference pressure
P_{SP} saturation point pressure
$\mathbb{P}$ time period
PBL   planetary boundary layer
PPI   plan position indicator
Pr   probability density
PRE-STORM   Preliminary Regional Experiment
    of STORM-Central
PS   process spectrum

Q

q   specific humidity (of water vapor)
q_G specific humidity at the ground surface
q_L liquid water specific humidity;
    local humidity scale
q_{Lf} local free-convection humidity scale
q_s surface specific humidity of air
Δq_s moisture difference between air at the
    surface and aloft
q_T total water specific humidity
q_2 moisture scale in the upper boundary layer
q_* generic turbulent moisture scale
q_*^{ML} humidity scale for mixed layer, based on w_*
q_*^{SL} humidity scale for surface layer based on u_*

Q   quadrature spectrum
Q_A effective bulk heat flux into the SBL
    associated with advection
Q_E latent heat flux
Q_g molecular heat flux within the ground
Q_G molecular heat flux within the ground at the
    surface
Q_H sensible heat flux
Q_R bulk heat flux
Q_T total heat flux acting on the SBL
Q_w effective bulk heat flux into the SBL
Q^* net radiation
Q_j^* component of net radiation in the j direction
ΔQ_S storage of internal energy

R

r   mixing ratio;
    correlation coefficient
r   spatial separation
r_a water vapor resistance through air
r_{AB} linear correlation coefficient between A, B
r_L liquid water mixing ratio
r_p total water vapor resistance of plant
r_T total water mixing ratio
r_{sat} saturation mixing ratio
r'θ'   moisture temperature covariance
rh   relative humidity

R   moisture flux
$\mathfrak{R}$ gas constant for dry air
R(L)   autocorrelation
$R_{AA}(L)$ autocorrelation of variable A with itself
R_B bulk Richardson number
R_c critical Richardson number for the onset of
          turbulence
R_f flux Richardson number
R_i Richardson number
R_T critical Richardson number for the
          termination of turbulence
R_v gas constant for moist air
Ra   Rayleigh number
Re   Reynolds number
RHI   range height indicator
Ri   gradient Richardson number
Ri^* convective Richardson number
RL   residual layer
Ro   surface Rossby number

S

s   dry static energy
s_{cc} change of saturation specific humidity with
          temperature
s_e moist static energy
s_{es} saturation static energy
s_i horizontal surface area occupied by
          roughness element i
s_L liquid water static energy
s_s average vertical cross-section area presented
          to the wind by one roughness element
s_v virtual dry static energy
s_A^2 variance of variable A
s^2 variance
S   spatial domain;
          solar irradiance (solar constant);
          spectral energy density
S_A spectral energy density of variable A
$S_A(n)$ spectral energy density of variable A
S_c net tracer body source term
S_G stability parameter
S_L ratio of total ground surface area to number
          of elements
S_q net moisture body source term;
          spectral energy density of moisture
S_{qT} net total water body source term
S_r area density of roughness element

S_{rT} net moisture body source term
S_T total area occupied by N roughness
          elements
S_X shape parameter for the lognormal
          distribution
$S(t,\kappa)$ spectral energy density
SBL   stable boundary layer
SCL   subcloud layer
SESAME   Severe Environment Storms and
          Mesoscale Experiment
SL   surface layer
SP   saturation point

T

t   time
t_o initial time
t_* time scale for the mixed layer
t_*^{ML} time scale for the mixed layer
t_*^{SL} surface-layer time scale (z/u_*)
T   temperature (usually absolute temperature);
          fundamental time dimension
T_d dewpoint temperature
T_G temperature of upper slab in force-restore
          model
T_K net sky transmissivity
T_{LCL} temperature of parcel at the lifting
          condensation level
T_M temperature of lower slab in force-restore
          model
T_{SP} saturation point temperature
T_v virtual temperature (usually absolute)
TAMEX   Taiwan Area Mesoscale Experiment
TDBU   top down/bottom up (diffusion)
TIBL   thermal (convective) internal boundary layer
TKE   average turbulence kinetic energy
Tr   turbulent transport across the spectrum
TS   process spectra
TS(k,m)   transport spectrum

U

u   eastward moving Cartesian wind
          component;
          optical thickness
u_L local (friction) velocity scale
u_* friction velocity
u_*^{ML} convective stress scale velocity

u'w'	kinematic flux of U-momentum in the vertical	w'r'	kinematic moisture flux in the vertical
U	eastward moving Cartesian wind component	w's'	kinematic vertical static energy flux
U_{eq}	equilibrium wind speed	w'T'	kinematic temperature flux in the vertical
U_g	eastward component of geostrophic wind	w'θ'	kinematic potential temperature (heat) flux in the vertical
U_{gj}	represents the components of geostrophic wind (U_g, V_g, 0)	w'θ_v'	kinematic virtual potential temperature (buoyancy) flux in the vertical
U_i	represents (U, V, W) for i = (1, 2, 3)	w'ρ_v'	kinematic vertical moisture flux
U_T	generic transport velocity	W	width of obstacle;
UTC	Coordinated Universal Time (virtually the same as GMT (Greenwich Mean Time) or Z time)		vertical velocity (same as w)
		W_H	width of the hill
		W_P	liquid water path
		W_T	length of obstacle disturbing the flow

V

		$W_{1/2}$	half-width of the hill
v	northward moving Cartesian wind component	W(k)	window weight
v'w'	kinematic flux of V-momentum flux in the vertical		

X

V	northward moving Cartesian wind component	x	Cartesian coordinate towards east (sometimes used in a coordinate system rotated such that x is aligned with the mean wind direction); location; generic distance; fetch
V_B	SBL buoyancy velocity scale		
V_g	northward component of geostrophic wind		
V_M	mechanical forcing scale		
V	velocity scale	x_d	coordinate parallel to mean wind
VAD	velocity-azimuth display	x_i	represents (x, y, z) for i=(1, 2, 3)
VIMEX	Venezuelan International Meteorology and Hydrology Experiment	X	generic distance
		X_s	relative humidity of the air near the surface
		X_G	relative humidity at the ground surface

W

		X^{ML}	convective horizontal distance
w	upward moving Cartesian wind component (vertical velocity, negative for subsidence)		

Y

w_c	average vertical velocity within the clouds at height z_i	y	Cartesian coordinate towards north (sometimes used as crosswind horizontal coordinate)
w_e	entrainment velocity		
w_i	longitudinal width of each roughness element i in the direction of travel	Y	mixing potential
w_L	mean large scale vertical motion acting at the top of the ML (subsidence)		

Z

w_{Lf}	local free convection velocity scale	z	Cartesian coordinate up, relative to local sea-level horizontal surface
w_{up}	average vertical velocity through active cloud bases	z_b	cloud base height (top of the subcloud layer)
w_*	convective velocity scale	z_B	height of cloud base
w'c'	kinematic tracer flux in the vertical	z_{hill}	height of the hill
w'e	vertical turbulent transport of TKE	z_i	average top of the mixed layer (base of the overlying temperature inversion)
w'p'	pressure correlation		
w'q'	kinematic moisture flux in the vertical	z_{LFC}	height of the level of free convection

z_{LOC}	height of the limit of convection	ε_{fT}	infrared flux emissivity from the effective
z_{LW}	depth of the column flowing over the hill		top of the atmosphere
z_o	aerodynamic roughness length	ε_{ijk}	alternating unit tensor
z_{o1}	aerodynamic roughness length upwind of	ε_{IR}	emissivity in the infrared
	border	ε_{mnq}	alternating unit tensor
z_{o2}	aerodynamic roughness length downwind of	ε_q	destruction rate of humidity variance by
	border		molecular processes
z_r	top of the residual layer	ε_R	destruction rate of temperature variance by
z_T	height of cloud top		radiative processes
Δz	height above local terrain	$\varepsilon_{u_i c}$	destruction rate of tracer concentration by
Z_s	scale of surface features or surface		molecular processes
	roughness	$\varepsilon_{u_i q}$	destruction rate of moisture flux by
zo_{eff}	effective roughness length		molecular processes
		$\varepsilon_{u_i u_k}$	destruction rate of momentum flux by
			molecular processes
		ε_{uw}	destruction rate of vertical flux of horizontal

Greek

			momentum by molecular processes
α	terrain slope;	$\varepsilon_{u_i \theta}$	destruction rate of heat flux by molecular
	various constants and parameters		processes
α_c	Charnock's parameter	ε_{wc}	destruction rate of vertical tracer
α_{gM}	angle between the surface and geostrophic		concentration flux by molecular
	wind		processes
α_k	Kolmogorov constant	ε_{wq}	destruction rate of vertical moisture flux by
α_{PT}	Priestly-Taylor parameter		molecular processes
α_{ws}	angle between the wind direction and the	$\varepsilon_{w\theta}$	destruction rate of vertical heat flux by
	surface stress		molecular processes
β	Bowen ratio;	ε_θ	destruction rate of temperature variance by
	various parameters and constants		molecular processes
γ	lapse rate ($\partial\theta_v/\partial z$);	ζ	dimensionless height in the surface layer
	psychrometric constant (C_p/L_v)		(z/L)
γ_E	Ekman spiral parameter ($[f_c/2K_m]^{1/2}$)	η	Kolmogorov microscale for turbulence;
γ_T	transilient mixing rate coefficient		radar reflectivity;
$\gamma(t,\kappa)$	spectral component of spectral heat flux		net volumetric flow rate into a volume
Γ_d	dry adiabatic lapse rate	η''	wave vertical displacement distance
δ	depth of the internal boundary layer	θ	potential temperature
δ_m	unit vector	θ_e	equivalent potential temperature
δ_{mn}	Kronecker delta	θ_{es}	saturation equivalent potential temperature
δ_s	solar declination angle	θ_G	potential temperature at the ground surface
ε	turbulence kinetic energy dissipation rate;	$\Delta\theta_h$	temperature jump at the top of the SBL
	ratio of dry air and water vapor gas	θ_L	liquid water potential temperature;
	constants		local temperature scale
ε_c	destruction rate of tracer concentration	θ_{Lf}	local free-convection temperature scale
	variance by molecular processes	θ_{ML}	potential temperature within the mixed
ε_f	infrared flux emissivity		layer
ε_{fB}	infrared flux emissivity from the bottom of	θ_s	surface potential temperature of air
	the boundary layer		

θ_v	virtual potential temperature	π_i	Pi group
θ_2	temperature scale within the upper boundary layer	ρ	density
		ρ_o	reference density
θ_o	initial potential temperature	ρ_v	absolute humidity
θ_*	generic turbulent temperature scale	σ	standard deviation
θ_*^{ML}	temperature scale for the mixed layer, based on w_*	σ_A	fraction of the sky covered by active cumulus clouds
		σ_c	fractional cloud cover
θ_*^{SL}	temperature scale for the surface layer, based on u_*	σ_{c_H}	fractional coverage of sky with high clouds
		σ_{c_L}	fractional coverage of sky with low clouds
$\Delta\theta_S$	SBL surface cooling (strength of inversion)	σ_{c_M}	fractional coverage of sky with middle clouds
$\Delta\theta_{seq}$	equilibrium SBL strength	σ_F	fraction of the sky covered by forced cumulus clouds
$\Delta_{EZ}\theta_v$	potential temperature jump across the entrainment zone	σ_M	standard deviation of variable M
κ	wavenumber	σ_L	total cumulus cloud cover
κ_b	large wavenumber bound on the buoyancy subrange	σ_P	fraction of the sky covered by passive cumulus clouds
κ_H	horizontal wavenumber	σ_{SB}	Stefan-Boltzmann constant
κ_{max}	wavenumber corresponding to peak in turbulence spectrum	σ_u	standard deviation of U-wind
		σ_w	standard deviation of the vertical velocity
κ_x	wavenumber in the x direction	σ^2	variance
κ_y	wavenumber in the y direction	τ	shear stress
κ_z	wavenumber in the z direction	τ_{ij}	shear stress
λ	wavelength	τ_{mol}	molecular stress
λ_e	longitude	τ_R	SBL response time
λ_{max}	wavelength corresponding to peak in turbulence spectrum	$\tau_{Reynolds}$	Reynolds stress
		ϕ	latitude
λ_R	wavelength of radar	ϕ_E	dimensionless moisture gradient in the surface layer
λ_{sol}	e-folding solar decay length	ϕ_H	dimensionless potential temperature gradient in the surface layer
Λ	empirical length scale parameter		
μ	dynamic viscosity	ϕ_M	dimensionless wind shear in the surface layer
μ_B	bulk viscosity		
μ^{ML}	mixed layer scaling parameter	ϕ_r	latitude of the Tropic of Cancer
μ^{SL}	surface layer scaling parameter	$\phi(t,\kappa)$	spectral component of momentum flux
ν	kinematic molecular viscosity	Φ	phase shift angle; phase spectrum
ν_c	kinematic molecular diffusivity for tracer constituent c in the air	χ	pollutant (tracer) flux
		Ψ	solar elevation angle
ν_g	soil thermal diffusivity	Ψ_H	surface layer stability correction term for heat
ν_q	kinematic molecular diffusivity for water vapor in air		
		Ψ_M	surface layer stability correction term for momentum
ν_θ	kinematic molecular diffusivity for heat in air		
ξ	any variable	ω	angular rotation rate of earth
Ξ	turbulent diffusivity transfer function	ω^*	convective mass flux
π	3.1415926535897932384626	Ω	angular velocity vector

Special Symbols and Operators

$\overline{(\ )}$	average operator
$\overline{\overline{(\ )}}$	average over a portion of the SBL depth
$\hat{(\ )}$	normal (dynamic) flux (flux without tilda is a kinematic flux)
$(\ )'$	deviation from mean value
$(\ )''$	wave perturbation
$(\ )^+$	just above or greater than the value in parentheses
$(\ )^-$	just below or less than the value in parentheses
$\backslash\,\backslash$	phase-averaging operator
$<\ >$	average over the depth of the boundary layer
$\cdot$	simple multiplication
$\bullet$	vector dot product
$\times$	vector cross product
∇	del operator
∇^2	Laplacian operator
∂	partial derivative
d	total derivative
Δ	difference
Δ_{EZ}	difference across the entrainment zone

Subscripts

g	within the ground; geostrophic
i,j,k,l,m,n,q	indices for summation notation (each index can take on the values 1, 2, or 3)
s	near surface air quantity
z_i	quantity at the top of the ML
B	bottom of cloud
G	at the ground surface
L	local scaling quantity; liquid water; low
T	top of cloud
$*$	scaling variable
o	initial or reference quantity
2	upper boundary layer scale

Superscripts

e	ensemble (average)
s	space (average)
t	time (average)
ML	mixed layer
SL	surface layer

Appendix C

Useful Constants, Parameters and Conversion Factors

Table C-1. Similarity "Constants"
(A range of values is also given for those cases where agreement has not been reached on the precise value.)

k	= 0.4 (range 0.35-0.41)	von Karman constant
α_k	= 1.53 - 1.68	Kolmogorov constant

Table C-2. Geophysical Parameters

g	= 9.8	$m \cdot s^{-2}$	acceleration due to gravity
r_e	= 6.37 x 10^6	m	average radius of earth
ω	= 2π radians / 24 h		rotation rate of earth
	= 7.27 x 10^{-5}	$(radians) \cdot s^{-1}$	
f_c	= $(1.46 \times 10^{-4}) \cdot sin(\phi)$	s^{-1}	Coriolis parameter as a function of latitude (ϕ)
$\mathbb{P}_{inertial}$	= $12/sin(\phi)$	h	inertial period

Table C-3. Radiation Parameters

S	= 1370	$W \cdot m^{-2}$	solar irradiance (i.e., solar constant)
	= 1.113	$K \cdot m \cdot s^{-1}$	(using sea-level air density for conversion)
σ_{SB}	= 5.67 x 10^{-8}	$W \cdot m^{-2} \cdot K^{-4}$	Stefan-Boltzmann constant

Table C-4. Parameters of Air

Values at sea level for a standard atmosphere:

ρ_{SL}	= 1.225	$kg \cdot m^{-3}$	standard density of air
	= 0.01225	$mb \cdot s^2 \cdot m^{-2}$	
	= 0.001225	$kPa \cdot s^2 \cdot m^{-2}$	
P	= 101.325	kPa	pressure
	= 1013.25	mb	
	= 1.013×10^5	$N \cdot m^{-2}$	
	= 82714	$m^2 \cdot s^{-2}$	(in kinematic units)
T	= 288.15	K	temperature
	= 15	°C	
μ	= 1.789×10^{-5}	$kg \cdot m^{-1} \cdot s^{-1}$	dynamic molecular viscosity
ν	= 1.461×10^{-5}	$m^2 \cdot s^{-1}$	kinematic molecular viscosity
k_θ	= 2.53×10^{-2}	$W \cdot m^{-1} \cdot K^{-1}$	molecular thermal conductivity
ν_θ	= 2.06×10^{-5}	$m^2 \cdot s^{-1}$	molecular thermal diffusivity ($=k_\theta/\rho C_p$)
C_{pd}	= 1004.67	$J \cdot kg^{-1} \cdot K^{-1}$	specific heat of dry air at constant pressure
	= 1004.67	$m^2 \cdot s^{-2} \cdot K^{-1}$	
C_p	= $C_{pd} \cdot (1+0.84 \cdot q)$	$m^2 \cdot s^{-2} \cdot K^{-1}$	specific heat of moist air, for q in g/g.
$\Re$	= 287.04	$J \cdot K^{-1} \cdot kg^{-1}$	gas constant for dry air
	= 287.04	$m^2 \cdot s^{-2} \cdot K^{-1}$	
	= 2.87	$mb \cdot K^{-1} \cdot m^3 \cdot kg^{-1}$	
	= 0.287	$kPa \cdot K^{-1} \cdot m^3 \cdot kg^{-1}$	

Dry air at other altitudes (standard atmospheric lapse rate assumed):

ρ_{100kPa}	= 1.212	$kg \cdot m^{-3}$	density of air at 111 m (where P = 100.0 kPa)
ρ_{1km}	= 1.112	$kg \cdot m^{-3}$	density of air at 1000 m (where P = 89.9 kPa)
ρ_{2km}	= 1.007	$kg \cdot m^{-3}$	density of air at 2000 m (where P = 79.5 kPa)

Air of other temperatures and humidities (at sea-level pressure):

ρ = 1.292 & 1.289	$kg \cdot m^{-3}$	density of dry & saturated air	at 0°C
ρ = 1.246 & 1.240	$kg \cdot m^{-3}$	density of dry & saturated air	at 10°C
ρ = 1.204 & 1.194	$kg \cdot m^{-3}$	density of dry & saturated air	at 20°C
ρ = 1.164 & 1.145	$kg \cdot m^{-3}$	density of dry & saturated air	at 30°C

Table C-5. Parameters of Liquid Water and Water Vapor

R_v $= 461.5$ $J \cdot K^{-1} \cdot kg^{-1}$ gas constant for water vapor

$C_{p\ vapor} = 1875$ $J \cdot kg^{-1} \cdot K^{-1}$ specific heat of water vapor

L_f $= 3.34 \times 10^5$ $J \cdot kg^{-1}$ latent heat of fusion (liquid:solid) at 0°C

L_s $= 2.83 \times 10^6$ $J \cdot kg^{-1}$ latent heat of sublimation (vapor:solid) at 0°C

L_v $= 2.50 \times 10^6$ $J \cdot kg^{-1}$ latent heat of vaporization (vapor:liquid) at 0°C.

L_v $= 2.45 \times 10^6$ $J \cdot kg^{-1}$ latent heat of vaporization (vapor:liquid) at 20°C.

L_v $\cong [2.501 - 0.00237 \cdot T(°C)] \times 10^6$ $J \cdot kg^{-1}$ latent heat of vaporization vs. T

ρ_{water} $= 1025$ $kg \cdot m^{-3}$ density of liquid water

C_{water} $= 4200$ $J \cdot kg^{-1} \cdot K^{-1}$ specific heat of liquid water

$(\rho \cdot C)_{water} = 4.295 \times 10^6$ $(W \cdot m^{-2}) / (K \cdot m \cdot s^{-1})$ heat capacity of liquid water

Table C-6.   Conversion Factors and Combined Parameters

$\rho \cdot C_p$	$= 1.231 \times 10^3$	$(W \cdot m^{-2}) / (K \cdot m \cdot s^{-1})$	for air at sea level (conversion factor between dynamic and kinematic heat fluxes for dry air)
	$= 12.31$	$mb \cdot K^{-1}$	
	$= 1.231$	$kPa \cdot K^{-1}$	
$g / \mathfrak{R}$	$= 0.0342$	$K \cdot m^{-1}$	
$\mathfrak{R} / g$	$= 29.29$	$m \cdot K^{-1}$	
$\mathfrak{R} / C_{pd}$	$= 0.28571$	-	
$C_{pd} / \mathfrak{R}$	$= 3.50$	-	
$\mathfrak{R} / R_v$	$= 0.622$	$g_{water} \cdot g_{air}^{-1}$	$(\equiv \varepsilon)$
$\rho \cdot g$	$= 12.0$	$kg \cdot m^{-2} \cdot s^{-2}$	at sea level
	$= 0.12$	$mb \cdot m^{-1}$	
	$= 0.012$	$kPa \cdot m^{-1}$	
g / C_p	$= 0.00975$	$K \cdot m^{-1}$	dry adiabatic lapse rate (Γ_d)
	$= 9.75$	$K \cdot km^{-1}$	
C_p / g	$= 102.52$	$m \cdot K^{-1}$	
C_p / L_v	$= 4.0 \times 10^{-4}$	$(g_{water} \cdot g_{air}^{-1}) \cdot K^{-1}$	psychrometric "constant" (γ)
	$= 0.4$	$(g_{water} \cdot kg_{air}^{-1}) \cdot K^{-1}$	
L_v / C_p	$= 2.5$	$K / (g_{water} \cdot kg_{air}^{-1})$	to convert from a kinematic moisture flux to a kinematic latent heat flux
$\rho \cdot L_v$	$= 3013.5$	$[W \cdot m^{-2}] / [(g_{water} \cdot kg_{air}^{-1}) \cdot m \cdot s^{-1}]$	to convert from a kinematic moisture flux to a latent heat flux (sea level, standard atmosphere)

Table C-7. Typical surface conditions (based on Anthes, R.A., E.-Y. Hsie, and Y.-H. Kuo, 1987: Description of the Penn State/NCAR Mesoscale Model, version 4 (MM4). NCAR Tech Note NCAR/TN-282+STR, Boulder, CO 80307. 66pp). Summer/winter values are listed.

Landuse	Albedo (%)	Moisture Availability (%)	IR Emissivity (% at 9 μm)
Urban land	18/18	5/10	88/88
Agriculture	17/23	30/60	92/92
Range-grassland	19/23	15/30	92/92
Deciduous forest	16/17	30/60	93/93
Coniferous forest	12/12	30/60	95/95
Forest swamp	14/14	35/70	95/95
Water or ocean	8/8	100/100	98/98
Marsh or wetland	14/14	50/75	95/95
Desert	25/25	2/5	85/85
Tundra	15/70	50/90	92/92
Permanent ice	55/70	95/95	95/95
Tropical forest	12/12	50/50	95/95
Savannah	20/20	15/15	92/92

Table C-8. Soil and ground properties (Lettau, personal communication), where: ρ = density (kg·m^{-3}), C = volumetric heat capacity (10^6 J·m^{-3}·K^{-1}), and ν = thermal diffusivity (10^{-6} m^2·s^{-1}).

Type	Composition	ρ	C	ν
Quartz sand	- dry	1500	1.24	0.24
	- 10% moisture	1650	1.54	1.22
	- 40% moisture	1950	2.76	0.91
Sandy clay	- 15% moisture	1780	2.42	0.38
Swamp land	- 90% moisture	1050	3.89	0.23
Rocks	- basalt	2800	2.34	0.66
	- sandstone	2600	2.30	1.13
	- granite	2700	2.13	1.28
	- concrete	2470	2.26	1.08
Snow	- new feathery	100	0.21	0.10
	- old packed	400	0.84	0.40
	- ice	920	2.05	0.92
Water	- still	1000	4.18	0.14

Appendix D

Derivation of Virtual Potential Temperature

Consider a volume (V) of cloudy air (saturated) with temperature T and total pressure P.

Let m_d = mass of dry air

   m_v = mass of water vapor

   m_L = mass of liquid water falling at terminal velocity

The density of the cloudy air is

$$\rho = \frac{m_d + m_v + m_L}{V}$$

$$= \rho'_d + \rho'_v + \rho'_L \qquad \text{"partial densities"}$$

where $\rho'_d = \dfrac{m_d}{V}$ $\rho'_v = \dfrac{m_v}{V}$ $\rho'_L = \dfrac{m_L}{V}$

Using the ideal gas law:

$$\rho'_d = \frac{P'_d}{\mathcal{R}\,T} \qquad \text{and} \qquad \rho'_v = \frac{e_{sat}}{R_v\,T} \qquad \text{where } e_{sat} = \text{saturation vapor pressure}$$

But $P'_d = P - e_{sat}$ Dalton's law of partial pressures

And $e_{sat} = \dfrac{r_{sat}}{r_{sat} + \varepsilon}\,P$ where $\varepsilon = \dfrac{\mathcal{R}}{R_v} = 0.622$

$$\rho_d' = \frac{P - e_{sat}}{\mathfrak{R}\,T} = \frac{P\left(1 - \dfrac{r_{sat}}{r_{sat} + \varepsilon}\right)}{\mathfrak{R}\,T} \qquad \text{where } r = \frac{m_v}{m_d} = \text{mixing ratio and}$$

$$r_{sat} = \text{saturation mixing ratio}$$

$$\rho_v' = \frac{e_{sat}}{R_v\,T} = \frac{\varepsilon\,e_{sat}}{\mathfrak{R}\,T} = \frac{P\varepsilon\left(\dfrac{r_{sat}}{r_{sat} + \varepsilon}\right)}{\mathfrak{R}\,T}$$

$$\rho_L' = \frac{m_L}{V} = \frac{m_L}{m_d}\frac{m_d}{V} = r_L\,\rho_d'$$

Hence

$$\rho = \frac{(1 + r_L)\,P\left(1 - \dfrac{r_{sat}}{r_{sat} + \varepsilon}\right)}{\mathfrak{R}\,T} + \frac{P\varepsilon\left(\dfrac{r_{sat}}{r_{sat} + \varepsilon}\right)}{\mathfrak{R}\,T}$$

$$\rho = \frac{P}{\mathfrak{R}\,T}\left(\frac{\varepsilon}{r_{sat} + \varepsilon}\right)\left[1 + r_L + r_{sat}\right]$$

Define a virtual temperature T_v such that T_v is the temperature that dry air must be in order to have the same density as the moist air.

$$\boxed{\rho \equiv \frac{P}{\mathfrak{R}\,T_v}}$$

Thus:

$$T_v = T\left(\frac{r_{sat} + \varepsilon}{\varepsilon}\right)\frac{1}{\left[1 + r_L + r_{sat}\right]}$$

Doing the long division:

$$1 + \left(\frac{1-\varepsilon}{\varepsilon}\right)r - r_L$$

$$\varepsilon + \varepsilon r + \varepsilon r_L \,\overline{\smash{\big)}\,\varepsilon + \quad r}$$

Eliminate ε:

$$\varepsilon + \quad \varepsilon r \quad + \varepsilon r_L$$

$$\overline{\quad r(1-\varepsilon) \quad - \varepsilon r_L}$$

Eliminate r:

$$r(1-\varepsilon) \quad + \quad \quad + \quad (1-\varepsilon)\, r_L r + (1-\varepsilon)\, r^2$$

$$\overline{\quad - \varepsilon r_L - (1-\varepsilon)\, r_L r - (1-\varepsilon)\, r^2}$$

Eliminate r_L:

$$- \varepsilon r_L - \quad \varepsilon r_L r \quad - \varepsilon r_L^2$$

Remainder:

$$\overline{\quad - (1+2\varepsilon)\, r_L r - (1-\varepsilon)\, r^2 + \varepsilon r_L^2}$$

All terms in the remainder are on the order of r^2.
Therefore neglect the remainder because $r \ll 1$, leaving $r^2 \ll r$.

$$T_V \cong T\left[1 + \left(\frac{1-\varepsilon}{\varepsilon}\right) r_{sat} - r_L\right]$$

$$\boxed{T_V \approx T\,(1 + 0.61\, r_{sat} - r_L)}$$

Similarly   ⎯⎯⎯ liquid water loading

$$\boxed{\theta_V \approx \theta\,(1 + 0.61\, r_{sat} - r_L)}$$

If unsaturated, then $r_L = 0$ and use r instead of r_{sat} :

$$\boxed{T_V \approx T\,(1 + 0.61\, r)}$$

Index

S

Z